NASA SP-4025

ASTRONAUTICS AND AERONAUTICS, 1985

A Chronology

THE NASA HISTORY SERIES

 Scientific and Technical Information Division   1988
National Aeronautics and Space Administration
Washington, DC

NASA maintains an internal history program for two principal reasons: (1) Sponsorship of research in NASA-related history is one way in which NASA responds to the provision of the National Aeronautics and Space Act of 1958 that requires NASA to "provide for the widest practicable and appropriate dissemination of information concerning its activities and the results thereof." (2) Thoughtful study of NASA history can help agency managers accomplish the missions assigned to the agency. Understanding NASA's past aids in understanding its present situation and illuminates possible future directions. The opinions and conclusions set forth in this book are those of the author; no official of the agency necessarily endorses those opinions or conclusions.

PREFACE

Astronautics and Aeronautics for 1985 continues the series of annual chronologies of events in aeronautics, aviation, and space science and exploration prepared by the History Office of the National Aeronautics and Space Administration. A compendium chronology covering the events between 1979 (the date of the most recently published Astronautics and Aeronautics chronology) and 1984 is in preparation.

The present volume introduces a new format designed to make the series more useful as an historical reference work. Events are arranged by major subject categories and subcategories, and chronologically thereunder. It was written by Bette R. Janson under the editorial supervision of the Director of the NASA History Office. Items for inclusion in this volume were collected by Eleanor H. Ritchie and Lee D. Saegesser of the NASA History Office.

As with any work of this nature, its extensive scope has made the use of primary sources prohibitive. Thus entries are based largely on normally reliable secondary sources. All entries are followed by their sources to provide serious researchers points of access for further investigation.

A chronology of this size necessarily involves considerable selection; readers are invited to notify the NASA History Office of any serious omissions, as well as errors, for possible correction in subsequent volumes.

Sylvia D. Fries
Director, NASA History Office
Washington, D.C.

CONTENTS

Preface	iii
Aerospace Industry	1
Aerospace Research and Development	3
Aeronautics	4
Balloons	16
Large Space Systems	17
Launch Vehicles	18
Military Applications	22
Simulation Technology	26
Transatmospheric Vehicle	28
Astronauts	31
Astronomy	37
Planetary Exploration	61
Atmospheric Research	71
Upper Atmospheric Research	77
Aviation	81
Civil Aviation	82
Human Powered Aircraft	96
Supersonic Transport	96
Awards	99
Colombia	105
Comsat	107
Conferences	109
Department of Defense	111
Antisatellite System	111
Budget	116
Missiles	117
NASA and the Department of Defense	120
Procurement	120
Satellites	121
Strategic Defense Initiative	125
European Space Agency	135
Federal Aviation Administration	141
France	145
Hermes Program	145
Geophysics	147
History	151
India	155
INTELSAT	157
Satellites	159

Japan	163
Launch Vehicles	163
Materials Processing	165
National Aeronautics and Space Administration	167
Budget	168
Data Tracking and Relay Systems	171
Management and Personnel	173
NASA Installations	186
Ames Research Center	186
Goddard Space Flight Center	189
Johnson Space Center	190
Kennedy Space Center	191
Langley Research Center	194
Lewis Research Center	196
Marshall Space Flight Center	198
Wallops Flight Facility	199
Office of Inspector General	200
Procurement	201
Public Affairs	203
Review Boards	204
Technology Transfers	206
Peoples Republic of China	213
Satellites	217
Search For Extraterrestrial Intelligence (SETI)	235
Smithsonian Institution	237
Space Science and Applications	243
Space Station Program	255
Space Transportation System (STS)	271
Civilian in Space Program	271
Crews	277
External Tank	285
Launch and Landing Facilities	286
Launch Schedules	294
Main Engines	306
Management	310
Military Applications	314
Missions	322
Revenues	343
Scientific Research	348
Shuttle Orbiter	358
Solid-fuel Rocket Boosters	366
Student Involvement Program	369
Spacelab	373

Union of Soviet Socialist Republics	385
Missiles	385
Satellites	386
Space Program	388
Spacecraft	396
United Kingdom	401
Satellites	401
United States	403
Congress	403
U.S. Air Force	404
Aircraft	404
NASA and the U.S. Air Force	409
U.S. Science and Technology	411
Policy	413
Resources	414
U.S. Space Policy	417
Civilian Programs	420
Commercialization	425
International	432
Military Programs	436
National Space Commission	440
Miscellaneous	443
Appendixes	447
A. Satellites, Space Probes, and Manned Space Flights, 1985	447
B. NASA Launches, 1985	499
C. Manned Space Flights, 1985	507
D. Abbreviations of References	513
Index	519

AEROSPACE INDUSTRY

March 18: In its aerospace forecast and inventory issue, *Aviation Week* reported that aerospace sales would continue to expand in 1985 as a result of business that was already on the books, with reviving commercial transport orders expanding on earlier rebounds in the military and space segments of the industry. However, as strong as the expansion was, sales were not as robust in some markets as had been predicted the previous year.

Among the factors affecting the sales picture was Congressional trimming of military funding in the FY 85 budget. But *Aviation Week* predicted that to trim outlays the Pentagon would go after operating funds rather than hardware money that was spent over periods longer than a year. Thus the aerospace industry experienced moderately slower but still firm growth. Also international and domestic competition was intensifying, particularly because the Pentagon sought to use competition as a primary tool to hold down weapons costs.

The magazine forecasted that total aerospace sales in 1985 would reach $99 billion and over $100 billion in 1986. Military aircraft would produce $32 billion in sales, compared with $27.5 billion in 1984; missile sales would rise to $15.7 billion from $13.8 billion in 1984; space technology, rebounding from weak sales in the 1970s, would reach $14.8 billion compared with $12.5 billion in 1984; commercial transport orders would revive from the $6-billion level in 1984 to $9.2 billion in 1985; and business flying sales would remain flat at about $1.8 billion.

Both transport and corporate aircraft had a common problem: the high cost of relatively small increments of new technology and the drop in fuel prices that had placed a premium on modest improvement in fuel efficiency. However, all-composite turboprop aircraft for the business market were flying and might begin to change that situation in 1986 and 87. Also, new ultra-high-bypass engines (in some situations a euphemism for propellers) were under test and might hit the commercial transport market in the early 1990s, as would increased use of composites, lighter and simpler subsystems, and more flexible cabins in terms of seating, galley, and lavatory layouts.

Military sales levels depended on what happened to President Reagan's defense budget in Congress. Although orders on the books would carry industry sales for 1985 to forecasted levels, a stalemate in Congress would affect the longer term outlook. (*AvWk,* Mar 18/85, 10)

August 8: The *Washington Post* reported that Fairchild Industries would sell its 50% interest in American Satellite Co. and Space Communications Co. for $105 million to Continental Telecom Inc., which owned the other half of the two joint ventures. Observers considered the two satellite communications subsidiaries Fairchild's leading growth opportunities.

American Satellite of Rockville (Amsat) provided private and general satellite communications services. It transmitted regional editions of *The Wall Street Journal*, *The New York Times*, and other newspapers and had won a large number of government contracts.

Space Communications of Gaithersburg (Spacecom) owned and operated the tracking and data relay satellite system for NASA, provided communications to the Space Shuttle, and was building a $122.8 million communications network for the Air Force.

Fairchild president Emanuel Fthenakis said the company decided to sell its interests in Amsat and Spacecom to consolidate its other businesses. "We are addressing the same market with more than one company," he said. "The main reason is to consolidate our activities and address these markets with our wholly owned subsidiaries."

Analysts said the move showed Fairchild was unwilling or unable because of troubles in its other operations to invest the large amounts of capital required to continue to expand the two companies. (*W Post*, Aug 9/85, B1)

August 19: Martin Marietta Corp. completed an agreement giving it a 25% stake in Equatorial Communication Co., a provider of satellite-based data communications networks, the *Washington Post* reported. Under the agreement, Marietta acquired about 3.6 million shares of Equatorial's common stock for $13.87½ a share and warrants to purchase an additional 1.8 million shares for $17.50 each. Marietta would assume two seats on Equatorial's board, which would expand to seven members.

Equatorial, which provided small, low-cost micro-earth stations, satellite transmission capacity, and other services that let companies construct and control their own private data communications networks, said it planned to use the funds to reduce debt and provide working capital.

Marietta and Equatorial said they planned to pursue joint ventures in the information systems market. (*W Post*, Aug 20/85, E4)

AEROSPACE RESEARCH AND DEVELOPMENT

April 15: NASA announced that Edward Taylor, a technician at Langley Research Center, modified an existing system, the uresco ultrasonic immersion testing system, by adding a holding device that permits accurate ultrasonic testing of small diameter polymeric tubes used in research programs, particularly small graphite fiber reinforced epoxy specimens, testing which formerly had to be done by hand.

The existing uresco system could not test curved or cylindrical shapes. Taylor's device rotated the tubes about their longitudinal axis while a high-frequency sound wave, transmitted through the tube wall, checked for flaws. This new method would permit nondestructive testing of tubular specimens without altering the electronic or mechanical configuration of the system.

Engineers checked graphite tubes for defects, trapped gases, void spaces, or foreign elements like a piece of metal. If they found a defect, engineers then decided if the tube would perform as one homogeneous material; if not, it was discarded. (NASA Release 85-57)

June 19: Dr. Raymond Colladay, NASA's associate administrator for aeronautics and space technology, said in an interview that NASA was developing a set of "technology roadmaps" that would provide the framework for the government, military, and industry to implement programs that would allow the U.S. to meet the national aeronautics research and development goals proposed in a White House report, "National Aeronautical R&D Goals," *Aerospace Daily* reported. Dr. Colladay said the report, written by the White House Office of Science and Technology, had charged NASA with developing the quantitative performance levels and technology steps that would be required to achieve three major goals in subsonics, supersonics, and transatmospherics. NASA hoped to have the roadmaps, being developed in cooperation with the Defense Department and industry, completed in July for review at the next planned interagency meeting.

The roadmaps "will represent a national consensus" on what needs to be done to achieve the aeronautics goals outlined in the report, Colladay said, although they would not spell out specific programs or funding levels.

The report had three goals: to advance technology for a new generation of fuel-efficient, affordable U.S. subsonic aircraft; to develop the technology for efficient, long-range supersonic cruise capability for military and civil aircraft; and to develop options in aeronautics and space technology for routine cruising and maneuvering into and out of the atmosphere with takeoff and landing from conventional runways. A fourth major national R&D goal, Colladay noted, was to maintain the basic research and technology that supported those three goals.

Colladay singled out the transatmospheric vehicle and hypersonic speeds as developments that would be "the biggest driver" with the "biggest payoff." Observers had identified hydroglen-fueled scramjets as holding greatest promise in producing substantially improved propulsive efficiency over current rocket technology. Colladay added that both NASA and the Defense Department were "very interested" in the hypersonic area.

Vehicles of that type depended on further technological developments in propulsion, materials, and intelligent integrated flight controls. Colladay pointed out that NASA had made substantial progress in all those areas over the last decade.

When questioned about future funding, Colladay indicated NASA might not face problems such as those created by the Office of Management and Budget when reportedly it had attempted to delete part of the NASA's aeronautics budget. Colladay said, "I think that the content of the program, when we focus on these national goals, will not be subjected to the same policy disagreements that we have had in the past," which centered primarily on the role of NASA in support of "civil aviation oriented research. In the case of supersonic cruise research and transatmospheric vehicles, there is no long-term commercial driver . . . that industry would do on their own . . . it's a clear example of where the government needs to take the lead to make it happen," Colladay concluded. (*A/D,* June 19/85, 1)

Aeronautics

During January: NASA announced it was testing a concept, called laminar (air)-flow control, which removed the layer of air molecules nearest an aircraft's wing surface to increase transport-aircraft flight efficiency. Since air/skin friction during a subsonic flight's cruise phase caused about one-half of the total drag on an aircraft, the laminar-flow control would use suction through porous wing surfaces to remove the turbulent air, reducing drag and fuel consumption.

NASA contractors had designed and fitted two laminar flow-control systems on the leading edge of each wing of a JetStar four-engine light transport. Gloves, perforated either by more than a million tiny suction holes or by narrow slots, encased the systems. Lockheed-Georgia Corp. had designed the test article on the JetStar's left wing; Douglas Aircraft Co. the one on the right wing.

Lockheed's construction was a sandwich of nomex-honeycomb core and graphite epoxy-face sheets, covered by a thin sheet of titanium bonded to the surface. Slots in the titanium sheets, .004 in., ran the length of the section and were the openings that drew air through the surface, directing it down ducts in the wing into the fuselage. The Douglas test article, instead of slotted, was perforated by an electron beam that drilled 800 precise, nearly microscopic holes per sq. inch of wing surface.

Both concepts had built-in systems to keep the wing surface insect and ice free. The Lockheed system pumped a cleansing fluid through the slots to the surface, making it too slippery for insects to adhere. The Douglas concept used a retractable insect shield deployed in front of the leading-edge panel during takeoff, climb, descent, and landing.

After tuning the leading-edge systems for best performance, NASA would flight test the airplane at various locations around the country at different times of year to record effects of weather and insect conditions.

Langley Research Center (LaRC) researchers managing the project pointed out that one energy-saving transport configuration, already efficient by today's standards, would be 22% more efficient with the addition of laminar-flow control. (NASA announcement, Jan 85, 7)

NASA announced that up to 20% of all fatal light-airplane accidents might be preventable with a LaRC- and ARC-developed wing modification, a carefully designed "glove" placed over the outer portion of a wing's leading edge and covering about the first 6 in. of the upper surface and first 18 in. of the lower surface. The glove was lightweight, had no moving parts, and required no maintenance. Wind-tunnel and flight tests of reshaped wings had revealed greatly increased resistance to airplane spins, which could result in aircraft stalls.

Before modification, three test airplanes would enter a spin about 18 of every 20 times pilots stalled wings and applied pro-spin controls. With LaRC's wing leading-edge design, the same planes had entered a spin only once in every 20 attempts. The spins that did occur had required improper airplane loading or extremely aggravated pilot inputs. And the pilot usually had three to four times as long (measured in seconds) to make a correction before the plane entered a spin.

NASA hoped ultimately to provide airplane designers with the ability to incorporate the modification as an integral part of a wing, rather than as an add-on, and the analytical tools to determine amounts of spin resistance for new planes. NASA was expanding the research beyond the series on unswept, low-wing airplanes to include high-wing planes and those with different airfoil shapes. (NASA anno Jan 85, 12)

March 1: NASA announced a demonstration at Lewis Research Center (LeRC) of a convertible gas turbine engine, which could operate in turboshaft or turbofan modes or both simultaneously, that would enable future rotorcraft to operate either as rotary or fixed-wing aircraft capable of speeds equivalent to commercial transports. The demonstration was part of a joint NASA/Defense Advanced Research Projects Agency (DARPA) program using a TF–34 engine modified under a General Electric Co. contract.

Such an engine would provide future X-wing (4-bladed) rotorcraft with the necessary shaft power to take off vertically. After reaching the conversion speed, near 200 knots, the pilot would disengage the transmission of the rotorcraft and lock the rotor blades in an "X" configuration for operation as a fixed-wing aircraft. The engine would convert from shaft to fan mode in 15 to 20 seconds to power the craft at speeds near .8 Mach. Landing would require reversing the conversion process.

During the past year, designers had evaluated performance characteristics of the engine and in Dec 1984 accomplished successfully the first transient operation simulating the conversion to and from rotary-wing to fixed-wing flight. The conversion had required 18 seconds, with the engine responding as predicted throughout. These tests had represented the first successful operation of a 5,000 hp-class convertible engine in both fan and shaft modes and the first dual mode operation for an engine.

The new engine would make possible a class of civil and military high-speed rotorcraft that could operate efficiently from hover to transonic speeds. (NASA Release 85–30)

March 21: NASA announced that it had scheduled for March 27 the first government flight in the joint Defense Advanced Research Projects Agency (DARPA)/U.S. and the Air Force/NASA X–29 flight research program. The X–29 was a high-performance research aircraft incorporating a wide variety of advanced technologies, each aimed at producing a better fighter-type aircraft. As a technology demonstrator, the X–29 was intended to provide engineering data for future design rather than to serve as a prototype for production.

The flight research program followed four recently completed contractor demonstration flights flown by Grumman Aerospace Corp., builder of the experimental craft for DARPA. All flights were at Dryden Flight Research Center (DFRC).

NASA, Air Force, and Grumman pilots would fly the first phase or initial concept evaluation of the government flight research program and then would gradually increase the performance of the forward-swept wing aircraft to speeds of about Mach 0.6 (almost 400 mph) at altitudes of about 30,000 ft. That phase would include as many as 30 flights and would conclude during the summer.

The first three flights would concentrate on handling qualities and stability and control aspects of the aircraft in its backup flight control mode, one of three different modes of the X–29's computer-controlled flight. These flights would also provide pilot familiarization and allow smooth program team transition from the contractor to the government. NASA test pilot Stephen Ishmael would pilot the first flight.

Aerospace Research and Development

Following the three flights, NASA would reconfigure the DFRC control room from its functional flight monitoring mode to a research mode that would allow the team to expand the aircraft's flight envelope for the second flight research phase scheduled for early fall 1985.

Prior to start of the second phase or full envelope concept evaluation, the contractor would modify the aircraft's flight control system to incorporate full envelope flight control system capabilities.

The second phase, lasting through October 1986, called for as many as 100 research flights by NASA and Air Force pilots and should extend the X-29's flight envelope to Mach 1.5, about 1,000 mph, and to 50,000 ft. in altitude. (NASA Release 85-40)

During May: Dr. George Keyworth, II, the President's science advisor, released a report of the Aeronautical Policy Review Committee that proposed national goals for research and development in aeronautics, NASA reported. The report stressed that, in the face of strong and growing challenges from foreign interests, the U.S. would maintain lasting aeronautical leadership only by vigorous application of the country's traditional strength—innovative technology. The report also pointed out that there were aeronautical opportunities that could make virtually all of that day's operational military and civil aircraft obsolete before the end of the century.

The report reaffirmed the government's role in supporting research and technology in aeronautics. "From our viewpoint, aeronautical research and technology remain an important investment. Our national interests, in particular our national security interests, dictate continued federal support."

The first goal outlined in the report addressed advancement of technologies applicable to a variety of subsonic aircraft including short-, medium-, and long-range transports; rotorcraft; high-speed turboprop engines; all-composite primary structures; high-lift aerodynamics systems, as well as new flight control and guidance improvements.

The second goal concerned higher-speed regimes of supersonic transports. The report noted that the Reagan Administration had made no commitment to supersonic transports, but was laying the groundwork, both in NASA and the Department of Defense, in the fundamental technologies essential for any future efforts in supersonic flight.

The third goal called for capturing the potential payoffs available from super high-speed flight—hypersonic speeds and transatmospheric vehicles. These vehicles would operate from conventional runways, maneuver at the fringe of the atmosphere, and ascend into space orbit when required.

The report concluded that the government could only provide part of the technological base and support to achieve those goals and that private industry must share in the development of new technology and in its application to new projects. (*NASA Activities,* May 85, 9)

June 14: With the aim of eventually conducting a piloted simulation study to assess airplane susceptibility to tumbling (an autorotative pitching motion where an airplane could theoretically go out of control end over end) and developing mathematical control laws for providing resistance to tumbling motions, Langley Research Center was testing in its spin tunnel a modern configuration, the X–29A forward-swept wing demonstrator, that had shown a tendency for the unaugmented airframe (without black box) to tumble, the *Langley Researcher* reported.

Certain modern aircraft design trends had established the need to study the phenomenon of tumbling. One such trend was relaxed static stability, where a displacement in pitch was not automatically corrected by the natural aerodynamics of the airplane; however, an electronic black box made such an airplane controllable. Other design trends contributing to tumbling were the elimination of aft-mounted horizontal tails and an unprecedented demand for pitch agility.

Langley's free-tumbling tests and "single-degree-of-freedom free-to-pitch tests" demonstrated tumbling motions that could not be controlled by the canard (forward horizontal wing) surfaces. The strake flaps were somewhat effective during tumbling. Researchers also obtained aerodynamic wind-tunnel data for the X–29A configuration over a 360° angle-of-attack range (to include flying backwards). (*Langely Researcher,* June 14/85, 4)

July 8: NASA announced that its Ames-Dryden Flight Research Center (DFRC) would begin a series of simulated airline flights to test two experimental laminar air flow control devices. Previous research had shown that laminar air flow could reduce aerodynamic drag from 25 to 40% and thus provide significant fuel savings under laboratory conditions. However, in actual flight insects, ice, and other obstructions adhering to the leading edges of an aircraft's wings could disrupt laminar flow.

Ames-DFRC had installed on its business-sized JetStar aircraft the two experimental laminar control flow devices and would simulate airline service flights in widely separated areas of the U.S. to experience a wide variety of contaminant conditions. NASA would conduct the simulated airline service flights just as an airline would under normal air traffic rules and regulations.

During the test, researchers planned up to four flights each day over a two-week period to gather as much information as possible on the test articles' performance.

The test article installed on the left wing used suction through 27 0.003-in. spanwise slots on the upper and lower surface to maintain laminar flow. A propylene glycol methyl ether (PGME)/water mixture discharged through several slots at the wing leading edge and flowing back over the wing provided ice and insect impact protection. Lockheed-Georgia manufactured the article for NASA.

The test article on the right wing used suction through approximately 1 million 0.0025-in. diameter holes in the titanium skin to maintain laminar flow on the upper surface of the article. For insect protection, a shield, which retracted at 6,000 feet altitude, extended much like a wing leading edge flap on commercial transports. Spray nozzles behind the shield could spray the PGME/water mixture on the test article for additional insect protection. For additional ice prevention, glycol was forced through the wing's porous metal section of the shield leading edge. McDonnell-Douglas manufactured the article for NASA.

The JetStar would carry a Knollenberg probe, mounted atop the aircraft, to measure the number and size of ice and water particles encountered in flight. A charge patch, located on the pylon that held the probe, would measure the static electric charge caused by particles in the air rubbing across the patch surface, thus giving a qualitative measure of ice and water particles. Correlation of the probe and patch data could calibrate the charge measurement in a simple cockpit display. (NASA Release 85-101)

July 26: In work done under a three-year grant in cooperation with Wichita State University (WSU) and aircraft and component manufacturers, Lewis Research Center (LeRC) developed the Electro-Impulse Deicing System (EIDI), which creates an electromechanical impulse at the leading edge of an aircraft's wing to produce a minute flexing of the metal skin that serves to shatter, debond, and remove ice buildup, the *Lewis News* reported.

At a LeRC symposium, more than 100 representatives from various companies and other government agencies met to review the results of the cooperative program to develop a technology base for EIDI. Cessna Aircraft Co. planned to apply for Federal Aviation Administration certification for their model 206 aircraft with the EIDI system.

Although several methods of deicing or anti-icing were available, all had undesirable energy requirements. The electro-magnetic impulse method, first suggested in 1937 by Rudolf Goldschmidt, a German national residing in London after World War II, offered a promising alternative.

During the 1950s and 1960s, industry used electro-impulse methods for metals forming in various industrial processes. In the 1970s, the USSR pursued further research including some aircraft applications. However, for various reasons the development stopped short of full implementation; the system still lacked a well developed underlying technology and known design parameters. Early EIDI system research offered promise of ice removal with low energy requirements, minimal maintenance (no moving parts), greater reliability, and a weight and cost competitive with existing methods.

In mid-1982, LeRC funded a six-month grant to WSU to work with two small plane makers and an aircraft electrical system manufacturer to do a feasibility study that resulted in a successful icing tunnel demonstration in

October/November 1982. LeRC then organized a consortium of companies for full development of the method for the whole range of civil aircraft, each company agreeing to contribute some services or equipment to the project in return for eligibility to submit its own products for deicing design and tests by the EIDI method.

Engine aircraft candidates for the EIDI system included commercial transport turbofans, business jet turbofans, commuter transport turboprops, and commercial transport propfans. (LeRC *News,* July 26/85, 2)

August 9: A flight today of an F–14 marked the completion of the latest phase in the plane's aileron-rudder interconnect (ARI) program, which was intended to coordinate turns, prevent wing rock, and resist spins at high angle-of-attach flight conditions, the DFRC *X-Press* reported. Dryden Space Flight Center (DFRC) modified the F–14 X's basic analog flight control system for ARI.

Although earlier configurations were successful in flight tests, pilots noted a deficiency in loss of roll power in high angle-of-attack flight regions. NASA then developed a "cross control" feature in the ARI system to correct the problem.

The cross control feature allowed the pilot to roll the aircraft opposite to lateral stick input but in the direction of the rudder pedal input. This gave the F–14 pilot roll power needed for tactical maneuvering while the aircraft was in a high angle-of-attack mode.

Flight tests of the cross control feature began in June 1985. Two NASA Ames-DRFC F–15 project pilots, three Navy pilots, and one Grumman pilot flew these initial evaluation flights. As part of the evaluation program, four other Ames-DRFC pilots not involved in the F–14 program flew the aircraft.

NASA Ames-DFRC pilot Ed Schneider said after his flight that "The new modification permits improved rolling performance at angles of attack above 30°, yet the system retains spin resistance and wing rock suppression, both ARI features."

Navy personnel taking part in the evaluation would make presentations on program results to NAVAIR. Navy pilot Lt. Commander Chuck Baucom would also present results at the Society of Experimental Test Pilots' September symposium. (DFRC *X-Press,* Aug 16/85, 2)

August 14: The deputy for tactical systems at the Air Force Aeronautical Systems Division (ASD) issued a request for proposals for demonstration and validation of the Air Force's planned Advanced Tactical Fighter (ATF), *Defense Daily* reported. ASD planned to award up to four fixed-price contracts running approximately 33 months.

Seven ATF concept formulation contractors would compete for the contracts. The companies were Boeing Military Airplane, General Dynamics,

Grumman Aerospace, Lockheed, McDonnell Douglas, Northrop, and Rockwell Internatl.

ASD said the demonstration/validation contractors would develop detailed concepts and designs of the ATF to meet operational needs, conduct testing to validate and assess proposed concepts, and identify and reduce overall program risks prior to start of full-scale development.

General Electric and Pratt & Whitney were competitively developing the joint advanced fighter engine for the ATF.

ASD was planning a FY 89 decision on full-scale development of the ATF, with initial operational capability set for 1995. The Air Force estimated the cost of developing and building 12 research and development ATF aircraft at $15.3 billion and believed it could build the ATF at a total unit cost only 20% higher than that of the F-15.

A top Air Force official had noted in 1984 that the ATF would have a basic Mach 1.5 speed and that it could employ technologies being demonstrated by the Grumman X-29 forward swept wing aircraft. (*D/D,* Aug. 14/85, 1)

August 23: The testing of the AFTI/F-111 mission adaptive wing program at NASA's Ames-Dryden Flight Research Center (DRFC) was an important milestone, the DRFC *X-Press* reported. It was the aircraft's first taxi under its own power in four-and-a-half years.

In the test, the control room was fully staffed, the air crew was aboard the aircraft just as they would be for the first flight, and the airplane operated as if it were ready to fly. Engineers checked and verified all aircraft and control room displays, including the mission adaptive wing color panel that duplicated cockpit displays and featured a caution panel that identified discretes, a trouble-shooting tool.

"Things went really well," said Ames DRFC AFTI/F-111 project manager Louis Steers. "There were some small problems on the aircraft, but no show stoppers." (DFRC *X-Press,* Aug 30/85, 2)

September 9: NASA announced that pilots flew the NASA/Army XV-15 tiltrotor research aircraft with its stability control augmentation system (SCAS) both operative and inoperative. The SCAS reduced pilot workload, allowing the pilot to devote attention to other tasks; however, if the SCAS failed, the pilot had to provide stability and control commands. A primary question answered by the NASA research was whether a pilot could fly a tilt-rotor aircraft with a sidestick controller when the SCAS failed.

The objective of the XV-15 tilt-rotor research aircraft program was to develop and evaluate tilt-rotor technology for civil and military applications. A tilt-rotor aircraft could lift off vertically like a helicopter and then, by rotating its rotor thrust from vertical to horizontal, fly like a conventional airplane at speeds up to 345 miles per hour.

The sidestick control system evaluation provided data for the V–22 Osprey, a tilt-rotor aircraft NASA was developing for the Department of Defense (DOD). DOD would consider the flight test when deciding whether to use a sidestick controller in the V–22.

Results of the flight evaluation indicated the sidestick controller provided a reasonable means of controlling the aircraft with or without the SCAS. However, when the third axis (yaw) was added to the sidestick controller, pilots noted a significant increase in workload, indicating that additional studies were necessary in this area.

The sidestick controller system used onboard digital computers and electronics, which replaced mechanical linkages that normally transmitted pilot signals to the aircraft's control surfaces. Ames Research Center developed the system and would refine and reevaluate it during future control system studies. (NASA Release 85–124)

September 10: The C–5's new B model made its first flight today from Dobbins Air Force Base, Georgia, climbing to 15,000 feet and cruising at 300 knots—an appropriate altitude and speed to determine airworthiness, the Air Force Systems Command (AFSC) *Newsreview* reported. Airborne for slightly more than three hours, the Galaxy began 55 hrs of production flight testing designed to evaluate its improved features. Evaluation would be shorter than usual for a newly designed aircraft because the C–5B only carried state-of-the-art features added to the already proven C–5A.

Col. Thomas Stover, acting C–5B director in Aeronautical System's Division (ASD) Deputy for Airlift and Trainer Systems, called the first flight a "key event . . . ushering in a significant improvement to our strategic airlift capability."

The nine-member crew alternately shut down and restarted each of the airlifter's four General Electric TF39–1C engines, which restarted without a hitch.

Production flight evaluation would include airborne testing of an improved onboard computer called MASAR II, the malfunction detection analysis and recording computer. Pilots would also evaluate the improved automatic flight control and landing gear actuation systems.

In October 1982 ASD awarded Lockheed Georgia a $50-million preliminary contract to begin C–5B production, and the first aircraft rolled out July 12, 1985. The Air Force had contract options to buy 50 C–5Bs, at total cost of $7.817 billion.

Military Airlift Command expected to receive its first B model by December, with delivery of the 50th expected near mid-1989. (AFSC *Newsreview,* Sept 27/85, 1)

Aerospace Research and Development

September 13: Researchers in the Workload and Ergonomics Branch of the Armstrong Aerospace Medical Research Laboratory were working to adapt machines to pilots rather than the traditional other way around, the Air Force Systems Command *Newsreview* reported. "The success or failure of a modern Air Force aircraft and mission depends on whether the pilot has the ability to handle the unpredictable situations typical in a wartime environment," said Maris Vikmanis, supervisory industrial engineer for the lab. He explained that laboratory researchers studied how pilots performed in different environments, including combat, to provide information to aircraft designers so that future aircraft would be better adapted to pilots.

But he pointed out there was more to it than simply assessing new aircrafts' cockpit designs. "Not only do we have to observe how the pilot performs in the aircraft cockpit, we also have to tap into the pilot himself," Vikmanis explained. "We must know how much effort the pilot is expending, his mental capability, and his workload limitation."

In analyzing cockpit design, researchers checked into how well a pilot could see and reach cockpit controls and displays.

Only in existence as an aerospace research field since 1979, workload research measured both the pilot's performance and his internal state of well-being. By combining experiments in the laboratory and in flight, lab researchers measured human performance/workload using three different methods.

First, researchers used a computer-based performance test to measure a pilot's physical and mental action. By having the pilot perform up to 11 different tasks, researchers could characterize various aspects of human behavior.

Second, pilots answered, after flying a simulator or an actual aircraft, subjective questions about their experience.

A third method measured human performance/workload from a physiological viewpoint by collecting and analyzing variable responses of the human brain, heart, muscles, and visual system.

Lab scientists and engineers used the test data to prepare a workload scale, which focused on problems the pilot, aircrew, and the aircraft itself might encounter in the future.

NASA was applying similar methodology to the Space Shuttle and its crew, particularly to examine the effects on astronauts of microgravity as an element of stress. (AFSC *Newsreview,* Sept 13/85, 7)

September 17: NASA announced the first flight at Ames Dryden Flight Research Facility, scheduled for late September, of the advanced fighter technology integration (AFTI) F–111 with a mission adaptive wing (MAW). The AFTI F–111 research aircraft was modified with a wing that could change its camber (fore and aft wing curvature) and its wing sweep to adjust to varied flight conditions.

NASA expected that MAW's variable camber capability would allow optimum efficiency at supersonic, transonic, and subsonic speeds by adopting the best wing shape for each speed. The MAW program was intended to demonstrate that smooth variable camber technology showed improvements over current aircraft wing lifting-altering devices in such areas as aircraft payload and range capability, maneuverability, fuel efficiency, and aircraft handling qualities.

The first flights would test the airworthiness and flight envelope expansion phase of the program, followed by the research phase. The plans for the initial phase called for about 15 flights that gradually increased the MAW aircraft to speeds of approximately Mach 1.05 and altitudes of about 35,000 feet. NASA expected that phase to be complete by January 1986; and the research phase, about 30 flights, by June 1986.

The MAW system had smooth-surfaced, variable-camber wing leading- and trailing-edge flaps, an actuation mechanism, hydraulics, and redundant computers to control flap positions. Six independent trailing-edge flaps (three per wing) and two leading-edge flaps provided a smooth continuously variable wing camber using flexible fiberglass skins on the upper surface and sliding panels on the lower surface.

Fully developed, the MAW system would result in an aircraft equipped with an automatic flight control system with several control modes. Cruise camber control mode would optimize the trailing-edge flap position for maximum cruise speed. Maneuver camber control mode would continuously position leading and trailing edge for optimum lift-to-drag ratio depending on the lift coefficient and speed of the aircraft. Maneuver enhancement/gust alleviation mode would enhance the aircraft reponse to pilot control inputs, while reducing the aircraft's sensitivity to wind gusts.

Designed for research rather than as a production prototype, the AFTI F–111 was a joint program of Ames-Dryden Flight Research Center and the U.S. Air Force's Flight Dynamics Laboratory. The Boeing Military Airplane Co. manufactured the MAW. (NASA Release 85–129)

October 4: As part of the continuing Lewis Advanced Turboprop Project (ATP), supported by some 50 contracts and 15 grants, two advanced propeller propulsion systems recently underwent testing, the *Lewis News* reported. Lewis Research Center's (LeRC) Advanced Turboprop Project Office was directing development of the new highly loaded, multi-bladed propellers for use at speeds up to Mach 0.85 and at altitudes compatible with commercial air support system requirements. Advanced turboprop engines offered the potential of 15 to 30% savings in aircraft fuel burned relative to advanced turbofan engines (50 to 60% savings over the current turbofan fleet). Investigators were developing both single- and counter-rotation propeller systems.

Researchers at Wright Patterson Air Force Base were testing in a static propeller test rig the first Large-Scale Advanced Propfan (LAP), designed and

Aerospace Research and Development

built by Hamilton Standard Div. of United Technologies Corp. under a NASA/ LeRC contract. Testing of the nine-foot diameter propfan, powered by an electric drive motor in the rig test, began in late August.

At approximately the same time, the General Electric Co. began static proof-of-concept tests of its Unducted Fan (UDF) demonstrator engine at the GE Peebles, Ohio, test site. LeRC partially funded development of this advanced counter-rotating pusher propellers propulsion system.

After completing a series of ground tests, both advanced propeller propulsion systems would undergo flight tests starting in mid-to-late 1986. The propfan would be tested in a wing-mount installation on a modified Gulfstream II testbed aircraft under a NASA-contracted program with Lockheed; the UDF would be tested as an aft-mounted pusher on a Boeing 727 as part of a GE/Boeing cooperative program.

The propfan of the LAP program was an example of the application of single-rotation tractor propulsion technology, whereas the UDF was a unique example of counter-rotation pusher propulsion technology. The UDF was unique in that its propellers were directly driven by the multi-stage power turbine without the need for gearbox speed reduction. Normal procedure in propeller installations was to drive the propeller through a reduction gearbox so that both the propeller and its driving turbine operated at close to their optimum speeds for peak efficiency.

The APT effort was also addressing more conventional counter-rotation propfan configurations, as well as associated gearbox designs. (LeRc News, Oct 4/85, 1)

October 9: NASA announced it signed with the Federal Aviation Administration (FAA) and the Department of Defense (DOD) a memorandum of agreement to conduct a two-year study of the possible benefits to the nation of the continued development of tilt-rotor aircraft. In particular, the study would assess the broader implications of V-22 aircraft development, including the potential for other versions and sizes, both civil and military; civil certification issues; civil production impact on the defense industrial base; and any indirect technology spin-offs.

Tilt-rotor aircraft combined the vertical take-off-and-landing ability of the helicopter with the relatively higher cruising speed and longer range of fixed-wing aircraft.

A small proof-of-concept aircraft, the XV-15, was tested successfully, leading to a larger tilt-rotor aircraft, the V-22 Osprey, which DOD was developing. DOD expected the V-22 to bring the tilt-rotor configuration closer to maturity, enabling lower risk development of other versions and sizes, for both military and civilian use.

The FAA would certify any civil versions of a U.S. tilt-rotor aircraft and would assess its impact on the national airspace system, airport and heliport development, and other air service needs.

NASA acquired a high level of expertise in tilt-rotor technology during development and flight testing of the XV-15, and the agency also had a charter to study use of NASA-developed technology and to assess future research needs. (NASA Release 85-144)

October 18: NASA announced the first flight of the NASA/Air Force/F-111 mission adaptive wing (MAW) [see Aerospace R & D/Aeronautics, Sept. 17] at the Ames-Dryden Flight Research Facility. The flight accomplished functional verification of the variable camber MAW at flight conditions up to MACH 0.6 and altitude of 15,000 feet.

The Boeing Military Airplane Co. manufactured the MAW. (NASA Daily Activities Report, Oct 21/85)

December 13: NASA announced that during 1985 it began a series of research flights under simulated airline conditions to evaluate the effects of an operational environment on new technology for smooth, or laminar, airflow over aircraft wings [see Aerospace R & D/Aeronautics, July 8]. Previous research had shown this technology could reduce aerodynamic drag from 25 to 40% under laboratory conditions and could provide significant fuel savings. However, insects, ice, and other obstructions could disturb laminar flow in actual flight service.

To test the system in an operational environment, NASA's Ames/Dryden Flight Research Facility's business-sized Jet Star, equipped with two experimental laminar flow control devices, was based in Atlanta and flown in and out of commercial airports in the southeast and midwest, including those in St. Louis, Cleveland, and New Orleans, to test the devices in humid summer climates.

While based in Pittsburgh to test the devices in early autumn east coast conditions, the plane flew in and out of airports in Boston; Chicago; Chattanooga; Cleveland; Charleston, W.Va.; Washington, D.C.; Detroit; Bangor, Maine; New York; Kalamazoo, Michigan; Oklahoma City; Albuquerque; and Denver among others.

The flight tests would continue in 1986 with the plane based in Cleveland during January. (NASA Release 85-170)

Balloons

February 11: Ben Abruzzo, who took part in the 1st balloon crossings of the Atlantic and Pacific oceans, died February 11 along with five others when his twin-engine Cessna 421 plane crashed after takeoff from Coronado Airport north of Albuquerque, the *Washington Post* reported.

In 1978 Abruzzo, along with Maxie Anderson and Larry Newman crossed the Atlantic in the Double Eagle II, landing near Paris after the 3,000-mile flight. Late in 1981 Abruzzo, Newman, and two other men flew the Double Eagle V across the Pacific. Lifting off November 9 from Nagashima, Japan, Abruzzo crash-landed four days later in northern California in one of the worst storms there in 20 years. Abruzzo and another crew member won the Gordon Bennett race in 1979. (*W Post*, Feb 13/85, C6)

Large Space Systems

January 24: Marshall Space Flight Center (MSFC) announced that a December 1985 experiment, EASE or the experimental assembly of structures in EVA (extravehicular activity), aboard Space Shuttle flight 61-C would provide information about humans working in space. In the experiment, two crew members would move six aluminum beams, each 12 feet long and four inches in diameter, into space from the Space Shuttle payload bay and assemble them into a tetrahedron shape and then disassemble them. This experiment would demonstrate human ability to build strong, practical structures in space. During the experiment videotape and still photographs would record crew member movements to link beams by special connectors at the beam tips.

This building approach was a technique in the Large Space Structures program called an "erectible" method, as opposed to a "deployable" method that would not be demonstrated until 1986 or 1987 when NASA would orbit a structural-assembly demonstration experiment. MSFC had conducted tests similar to EASE in its neutral buoyancy simulator, a million-gallon water tank that gave a sense of weightlessness to submerged people working in space suits. Researchers would compare assembly and disassembly times recorded in the space test with results found in the neutral buoyancy tank. However, MSFC's Ed Valentine, experiment manager, noted, "We'd like to be able to measure a human's productivity in space scientifically, and we're not really sure if time measurements are all that are needed to guage that, so we'll be examining other possible criteria to use in the future."

A Langley Research Center structural assembly experiment would accompany the MSFC experiment. (MSFC Release 85-3)

During April: NASA announced it awarded to Howard University's School of Engineering Large Space Structure Institute (LSSI), Washington, D.C., an ongoing grant expected to total $1.9 million to develop theoretical knowledge supporting the design and construction of very large space systems. NASA expected the knowledge to make a significant contribution to the establishment of large space stations and space settlements.

The LSSI research project would help demonstrate the practical benefits of U.S. space capabilities, including construction of a solar power satellite that would convert the sun's energy into electricity for transmission back to earth and construction of large platforms that would allow scientists to observe the earth for improved understanding of the weather and the ground water table.

The LSSI team would consist of Dr. Taft Broome, chief project investigator and chairman of the Department of Civil Engineering, and other Howard University faculty members; and one faculty member each from the Massachusetts Institute of Technology, Cambridge University, and Rensselaer Polytechnic Institute. Graduate students from the latter institutions would also participate.

NASA was primarily interested in LSSI doing the concept development and the research and demonstration analysis. Once Broome and his team provided the agency with the needed data, NASA could turn the project over to private industry for implementation. (NASA Activities, Apr 85, 8)

Launch Vehicles

January 9: NASA announced that Lewis Research Center awarded a $32,346,000 follow-on contract to General Dynamics's Convair Division (GD/C) to erect, test, and launch (with accompanying launch facility modifications) Atlas-Centaur launch vehicles and to perform support testing and launching of Centaur upper stages from the Space Shuttle in 1986; GD/C would work at Kennedy Space Center. The cost-plus-award-fee contract, beginning January 1, 1985 and running through December 1986, was a continuation of work GD/C had performed under earlier contracts. The Centaur should increase the Space Shuttle's capability to carry heavier payloads from low-earth orbit to geosynchronous orbit or interplanetary trajectories. (NASA Release 85–5)

During March: In a report of 1984 accomplishments, Arnold Engineering Development Center (AEDC) commander, USAF Col. Philip Conran, said that AEDC conducted more than 200 separate test and evaluation programs during FY 84 and demonstrated the importance of the center's full-spectrum test support to development and operational aerospace programs.

In that year, propulsion systems testing dominated test and evaluation efforts, turbine engine testing being the busiest in the history of the center. Much of that had resulted from the new start for the F109 next-generation trainer engine and the fly-off competition between the F110 and F100 engines for the USAF's F–16 Fighting Falcon and the subsequent product verification of the F110.

Aerospace Research and Development

In rocket motor testing, AEDC conducted three altitude test firings of the inertial upper stage (IUS) in an effort that helped Space Division and Boeing Aerospace Co. successfully understand and fix the April 1983 upper stage in-flight anomaly. After failure of the *Palapa B-2* and *Westar 6* commercial satellites to reach desired orbits, AEDC supported the conversion of defense and commercial satellite perigee and apogee booster motors from carbon-carbon to carbon phenolic rocket nozzles.

AEDC considered completion of construction of the aeropropulsion systems test facility (ASTF) its outstanding achievement for 1984 and was proceeding with activation and acceptance testing to prepare the ASTF for operations in September 1985.

In the flight dynamics area, AEDC completed in a specially modified vacuum chamber the initial ground testing phase of the antisatellite pathfinder sensor that required the use of a spin mount rig, which had been three years in development, and an optical alignment system that provided accuracies on an order of magnitude better than any other available system. AEDC also developed and successfully demonstrated a new test technique of spinning a nose cone to evaluate reentry erosion.

One of the most productive programs at AEDC in 1984 was for the C-17 advanced cargo aircraft. AEDC provided more data in less time than usual and saved the sponsors about $1 million in wind-tunnel test time.

The in-house technology program, which supported testing and evaluation missions, delivered needed test techniques for testing of advanced reentry-vehicle materials, engine icing, and rocket motors. For example, AEDC applied a flash X-ray technique and worked with the University of Tennessee Space Institute to enhance the image of three test firings of the inertial upper stage, resulting in successful completion of the anomaly investigation. (*AEDC Test Highlights*, Spring/85, 2)

July 12: Arnold Engineering Development Center completed the fourth and final qualification test firing of the Payload Assist Module-D (PAM-D) II rocket motor, the Air Force Systems Command *Newsreview* reported. The PAM-D II featured increased propellant capacity (7200 lb.) over its predecessor, the PAM-D (4400 lb.), and could boost both military and commercial communications satellites into geosynchronous orbit about 22,300 miles above earth.

A Space Shuttle would deliver the motor and its payload into space at an altitude of 160 to 180 miles, where the PAM-D II satellite package, spinning at 30 to 85 rpm, would spring-eject from the orbiter's open bay doors.

At a safe distance from the orbiter, the PAM-D II, which served as a perigee kick motor, would ignite and propel its payload more than 22,000 miles higher. PAM-D II would then separate from the payload, which was positioned by its own apogee motors activited by signals from earth.

AEDC test-fired the motor in one of its high-altitude rocket development test cells at a simulated pressure altitude of approximately 100,000 feet while spinning it about its thrust axis at 70 rpm. Test objectives were to determine ballistic performance and to take plume (exhaust) radiation measurements, which were important in determining if the exhaust radiation would damage the motor case or satellite.

With the fourth test completed, the Air Force considered the motor ready for production and flight. Morton Thiokol's Wasatch Division was the rocket's manufacturer. NASA had tentatively scheduled PAM-D II's first flight for November 1985 aboard the Space Shuttle Challenger to place the RCA SATCOM satellite into orbit. (AFSC *Newsreview,* July 12/85, 7)

July 26: NASA announced the Lewis Research Center awarded a $10,169,016 contract to General Dynamics Corp. for management and engineering services in support of the Atlas/Centaur launch vehicle program. The contract also called for the company to provide technical management, engineering design, analysis and development, engine testing and development, and reliability and quality assurance. The cost-plus-award fee follow-on contract would begin July 1985 and continue through December 31, 1986.

Centaur was a high-energy, upper-stage used with expendable boosters to deliver large payloads to geosynchronous orbit. NASA was modifying it for use as an upper stage on the Space Shuttle. (NASA Release 85-110)

August 1: Lewis Research Center (LeRC) announced it awarded a $12,386,535 sole source fixed-price-incentive-fee contract to Teledyne Industries, Inc. for the fabrication, assembly, and testing of digital computers and remote multiplexer units in support of the Space Shuttle/Centaur program. Work began in July 1985 and would run through October 1987.

Digital computers would serve as the focal point for the Centaur flight-control system, processing data from the other flight boxes including the multiplexer units. They would also provide the hardware capability to interface with the orbiter and ground system.

NASA scheduled the first Centaur launch—the Ulysses mission—for May 15, 1986, from the Space Shuttle payload bay followed on May 21, 1986, by the Galileo mission. (LeRC Release 85-58)

September 6: NASA announced that its Lewis Research Center (LeRC) awarded a $79,194,665 contract to General Dynamics Corp.'s Convair Division to support the Space Shuttle/Centaur program from August 30, 1985 through June 1987. Under the cost-plus-award-fee contract, General Dynamics would supply labor, materials, and resources necessary for management and engineering services and launch operations services to implement the program.

Included in the multi-contract concept were services such as overall program management; administration and engineering support; technical management; operations, planning, and integration; and launch complex modifications.

NASA planned that the Space Shuttle/Centaur combination in May 1986 would launch Ulysses, the first and only planned mission to observe the polar regions of the sun, and a week later Galileo to explore Jupiter's atmosphere. In spring 1988 the combination would launch the Venus radar mapper mission to investigate the surface and interior of Venus. (NASA Release 85–125)

September 9: General Dynamics would delay for more than two weeks the delivery of the first Space Shuttle/Centaur upper stage to the company's facility at Cape Canaveral Air Force Base, adding an increased schedule risk to already tight timing for launch in May 1986 of the Galileo and Ulysses missions on trajectories toward Jupiter, *Aviation Week* reported. Martin Winkler, General Dynamics Space Shuttle/Centaur program manager, said the stage remained in San Diego to complete acceptance testing and also to conduct some tests that were to have been done in Florida. Winkler said that keeping the stage in San Diego to complete assembly and testing would not affect the testing and processing schedule in Florida, since it would be delivered in time for the scheduled mating with the Centaur integrated support structure (CISS).

Winkler said the schedule called for completion of acceptance testing by September 11 and that there were no unresolved flight safety issues, although it was possible some might come up during future safety reviews and subsystem qualification testing.

Problems fixed prior to shipment included a mislocated forward bearing bracket that held each of the 12 springs on the stage, installation difficulties resulting from a minor tooling problem, and strut reengineering required for installation of a component in the tank.

The CISS, which was then in Florida for the first Space Shuttle/Centaur stage, was flight hardware and was a complete system with only some small parts yet to be delivered by mid-September. NASA had scheduled a major test for September 16, when the assembly would rotate with all of the installations in place to assure clearances. A November milestone would be four tanking tests.

The second Space Shuttle/Centaur was on schedule; contract date for delivery of the second stage and second CISS was the end of December. General Dynamics estimated it would deliver the second Space Shuttle/Centaur on November 22 and the second CISS on October 29. (*AvWk*, Sept 9/85, 24)

October 7: NASA announced that it signed a 21-month agreement with Scott Science and Technology, Inc. (SST) to provide technical expertise to SST in the development of a commercial liquid-fuel upper stage for boosting Space Shuttle-deployed satellites to geosynchrous orbit.

SST, whose president was former astronaut David Scott, had worked for two years on developing the upper stage, called the Satellite Transfer Vehicle, which would be able to boost satellites ranging from 2,000 to 19,000 lb. On some missions, the stage would be recoverable.

Engineers at NASA's Johnson Space Center would monitor the stage's development and consult with SST staff on technical problems. SST would reimburse NASA for the use of any test facilities and the salaries and travel of JSC personnel working on the project. (NASA Release 85-143)

November 21: NASA's Marshall Space Flight Center (MSFC) requested proposals for a 12-month study of an advanced high-thrust, high-performance liquid oxygen/liquid hydrogen rocket engine that could be used for the advanced launch vehicles being examined in NASA/Air Force Space Transportation Architecture Studies (STAS), which included heavy lift launch vehicles, *Defense Daily* reported. NASA considered the advanced LOX/hydrogen cryogenic engine and a new LOX/hydrocarbon engine to be key elements to advanced transportation systems. MSFC said it planned to award up to three parallel firm-fixed-price contracts for the study of the Space Transportation Main Engine (STME).

During the study, the contractors would prepare parametric trade-off data for engine requirements and configurations that might be employed in future launch vehicles. NASA would compare these engine configurations to the Space Shuttle main engine (SSME) for application to new launch vehicles. The contractors would identify candidate engine configurations, compare their features, and then recommend engine configurations that would best satisfy the requirements of future launch vehicles.

The three principal contenders for the STME study were Rockwell's Rocketdyne Div., prime contractor for the SSME; and Aerojet Technologies and United Technologies's Pratt & Whitney Aircraft, who conducted component improvement studies for the SSME. (*D/D,* Nov 21/85, 105)

Military Applications

April 26: Although the Defense Department's fourth edition of *Soviet Military Power* described an unfavorable imbalance between U.S. and Soviet strategic forces and research programs [see USSR/Missiles, Apr. 1], the annual report of the Under Secretary of Defense for Research and Engineering stated that the U.S. would lead the Soviet Union in virtually every basic technology "critical to defense" over the next 10 to 20 years, *Science* reported, including

Aerospace Research and Development

many that would be crucial to the success of a sophisticated ballistic missile defense.

U.S. technology was superior in computers and software, electro-optical sensors, guidance and navigation, materials, microelectronics, optics, propulsion, radar, robotics, signal processing, telecommunications, and signature reduction (or stealth), the report said, and also in life sciences, production/manufacturing, and submarine detection. The Soviets, in contrast, led in no areas and matched the U.S. only in aerodynamics, conventional and nuclear warheads, and power sources.

In the area of weapons systems, the report said the U.S. was superior in 25 of 30 categories; the only exceptions were in categories where the U.S. had not chosen to deploy any weapons or where the U.S. was ready to deploy what experts believed were vastly superior weapon systems. These categories included deployed antiaircraft missiles, a deployed ballistic missile defense, deployed chemical weapons, and deployed antisatellite weapons.

The report concluded that the "strengthening of the U.S. military R&D commitment will make it more difficult for the USSR to close existing technology gaps." (*Science,* Apr 26/85, 478)

May 10: Deputy Secretary of Defense William Howard Taft gave approval to the U.S. Air Force to begin development of higher thrust derivatives of the Pratt & Whitney F100 and General Electric F110 engines that powered the F–15 and F–16 fighters, *Defense Daily* reported. The development program, expected to cost between $400 and $500 million, would boost the thrust of the 24,000-lb. thrust F100 and 27,000-lb. thrust F110 to 29,000 lb., about the maximum possible.

Air Force Assistant Secretary Thomas Cooper said durability, not thrust, was the key problem for fighter engines and that the new engines would have the same durability as current engines. He said the Air Force planned to start production of the engines in 1989 and to complete annual buys on a shared basis. The Air Force would seek warranties on the derivatives as on current engines, access to proprietary data, and pre-priced production options.

Cooper said the new engines, costing slightly more than current engines that were between $2 and $3 million apiece, would have the same dimensions as present engines, could be used in all models of the F–15 and F–16, and could be retrofitted. He pointed out the Navy could also use either derivative engine for its fighter program.

Under questioning, Cooper declined to say whether the Air Force planned to buy the Northrop F–20 fighter in FY 87, but said the Air Force was interested in the aircraft and needed more detailed cost data to make a decision. Northrup had offered to supply 396 F–20s over four years at a fixed price of $15 million per plane; however, Cooper said the Air Force required more detailed data on the "bits and pieces" of the program in order to make the cost comparison that people were asking for. (*D/D,* May 10/85, 57)

June 26: U.S. Rep. George Brown, Jr. (D–Calif.), author of an amendment to prohibit tests of the antisatellite miniature homing vehicle (MHV–ASAT) against objects in space, suggested that in light of the successful ground-based laser test with the orbiting Space Shuttle [see Space Transportation System, Military Applications, June 21], the U.S. should drop the MHV in favor of a laser antisatellite system, *Defense Daily* reported.

Brown said that ". . . This very successful demonstration raises the question of whether the United States should spend several billion dollars on the technologically less-advanced MHV ASAT system, still in the testing stage, when we could have a more technologically advanced, far more versatile system available for deployment in the same timeframe . . . and probably at a lesser cost."

Brown emphasized the significance of the laser test, which he noted had hit a target on the Space Shuttle "less than a foot in diameter, much smaller than an ICBM, traveling faster than a missile, and there is no question but that a missile would be just as easy to hit."

Although lethality was not demonstrated in the test, the U.S. had "demonstrated lethality more than 10 years ago when we knocked down a drone airplane with a laser, and we have now in the inventory lasers with about a million times the energy of the one that was used" in the Space Shuttle test, Brown added. "There is no question that lethality is achievable with existing technology." (*D/D,* June 26/85, 318)

July 12: The U.S. Air Force rolled out the C–5B today, the first B model of the free world's largest military cargo aircraft, the Air Force Systems Command *Newsreview* reported. Although it looked like a C–5A, the new C–5B had many improved features including state-of-the-art materials and avionics. It also had the new wings, which would boost the service life of the aircraft to 30,000 flight hours, that the Air Force was retrofitting to the A model.

The Air Force planned to buy 50 C–5Bs, nearly doubling its capability to airlift non-stop anywhere in the world large equipment such as the Army's 74-ton mobile scissors bridge, Abrams M–1 tanks, and helicopters. When the purchase of the C–5Bs was completed, the Air Force would be about 18 million-ton-miles per day short of its goal of 66-million-tons per day of outsized airlift capability, the minimum capability recommended in a 1981 Congressionally mandated mobility study. The Air Force planned to make up the difference by acquiring the C–17.

Four improved General Electric TF–39 engines, rated at 41,000 lb. of thrust each, powered the C–5B. The aircraft seated 75, and its seats were 30% lighter than the A model's, decreasing total aircraft weight by 725 lb and cutting fuel consumption by an estimated $13 million over the life of the C–5B fleet.

The C–5B's 24 main landing gear wheels had carbon brake assemblies with multiple discs. More than 400 lb. lighter per aircraft, the new brakes

Aerospace Research and Development

doubled the life of the A model's beryllium system and should bring about an estimated 20-year savings of $20 million.

In October 1982 the Air Force's Aeronautical Systems Division had awarded a $50-million preliminary C–5B production contract to Lockheed-Georgia. Plans called for early September flight testing of the first production aircraft at Dobbins AFB, Georgia. The Air Force had contract options to buy 50 C–5Bs for $7.817 billion. (AFSC *Newsreview,* July 29/85, 1)

August 23: In the first test of the precision location/strike system (PLSS), three TR–1 aircraft picked up a radar signal like one an enemy might use to guide a missile attack and passed the signal direction and time data to a ground station computer. The computer analyzed the data, compared inputs from the three aircraft, and pinpointed where the radar signal was coming from, the Air Force Systems Command *Newsreview* reported. If the source had been an enemy site for radar-directed missiles, artillery, or an anti-aircraft system, the U.S. Air Force could quickly direct an F–16 to it for an attack. The PLSS passed the test by zeroing in on a prepositioned radar emitter the moment its signal started.

Lt. Col. Dayl Donahey, PLSS program director for the Aeronautical Systems Division, called it "a giant leap forward" for future wartime capabilities of U.S. tactical forces. "The system performed better than expected at this point in its development testing," he said.

Lockheed Missiles and Space Co. was developing the PLSS, which featured an integrated, lethal mix of ground-based and airborne electronics.

In an operational scenario, the three highly instrumented, high-altitude TR–1s would fly race-track patterns behind the forward edge of a battle area. Around the clock, in all kinds of weather, the TR–1 listening platforms would detect enemy electronic transmissions. Despite dense electromagnetic interference, the TR–1s could pick up individual pulses of enemy radar. (AFSC *Newsreview,* Aug 23/85, 1)

December 6: Lockheed Georgia, manufacturer of the C–5B Galaxy, resumed flight testing of the aircraft after it replaced about 12,000 fastener nuts on the plane, the Air Force Systems Command (AFSC) *Newsreview* reported. Lockheed-Georgia grounded the plane after its September maiden flight when quality control inspectors found that construction crews had installed aluminum, instead of steel, nuts on the first eight planes. The mistake slowed production and temporarily halted certain aircraft flight testing.

Col. Thomas Stover, acting C–5B program director for Aeronautical Systems Division, said, "It is important to note that flight safety was never an issue. The concern was for development of maintenance problems 10 to 15 years down the road." Stover explained that steel nuts were stronger and lasted longer than aluminum nuts and that, although there were about 3.7 million nuts for each C–5B, only 35,000 had to be steel.

Lockheed-Georgia spokesman Dick Martin said workers installed the wrong nuts because some engineering drawings, used by supervisors and assembly line workers, were not specific about nut type. He said about 35% of the nuts had to be replaced on the first aircraft.

The Air Force had scheduled the first C-5B for delivery by January 1, and Lockheed-Georgia still expected to deliver the aircraft on time because the company had been four weeks ahead of schedule before the delay caused by the replacement. (AFSC *Newsreview*, Dec 6/85, 4)

Simulation Technology

March 20: NASA announced that a five-year project at Lewis Research Center (LeRC) resulted in development of the real-time multiprocessor simulator, a low-cost easily programmable computing system that simulated or monitored complex physical processes in real time. The simulator could offer considerable savings, for example, in the development and testing processes for producers of airplane engines.

Real-time simulations were particularly important in studying the internal aerodynamics of a jet engine, including such variables as the performance of turbine and convergent-divergent nozzles, bypass valves, injectors, and fan and compressor geometry and the coordination of these variables with fuel flows and flight controls. In a real-time simulation of a helicopter engine, the NASA system solved all 63 engine equations within 1/100th of a sec.

The system had hundreds of real-time applications where there was a critical need to simulate or monitor many processes at once or when the reaction of one process could be vital to the performance of another, such as monitoring the structural dynamics of a space station at several locations, with a processor dedicated to each location.

The essence of the NASA system was a network of computers tied together and operated simultaneously but with different programs, each one a portion of the overall task. The system could detect a change in one aspect of a dynamic process and report the effect of that change on other parts of the process. Using state-of-the-art, high-performance microcomputers, NASA scientists had designed hardware configurations that enabled a person at a single terminal to control up to 10 microcomputers operating simultaneously, each handling a portion of a total program. NASA scientists had also developed a companion form of computer language, taken from the PASCAL format, that enabled an engineer who was not a computer specialist to program his or her own multiprocessor for experimentation and monitoring. (NASA Release 85-39)

December 6: NASA announced that a flight simulation laboratory, the Man-Vehicle Systems Research Facility (MVSRF), was in operation at its Ames

Aerospace Research and Development

Research Center to give scientists a new tool to study human factors in operating commercial aircraft.

The MVSRF contributed to flight safety by allowing researchers to simulate the complex set of relationships among pilots, their crews, converging aircraft, the weather, and air traffic controllers. Using the facility, scientists could study the effects of new and increasingly automated technologies on flight crew performance. NASA's human factors researchers worked directly with Federal Aviation Administration (FAA), airplane manufacturers, and commercial airlines to improve flight safety. For example, with the 727 simulator, NASA scientists researched airplane landings to learn methods that minimized fuel consumption while allowing pilots to maintain safe distances between their planes and other air traffic.

The MVSRF included an exact replica of a Boeing 727 cockpit, a mock air traffic control station, and a cockpit known as the advanced concept flight simulator (ACFS). In both the 727 replica and the ACFS, a computer visual system provided out-the-cockpit window displays of approaching aircraft and landscapes of many U.S. airports. The computer system also depicted clouds and other weather conditions, the glow of the horizon at dusk and dawn, and the varied texture of the area illuminated by the simulated aircraft's landing lights.

Researchers at the MVSRF put the pilots and controllers through created situations, during which the researchers monitored communications among flight crew members and between the crew and traffic controllers. They also read the aircraft's instruments and observed flight control settings while making audio and video recordings that included tracking a pilot's eye movements. (NASA Release 85-162)

December 20: Scientists at the Air Force Geophysics Laboratory's (AFGL) Space Physics Div. developed a computer program that predicted the electrical effects of the highly charged polar space environment on spacecraft. These effects could range from the production of noise in circuits to system failure in extreme cases, the Air Force Systems Command's (AFSC) *Newsreview* reported. Called Potential of a Large Object in the Auroral Region (POLAR), the code simulated in three dimensions the electrical interaction of the Space Shuttle or any other space vehicle with the polar space environment in the region from 100 to 600 miles above earth. Space Shuttle launches from Vandenberg Air Force Base would orbit through the polar region, and the polar charging code could provide important data for the design of large space structures such as space stations.

Lockheed Corp. used POLAR to predict the effect of charging on the Space Shuttle in low-altitude polar orbits with payloads funded by the Air Force Space Test Program Directorate and other Air Force and NASA programs. RCA performed calculations with POLAR for the Defense Meteorological Satellite Program vehicles.

NASA's Spacelab 2 mission scientists used POLAR to predict the electrified gaseous environment, or plasma, of the Space Shuttle when the Plasma Diagnostics Package, a free-flying satellite, deployed. And NASA's LAGEOS Satellite Project Office would use POLAR to study charged particle drag on this laser reflecting satellite.

POLAR was a successor to NASCAP (NASA Charging Analyzer Program), the computer code AFGL developed with NASA to model charging of Air Force satellites at geosynchronous altitude. The European Space Agency (ESA) had used that code to design weather and TV relay spacecraft, and the Jet Propulsion Laboratory used it to design the Galileo satellite.

POLAR would incorporate data derived from future spacecraft flights and would become a computer-aided design tool for the construction of the advanced extravehicular mobility unit, the manned maneuvering unit, the NASA-proposed space station, and the Strategic Defense Initiative. (AFSC *Newsreview*, Dec 20/85, 7)

Transatmospheric Vehicle

April 2: Robert Cooper, director of the Defense Advanced Research Projects Agency (DARPA), said in testimony delivered in late March before the U.S. House Armed Services Committee research and development subcommittee that a space plane capable of flying to anywhere on the earth within half an hour to intercept enemy bombers or perform space spy missions might be possible within the next 10 years, the *Washington Times* reported.

The U.S. Space Shuttle, which had limited ability to maneuver in space, was put in orbit only after a lengthy and complicated rocket launch. So the U.S. Air Force wanted a "transatmospheric vehicle" that would be able to quickly reach and return from space like an ordinary aircraft. Cooper said a recent breakthrough in so-called ramjet technology might make possible an aircraft capable of taking off from a runway and reaching speeds of Mach 25 (times the speed of sound), the velocity necessary to escape earth's gravity.

Cooper noted the plane would weigh about 100,000 lb., take off from and land on a 10,000-foot runway, and probably use hydrogen as a fuel. Only a small amount of rocket power would be required to "deorbit" the plane, he said.

DARPA proposed to build two full-scale model ramjet engines for testing on the ground. "There are a number of very interesting technical problems here . . . none of them insurmountable. They're all within our grasp," Cooper said.

Cooper explained that the ramjet combustion providing the extra boost took place only when the air flowing through the jet engine chamber was at supersonic speeds. Ordinarily, to get air flowing at this speed through the entire length of the chamber, the aircraft itself had to be moving at Mach 6. However, "a way has been found," he said, "to use a subsidiary scheme for driving the air through the engine up to supersonic speeds, even though the vehicle on which the engine is attached is sitting on the runway."

By slightly varying the geometry of the engine, it might be possible to fly the aircraft from takeoff to Mach 6, and "with slight variations" fly from Mach 6 to 14 and then to Mach 25, Cooper said.

The Air Force and Navy were interested in the engine, development of which marked a resurgence in hypersonic speed research that had been "put on a back burner" in the late 1960s, Cooper commented. (*W Times*, Apr 2/85, 3A)

November 21: U.S. Air Force Maj. Gen. Donald Kutyna said today that the Defense Department had decided to proceed with a $500 million program to design a hypersonic plane capable of flying around the globe in less than two hours and of flying in the highest reaches of the atmosphere and to provide a low-cost method for launching satellites and other equipment critical to President Reagan's Strategic Defense Initiative, the *NY Times* reported. Speaking at a Colorado Springs symposium on space technology, Kutyna said construction of the plane itself could cost $2 to $3 billion.

"It is something we are very serious about, and we think the technology is now within reach," said Kutyna, who coordinated the Air Force's research and development of space systems. "We have been examining the principles for some time, and now we are ready to head into the next phase."

However, in Washington an aide to George Keyworth II, the President's science advisor, said, "Funding for the project's next phase is not assured." He noted Congress would have to approve the program. "This is all a little premature," the aide noted. "We have not yet presented the plan to President Reagan."

But on November 20 Keyworth and top officials on the Defense Advanced Research Projects Agency (DARPA) and NASA presented details of the plan to members of Congress interested in aerospace issues.

Unlike the Space Shuttle, which was launched vertically with rocket boosters, the aerospace plane would take off like an ordinary airplane, but would not have jet engines with heavy compressors, making the aerospace plane a light-weight vehicle. "Once you reach about Mach 3," or three times the speed of sound, Kutyna said, "the air streaming through the engines begins to compress itself. Some studies indicated that the plane could attain speeds of Mach 20, although there were experts who said half that speed would be more reasonable.

If successful, the project could solve a key problem for the military: drastically reducing the cost of placing satellites and defensive space weapons in orbit. (*NYT*, Nov 22/85)

ASTRONAUTS

February 6: President Reagan and USSR Chairman Konstantin Chernenko had approved the idea of establishing ties between the U.S.'s Young Astronauts and the Soviet's Young Cosmonauts, the *Washington Post* reported, as a step toward better U.S.-Soviet relations and peaceful cooperation in space.

The program would begin with a Young Cosmonauts' delegation visiting the U.S. to view a space launch and tour space facilities. Young Astronauts would then write to Young Cosmonauts, explaining their interest in space and what the groups might accomplish together. Those writing the best letters would visit, all expenses paid, their letters' recipients. In addition, the International Aeronautical Federation had adopted the Young Astronaut program and planned to offer it to other nations to establish ties between Young Astronauts and Cosmonauts around the world.

In his *Washington Post* column, Jack Anderson wrote that the Young Astronaut Council was recruiting 6-to-16-yr-olds for future space exploration and that those joining would be eligible to participate in the "Letter to a Young Cosmonaut" contest. (*W Post*, Feb 6/85, F11)

February 8: Johnson Space Center (JSC) announced that Warren North, JSC specialist in astronaut selection and training, retired February 1 after 38 years with NASA.

North had joined the National Advisory Committee for Aeronautics (NASA's predecessor agency) in 1947 as a test-pilot engineer at LeRC. In 1959 North had transferred to NASA Headquarters, where he participated in early planning for Project Mercury, including selection and training of the seven original Mercury astronauts; in 1962 North assumed leadership of the JSC division responsible for training astronauts for Gemini rendezvous and docking development flights and for the Apollo lunar landing program. North was currently special assistant to the JSC flight-operations director for planning Space Shuttle crew training and flight-simulation equipment and techniques.

North received a bachelor's degree from Purdue University and master's degrees from Case Institute of Technology and Princeton University. (JSC Release 85-7)

May 9: NASA announced that astronauts Paul Weitz, Dr. Kathryn Sullivan, and Dale Gardner would support NASA activities at the 36th Paris Air Show at Le Bourget from May 31 through June 9 [see Space Station Program, Mar. 22].

Weitz, then chief of astronaut appearances at Johnson Space Center, was a member of the first Skylab crew in 1973 and later commanded the first flight of the Space Shuttle orbiter Challenger in April 1983. He had logged 793 hours in space.

Sullivan, an astronaut since 1978 with 197.5 hours in space, served as a mission specialist on the October 1984 Space Shuttle mission 41–G and was the first American woman to perform an extravehicular activity or space walk, during which she and astronaut David Leestma conducted a 2.5-hour demonstration of the potential of refueling satellites in space.

Gardner, an astronaut since 1978 with 337 hours in space, had flown on the August 1983 Space Shuttle mission STS-8, the first night launch and landing using the orbiter Challenger, and in November 1984 on the 51–A mission, during which the crew retrieved two communications satellites that had failed to achieve proper orbits after launch from an earlier Space Shuttle flight. NASA refurbished the satellites for reflight.

During the Air Show, the astronauts would make numerous public appearances, participate in media interviews, and assist NASA's marketing efforts at its exhibit in the U.S. pavilion. (NASA Release 85–72)

May 22: The *Congressional Record* reported the remarks of Cong. Frank Annunzio (D-Ill.) during a U.S. House session on May 22.

". . . Today I am introducing legislation to strike medals in commemoration of the Young Astronaut Program. The medals would be sold by the Young Astronaut Council to raise money to help fund the program.

". . . The Young Astronaut Program is designed to use the excitement of the United States space program to increase student interest and skills in math and science through a comprehensive program in elementary and junior high schools.

". . . This September, the council will launch its 'Decade of Discovery.' Already more than 20,000 schools have responded to the program. The council estimates every elementary and junior high school in the nation will be signed up for the program by the beginning of the next school year.

". . . To assist the council, and commemorate the 'Decade of Discovery,' the Young Astronaut Program Medal Act will authorize the United States Mint to strike up to 750,000 gold, silver, and bronze medals for the council to sell to raise money for this exciting educational endeavor. The council will pay the mint the cost of manufacture plus a 10 percent surcharge.

". . . I urge my colleagues to join with me in cosponsoring this worthwhile legislation." (*CR-House,* May 22/85, H3585)

June 4: NASA announced selection of 13 new astronaut candidates, six pilots and seven mission specialists, who would report to the Johnson Space Center in late summer to begin a one-year period of training and evaluation. Upon successful completion of the training, they would be eligible for assignment to Space Shuttle flights.

The candidates would join 90 current astronauts. Including the new group, NASA had named 157 astronauts since the beginning of the program. In making the latest selections, NASA considered 33 civilians from the selection rosters developed during the 1984 selection process and 133 nominees from the military services.

Those selected as pilot candidates were LCdr. Michael Baker (USN), Maj. Robert Caban (USMC), Capt. Brian Duffy (USAF), Stephen Oswald (civilian), and LCdr. Stephen Thorne (USN). Mission specialist candidates were Jerome Apt (PhD—civilian), Capt. Charles Gemar (US Army), Linda Godwin (PhD—civilian), Richard Hieb (civilian), Tamara Gernigan (civilian), Capt. Carl Meade (USAF), and Lt. Pierre Thuot (USN). (NASA Release 85-84)

June 7: NASA announced it would alter its process of soliciting astronaut applications for pilot and mission specialist positions to one where it would accept applications from civilians on a continuing basis beginning August 1, 1985, and nominees from the military on an annual basis. NASA would make selections in the spring of each year with successful candidates reporting in the summer. Mission requirements and attrition rate of the current astronaut corps would determine number of candidates selected. (NASA Release 85-89)

June 19: NASA announced that Dr. Joseph Allen, who in November 1982 was a mission specialist on the four-member crew of mission STS-5, the first fully operational flight of the Space Shuttle, would resign July 1, 1986 from NASA to become executive vice president of Space Industries, Inc., a Houston-based firm pursuing commercial ventures in space. Allen also flew on mission 51-A, the second flight of Discovery, during which he participated in the deployment of two communications satellites and, in the first space salvage mission, helped retrieve the Palapa B-2 and Westar VI satellites for return to earth.

In addition to his astronaut work, Allen was a mission scientist for Apollo 15, staff consultant on science and technology to the President's Council on International Economic Policy, and NASA assistant administrator for legislative affairs.

Allen had received NASA's Exceptional Scientific Achievement Medal, Exceptional Service Medal, and Superior Performance Award. (NASA Release 85-91)

September 13: During a recent ceremony, the U.S. Naval Academy named a new Space Systems Research Chair in memory of two naval astronauts—Roger B. Chaffee, who was a Lt. Commmander in the Navy, and Clifton C. Williams, who was a Major in the U.S. Marine Corps. The faculty chair was designed to enhance space education programs for the Navy. The astronauts' widows, Jane Williams and Martha Chaffee, joined Rear Admiral Charles Larson, Academy superintendent, and Commodore Richard Truly, commander of the Naval Space Command, for the plaque unveiling ceremony.

Both Williams and Chaffee were among the third group of astronauts NASA named in October 1983. Williams served as backup pilot for the Gemini 10 mission. He was killed in a T-38 crash October 5, 1967. Chaffee was assigned as one of the pilots for the first three-man Apollo flight. He was killed in the Apollo 1 spacecraft fire January 27, 1967, at Kennedy Space Center along with astronauts Virgil Grissom and Edward White.

The Naval Space Command and the Naval Academy signed earlier in the year an agreement to establish the Space Research Chair. The principal objective of the new Academy professorship, as stated in the agreement, was to provide a means for future officer candidates at the Academy to further their understanding of space systems and their naval applications. "It will provide the mechanism by which midshipmen and faculty members will become exposed, involved, and committed to scientific activities at the forefront of the technologies related to space systems," the agreement said.

Rear Admiral William Ramsey, director of the Naval Space Systems Division, also participated in the Space Chair dedication ceremony. He presented the Defense Superior Service Medal to three Navy astronauts present at the dedication—Commodore Truly, Captain Robert Crippen, and Commander Robert Gibson—commending them for "exceptionally superior achievement" as astronauts. Truly served as pilot on STS-2 and commander of STS-8; Gibson served as pilot for STS 41-B; Crippen had flown on STS-1, STS-7, STS 41-C, and 41-G. (JSC *Roundup,* Sept 13/85, 2)

October 28: Astronaut John Fabian resigned from the NASA astronaut corps in the summer of 1985. It was the eighth astronaut resignation over the previous 16 months, the *Washington Post* reported, suggesting the way of life that was the hallmark of astronaut service might have begun to wear thin. Besides Fabian, NASA lost over those months Joseph Allen, Terry Hart, William Lenoir, Jack Lousma, Thomas Mattingly, Donald Peterson, and Richard Truly, all veterans in the prime of their careers whose combined experience covered 12 of the 21 Space Shuttle missions flown.

"Are these resignations something we worry about?" said NASA Administrator James Beggs. "Yes, especially the younger ones like Allen [45] and Fabian [44], who still have a lot of tread left on them. We've now begun to lose the guys we've educated and trained to the most difficult things we do, like spacewalks, and if this trend starts to increase, it's going to disturb me," he concluded.

Stripped of its glamor, astronaut service was demanding work that kept its members away from home, the *Post* said. One female astronaut, who just began training for a flight late in 1986 said she had seen her husband twice in the previous two months, "The last time it was for 12 hrs," she said, "and for six of those we were both asleep."

Astronaut training ranged from studying the physics and biology of spaceflight to making parachute jumps and three-day survival visits in Panama's jungle. Classroom hours were equivalent to time spent by Ph.D. candidates. Physical fitness was a must, but astronauts had to exercise on their own time.

Once assigned a mission, the astronaut's training pace quickened. At Johnson and Kennedy Space Centers there were around-the-clock computer simulations of astronaut tasks and potential problems in space. Astronauts also visited other NASA centers and contractors' facilities to become familiar with the equipment they would use on their missions.

NASA officials said there was nothing they could do to slow the training pace, because they believed it was the reason the U.S. had not had a fatal accident in space. NASA Administrator Beggs indicated that the way to keep astronauts happy and in the corps was to keep them busy and assigned to a mission.

That didn't keep Fabian from resigning, however, the *Post* pointed out. "I'll miss flying and I'll miss the people, but I don't want to be known to posterity as the oldest astronaut to fly in space," Fabian said. (*W Post*, Oct 28/85, A1)

ASTRONOMY

January 7: NASA was accelerating design of an unmanned U.S. space mission that would rendezvous with a comet and possibly drop a penetrator to return data directly from the comet's surface, *Av Wk* reported. NASA officials believed that this first U.S. comet-rendezvous mission using the new Mariner Mk. 2 spacecraft would be a high-priority item in the FY 87 budget.

NASA planners would probably select the Comet Kopff or Comet Wild-2 as mission target; FY 87 new-start funding could make a 1991 launch possible.

Unlike Halley's Comet flybys, the U.S. spacecraft in a rendezvous mission would spend many months flying in formation with a comet. Mission concept proposed rendezvous well before closest approach to the sun, so the spacecraft could observe the comet while it was in its asteroidal stage and before solar wind would blow large amounts of dust off its surface. The spacecraft would remain in formation during the comet's active phase and possibly the post-solar-encounter phase.

GSFC and ARC were examining comet-penetrator concept systems that would return compositional data using a gamma-ray spectrometer. (*AvWk*, Jan 7/85, 18)

January 8: NASA announced it signed with the University of Arizona, Tucson, a memorandum of understanding to study, plan, and, if approved by Congress, develop an orbiting astrometric telescope facility, which would be mounted in the mid-1990s on the planned space station, to detect and study planetary systems around other stars. Astrometric telescopes measured with extraordinary precision positions of stars and other celestial objects.

Eugene Levy, director of the University's Lunar and Planetary Laboratory, and David Black, head of Ames Research Center's (ARC) Space Sciences Division, would be project principal scientists. Black described the effort as "the beginning of a new scientific discipline. The results of the search, whether positive or negative, will profoundly alter our view of the universe."

The gravitational pull of planets orbiting a star causes it to wobble as seen against the sky. By measuring this wobble, an astrometric telescope could determine the presence of planets and perhaps determine the properties of such a planetary system. Although ground-based telescopes use the wobble-measuring technique to measure properties of double stars in orbit around one another, an astrometric telescope could measure the disturbance in a star's motion caused by objects as small as planets. Until researchers find other planetary systems, they cannot fully test theoretical models used to explain the origin of our solar system or the formation of stars. The joint project would be coordinated with development of photometric and spectroscopic techniques and theoretical modeling of planetary system formation.

ARC would clarify final telescope design and manner of mounting on the space station as well as oversee telescope and instrument construction and mounting. The University would establish operating facilities on the ground and coordinate continuing scientific investigations that would extend more than a decade.

The project would include a core program of planet detection and study, and guest investigators could use the telescope for astrometric studies of other important celestial objects. Initial study and definition work would begin shortly; ultimate implementation would require Congressional approval and funding. (NASA Release 85-3; ARC Release 85-01; ARC *Astrogram*, Jan. 17/85, 1; *A/D*, Jan 17/85, 95)

January 15: Naval Observatory astronomer Richard Walker proposed a simple explanation for the inclination of a descending passageway in the Great Pyramid of Cheops in Egypt, the *Washington Post* reported. Walker had first examined the suggestion of John Herschel, an early 19th-century astronomer, that the 337-foot passageway was at an angle of 26.5230° to point at the North Star, allowing the pyramid to serve an as astronomical observatory as well as tomb for the pharaoh. Walker noted, however, that due to the wobble of earth's axis in its orbit around the sun, no prominent star could have been seen from the base of the passageway in 2800 B.C.

Walker suggested the angle resulted from the construction technique. By placing three stones of equal length horizontally and another of equal size on top of the 3rd horizontal stone, the angle from the top stone to bottom stone at the other end was 26.5°. (*W Post*, Jan 15/85, A10)

January 17: The *Washington Times* reported on the "Planet X" theory of Daniel Whitmire and John Matese, physics and astronomy professors at the University of Southwestern Louisiana, in which they had suggested that an undiscovered planet triggered a rain of comets upon earth, resulting in extinctions of dinosaurs and other species at roughly 26-million-year intervals. "Planet X, if it's out there," Whitmire said, "lies in an orbit outside Neptune and, every 30 million years or so, passes through a ring or disc of comets that also lie out there. When it does, gravity diverts some of the comets toward earth, kicking up huge clouds of dust on impact that disrupt the weather enough to kill off many species of plants and animals."

Whitmire noted the theory would explain several phenomena that had puzzled scientists: fossil evidence of extinctions at approximately 26-million-year intervals and the slight variations in Neptune and Uranus's orbits by Planet X's gravitational pull. (*W Times*, Jan 17/85, 3A)

January 18: Observations by the *International Ultraviolet Explorer* (*IUE*) revealed what appeared to be a variable gas cloud surrounding the star Beta

Pictoris. This was the first evidence of gas around the star as well as the large dust particles previously observed from the *Infrared Astronomy Satellite* (*IRAS*) and ground-based telescopes, the JSC *Roundup* reported. Many astronomers considered Beta Pictoris, in the constellation Pictor in the southern hemisphere, a leading candidate for another planetary system.

Dr. Yoji Kondo, Goddard Space Flight Center (GSFC) *IUE* project scientist, and Dr. F.C. Bruhweiler of Catholic University jointly made the observations in December and said the coexisting dust and gas clouds were consistent with scientific theories of planetary system formation. "Many stars like Beta Pictoris show evidence of being surrounded by clouds of gas at ultraviolet wavelengths," said Kondo. "If such data imply that these stars also are surrounded by a protoplanetary gaseous cloud with a disk of dust particles, then many nearby stars may have evolving planetary systems. That's exciting," he concluded, "because these stars would offer a unique opportunity to see the way a solar system might appear in formation." Theories of planet formation had generally predicted that planets evolved from large stellar clouds of dust and gas that gradually condensed into orbiting planetary bodies, a theory strengthened by the *IUE* observations. (JSC *Roundup*, Jan 18/85, 3)

March 3: The *NY Times* reported the death in Moscow of Iosif S. Shklovsky, internationally known Soviet astrophysicist and early champion of the search for intelligent extraterrestrial life. Shklovsky was 68.

Dr. Herbert Friedman of the National Academy of Sciences in Washington said Shklovsky's contributions were diverse and, in some cases, basic. For example, in 1953 Shklovsky had proposed that the strange light from such features as the Crab Nebula, puzzling because it occurred at all wavelengths instead of being concentrated into spectral lines like ordinary light, was synchrotron radiation generated by electrons whirling in extremely strong magnetic fields. Researchers later discovered that the Crab Nebula and other remnants of great stellar explosions, or supernovas, emitted synchrotron radiation across the full width of the spectrum.

Shklovsky also proposed the widely held concept that periodic X-ray emissions from space came from superdense stars circling one another at close range.

In 1962 Shklovsky published a book arguing that intelligent creatures might exist in other worlds; it had gone through four Soviet editions and was translated by Carl Sagan. Later, Shklovsky began to doubt the existence of nearby civilizations, but had urged the creation of hugh space colonies.

He received the Lenin Prize in 1960 and was made a corresponding member of the Soviet Academy of Sciences in 1966. (*NYT*, Mar 6/85, B10)

March 6: NASA announced selection of the scientific investigators whose projects would be included on the proposed Advanced X-ray Astrophysics

Facility (AXAF), an observatory that would operate in low-earth orbit for at least 15 years. A spacecraft carrying an array of instruments and providing power, precise pointing, and data transmission would house the observatory containing a 1.2m diameter, 10m focal length telescope. AXAF construction could begin around 1987–88 with launch approximately five years later.

In orbit, AXAF would join the Hubble Space Telescope, scheduled for launch in 1986, and the Gamma-Ray Observatory, planned for a 1988 launch. The three observatories, together with the planned Space Infrared Telescope Facility, would provide simultaneous observations of cosmic sources over infrared, visable, ultraviolet, X-ray, and gamma-ray wavelengths.

Those selected included instrument principal investigators to design and fabricate scientific instrumentation for placement in the telescope focal plane, interdisciplinary scientists to provide expertise on X-ray astrophysics and other fields of astronomy, and a telescope scientist to guide telescope fabrication.

Instrument principal investigators selected were Dr. Gordon Garmire, Pennsylvania State University (charged coupled device imaging spectroscopy); Dr. Steven Murray, Smithsonian Astrophysical Observatory (high-resolution camera); Dr. Stephen Holt, Goddard Space Flight Center (GSFC) (X-ray spectroscopy investigation for the AXAF); Dr. Claude Canizares, Massachusetts Institute of Technology (high-resolution X-ray spectroscopy using AXAF); and Dr. Albert Brinkman, University of Utrecht, Netherlands (high throughput transmission grating for cosmic X-ray spectroscopy).

Interdisciplinary scientists selected were Dr. Riccardo Giacconi, Space Telescope Science Institute (advanced X-ray astrophysics facility mission); Dr. Jeffrey Linsky, National Bureau of Standards (the coronal structures of selected cool stars and close binary systems); Dr. Richard Mushotzky, GSFC (a program to measure the mass of galaxies and clusters of galaxies with AXAF); Dr. Andrew Wilson, University of Maryland (studies of radio jets and the narrow line regions of active galaxies with AXAF); and Dr. Andrew Fabian, Cambridge University, United Kingdom (cooling flows in clusters and galaxies).

Dr. Leon Vay Speybroeck, Smithsonian Astrophysical Observatory, would serve as telescope scientist.

The investigators, as members of the AXAF science working group, would provide scientific and technical guidance throughout the project and would receive a specified amount of time to use the telescope during its first 30 operational months. Astronomers from the U.S. and around the world would use the majority of observing time.

Marshall Space Flight Center managed the AXAF project for NASA's astrophysics division, Office of Space Science and Applications. (NASA Release 85-33)

March 12: Soviet and French scientists would soon launch into orbit a space observatory, the Gamma 1 Project, that would carry a large gamma-ray telescope to help determine sources of cosmic rays, streams of highly charged particles that bombarded the earth from space, the *NY Times* reported the newspaper *Izvestia* as saying. Researchers hoped to obtain a detailed picture of the sky in gamma rays in order to understand the mysterious nature of gamma stars and their possible relationship with known astrophysical objects. *Izvestia* had gone on to say that testing was underway on the observatory but did not report a launch date.

The *Izvestia* article said the two countries had cooperated in designing a special spacecraft, to be placed in an orbit 216 miles above earth, capable of carrying nearly two tons of scientific equipment for the Gamma 1 experiment, including main, small gamma-ray and X-ray telescopes.

Franco-Soviet space cooperation dated back to an agreement signed in 1966; France was a major partner in the Vega missions the Soviets had launched in December to examine Venus and Halley's Comet and, in June 1982, the French cosmonaut Jean-Loup Chretien had been the first and was the only man from a country outside the Soviet block to go into space aboard a Soviet spacecraft, the Soyuz T-6. (*NYT*, Mar 12/85, C3)

April 10: NASA announced that observations by the *International Ultraviolet Explorer* (*IUE*) satellite revealed what appeared to be a variable gas cloud surrounding the star Beta Pictoris in the constellation Pictor in the southern hemisphere, a star that many astronomers considered a leading candidate for a planetary system resembling earth's. Scientists using the *Infrared Astronomy Satellite* (*IRAS*) and ground-based telescopes previously had reported large particles of dust around the star.

The *IUE* findings of coexisting dust and gas clouds were consistent with scientific theories of planetary system formation, noted Dr. Yoji Kondo, *IUE* project scientist at Goddard Space Flight Center. In general, theories predicted that planets evolved from large stellar clouds of dust and gas that gradually condensed into orbiting planetary bodies. Observations with the *IRAS* strengthened this theory by suggesting that planetary disks surrounded a number of nearby stars. Also, astrometric measurements of the movement of several stars indicated the possible presence about each star of a planetary body several times more massive than Jupiter.

Kondo and his associate Dr. F. C. Bruhweiler of Catholic University, Washington, D.C., used the *IUE* telescope to obtain the ultraviolet spectra of Beta Pictoris to determine the manner in which ultraviolet light was absorbed by the gas cloud and the extinction of the light by the dust cloud. The ultraviolet spectra showed an absence of selective extinction of the light at shorter wavelengths, evidence that the star's light was being blocked evenly at all wavelengths by dust particles greater than one micron (millionth of a meter), as infrared data had reported.

"The *IUE* observation," Kondo said, "indicates the gas is either clumpy or is varying under the pressure of the stellar wind and radiation. If so, this would tend to complement *IRAS* and ground-telescope observations that a substantial dust cloud of protoplanetary material exists around Beta Pictoris."

Kondo cautioned that the lack of observed extinction could also be due to viewing the star just a little off-plane, so that the telescope's line-of-sight was not aligned directly through the disk of orbiting dusty matter. However, Bruhweiler added, "that would be unlikely. All evidence seems to point to a belt too thick around Beta Pictoris for the line-of-sight to be missing the dust belt." (NASA Release 85-52)

April 16: NASA announced it selected the first participants in a program to assimilate discoveries from space astronomy experiments into a comprehensive modern astrophysical theory in order to use more effectively future experiments such as the Hubble Space Telescope, the Gamma-ray Observatory, the Solar Optical Telescope, and similar planned projects. After evaluating 49 proposals from 300 individuals at 80 institutions, NASA selected seven groups to participate in the astrophysical theory program.

Dr. Richard McCray and a scientific team from the University of Colorado would study detailed ways in which spectra form in a variety of astrophysical sources such as stars, supernovae, and active galactic nuclei.

Dr. Jeremiah Ostriker and colleagues from Princeton University would analyze the most distant observable parts of the universe to determine possible effects of the theoretical "big bang".

Dr. Pierre DeMarque and a team from Yale University would develop computer models of the sun to examine the internal motions and magnetic fields that caused sunspots, solar flares, and solar wind.

Dr. Roger Chevalier and his team from the University of Virginia would study the hot gases that appeared to surround clusters of galaxies, individual galaxies, and supernova remnants to determine the origin and motion of the gases as well as their effect on the development of galaxies.

Dr. Ronald Taam and colleagues from Northwestern University would study the origin of rapid bursts of high-energy radiation from neutron stars and globular clusters in order to model the ignition, nuclear evolution, and propagation of burning fronts on surfaces of neutron stars and to understand the nature of hot plasma confinement in magnetic fields near such subjects.

Dr. David Black, heading a team from Ames Research Center and the University of California, Santa Cruz and Berkeley campuses, would investigate various physical and chemical processes taking place in the formation of stars, concentrating particularly on the stars' origins in giant molecular clouds in interstellar space and the protostellar/protoplanetary disk-shaped nebulae surrounding fledgling stars.

And Dr. Simon White and a team of scientists from the University of Arizona would study questions regarding star and planet formation, including properties of star-forming regions, galaxy formation, and evolution.

NASA would bring the researchers together in 1986 to exchange information and to encourage a better understanding of the advances that science has made with data acquired from space research. (NASA Release 85-59)

May 31: When a gap occurred in ancient records dealing with the appearance of Halley's Comet—its appearance in 164 B.C.—pertinent information was found on ancient clay tablets originating in Babylonia and residing in the British Museum, the Jet Propulsion Laboratory's (JPL) *Universe* reported.

The museum had about 1,200 astronomical texts from Babylonia, but they were in disarry and few scholars had looked at them. Museum representative C.B.F. Walker said that information from the tablets relating to Halley constituted "the first significant addition to our knowledge of past history of the comet" in more than a century. The information was impressed into the clay in the cuneiform alphabet, an ancient system that employed wedge-shaped symbols.

JPL's Dr. Don Yeomans said that since recorded history mentioned all ancient sightings of the comet except the 164 B.C. appearance, archeologists and astronomers had joined forces to discover such a reference, resulting in the find at the British Museum.

"It filled in a blank," Yeomans said. "The information showed that we were extremely close to where we thought it [Halley's 164 B.C. visit] would be."

For the International Halley Watch, of which JPL was the lead organization, Yeomans was discipline specialist in the area called "astrometry," the science of predicting where, when, and how the comet would appear and what the world could expect from it.

Yeomans served in a similar capacity for the planned mission to the comet Giacobini-Zinner with the international cometary explorer (ICE) spacecraft. (JPL *Universe*, May 31/85, 1)

June 5: NASA announced that flight controllers for its International Cometary Explorer (ICE) performed the first and largest of four midcourse corrections to direct the spacecraft toward a September 11 intercept of Comet Giacobini-Zinner. Conducted from ICE mission control at Goddard Space Flight Center, these corrections used two thousand pulses of the spacecraft's two-lb. hydrazine thrusters to change its heading so it would fly through the comet's tail 16,200 miles from the comet's nucleus. The thruster burns lasted four-and-a-half hours.

In 1982, NASA redirected the International Sun-Earth Explorer, which was launched in 1978, toward the comet for what would be the first satellite/comet encounter in history. This encounter would provide scientists with their first look at the makeup and dynamics of a comet's tail.

The ICE trajectory adjustment would place the satellite/comet encounter within the coverage window of the world's largest single dish radio astronomy telescope located at the Arecibo Observatory, Puerto Rico. The 300-m dish, managed by the National Science Foundation, would supplement coverage by three NASA deep space network stations at Goldstone, California; Madrid, Spain; Canberra, Australia; and the recently completed 64-m antenna of the Japanese Space Agency at Usuda, Japan.

NASA based ICE's course adjustment on telescopic observations of Giacobini-Zinner as the comet emerged in April from behind the sun, the first observation since July 1984. The comet was approximately 149 million miles from the sun on the inbound leg of its orbit between the sun and the neighborhood of Jupiter.

NASA would direct three smaller orbit maneuvers as ICE approached the comet, the last scheduled for three days prior to encounter. Course corrections were necessary because material spontaneously outgassed from the comet's core, acting as jet thrusters on the comet's body, slightly altered the its orbital path.

Astronomers in 1900 discovered Giacobini-Zinner, which returned to earth's neighborhood every six-and-one-half years. It would not be visible to the naked eye, but amateur astronomers with small telescopes would be able to see it. (NASA Release 85-86)

June 7: Two groups of scientists, who were making observations at infrared and radio wavelengths, reported the strongest evidence so far of a supermassive black hole in the center of the Milky Way galaxy, the *NY Times* reported, strengthening the theory that very massive black holes existed in the cores of all galaxies. Astrophysicists had already deduced that black holes might be at the cores of some distant galaxies.

Eight physicists and astronomers from the University of California, Berkeley, reported in the British journal *Nature* their infrared observations showing gas whipping around a core at 100,000 to 400,000 miles per hour. They concluded the gas was in the grip of a black hole four million times the mass of the sun.

The group measured the velocities near the center of the core with sufficient precision to show that they decreased further out in a way to be expected if controlled by gravity from a single central object. They believed this ruled out a dense cluster of stars, proposed by some astronomers, because that gravity field would be more uniform.

Radio astronomers using coordinated observations with antennas in Westford, Massachusetts; Green Bank, West Virginia; Fort Davis, Texas; and Goldstone, Hat Creek, and Owens Valley, California; and the 27 antennas of the very large array in New Mexico aimed the antennas at a radio source, a suspected black hole, almost directly in the center of the Milky Way called

Sagittarius A (in the constellation Sagittarius). The results indicated that the source, despite its enormous mass and output of energy, was smaller than the orbit of Saturn. Donald Lyndon-Bell of Britain's Royal Greenswich Observatory and Martin Fees of Cambridge University had proposed 14 years before that there might be such an object there.

Astronomers using improved radio and infrared techniques had mapped in more and more detail the Milky Way core, which was 30,000 light years from earth. They could not see the core with visible light because of dense clouds of dust and gases. (*NYT* June 7/85, A15)

June 22: NASA announced that the International Ultraviolet Explorer (*IUE*) observed for the first time the comet Giacobini-Zinner while the comet was approximately 87 million miles from earth on the inbound leg of its orbit between the sun and the neighborhood of Jupiter. *IUE* would continue observations of the comet as it moved closer and ultimately became the target of the world's first intercept, on September 11, 1985, of a comet by a satellite, the International Cometary Explorer (*ICE*) spacecraft.

The June 22 *IUE* observations of Giacobini-Zinner produced ultraviolet spectrograms—graphic depictions of the intensity of light over a range of ultraviolet wavelengths—revealing emissions of two molecules, carbon monosulfide and another known as the hydroxyl radical. Preliminary evaluation indicated that the ultraviolet emissions from both molecules were somewhat brighter than expected at this stage of the comet's development.

The *IUE* had taken the ultraviolet "pulse" of not only nearby solar system objects such as comets and planets, but also extragalactic objects such as quasars, providing research data to more than 1,000 astronomers around the world.

Scientists, engineers, and technicians at NASA's Goddard Space Flight Center designed and integrated the spacecraft, which was launched in 1978 as a cooperative venture of the British Science Research Council, the European Space Agency, and NASA. (NASA Release 85–103)

July 1: The European Space Agency (ESA) announced that the European X-ray Observatory Satellite, *EXOSAT*, had been in orbit over two years, during which time it had performed nearly 2,000 observations, returning data to an observatory based at the European Space Operations Centre, ESOC, in West Germany.

The *EXOSAT* telescopes had provided among other findings detailed images of the remnants of dead stars, called supernova, which had consumed their nuclear fuel (hydrogen and helium) and ended their lives in a spectacular explosion. The images enabled astronomers to deduce the amount of energy in the original explosion and the physical conditions in space in the neighborhood of the explosion.

Some astronomers speculated that a supernova might end as neutron stars or black holes, and *EXOSAT* was helping astronomers study a number of these "exotic" objects in the Milky Way.

Early in 1984 astronomers proposed using *EXOSAT* to search for very fast rotating neutron stars in X-ray binaries. In mid-July 1985, ESA announced that these observations had led to the discovery of a completely unexpected phenomenon, that of quasi-periodic oscillations that might indicate the presence of fast rotating neutron stars in very old X-ray binary systems. The discovery of the oscillations opened up an entirely new area of investigation in the study of neutron stars that, because of the stars' extreme gravitational conditions, could make it possible to test to their limits some of the fundamental laws of physics. (ESA release July 1/85; ESA release July 19/85)

July 2: The European Space Agency (ESA) announced that it launched at 11.23 hours GMT July 2 Giotto, ESA's interplanetary space probe that would after a 700 million km voyage encounter Halley's Comet in March 1986. ESA launched Giotto by an Ariane 1 launcher from the Kourou Space Center in French Guiana, the ninth straight success for the agency's unmanned Ariane 1 rocket.

The rocket placed Giotto into a geostationary transfer orbit with parameters of 35,966 km, apogee; 199.2 km, perigee; and 7° inclination. Following separation from the launcher's third stage 22 minutes after liftoff, ESA's Space Operations Centre (ESOC) at Darmstadt, West Germany, took control of the spacecraft. During the following 32 hours, ESOC spacecraft controllers worked around the clock to reconfigure the spacecraft in its geostationary transfer orbit and carry out spin-up and attitude maneuvers. After Giotto made three revolutions in its transfer orbit, ESOC at 19.24 hours GMT July 3 injected the spacecraft into an earth-escape trajectory.

Giotto, named for the 14th-century artist Giotto de Bondone, who in 1301 witnessed what probably was Halley's Comet, should come within 310 miles of the comet's five-mile-diameter nucleus. When the spacecraft reached its rendezvous, ESA scientists expected it to be destroyed by billions of high-speed dust particles pouring off the comet's nucleus. Before its destruction, Giotto would perform approximately 10 experiments examining the image, chemistry, magnetism, and other aspects of the comet and its tail.

Other spacecraft participating in the Halley's Comet investigation were the USSR's Vega 1 and 2, launched December 15 and December 21, 1984, respectively, to encounter the comet at a distance of 62,000 miles; Japan's Sakigake launched January 8, 1985, and Planet A, scheduled for launch in August 1985 to photograph the comet's tail at a distance of 160,000 miles; and NASA's International Comet Explorer, ICE (formerly ISEE–3), launched December 22, 1983, and diverted from its previous earth orbiting position for the Halley's Comet encounter. (ESA release July 2/85; ESA release July 4/85; *W Post,* July 3/85, A3)

August 2: Crew aboard the Space Shuttle Challenger on mission 51–F made major progress in fixing the instrument pointing system (IPS) for a $60 million solar telescope, the *Washington Post* reported; and flight planners moved closer to extending the mission one day to make up for lost sun-watching time. The extension would have Challenger landing in California after eight days in orbit.

The IPS's improved operation enabled the astronauts to increase the pace of their observations of the sun, the only star close enough for detailed study. Although the balky telescope mount late in the evening again failed to work properly, the Space Shuttle scientists manually locked the telescope on target. New computer commands radioed earlier from earth had helped solve some initial problems with the IPS, and it aimed its three solar telescopes precisely at specific spots on the sun, holding them there for a few minutes each time. However, the steadiness did not last.

During one long sun-sighting period, astronaut-physicist Loren Acton said the solar chronosphere, a layer of gas that resembled a flaming forest, appeared more active than he had suspected. The phenomenon was dubbed the "Acton effect."

A decision to extend Challengers's mission depended largely on the orbiter's supply of hydrogen for generating electricity. (*W Post,* Aug 3/85, A10)

August 2: NASA Administrator James Beggs met with planetary science experts to discuss a plan to develop a spacecraft capable of flying in formation with a comet known as Wild-2 and dropping a probe down onto its surface in hopes of bringing back samples, the *Washington Times* reported.

Space scientists said NASA wanted to launch the $400 million mission in the early 1990s. NASA hoped to get about $30 million in start-up money earmarked for the project in the agency's FY 87 budget.

The mission would be an important space first, because none of the previous spacecraft launched toward Halley's Comet would fly alongside that comet, launch a probe, or return to earth after rendezvous. (*W Times,* Aug 2/85, 10A)

August 5: Astronauts aboard the Challenger on Space Shuttle mission 51–F spent an extra day in space gazing at the sun after the instrument pointing system was back in order, the *Washington Post* reported. The sudden recovery on August 4, as a result of a lucky radio command transmitted on a whim after a jolting rocket firing, of a fourth solar telescope called a solar optical universal polarimeter added to the science data. Most of the August 5 sun observations used that telescope after NASA had considered it a failure.

The telescope initially recorded images of sunspots, dark splotches that erupted through the visible surface of the sun when underlying magnetic fields became convoluted, the *Washington Times* reported. The astronauts turned the telescope on the sunspot again on August 5, and the TV image was so clear scientists on the ground reported that they could see changes in its penumbra, a surrounding halo-like feature made up of jets of gas streaming out from the dark central area. (*W Post,* Aug 6/85, A5; *W Times,* Aug 6/85, 2A)

August 15: Lockheed Missiles and Space Co. personnel fitted the Jet Propulsion Laboratory's (JPL) wide-field/planetary camera (WFPC) into the Hubble Space Telescope, the JPL *Universe* reported. They then removed and returned it to JPL for modifications and testing. "Everything went fine," Dave Rodgers, JPL's WFPC project manager, said. "There were no problems with any of our equipment."

The WFPC consisted of two complete camera systems with different focal lengths. There were four cameras in each system, all sharing the same housing and electronics. The wide-field camera would obtain a wide perspective of a region in the sky, while the planetary camera would make high-resolution studies of individual objects. "It's called a planetary camera, but you can use it for any object that you want to study in detail," Rodgers explained.

Besides planets in earth's solar system, the planetary camera would study galaxies and star-like objects such as quasars. Astronomers could also use the camera system to search for clues suggesting the presence of other planets orbiting nearby stars. Because the space telescope would be above the earth's obscuring atmosphere, it would be able to detect objects 100 times dimmer than those visible from earth with about 10 times greater resolution.

The orbiting observatory was a major part of NASA's astronomy effort through the end of the century. Space Shuttle crew would service and replace in orbit the space telescope instruments, keeping the telescope operating for at least 15 years. (JPL *Universe,* Aug 23/85, 1)

August 19: Officials at Japan's Institute of Space and Astronomical Research announced that today Japan launched from the Kagoshima Space Center in Kyushu its second spacecraft to observe Halley's Comet the following March, the *NY Times* reported. The spacecraft, Planet A, was on a trajectory that would take it within about 60,000 miles of the comet. The USSR had two craft bound for Halley and the European Space Agency had one.

Planet A had a solar-wind analyzer to measure the distribution and direction of high-energy particles from the sun as they encountered the vicinity of Halley's Comet and an ultraviolet telescopic camera to examine the cloud of atomic hydrogen, glowing in the ultraviolet, that extended tens of thousands of miles out from the comet. (*NYT,* Aug 20/85, C3)

September 3: NASA announced that, when its International Cometary Explorer (ICE) intercepted the dust-filled tail of Comet Giacobini-Zinner on

September 11, it would be the world's first spacecraft encounter with a comet. When it moved into the comet's bow shock approximately 45 minutes before the predicted rendezvous with the tail at 7 a.m. EDT, ICE would have traveled for more than seven years and covered more than 1 billion, 500 million miles, participating in three distinct scientific missions instead of the one planned.

When launched August 12, 1978 from Cape Canaveral, ICE was known as the International Sun-Earth Explorer (ISEE–3). The 1054-lb. 16-sided spacecraft became the first spacecraft to orbit at the sun-earth libration point (that point in space where a satellite is suspended in a gravitational equilibrium between the sun and the earth-moon system) and the first to traverse the earth's distant geomagnetic tail. ISEE–3 also was the first spacecraft to make multiple swings by the moon and the first to use a lunar gravity-assist maneuver for targeting-escape trajectory. It made more gravity-assist maneuvers (five) than any other spacecraft.

After the spacecraft completed most of the objectives of its original mission, it remained in good operational condition and had approximately 75% (150 lb.) of its total propellant reserves (200 lb.). Engineers at Goddard Space Flight Center (GSFC), under the leadership of Dr. Robert Farquhar, developed a plan that would divert ISEE–3 from its libration point orbit, take it past the moon five times, and propel it out toward Comet Giacobini-Zinner. Although the spacecraft was not designed for cometary investigation (it had no cameras or dust detectors), GSFC scientists believed it was well-suited for measurements of a comet's plasma properties, a chief objective of cometary exploration. In addition, the trajectory developed by Farquhar and his colleagues brought the spacecraft within range of a March 1986 upstream intercept of Comet Halley.

That 1986 upstream pass, following an October 31, 1985 distant pass, was important because ICE would provide data on the solar-wind state upstream from Halley and earth-based telescopes would then observe the effect of the solar wind on Halley's tail.

The spacecraft would return in July 2012 to the vicinity of earth, where a lunar gravity-assist maneuver could place the spacecraft into an earth orbit from which it might be retrievable. (NASA Release 85–121)

September 11: At 44 million miles from earth, the International Cometary Explorer (ICE) spent 20 minutes traveling through the tail of Comet Giacobini-Zinner, the *Spaceport News* reported. During the precedent-setting encounter, ICE was traveling at 46,000 miles per hr. when it entered the 14,000-mile wide tail at a point 4,900 miles behind the comet's nucleus; when it emerged, ICE had no apparent damage.

Dr. John Brandt, head of Goddard Space Flight Center's Laboratory for Astronomy and Solar Physics, summarized scientists' tentative conclusions at a press conference, saying that mission discoveries added significantly to and confirmed many predicted features of the scientific portrait of a comet.

Scientists expected to see a sharply defined bow shock, but instead they observed that a broad U-shaped turbulent interaction region preceded the comet as it moved through the solar wind of interplanetary space.

Some data, however, confirmed what cometary scientists postulated about the comet's plasma tail: it was threaded by hairpin-shaped magnetic field lines captured from the solar wind and included electrified gases both denser and colder than those of the surrounding solar wind.

The flight also showed that the hazard of flying through the dust in Comet Giacobini-Zinner's tail was less than expected.

There was one new-found phenomenon that puzzled the ICE scientists. At least 300,000 miles before the ICE reached the comet, it detected high-speed heavy ion beams never before found in space. ICE scientists theorized that these beams were actually low-speed molecules that escaped from the comet, were ionized by solar ultraviolet light, and then turned around by the supersonic solar wind and accelerated back toward the comet as particle beams.

ICE was then continuing on to take solar wind measurements upstream of Halley's Comet on two occasions—the second, March 28, 1986, to be within 19.5 million miles of the comet. (*Spaceport News*, Sept 27/85, 1)

September 12: Planetary scientists at the Jet Propulsion Laboratory outlined plans for two future $400 million missions, one of which NASA would select for funding, the *Washington Post* reported.

One mission entailed a spacecraft that in September 1991 would fly by the asteroid Hedwig and rendezvous in 1995 with Comet Wild 2. The other called for a spacecraft in 1995 to pass by the asteroid Freia and rendezvous in 1996 with Comet Tempel 2.

JPL scientists preferred the Hedwig and Wild 2 mission, in part because Wild 2 was a newer and brighter comet than Tempel 2. Plans for that mission called for a March 1991 spacecraft launch from the Space Shuttle with an ultimate encounter January 8, 1995, with Wild 2, when the comet was at the far end of its orbit 465 million miles from earth.

Scientists would program the spacecraft to match the comet's speed as it turned in toward the sun, which would then draw the comet and spacecraft toward it at the same speed for the next 850 days while the spacecraft's instruments measured the comet. (*W Post*, Sept 31/85, A23)

October 2: NASA announced that U.S., French, West German, and United Kingdom scientists completed preliminary findings from the September 11 [see Astronomy, Sept. 11] encounter of NASA's International Cometary Explorer (ICE) spacecraft with comet Giacobini-Zinner, which took place 44 million miles from earth and was the first comet intercept in history.

In the precedent-setting encounter, ICE, traveling at 46,000 miles an hour, entered at approximately 6:50 a.m. EDT, September 11, the 14,000-mile-wide tail of the comet 4,900 miles behind the cometary nucleus and emerged from the tail about 20 minutes later. Midpoint of the tail encounter came at 7:02 a.m., when the spacecraft passed through a narrow region called the neutral sheet.

ICE was then on its way to its fourth space exploration assignment to record on October 31 and March 28, 1986 solar wind measurements upstream of Halley's Comet. (NASA Release 85-138)

October 8: The National Academy of Sciences' Astronomy Survey Committee, the NASA Solar System Exploration Committee, and the NASA Committee on Solar and Space Physics made recommendations for future NASA space exploration and science programs, which Rep. Wyche Fowler, Jr. (D-Ga.) was asking the U.S. House to endorse in concurrent resolutions, *Defense Daily* reported. Advisory committee recommendations were often the basis for programs NASA sought to implement.

The Astronomy Survey Committee recommended the following major new programs "in order of priority" for total funding of up to $950 million (in constant 1980 dollars) over the next ten years: (1) the advanced x-ray astrophysics facility (AXAF) to be authorized in FY 87, (2) a very-long baseline (VLB) array of radio telescopes, (3) a new technology telescope (NTT) of the 15-m class operating from the ground at optical and infrared wavelengths, and (4) a large deployable reflector (LDR) in space.

The Solar System Exploration Committee recommended a program funded at a sustained annual level of $300 million per year (in FY 84 dollars). Its four initial core program recommendations were the Venus radar mapper (authorized in 1984), the Mars observer (authorized in FY 85), a comet rendezvous/asteroid flyby to be authorized in FY 87 for a 1990-92 launch, and the Titan probe/Saturn orbiter to be authorized for a 1988-92 launch as a joint NASA/European Space Agency mission.

Other committee recommendations included the Mars aeronomy orbiter, the Venus atmospheric probe, the lunar geoscience orbiter, the Mars surface probe, the comet atomized sample return, the multiple mainbelt asteroid orbiter/flyby, the earth-approaching asteroid rendezvous, the Saturn flyby/probe, the Uranus flyby/probe, the Neptune flyby/probe, and the Pluto flyby.

The Committee on Solar and Space Physics recommended an average annual funding of $400 million (in FY 84 dollars) "which would be sufficient to achieve the highest objectives in solar and space physics between the year 1988 and the year 2000." Major missions recommended were the upper atmosphere research satellite (authorized in FY 85), solar optical telescope for launch in 1990, the international solar terrestrial physics program for launch between 1989 and 1993, a solar probe for launch in 1995, and a solar polar orbiter to be launched in 2000. The committee also endorsed a series of moderate Explorer-class missions of about one launch per year. (*D/D*, Oct 8/85, 197)

October 17: NASA Administrator James Beggs confirmed this week that NASA cancelled a mission to send an unmanned spacecraft to comet Wild 2, citing budgetary constraints, the *Washington Post* reported. "However, the comet rendezvous proposal is a very promising candidate for a new start" in FY 88, Beggs said. NASA had expected the mission from FY 87 to 1992 to cost $400 million, which was within the budgetary limits NASA established in 1981 for new planetary missions. The NASA budget the Reagan Administration was drafting eliminated the mission.

Delaying a decision on a comet rendezvous by even one year meant the spacecraft could not fly to Wild 2, the favorite candidate of space scientists because it was a bright, fresh comet. Discovered in 1978, the comet would make its third and fourth appearances in 1990 and 1996. In order to rendezvous with the comet when it passed near the earth, NASA in 1986 would have to begin work on the mission to have the spacecraft ready by 1991 for launch from the Space Shuttle. The announcement followed NASA decisions in 1980 and 1981 not to attempt a mission to fly by Halley's Comet.

"Once again, the United States is passing up an opportunity to initiate the exploration of the comets, which contain the most unblemished record of the origins of our solar system," the *Post* quoted Dr. Laurel Wilkening, dean of the graduate school at the University of Arizona and vice chairman of President Reagan's Commission on Space. "The decision is a major setback for the U.S. planetary exploration program," he concluded.

Space scientists were also upset because NASA's science advisory groups recommended that a comet rendezvous be the next new misson in planetary exploration [see Astronomy, Oct. 8]. Scientists pushed for a mission to Wild 2 because the spacecraft could travel with the comet for 850 days, observing it before its long dust tail formed and obscured the comet's nucleus or damaged the spacecraft. It would have been NASA's most ambitious mission to explore the solar system.

The next comet that NASA could approach would be Tempel 2, which would pass by earth in 1998. Such a mission, also costing $400 million, required a NASA decision by the middle of 1986. (*W Post*, Oct 17/85, A21)

October 24: The European Space Agency announced today that the European Space Operations Centre (ESOC) on October 14 and 16 simulated the spacecraft Giotto's encounter with Halley's Comet (scheduled for about midnight March 13, 1986) by testing the spacecraft's systems, the different scientific instruments, and the associated ground systems in the first of a series of rehearsals that would take place up until a few days before the encounter. The first rehearsals showed satisfactory operation of the spacecraft and its payload and helped experimenters and engineers obtain better insight into the type of situations they might face during actual encounter.

Before the formal rehearsals, engineers had switched on and tested all scientific instruments. The two plasma analyzers and the ion mass spectrometer produced measurements of the solar wind and its composition; the magnetometer, of the interplanetary magnetic field.

The onboard camera made observations of the star Vega, the planet Jupiter and, on October 18 and 23, of earth. Taken at a distance of about 20 million km, images of the Pacific Ocean region showed a pattern of darker and brighter structures, the latter being associated with cloud formations. The quality and resolution of the images showed that the camera was functioning according to its design specifications. (ESA release Oct 24/85)

October 28: NASA announced that the Infrared Astronomical Satellite (IRAS), launched January 25, 1983, worked almost perfectly for 300 days until it depleted its liquid helium coolant and that it had achieved mission objectives, which were to produce an unbiased all-sky survey using a number of broad infrared photometry channels, to study selected galactic and extragalactic sources, and to map extended sources.

For infrared astronomy, IRAS represented as great an improvement over ground-based telescopes as the Palomar 200-in. telescope had over Galileo's telescope. Because fewer than 1,000 infrared sources were cataloged before the IRAS launch, the sensitivity of IRAS produced a survey of a large unexplored area in the electromagnetic spectrum. The IRAS All-Sky Survey catalog contained about 250,000 sources, including some 20,000 new galaxies and 16,000 small extended sources. IRAS surveyed more than 96% of the sky, 16% more than was required for mission success. Information in the IRAS databased revolutionized the infrared study of the sky and provided a rich data source to observers in other wavebands.

The IRAS program resulted from a cooperative effort among the U.S., The Netherlands, and the United Kingdom in the areas of operations, spacecraft design, infrared technology, and scientific collaboration. During the next four years, scientists would continue to examine and analyze the more than 200 billion data bits in the IRAS database and, by developing specialized processing techniques, possibly increase the sensitivity of the survey. (NASA MOR E-885-83-01 [postlaunch] Oct 28/85)

During October: NASA announced that its Marshall Space Flight Center (MSRC) awarded a 14-month $140,000 contract to Perkin-Elmer Corp. to study concepts and technologies needed for a next generation telescope with ten times the capability of the Hubble Space Telescope. This study, the first step toward preparing to launch a space telescope array in the first decade of the next century, would define a technology development plan for the array and provide NASA with cost and schedule data.

Perkin-Elmer would initially examine the probable science such a telescope would perform as well as the concept of an array of telescopes working together as one. Perkin-Elmer said it would develop two telescope concepts based on the best use of the proposed space station and technology advancements needed to implement them. The first would emphasize assembly and optical alignment in orbit using the space station as an assembly base. The second concept would focus on modularization, ground assembly and checkout, and minimal on-orbit assembly.

The study would also explore the possibilities of adapting, for visible light purposes, interferometer techniques used by radio astronomers to obtain high-resolution maps of deep space objects. In addition, Perkin-Elmer would define and recommend technologies needed for maximum use of the space station for assembling, testing, and servicing the next generation facility for optical astronomy.

Perkin-Elmer built the Hubble Space Telescope scheduled for launch August 1986 on Space Shuttle flight 61–J. (*NASA Activities,* Oct 85, 14)

November 1: Jet Propulsion Laboratory (JPL) planetary scientist Steve Ostro recently completed a five-year study of asteroids located in a belt between the orbits of Mars and Jupiter, the JPL *Universe* reported, and then turned his attention to observing near-earth asteroids using the world's most powerful radar telescope at the Arecibo Observatory in Puerto Rico.

"Some asteroids are among the most primitive planetesimals—the matter which accreted to form the solar system," Ostro said. "Asteroids are every bit as exciting as comets and potentially the most important objects in our solar system," he pointed out.

Asteroids constituted an enormous and diverse population, ranging in size from a few hundred meters to 1,000 km, and varying in composition, spin rate, and shape. Most asteroids were in the belt Ostro had studied and were thought to be the remnants of a planet that failed to form due to Jupiter's gravitational influence. However, many other asteroids tumbled through space close to and sometimes across earth's orbit.

Some scientists believed near-earth asteroids could be the nuclei of burned-out comets. "We really don't know what the nucleus of a comet is," Ostro said. "Comets might eventually lose all of their volatile compounds and then travel as asteroids." And scientists believed asteroids were some of the most primitive objects in the solar system. "We really have a spectrum of asteroids—some of the primitive material left from the solar nebula all the way up to those that are highly evolved," Ostro commented. "So asteroids are a laboratory for studying the evolution of planets and other planetary objects."

Astronomy

No spacecraft had ever explored an asteroid, nor had scientists obtained an image of an asteroid's surface. That was why many planetary scientists believed JPL's proposed Comet Rendezvous/Asteroid Flyby mission was critical to understanding not only comets and asteroids, but also the history of the solar system.

Another JPL planetary scientist, Eleanor Helin, recently discovered a near-earth asteroid with a 36-inch telescope located in Caussols, France. The asteroid, designated 1985-PA and the 22nd asteroid discovered by JPL researchers, had a small elongated orbit highly inclined to the ecliptic plane; scientists had discovered only two other near-earth asteroids with orbital inclinations higher than 1985-PA. Helin said the asteroid's inclination suggested it could have had a close encounter with Mars and been knocked into the steeply inclined orbit.

At discovery the asteroid was moderately bright—measuring at 16.5 magnitude—and reddish, indicating the asteroid was probably a stony, silicaceous body. (JPL *Universe*, Nov 1/85, 1 and 2)

November 15: Nobel Laureate Charles Townes, speaking recently at a Smithsonian Institution Associates lecture, discussed how he and his colleagues at the University of California-Berkeley proved, after years of hypothesizing and researching, that there was a black hole, twice the size of the sun and four million times as heavy, in the center of our galaxy, the Milky Way, the *Washington Times* reported.

A black hole is a body of such enormous gravity that nothing escapes it—no light waves, no radio waves, no waves of any kind—and it cannot, by definition, be seen or heard. Townes said he and his colleagues identified the black hole by deductive reasoning and picking up clues here and there.

The first clue came in the 1940s when a scientist at Bell Laboratories picked up a steady stream of radio noise from roughly the center of the Milky Way. Electrons bumping into protons and heating them produced radio waves, but great permanent clouds of silicate dust obscured the radio wave source. Several decades later scientists at Berkeley detected infrared waves coming from the same place. The waves were just long enough to get through the dust clouds, seeming to indicate they came from a cluster of stars.

About ten years before, the scientists picked up gamma rays—a sure sign that, somewhere near the galactic center, antimatter (protons) was destroyed at a rate of about 10 million tons a second. Black holes sucked in stars, causing them to spiral in a whirling cluster, moving faster and faster until they reached the speed of light and were ripped apart and swallowed, a process that created great waves of energy. "That suggested that something very strong and violent was going on," Townes said.

"There were three possibilities," Townes said. "A large cluster of stars, alone, which pulled everything into it, a black hole, or some combination of the two." If it was a cluster of stars, they would all travel at the same speed. If it was a spiral of stars in the process of falling into a black hole, the ones nearest the hole would move a lot faster than those on the outer edge.

Since each element had its own spectrum, an astronomer using a radio telescope could search and find the identifiable wave emitted by a particular element. The astronomers at Berkeley picked out molecules of neon along the line of stars around the galactic center; those furthest from the center were going at the slowest rate, those nearest at the fastest. That proved the presence of a black hole.

Yet to be definitively answered, Townes said, were such questions as whether black holes came in different sizes; whether every galaxy had a black hole at its center; and how it all would end, since black holes were destroying stars. "It's possible," Townes said, "that eventually the black hole will pull in all the stars in our galaxy, if something else doesn't happen first. It would take an enormous time, so far in the future, that all kinds of other things would be more likely to happen. Such as our sun cooling off. We're more likely to fall into the sun than into the black hole. I don't think we have to worry about that immediately." (*W Times*, Nov 15/85, C1)

November 19: NASA announced it had added an educational experiment package sponsored by the Charleston County, South Carolina school district to the Astro-1 payload aboard the Space Shuttle Columbia scheduled for launch March 6, 1986. The Astro-1 complement of three ultraviolet astronomical instruments and a special visible-light, wide-field camera system would study Halley's Comet and other celestial objects during its nine-day flight in space. The primary objective of the student project, called "Can Do," was to photograph Halley's Comet; middle school students (grades 6 through 8) would then study and interpret astronomical photos obtained from the flight.

The photographic equipment—four 35mm still cameras equipped with lenses addressing a particular aspect of cometary science—would be in a canister normally used for Get Away Special experiments. The camera package, developed with the participation of the National Geographic Society, would use a newly developed color film with unprecedented film speeds to photograph the dim tail of Halley's Comet. The color photos would complement the black-and-white images obtained by Astro-1's own image-intensified, scientific wide-field camera.

In addition, the canister would contain a separate package of passive student experiments including biological samples, magnets, and two kinds of accelerometers to study effects of microgravity.

The Charleston County Schools, in cooperation with NASA's Educational Affairs Division, developed a broad program of astronomical science and related educational activities that they would offer to other students throughout the nation. These included concurrent ground-based photography by student groups to compare with results from the Space Shuttle-based photos, teaching packets for classroom use, and comparative studies of Comet Halley through interviews with people who recalled the comet's passage in 1910. (NASA Release 85-154)

November 24: Astronomers tracking Halley's Comet with the recently built radio telescope called the Very Large Array found that water streamed off the comet at two to three times the rate expected from calculations made after the 1910 passing, the *Washington Post* reported. Comets were mainly ice balls embedded with dust and rock; they remained frozen until they neared the sun, when their surfaces began to sizzle. On October 19, when the comet was about 190 million miles from the sun, Halley was giving off two tons of water per second.

Looking at the watery stream through a radio telescope showed only a "hole" extending 30,000 miles around the comet. This appearance occurred because earth-based telescopes could not pick up the wavelengths of light given off by water—they were obscured by the earth's soggy atmosphere. Past the 30,000-mile hole, astronomers detected the water broken up into oxygen and hydrogen at a distance extending out to 60,000 miles around the comet.

Astronomers were surprised to find the oxygen-hydrogen emission in "clumps" about the size of earth rather than in a continuous cloud. "No one had even hypothesized such clumps," said Imke de Pater of the University of California. Researchers had no explanation for the clumps.

To observe the comet, de Pater and colleagues Patrick Palmer, University of Chicago, and Lewis Snyder, University of Illinois, were the first to use the two dozen "dishes" arranged in a grid at the Very Large Array, which provided resolutions 20 times better than that of a single telescope. (*W Post*, Nov 24/85, A18)

December 2: NASA issued a revision of its October 28 report on the Infrared Astronomical Satellite (IRAS) [see Astronomy, Oct. 28], in which it noted a major IRAS survey discovery that many galaxies emitted far more energy in the infrared than in the optical band where normal stars emitted the bulk of their luminosity.

Astronomers considered these galaxies to be of three distinct kinds. The first was "starburst galaxies," those that were presumably undergoing an active period of star formation leading to a large infrared luminosity and in which as much as 10 to 20 solar masses of gas and dust were being converted into new massive stars each year.

Another class was sometimes called Seyfert galaxies, which astronomers believed were powered by matter falling into an accretion disk surrounding a black hole. The active regions of these galaxies had a great deal of interstellar dust, which absorbed the optical and ultraviolet radiation from the accretion disks, accounting for the large infrared luminosity.

The third and most mysterious class of extragalactic objects discovered by IRAS was the extremely luminous infrared galaxies corresponding to very faint optical galaxies. Such galaxies produced 100 to 300 times as much energy in the infrared as in the optical, and the total luminosities of these galaxies were approximately 100 times those of ours. Astronomers believed the infrared radiation in these galaxies was probably from thermal dust emission, where the dust was reradiating energy absorbed from shorter wavelengths.

Astronomers found a large fraction of the active infrared galaxies were interacting galaxies, that is, galaxies that were undergoing a collision or near collision with a neighboring galaxy. The IRAS survey demonstrated the profound effect such collisions had on galaxies, and it could well be that such events dominated the evolution of most galaxies. (NASA MOR E–885–83–01 [postlaunch] Oct 28/85, revised Dec 2/85)

December 5: NASA announced it selected three scientific investigations for the Hubble Space Telescope that would lead to the design and development over the next several years of one or two advanced scientific instruments for flight readiness during the period 1992 to 1994 for that orbital observatory.

The three investigations selected for the definition phase were an imaging Michelson spectrometer, by Donald Hall, University of Hawaii, Honolulu; near-infrared camera and multi-object spectrometer, by Dr. Roger Thompson, University of Arizona, Tucson; and space telescope imaging spectrograph, by Dr. Bruce Woodgate, NASA Goddard Space Flight Center. All the proposed instruments would be capable of taking photos of celestial objects and breaking down the light from these objects into spectral colors.

NASA would then select one or two of the investigations for flight hardware development about one year after telescope launch, scheduled for the last half of 1986. Space Shuttle astronauts would install the selected instrument(s) on the telescope while it was in earth orbit. (NASA Release 85–164)

December 6: NASA announced that, during mission 51–L scheduled for launch no earlier than January 22, 1986, it would release from the Space Shuttle a free-flying payload called Spartan-Halley to observe Halley's Comet. Scientists believed that comet exploration was a search for the beginnings of earth, because comets were the most unaltered samples of the early solar system.

Spartan-Halley was the only NASA satellite dedicated solely to observing Comet Halley. During these observations, January 24 through 31, the comet would be in its most active period as it headed towards its closest approach to the sun on February 9. The objective of the Spartan-Halley mission was to measure from earth orbit the ultraviolet spectrum of Comet Halley using twin spectrographs designed and built by the University of Colorado's Laboratory for Atmospheric and Space Physics.

Spartan-Halley's cameras and instruments would analyze the active comet, looking for hydrogen and oxygen atoms and carbon, nitrogen, and sulfur molecules. If the spacecraft detected these elements, it might indicate more complex compounds could be present in the ice and dust that made up the comet's nucleus.

Goddard Space Flight Center developed Spartan-Halley, a modified Spartan-2 spacecraft, to be a research carrier. The spacecraft consisted of the spectrographic instrumentation, attitude control cameras (which also provided science photography), a microprocessor, associated electronics, a sunlight baffling system, and a service module. The January mission would be the first for the Spartan-2 type carrier, a completely reusable spacecraft to be reflown later on several other missions. Following completion of the January mission, the Space Shuttle's robot arm would retrieve Spartan-Halley and store it in the cargo bay. (NASA Release 85-161)

December 10: NASA announced that scientists from the U.S., France, the United Kingdom, and West Germany released their findings from the rendezvous on September 11 of the International Cometary Explorer (ICE) spacecraft with Comet Giacobini-Zinner. Although some ICE data confirmed the traditional portrait of a comet, other information was unexpected.

Perhaps most surprising was detection of electrical wave (plasma) disturbances and high-speed molecular species coming from the comet more than a day before rendezvous. The spacecraft's Plasma Wave Experiment, developed by TRW's Dr. Frederick Scarf, detected the electrical waves while ICE was 1,429,200 miles from the comet. Scientists had theorized that first detection might occur just a few hours before the spacecraft crossed the comet's tail.

A few hours after initial detection, but still one day before intercept, two of ICE's instruments discovered electrically-charged particles (ions) as far as 1,130,000 miles from the comet. The Energetic Proton Experiment, directed by Dr. Robert Hynds, Imperial College, London, and the Low-Energy Cosmic Ray Experiment, directed by Dr. Dieter Hovestadt of the Max Planck Institute for Extraterrestrial Physics, Garching, West Germany, detected these ions. According to Hynds, it was believed solar ultraviolet light ionized gas molecules escaping from the comet's nucleus, then the solar wind picked up and accelerated the ions back toward the comet.

In another finding, Dr. Samuel Bame of the Los Alamos National Laboratory reported that, in contrast to the hot electrons on the outskirts of the comet, its tail consisted of a dense narrow structure of cool plasma. The Radio Wave Experiment of France's Meudon Observatory made this same finding.

The ICE's Ion Composition Experiment, directed by Dr. Keith Ogilvie of Goddard Space Flight Center, made the first direct measurements of molecules in a comet. The experiment found mainly water vapor ions, confirming the "dirty snowball" model of comets.

Astronautics and Aeronautics, 1985

A major prediction confirmed by the ICE data was that the magnetic structure of the comet's plasma tail consisted of two parallel lobes, each threaded by a magnetic field of opposite polarity. ICE's Magnetometer Experiment of Dr. Edward Smith of the Jet Propulsion Laboratory mapped this structure, which was predicted in 1957 by Nobel physicist Dr. Hannel Alfven.

Putting conflicting scientific conjectures to rest, the ICE mission revealed that no clear-cut bow shock—a surface moving ahead of the comet like that through which a jet aircraft passes as it breaks the sound barrier—accompanied the comet. Instead, researchers found around the head of the comet what they called "a transition region" in which the solar wind was heated, compressed, and slowed.

Neither the ICE spacecraft nor its instrument payload suffered any detectable damage as a result of the impact with the comet's dust. (NASA Release 85-166)

December 17: NASA announced that its Comet Halley Active Monitoring Program (CHAMP), a middeck payload to obtain wide field visible and ultraviolet data on Comet Halley using handheld equipment, would have its primary flight January 1986 on the Space Shuttle (STS) mission 51-L. It was during that time that the comet would be most active but viewing from earth would be most difficult due to the narrow viewing angle between the comet and the sun.

The first flight of CHAMP would be on STS mission 61-C in December 1985; the third, on STS mission 61-E March 1986. The December and March flights would provide important pre- and post-perihelion baseline data that would assist with the interpretation of the January flight data.

NASA considered CHAMP data supplemental to the extensive and detailed space-based data to be obtained from the flight of the ASTRO payload on STS mission 61-E, the Spartan-Halley payload on STS mission 51-L, the UVX payload on STS mission 61-C, and the International Cometary Explorer (ICE) spacecraft. However, the CHAMP data would be unique in its nature and ability to provide a single context for all of these activities, because it would make observations on three Space Shuttle missions.

Before each observation, the assigned mission specialist would set up the experiment by mounting the CHAMP equipment in the appropriate window. On the January flight, the mission specialist would use the orbiter Challenger's UV transmissive side hatch quartz window. On the other two flights, data probably would be acquired through the aft flight deck windows. During observations, the orbiter would be oriented so the appropriate window was pointed to within 5° of the celestial coordinates of the comet, with attitude drift rates of less than 1.002° maintained.

Each observation required exposing a series of one to four calibration shots plus six to eight frames of film for images or spectra during the approximately five-minute period occurring in each orbit when the comet was in view and the sun was at least 8° below the earth's horizon.

NASA hoped there would be an observation for each flight day during the seven-day December and March missions, and up to 20 observations during the January flight. However, conflicts with other operational requirements might reduce the actual number of observations, although unusual cometary behavior could require additional time allocated to the experiment. (NASA MOR E-420-61-C-20 [prelaunch], Dec 17/85)

December 31: NASA announced that on December 26 its Venus orbiting Pioneer spacecraft began six weeks of observations of Halley's Comet during the comet's most active period closest to the sun (perihelion). Near the time of perihelion, the comet, Venus, and Pioneer were located on the opposite side of the sun from earth, making observations difficult from earth (160 million miles away).

Pioneer was the only spacecraft close to Halley's Comet during perihelion, and NASA expected that Pioneer's ultraviolet observations would provide valuable insights into the state of the comet in advance of flybys of Halley's by European, Soviet, and Japanese spacecraft beginning March 6, 1986.

Pioneer's first phase of observations would end January 4 when both Venus and Pioneer passed behind the sun for almost a month, cutting off effective communications between the spacecraft and ground controllers at Ames Research Center. Observations would resume about February 3, six days before perihelion, and would continue until March 6.

NASA scheduled Pioneer to produce two images of the comet in ultraviolet light. One would show the hydrogen cloud surrounding the comet, an image expected to be 20 times larger than a photo of the comet in visible light. The image would come from 20,000 scans of the comet as it drifted for three days through Pioneer's ultraviolet spectrometer's field of view. A second image would show both Venus and Halley's Comet in a single view. (NASA Release 85-181)

Planetary Exploration

January 18: NASA Administrator James Beggs approved the addition of an asteroid 28 Amphitrite-flyby option to the Galileo mission, permitting a December 1986 final flyby decision and changing the Jupiter arrival date from August to December 1988, the JSC *Roundup* reported.

Approval had followed a two-year study by scientific groups, mission designers, and program officials, in which the National Academy of Sciences and NASA's Solar System Exploration Committee identified asteroids investigations as an essential element of a balanced planetary exploration program.

The asteroid flyby could not compromise or risk mission objectives. NASA would make the final decision for a flyby after launch based on analysis of spacecraft health, particularly the attitude-control and mission-operations systems.

Amphitrite, about 200 km in diameter and one of the larger of the minor asteroids, was in a near-circular solar orbit in the middle of the asteroid belt at 2.5 Astronomical Units (AU) from the sun (the earth's distance was 1 AU or about 150,000,000 km).

A specially convened hazards workshop had concluded that, at a 10,000- to 20,000-km flyby distance, hazard to the spacecraft was no greater than flying through the asteroid belt as 2 Pioneer and 2 Voyager spacecraft had done, and measurements analysis and Doppler-tracking data could achieve significant scientific objectives. As Amphitrite had a rotation period of about 5.39 hours, Galileo's mapping spectrometer could photograph and scan most of its surface. Data analysis would reveal size, shape, mass, density, exact rotation rate, pole orientation, detailed surface morphology, and mineral composition, thus indicating whether Amphitrite was a primitive accumulation of solar nebulae condensates or an evolved body that was a fragment or perhaps a core of a broken-up minor planet. With this data, scientists could possibly confirm or refute a hypothesis that asteroids were sources of many of the meteorites that had fallen on earth.

NASA had developed a new trajectory containing both Amphitrite and Jupiter, constrained by launch-vehicle energy and the existing launch window, that would result in a Jupiter arrival date delay from August 29, 1988 to December 10, 1988. Since the flyby would require early added-propellant expenditure, NASA would decrease the number of Jupiter-tour orbits from 11 to 10 and lengthen the tour from 20 to 22 months, permitting achievement of all major 11-orbit tour objectives. The delayed arrival and increased tour time would add five months and about $20–25 million in costs to the mission. (JSC *Roundup*, Jan 18/85, 1)

January 25: NASA and the European Space Agency (ESA) would launch in 1986 Ulysses, previously known as the International Solar Polar Mission, to explore behind and around the sun beginning in July 1987, the JPL *Universe* reported. JPL managed the U.S. mission phases and was responsible for part of the magnetometer experiment. The Space Shuttle would take Ulysses into orbit; a modified Centaur upper stage rocket would send it on a trajectory toward Jupiter, where it would gain energy from Jupiter's gravity field to pull it out of the ecliptic plane (the plane of the earth's orbit) and back over the sun.

Researchers would incorporate into the spacecraft bus instruments shipped to the European Space Research Technology Center near Amsterdam, Holland; after extensive testing scheduled for completion in August, an international board would review the spacecraft's readiness. After a mission operations review, ESA would ship Ulysses to KSC for flight tests and preparations. At a June meeting in Switzerland, a team of scientists would decide whether Ulysses would approach the sun over its north or south pole. (JPL *Universe*, Jan 25/85, 1)

January 28: NASA announced it had received signals from the USSR's Venera-Halley space probes at its Deep Space Network's (DSN) 210-foot antenna at Goldstone, California. The USSR had agreed that *Vega-1* and *2*, launched in December 1984, would carry experiments for joint U.S./French research.

For the June 1985 experiment, each spacecraft would drop into the Venusian atmosphere an instrument-laden balloon, which would float free in the middle, most active layer, of Venus's three-tiered cloud system. Transmitters on the balloons would continually send signals to earth during the two-day lifetime of each balloon. Three globally-placed, hypersensitive dish antennas of NASA's DSN, part of an international network of radio astronomy antennas organized by the Centre National d'Etudes Spatiales for the study, would receive the signals. Using data received from the combined array of at least 10 antennas worldwide, in addition to that from the USSR's network, scientists could calculate in detail the balloons' locations and motions.

Scientists would use a radio astronomy technique known as Very Long Baseline Interferometry (VLBI) to measure balloon velocity and hence Venus's wind velocity with a precision of approximately two miles per hr. at a distance of 67 million miles from earth. Scientists hoped data obtained would help further understanding of the complex Venusian weather system.

The DSN antennas would also receive data from the other scientific instruments on the balloon gondola, including instruments to measure lightning flash frequency, wind gusts' vertical velocity, atmospheric temperature and pressure, and cloud location and density.

Understanding turbulence and wave-type motions in clouds was important because scientists believed Venus's cloud layers to be the driving gear of the planet's multilayered weather machine. For example, scientists hoped data would explain why the atmosphere at Venus's cloud tops circled the planet at 250 miles per hr. as compared to near-calm on the surface.

By studying the atmospheric dynamics on Venus and other planets, scientists hoped to understand atmospheric processes in general in order to characterize all planets, including earth, in terms of a general set of physical laws. (NASA Release 85–13)

March 8: NASA announced that astronomers at the Jet Propulsion Laboratory (JPL) and the Universities of Hawaii, Arizona, and Texas were observing Pluto and its only known satellite, Charon, as the two alternately moved in front of one another in a series of eclipses that occurred every 124 years or twice in each orbit of the sun. Each time Charon passed between Pluto and earth, a portion of Pluto's surface was blocked from view, resulting in a dimming of the combined light from both bodies. When Charon moved behind Pluto, their roles reversed. Measurements of the times, durations, and changes in brightness of the events would allow astronomers to calculate the masses, diameters, and densities of Pluto and Charon, permitting development of models of the two bodies' composition.

Current estimates of Pluto's density had an uncertainty of 50%, not accurate enough to derive information on its composition. However, researchers thought Pluto's density was about that of water, making it the lowest-density planet known that had a solid surface.

The new measurements indicated that the combined brightness of Pluto and Charon diminished by 4% during the eclipses, a dimming that lasted about two hours and was superimposed on a 30% brightness change that occurred over a 6.4-day period. The longer change in brightness happened because one hemisphere of Pluto was 30% brighter than the other.

Very little was known about Pluto and even less about Charon, including when or even if the five-year-long series of eclipses would begin. In order not to miss any of the earliest events, the astronomers had established an observing network at McDonald Observatory in Texas, the University of Arizona observatories, Palomar Observatory in California, and Mauna Kea Observatory in Hawaii. Dr. Edward Tedesco of JPL had been the first to see and measure an eclipse of Pluto by Charon January 16 from Palomar; Richard Binzel observed another eclipse February 17 from the McDonald Observatory; and Dr. D. J. Tholen observed a third eclipse February 20 from the Mauna Kea Observatory. Pluto and Charon's great distance from earth and relatively small sizes made them the solar system's most difficult objects to observe. (NASA Release 85–36)

March 8: Galileo project team members would soon complete the two-month-long environmental test, phase 2, of the spacecraft, during which they put Galileo in a 25-ft. space simulator, an evacuated chamber cooled to −250 to 270° F, in order to observe how Galileo functioned in a space-like environment, the JPL *Universe* reported. Engineers also exposed the spacecraft to an array of high-intensity quartz lamps, which simulated the solar effects Galileo would experience in space.

Jet Propulsion Laboratory (JPL) personnel would move Galileo in about two weeks to the spacecraft assembly facility for the system test, phase 3. Following spacecraft disassembly, researchers would simulate different operational phases of the mission for spacecraft testing.

Galileo was composed of a planetary orbiter and an atmospheric entry probe for investigations of the planet Jupiter's magnetosphere's chemical composition and its physical state of structure and physical dynamics. (JPL *Universe*, Mar 8/85, 1)

Astronomy

June 1: After its 1979 Jupiter encounter and 1980 Saturn encounter, Voyager 1 at 12 noon EDT June 1 was 2,095,063,224 miles from earth and 2,170,849,615 miles from the sun, *Omni* reported. Voyager 2, after Jupiter and Saturn encounters in 1979 and 1981, respectively, on June 1 was 1,499,732,272 miles from earth, 1,593,867,481 miles from the sun, and 187,782,284 miles from Uranus.

At 12 noon EDT June 14, Pioneer 10, which encountered Jupiter in 1973 and was the first spacecraft to leave the solar system, was 3,410,823,980 miles from earth and 3,319,306,400 miles from the sun. (*Omni*, June 85, 28)

June 15: An international team of planetary scientists gathered at the USSR's Space Research Institute to catch the first return of data from the Vega-2 spacecraft plunging into the atmosphere of Venus, the *NY Times* reported. At 6:06 a.m. Vega-2 began its descent. Among the roughly 100 participants were representatives from eight European nations and the French Ambassador.

Two days before, exploding bolts had freed an eight-foot sphere that carried one-and-a-half tons of payload bound for Venus's night side. The main craft would miss Venus entirely and use energy provided by the planet's gravity to propel it toward an encounter with Halley's Comet. Inside the rapidly decelerating sphere a timing device programmed months before switched on the scientific packages around it, then set loose a 15-lb. stack of miniaturized weather instruments built by French, Soviet, and U.S. scientists. An 11-foot-diameter teflon balloon filled with helium kept the instruments from reaching Venus's surface.

An hour after the balloon inflated, the Vega-2 lander had reached the surface safely. The lander, in an atmospheric pressure 80 times the earth's at sea level and in a nighttime temperature of 855° F, was relaying information on surface conditions and excavating a small sample of the surrounding soil for crude analysis of its composition.

Vega-1's balloon had reached the planet June 11 and had covered nearly 6,000 miles through the Venusian clouds in 46 hours, passing into the planet's sunlit hemisphere before exhausting its battery.

The new Vega results suggested the planet had five discrete cloud layers whereas earlier missions detected only three. (*NY Times*, June 18/85, C8, June 16/85, 24; *W Post*, June 14/85, A10)

September 6: Jet Propulsion Laboratory (JPL) engineers and technicians recently practiced the exact technique that would be used to fuel the Galileo spacecraft, simulating the Kennedy Space Center (KSC) environment where the spacecraft would actually be fueled, the JPL *Universe* reported. This was necessary because Galileo's propulsion system, designed and built by West Germany and unlike previous JPL spacecraft in which the propulsion system was a separate module, would be sandwiched in between the bus and the despun section of the spacecraft. This would require fueling when Galileo was fully assembled, and any resulting problems could jeopardize the sensitive cargo.

"This four-day exercise is basically an 'undress rehearsal'," Hank Delgado, propulsion and pyro devices group leader, said. "We want to acquaint everyone with the procedures and equipment we plan to use to fuel Galileo."

Instead of using the actual propellants—mono-methylhydrazine and nitrogen tetroxide—that Galileo would use, the team practiced with less hazardous fluids—isopropyl alcohol and freon—allowing them to work without special suits. When Galileo was fueled at KSC, technicians would wear 55-lb. SCAPE (self-contained atmospheric protective ensemble) suits to protect them from the highly toxic propellants. Although the suits had an air supply and thermal control system, the limited air supply allowed only about an hour of work, requiring technicians to work in shifts.

JPL would send Galileo to KSC early in 1986 for final launch preparations. Technicians would then fuel the spacecraft, taking a day for each of its four tanks. For the following four months, the technicians would monitor the propellant closely for leaks or changes in temperature. (JPL *Universe*, Sept 6/85, 1)

October 8: NASA announced that a panel of scientists at its Ames Research Center (ARC) said today that, although Mars presently was a frigid desert planet, it once had enough water to cover its entire surface with an ocean more than 300 feet deep. Dr. Michael Carr of the U.S. Geological Survey said at the meeting, which grew out of research discussed the previous winter at ARC's Water on Mars Workshop, that "The pictures taken by the two Viking spacecraft in orbit around Mars tell us that Mars had as much water in geologic history as earth did."

Carr said the 20,000 close-up pictures taken since 1976 by the two Viking spacecraft revealed canyons that were deeper, wider, and longer than the Grand Canyon and could be made only by rushing rivers. The photos also showed thousands of gullies formed only by water or snow and ice slicing into the surface or forcing their way up to the surface.

"If all the water that existed on Mars to form these channels covered its surface today," commented ARC's Dr. James Pollack, "it would be enough to form a global Martian ocean tens of hundreds of meters deep."

Carr pointed out that the Viking photos also suggested that water was below the surface at latitudes near the Martian equator where the planet's interior heat could keep it from freezing just as do underground rivers in most temperate latitudes on earth. "There is a softening of the terrain, a rounding off of the edges . . . that is evidence of underground water flow and the creep of ice near the surface," he said.

"This terrain is also located where most of the Martian gullies are today, where snow, ice, and water burst out of the ground to cause the colossal floods that formed the channels we see today," Carr said.

And Pollack added that Mars long ago lost most of its carbon dioxide, which helped trap the sun's heat, causing in effect a permanent Martian ice age. (NASA Release 85–140)

November 29: New observations supported the prediction that Saturn's satellite Hyperion was tumbling wildly rather than rotating with a regular predictable period, *Science* reported.

In 1983 Jack Wisdom, who was currently at the Massachusetts Institute of Technology, and his colleagues predicted that the combination of Hyperion's odd potato shape and the stretching of its orbit by the gravitational tugs of the larger satellite Titan would prevent Hyperion from rotating regularly. Instead of keeping one face toward Saturn, as the moon faced earth, Hyperion would rotate chaotically, tumbling one way then another, slowing down and speeding up, in a fashion impossible to predict in any detail from its preceding behavior.

Since then, Peter Thomas of Cornell University and his colleagues reported analyses of the brightness of Hyperion in Voyager images that indicated a regular 13-day period of rotation during the 61 days of Voyager-2's encounter with Saturn.

Jack Wisdom and Stanton Peale of the University of California, Santa Barbara, countered that determining a period from 14 brightness observations scattered over several supposed rotations could not determine whether the rotation was chaotic or not. They found many periods by similarly sampling a numerically generated chaotic light curve.

The University of Texas's Richard Binzel, Jacklyn Green, and Chet Opal reported at a recent meeting of astronomers that the light curve of Hyperion they observed in April was "highly inconsistent with a 13.1-day rotation period." On April 16 and 17, Hyperion's magnitude was at a maximum for the 14-day observing period; it faded by 1 magnitude by April 21 and was no brighter 13 days after the first observations. It should have become 1 magnitude or 2.5 times brighter if it had a 13-day period. Although the observations did not prove chaotic rotation, the group said, "they provide strong evidence in favor of the hypothesis."

To say much more about Hyperion's rotation, astronomers would have to observe Hyperion nightly for many weeks, something that those assigning telescope time had been reluctant to permit. (*Science*, Nov 29/85, 1027)

December 3: Dr. Michael Kaiser of the Goddard Space Flight Center said at a news conference today that, "We see no radio emissions from Uranus that would tell us it has a magnetic field, and we're less than 46 million miles from the planet" with the Voyager 2 spacecraft, the *Washington Post* reported.

"We still pick up Jupiter's radio noise, Saturn's radio noise and even the sun's radio noise but we're not hearing any radio events at Uranus," he added. If Uranus had no magnetic field, it would make it only the second known planet in the solar system without one, Venus being the other.

Kaiser said the sign of a planet's magnetic field was its radio noise, generated when protons and electrons poured off the sun and collided with the planet's magnetosphere. This collision triggered the one billion-watt radio signal from earth called the Northern Lights, a 100 billion-watt signal from Saturn, and a signal so loud from Jupiter that it dwarfed every other radio signal in the solar system.

"You don't need much to form a magnetosphere that would generate noise out at Uranus's distance," Kaiser said, "so we figure we're either dealing with a planet that has no magnetic field at all or is so bizarre we don't even know what to look for."

Kaiser pointed out that a missing magnetic field suggested that Uranus had no internal heat source—no radioactive core such as that which made the rotating earth behave like a dynamo, and no internal heat source such as those that gave the rapidly rotating Jupiter and Saturn a strong magnetic field. "You need an internal heat source to drive a magnetic field," Kaiser said. "A planet's rotation is not enough by itself to create one."

Voyager 2, which had been in space for eight years and had passed Jupiter and Saturn, on January 24 would fly within 51,000 miles of the cloud tops of Uranus. Voyager 2 would in August 1989 encounter Neptune, but its flight would not take it near Pluto, the last known planet from the sun. The spacecraft would then pass out of the solar system. (*W Post*, Dec 4/85, A20)

December 4: NASA announced that Voyager 2 began today its encounter with Uranus, which would continue through February 25, 1986. During the period, the spacecraft's 11 instruments would perform close-range studies of the planet, its five known satellites, and nine rings. Voyager 2 would also search for a planetary magnetic field, new satellites, and new rings.

The spacecraft would make its closest approach to Uranus, flying 81,500 km above the cloud tops of the seventh planet, at 1:00 p.m. EST January 24, 1986. Because Voyager 2 was the first spacecraft to reach the planet, the encounter would provide scientists with more information about Uranus and its satellites and rings than had been learned since William Herschel discovered the planet March 31, 1781.

In addition to two cameras, a photopolarimeter, and a spacecraft radio, Voyager 2 carried an infrared interferometer/spectrometer and radiometer, an ultraviolet spectrometer, a cosmic-ray detector, a plasma instrument, a low-energy charged-particle detector, magnetometers, a planetary radio astronomy receiver, and a plasma-wave instrument. Three radioisotope thermoelectric generators supplied the spacecraft's electric power, a system

Astronomy

necessitated because solar cells could not receive sufficient solar energy at such a great distance from the sun.

NASA's Deep Space Network (DSN) antenna complex at Canberra, Australia, would receive most key data obtained during the Uranus encounter and all of that during the closest approach. (NASA release Dec 85, NASA Voyager Bulletin, Dec 4/85)

ATMOSPHERIC RESEARCH

January 14: NASA announced that Goddard Space Flight Center's (GSFC) Wallops Island, VA, facility in cooperation with the Air Force Geophysics Laboratory (AFGL) at Hanscom AFB, Massachusetts, and the Danish Meteorological Institute (DMI), Copenhagen, Denmark, would carry out that winter the 1985 Cooperative Observations of Polar Electrodynamics (COPE) project of sounding rocket research in Greenland to gather knowledge about the solar-earth relationship. The project would include studies of polar-cap turbulence and electrodynamics, auroral-zone electrodynamics, auroral-electrojet turbulence, neutral-atmosphere coupling, and polar-ionospheric irregularities.

Researchers selected Greenland for the investigations because of its access to the auroral oval, polar cap, and polar cusp; an existing rocket range; support from the Sondre Stromfjord incoherent-scatter radar, the Hi Lat Spacecraft, and the extensive array of scientific ground-observing stations in Greenland, Scandanavia, and North America; and a broad choice of launch azimuths. GSFC/Wallops Flight Facility personnel had spent several weeks during the previous summer at the Greenland Sondre Stromfjord facility installing additional launchers and related ground-support equipment, radar, telemetry, and communications systems and erecting vehicle and payload assembly structures.

Project experimenters scheduled nine suborbital rocket launches of which two would release chemicals creating artificial vapor clouds 250 km high. NASA scheduled seven flights: two Black Brant Xs, two Terrier-Malemutes, a Taurus-Orion, a Nike-Tomahawk, and a Taurus-Tomahawk; AFGL scheduled two missions: a Black Brant VIII and a Black Brant IX. (GSFC Release 85–5)

February 22: Goddard Space Flight Center (GSFC) announced that on February 4 technical/scientific teams from Australia, India, Finland, and the U.S. had begun a six-week meeting at Wallops Flight Facility, Va., to compare radiosonde instruments used worldwide. Measuring only a few inches square and weighing less than 2 lb., the radiosonde was a low-cost, mass-produced instrument for measuring atmospheric pressure, temperature, relative humidity, and wind while ascending on a small balloon up to about 25 km (15 miles), or more frequently 30 km (18 miles), above earth. Worldwide weather services used radiosonde data for forecasts to aid aviation and for research.

There were currently 17 radiosonde manufacturers in a number of countries, resulting in occasional differences in instrument measurements and the need to make periodic comparisons. A working group on the quality of meteorological data meeting in 1982 at the World Meteorological Organization's (WMO) headquarters, Geneva, Switz. had agreed that two phases of tests were required.

The United Kingdom's Meteorological Office had hosted the 1st phase in 1984, when researchers released 106 ballons. Phase-2 participants at Wallops launched four balloons a day, five days a week, weather permitting, for an anticipated total of about 100 flights, each balloon carrying four or more instruments. Researchers would then correlate Phase-2 data with Phase-1 results. (GSFC Release 85-8)

April 16: NASA announced the postlaunch status of the NOAA–A, –B, and –C missions carried out for the National Oceanic and Atmospheric Administration. In addition, NASA announced it scheduled NOAA–D for launch in 1987. NASA submitted on January 24, 1984 a postlaunch report for NOAA–E (launched March 18, 1983) and would submit a postlaunch report for NOAA–F (launched December 12, 1984) after a full year of data collection and evaluation.

NASA successfully launched NOAA–A (*NOAA–6*) June 27, 1979, from the Western Space and Missile Center into a sun-synchronous orbit. The satellite ceased providing usable atmospheric soundings on October 22, 1983, when the filter wheel assembly became inoperable. NASA still used the primary imaging system. NOAA–G would replace *NOAA–6* in November 1985.

During NOAA–B launch on May 29, 1980, one of the two Atlas booster engines had a thrust reduction. This premature booster engine cutoff caused a deficit in velocity and altitude of the Atlas, making the sustainer engine burn 55 seconds longer than planned. As the sustainer engine was still thrusting when the spacecraft attempted to separate from the launch vehicle, the spacecraft consumed most of its attitude control gas as it fought the booster for its attitude control. The apogee kick motor burn put NOAA–B into a highly elliptical, 269 × 1466 km, 92.5° inclination orbit. As a result of the combination of incorrect orbit and loss of attitude control, NASA terminated the mission. The aborted NOAA–B mission brought about changes in subsequent NOAA launches, and the next three launches were successful.

On June 23, 1981, NASA successfully launched NOAA–C (*NOAA–7*) from the Eastern Space and Missile Center into a sun-synchronous orbit. Failure of the filter wheel assembly on February 7, 1985, terminated the collection of usable atmospheric soundings. On February 25, 1985, NASA placed the primary imaging system on standby when *NOAA–9* became the operational spacecraft for afternoon observations.

NASA achieved the mission objectives of NOAA–A and –C, and the spacecraft provided useful sounding and imaging data far in excess of their two-year design life. NASA certified these missions successful. Due to launch vehicle malfunction, NASA terminated the NOAA–B mission and reported that mission unsuccessful. (NASA MOR E–615–79–01 [postlaunch], E–615–80–02 [postlaunch], E–615–81–03 [postlaunch] Apr 16/85)

Atmospheric Research

During May: The Canadian government had cut $5 million (Canadian) from the National Research Council of Canada's space sciences budget, eliminating the sounding rocket program and most of the balloon research, *Spaceflight* reported. The cuts meant the end of nearly a quarter-century of space research with the indigenous Black Brant family of sounding rockets. The council did not cut major satellite and Space Shuttle research projects. (*Spaceflight,* May 85, 201)

June 6: NASA announced it selected Aerojet ElectroSystems Co. to negotiate a cost-plus-award-fee contract for three advanced microwave sounding units-A for flight aboard National Oceanic and Atmospheric Administration (NOAA) spacecraft. The contract work statement called for development and delivery of flight instruments and associated ground support equipment as well as instrument/spacecraft integration and test support.

The contract, with an estimated value of $29 million, would take effect about November 11 and continue for 51 months. (NASA Release 85–88)

June 14: Goddard Space Flight Center announced that a team from its Wallops Flight Facility completed studies of the atmospheric ozone profile at Natal, Brazil. The team acquired data from 9 to 40 miles above the earth at an equatorial site using the improved "A" version ROCket OZonesonde (ROCOZ–A).

The team took seven profiles during ozone-measuring satellite overpasses to compare the ROCOZ–A data with other satellite data for the same time and place. Although the primary mission was to support the SAGE II ozone measurements acquired by the *ERBS/SAGE II* satellite, the team would also compare the data with that from the solar backscatter ultraviolet spectrometer on the *Solar Mesopheric Explorer (SME)* satellite.

Each ROCOZ–A launch included the supporting launch of a super-loki datasonde dart (a small meteorological rocket) to collect temperature data from 9 to 43 miles in the atmosphere and an electrochemical concentration cell (ECC) balloon-launched ozonesonde to collect lower-atmospheric (below 18 miles) ozone, temperature, and pressure data. The team would use the supporting data to define the atmosphere at the site for comparison with the appropriate satellite.

The team provided ROCOZ–A coverage during five dual and two single satellite overpasses. There were at least two ROCOZ–A profiles for each satellite with four profiles for the *NOAA-9*. (GSFC Release 85–17)

June 26: Goddard Space Flight Center (GSFC) announced that its Wallops Flight Facility had launched at 1:46 p.m. EDT June 24 and 1:46 p.m. EDT June 26 Nike-Orion sounding rockets for the University of Illinois to investigate the daytime mid-latitude ionosphere between 96 and 241 statute miles above earth. Specific mission objectives were to investigate differential absorption of radio waves; determine the electron density profile; and investigate the irregular structure of the electron density profile in terms of neutral turbulence, plasma instabilities in the mid-latitude region, and gravity wave effects. The 120-lb. payloads reached peak altitudes of 110 statute miles, and preliminary results indicated all instrumentation performed satisfactorily and obtained good data.

The Nike-Orion was a two-stage, solid-propellant unguided sounding rocket about 30 feet long.

Researchers would correlate scientific data from the missions with information from a similar study of the daytime equatorial ionosphere conducted March 1983 in Peru as part of Project CONDOR. Dr. L.G. Smith, University of Illinois project scientist, noted the studies were important "because turbulence is the major unsolved problem of the atmosphere and ionosphere." (GSFC Release 85–19)

September 9: Goddard Space Flight Center (GSFC) announced that its Wallops Flight Facility conducted today a coordinated series of rocketborne experiments to study effects of lightning in the earth's troposphere. Three rocket payloads at different altitudes simultaneously collected lightning measurement data, and a series of ground-based instruments designed to detect the location and characteristics of lightning also made coordinated measurements.

Wallops researchers launched at 8:01 EDT a single-stage Orion sounding rocket, 70 seconds later a two-stage Taurus-Orion, and then 57 seconds later a Nike-Orion; all three were solid-propellant sounding rockets.

The Taurus-Orion lofted its 297-lb. payload to a peak altitude of 120 statute miles, and the Nike-Orion carried its 303-lb. payload to a peak altitude of 71.5 statute miles. The Taurus-Orion experiment package contained a water recovery system using an eight-foot parachute with flotation bag, flashing strobe light, and recovery beacon.

Wallops personnel, following the three rocket series, launched a meteorological data sonde on a small rocket to determine the meteorological characteristics in the upper atmosphere near the time of the other measurements.

The Wallops P–3 search aircraft located the Taurus-Orion payload package at 9:45 p.m. that same day about 40 miles offshore, and the next day the Coast Guard cutter "Point Brown" recovered and returned it to the Wallops Flight Facility for refurbishment and future spaceflights. (GSFC Release 85–26)

October 16: NASA's Goddard Space Flight Center announced the launch at 6:01 a.m. DST today of a single-stage Orion sounding rocket from its Wallops

Flight Facility to evaluate electrical fields and charged particles in earth's middle atmosphere.

The Orion, about 16 feet long and 14 inches in diameter, carried the 90-lb. payload to a peak altitude of 52 statute miles, after which a Wallops Skyvan aircraft recovered the payload in mid-air.

NASA launched the experiment from the Poker Flat Research Range (PFRR) in Alaska on April 12, 1985; however, snow cover hindered payload recovery. On August 26 a sheep hunting party found the payload and its parachute floating on Beaver Creek in the White Mountains, Alaska and returned it to the PFRR. Range officials on September 23 sent the payload to Wallops for refurbishment for today's launch.

NASA tentatively scheduled for July 1986 another launch of the experiment in Sweden. (GSFC Release 85-29)

November 7: Scientists analyzing data recently sent from two monitoring devices aboard the *Nimbus-7* satellite said the observations confirmed a progressive deterioration in the earth's ozone layer above Antarctica, the *NY Times* reported. Since 1974, the satellite data showed that a "hole" appeared each October in the ozone layer there, meaning the layer in that area became less able to shield the earth from damaging solar ultraviolet rays. This had caused scientists to predict that increased atmospheric pollution was causing the gradual depletion of stratospheric ozone; the new data seemed to show researchers that the ozone loss was proceeding much faster than expected.

However, Goddard Space Flight Center's Dr. Donald Heath, who had monitored the satellite recordings for several years, said he was uncertain of the reason for the ozone decline. In addition to the theory that fluorocarbons caused ozone depletion, some scientists blamed the depletion on the sulfur compounds and other particles ejected into the stratosphere in the 1982 eruption of El Chichon in Mexico. And Heath said there were other possible explanations, such as the sunspot cycle, which was then near a minimum. According to a study by NASA scientists, the chemical reactions that produced stratospheric ozone were stimulated by a form of ultraviolet radiation that became weak when sunspots were fewest.

Even under normal conditions the ozone layer was subject to wide variations, so it was difficult to establish that the recent depletion was part of a long-term trend. Heath also pointed out that it was not clear whether the Antarctic readings manifested a local change in atmospheric circulation or a global depletion, since the condition of the winter atmosphere over Antarctic was unique.

Government officials in 1977 imposed a ban on fluorocarbons as spray-can propellants, but it became evident that the ozone varied in response to a variety of interacting natural and human influences. By 1984 a National Academy of Science report estimated fluorocarbon-caused ozone reduction at only 2 to 4%. (*NYT*, Nov 7/85, B21)

November 19: The TOMS, or Total Ozone Mapping Spectrometer, aboard *Nimbus-7* made measurements the previous week indicating that the gas cloud ejected from the Nevado del Ruiz volcano in Columbia released twice as much sulfurous gas into the sky as the 1980 eruption of Mount St. Helens and five times as much as the 1982 eruption of Galung-gung in Indonesia, the *NY Times* reported. Dr. Arlin Krueger of the Goddard Space Flight Center said that within 14 hours of the November 13 eruption, the cloud spread over an area of 600,000 sq. km (230,000 sq. miles).

NASA researchers designed TOMS to monitor changes in earth's protective ozone layer, and it scanned the stratosphere at a succession on ultraviolet wavelengths whose relative intensities provided an index of ozone abundance. But the TOMS readings could also identify sulfur dioxide.

The sulfur cloud, 14 hours after the main eruption, spread chiefly to the east, covering most of Colombia and part of Venezuela. One arm of the cloud reached as far south as the equator in Colombia and as far north as Lake Maracaibo in Venezuela.

Because TOMS was not used for day-to-day weather forecasting, it was not ordinarily subjected to a rapid-analysis process. As soon as the Colombian eruption occurred, however, Dr. Krueger and his colleagues began to study its readings. The sulfur dioxide, he said, "shows up very well."

Although estimates of the cloud's volume were preliminary, Krueger said it appeared much smaller than that formed by the eruption in 1982 of El Chichon in southern Mexico. Visible and infrared data from National Oceanic and Atmospheric Administration satellites had traced El Chichon's dust, which formed a narrow plume, through one complete circuit of the globe from April 5 to April 15, 1982.

Although Nevado del Ruiz's sulfur cloud was extensive, it did not appear sufficient, with its associated ash cloud, to have an effect on climate. Whereas El Chichon's sulfur cloud was traced around the world, researchers expected del Ruiz to dissipate within days without any severe environmental effects. (*NYT*, Nov 19/85, A14)

November 20: NASA's Wallops Flight Facility announced it launched at 5:19 a.m. today a three-stage, Taurus-Nike-Tomahawk sounding rocket that left several colored clouds high over the mid-eastern U.S. coastline. The objective of the experiment was to provide baseline data on expected yields, vapor expansion velocities, and other parameters for the chemical release canisters for the Combined Release and Radiation Effects Satellite (CRRES), scheduled for launch in July 1987 from the Space Shuttle.

The 157-lb. payload ejected balls of greenish-white titanium-boron-barium and barium-cupric-oxide at an altitude of 230 statute miles, one as the rocket ascended and one as the payload descended. The barium clouds rapidly expanded, while ejecting barium ions along the earth's magnetic field to form a visible streak more than 62 miles long. Another red titanium-boron-lithium cloud released at 325 miles expanded to a diameter of several hundred kilometers in less than a minute.

Goddard Space Flight Center's (GSFC) Wallops Range Control Center received reports of cloud sightings from as far north as Waterville, Maine, as far south as Wilmington, North Carolina, and as far west as South Bend, Indiana. (NASA Release 85–148; GSFC Release 85–39)

Upper Atmospheric Research

January 11: Observers had anticipated seeing the world's first man-made comet 72,000 miles above earth (over the Pacific Ocean west of Lima, Peru) the morning of December 27, 1984, the JPL *Universe* reported. Although clouds obscured viewing from official ground-observation sites, scientists called the artificial comet a success, marking another milestone in the active magnetospheric particle tracer explorers (AMPTE) program sponsored by the U.S., West Germany, and the United Kingdom. Mission goal was to inject, from a satellite tracer, ions of lithium and barium inside, outside, and just within the earth's magnetosphere (creating the comet appearance) and to detect and monitor these ions with two other satellites as the ions convected and diffused through the inner magnetosphere. Scientists hoped that this new data would improve understanding of the influence and mechanisms of interaction of the solar wind with earth's magnetosphere, including formation of the van Allen Belts.

A NASA Convair 990 reported a six-minute viewing; an Argentine Boeing 707 an eight-minute view. Overall intensity of the "comet" was lower than expected; the 12,000- to 20,000-km tail was somewhat shorter than expected. Experimenters withheld other canisters of chemicals aboard the German ion-release module (IRM) for AMPTE experiments later in the year.

GSFC managed the U.S. portion of the project; a JPL mission control team operated the charge composition explorer (CCE) spacecraft to observe the cloud. (JPL *Universe*, Jan 11/85, 1; GSFC *News*, Jan 85, 3)

March 6: NASA announced it awarded a $145.8 million contract to General Electric Co.'s Valley Forge Space Center, Philadelphia for development of the upper atmosphere research satellite (UARS) observatory. Scheduled for October 1989 deployment from the Space Shuttle, the satellite would carry 10 scientific instruments into a 373-mile circular orbit. Valley Forge Center would design the observatory system, design and fabricate an instrument module compatible with the NASA standard multimission modular spacecraft (MMS), integrate the instrument module with the MMS and flight instruments, test the observatory system, integrate the observatory into the Space Shuttle, and support postlaunch flight operations.

The UARS, with its remote sensing instruments providing essentially global coverage, would for the first time generate data for understanding the composition and dynamics of the upper atmosphere, important to solving many questions about the earth's weather and climate. For example, the observatory would provide previously unavailable data on the nature of natural and human effects on earth's ozone layer.

GSFC would provide UARS project management for NASA's office of space science and applications. (NASA Release 85–32)

March 22: NASA announced that the third phase of the active magnetospheric particle tracer explorers (AMPTE) magnetotail probe, a U.S., West German, and United Kingdom scientific experiment to determine how the solar wind interacted with the earth's magnetosphere, was underway.

Gilber Ousley, AMPTE project manager of Goddard Space Flight Center (GSFC), said that the first AMPTE magnetotail release of barium occurred March 21 at 4:20 a.m., EST and that the release conditions had been well within established criteria. The German ion release module spacecraft immediately detected a magnetic field change as expected. The satellite would make one additional release of barium and two of lithium (the tracers) into the earth's magnetotail. The U.S. satellite would analyze the charged-particle space environment around the magnetotail to determine the effects on the environment of the injected elements.

Ousley also said that all ground stations except the one in Argentina reported clear weather and that the observatory at Kitt Peak, Arizona, reported visual sighting with the naked eye for about 20 minutes. The airborne NASA Convair 990 and Argentine Boeing 707 observation aircraft had recorded the event.

The first phase of the AMPTE project had consisted of lithium releases September 11 and 20, 1984, into solar wind outside the earth's magnetosphere about 70,000 miles above the Pacific Ocean off the coast of Lima, Peru. Results of the experiment indicated that the artificial comet was "eroded" by the solar wind much faster than previously anticipated.

The only scheduled AMPTE experiment remaining after the March-April releases was formation July 13 or 14, 1985, of another artificial comet on the opposite flank of the magnetosphere from the December experiment.

All releases had to satisfy several criteria, including clear voice communications between the U.S. AMPTE science data center located at Johns Hopkins University's Applied Physics Lab and the German Space Operations Center in West Germany, where German scientists gathered to view the data. The latter sent the command for release through their center and NASA's Spaceflight Tracking and Data Network.

Goddard Space Flight Center managed the U.S. portion of AMPTE for NASA's office of space science and applications; Johns Hopkins Applied

Physics Lab under contract to NASA had built the U.S. spacecraft. (NASA Release 85–42)

July 11: Dr. Mario Acuna, project scientist at Goddard Space Flight Center (GSFC), announced that the Active Magnetospheric Particle Tracer Explorers (AMPTE) project, an international scientific experiment to determine how the solar wind interacted with the earth's magnetosphere, would produce on July 18 or 20 at midnight EDT the world's second artificial comet. In the fourth and final phase of the project, a West German satellite would release at 70,000 miles above earth two barium canisters into the solar wind on the flank of the earth's magnetosphere. The release would create an artificial comet expected to be visible in the southwestern United States to the unaided eye for approximately four minutes.

In the first phase of the AMPTE project, the West German satellite released September 1984 lithium into the solar wind outside the earth's magnetosphere. Preliminary results of that experiment indicated that less than 1% of the solar wind gained access to the magnetosphere under the conditions in which the releases took place.

In the second phase, the German satellite created on December 27, 1984, a barium cloud on the flank of the earth's magnetosphere about 70,000 miles above the Pacific Ocean off the coast of Peru. Data from the artificial comet indicated that the solar wind eroded it much faster than scientists had anticipated.

Releases on March 21 and May 13, 1985, of barium and lithium into the earth's magnetotail region ended the third phase of the AMPTE experiments. Acuna said of those releases, "Although, disappointingly, no tracer ions were detected by the U.S. satellite (Charge Composition Explorer) located inside the magnetosphere, this fact is in itself a very significant result. It implies that fundamental revisions to our current models of the magnetosphere need to be made to account for these negative observations." (NASA Release 85–105)

November 4: A team of Jet Propulsion Laboratory (JPL) researchers launched on November 4, as part of the Balloon-borne Laser In-situ Sensor (BLISS) experiment, a 10.2-million-cubic-foot balloon carrying a microprocessor-controlled payload that produced vertical profile measurements of nitrogen dioxide, water vapor, and nitric oxide concentrations at 115,000 and then 90,000 feet in altitude, the JPL *Universe* reported. When researchers left JPL for Palestine, Texas, to conduct the research, they expected to return home in three weeks. But two hurricanes, a cancelled insurance policy, and a balloon failure—which caused JPL's instrument-laden gondola to be dumped in the mud in a prison camp—extended the trip to 11 weeks.

The researchers used two balloons, each large enough to enclose an entire football field, to lift scientific payloads into the atmosphere. At launch, the balloons stood about 600 feet high and were about 70 feet across; at cruising altitude they expanded to about 420 feet.

A balloon's 3000-lb. payload included two types of lasers, a TV camera, and a retroreflector suspended by a steel cable one-third mile below the gondola after the balloon reached cruising altitude. The optically tracked retroreflector—swinging like a pendulum—bounced infrared laser beams through the atmosphere and back to the payload's receiving optics. "We use liquid helium to cool the lasers and our supply only lasts a day," said Dr. Christopher Webster, principal investigator for the experiment. "So over the 11 weeks we had to fill the cryostats 220 times."

Balloon failures in scientific research occurred frequently, mainly due to defects in the thin plastic balloon material or because seams ripped out, Webster noted. "This is a serious problem and scientific research is definitely suffering because of it," he said. "However, the payload recovery is excellent, so JPL does not lose its instruments."

"The greatest advantage of a balloon-borne laser instrument," he added, "is that we can obtain continuous spectroscopic analysis of the upper atmosphere in situ with high sensitivity." (JPL *Universe*, Nov 29/85, 3)

AVIATION

August 1: Fourteen-year-old Todd Holmes of Fort Smith, Arkansas, unofficially became the youngest person to solo in a glider, the *Washington Times* reported.

According to Federal Aviation Administration regulations, a pilot had to be 14 before he or she could solo in any kind of aircraft and 16 to be licensed in a passenger-carrying glider or to fly a powered aircraft. The agency did not keep statistics on who might have been the youngest person ever to solo.

Todd, who got his glider license on his 14th birthday, said, "I could've done it two weeks ago, but I had to wait until I was 14." (*W Times*, Aug 1/85, 4B)

August 6: Arthur Smith, 74, retired chairman of United Technologies Corp. and the originator of the concept for water injection in piston engines that made major contributions to the performance of World War II combat planes, died August 6 after a long illness, the *Washington Times* reported.

Smith retired as chairman of United Technologies—then known as United Aircraft—in 1973. He had assumed the post 15 months earlier after serving as executive vice president, president, and chairman of the executive committee.

In 1935, after two years in the automotive industry, Smith joined Pratt & Whitney as an experimental test engineer. As a project engineer, he and a coworker invented and patented a system for water injection to control detonation and increase horsepower in aircraft piston engines. The 1938 invention increased the speed of U.S. fighter planes used during World War II by 40 mph.

In 1949 he became chief engineer at Pratt & Whitney and helped develop the J57, which powered the F-100 fighter, the first production aircraft to exceed the speed of sound in level flight. Smith became president of the company in 1967. (*W Times*, Aug 8/85, 7B)

December 20: The U.S. Air Force's Aeronautical Systems Div. (ASD) issued in December a request for proposals to supply a wide-body jet aircraft to replace two aircraft known as Air Force One, which was used by the president and his staff, the Air Force Systems Command's (AFSC) *Newsreview* reported. Current candidates included the Douglas DC-10 and Boeing's 747SP and 747-300. The Air Force planned to award a contract for the plane by May 1986. Military Airlift Command's 89th Military Airlift Wing at Andrews Air Force Base would receive the first jumbo jet in late 1988; the second in 1989.

Existing Air Force One aircraft were deficient in three basic ways, Air Force officials said. They were getting more difficult to maintain; they were so crammed with equipment that there was no room for new communications equipment, an emergency medical treatment facility, or improved work areas for the president and staff; and they did not meet Federal Aviation Administration (FAA) standards and had limited performance, especially range.

The Air Force said the new aircraft must be certified by the FAA, have three or more engines, and be a model that had at least two years of airline service time. "The two-year in-service requirement guarantees a performance record will be available to evaluate the maturity of aircraft design relative to safety, reliability, and maintainability," said Col. Robert Black, program manager of ASD's Deputy for Airlift and Trainer Systems. Other Air Force One performance requirements under evaluation would call for the aircraft to take off from a 9,300-foot runway and fly 6,000 nautical miles nonstop. The new plane had to have a minimum cruising speed of about 528 mph while flying between 36,000 and 45,000 feet and a high-speed cruise capability of about 575 mph.

On the new Air Force One the president and staff would have state-of-the-art communications complete with secure voice terminals and cryptographic equipment for writing and deciphering classified messages. The cabin areas would provide seating for 80 passengers and 23 crew members; and presidential accommodations would consist of an office, stateroom, and adjacent dressing room and lavatory.

Other features would be a conference room; guest, staff, press, and Secret Service compartments; and a complete medical treatment facility. Onboard galleys would allow stewards to prepare and serve about 50 meals from each galley. (AFSC *Newsreview*, Dec 20/85, 4)

Civil Aviation

January 24: NASA announced that its anonymous and voluntary Aviation Safety Reporting System (ASRS), managed by Ames Research Center, had evaluated some 42,000 incident reports in eight years and had issued 805 alert bulletins and 28 research reports to improve airway safety. Designed and implemented by NASA in 1976 at the request of the Federal Aviation Administration (FAA), ASRS gave pilots and flight controllers a means of reporting incidents that would otherwise have passed unnoticed, resulting in revisions to both air traffic control procedures and FAA regulations. NASA served as a neutral third party in operating the system, protecting the confidentiality of those who reported. In addition, FAA could grant participants limited immunity from disciplinary action except in cases of accident or criminal conduct.

The system had identified, for example, the need for more flight-controller help to general aviation pilots operating at night over unfamiliar terrain and restriction of cockpit conversation and activity in transport aircraft flying below 10,000 ft. The system had also produced a number of changes in airline, military, and general aviation pilot flight training based on real flight data. Reports in the ASRS data base had become a major resource for human-factors research and behavior models.

ASRS received about 500 incident reports a month; experienced pilots and controllers screened these to identify existing safety problems and forecast future problems and trends, and NASA used the data for safety research for FAA, the National Transportation Safety Board, Department of Defense, and other government agencies. ASRS issued a monthly safety bulletin and also alert bulletins for hazards needing rapid response. (NASA Release 85–12)

During January: NASA announced that its new findings on the nature of high-altitude, clear-air turbulence (CAT), based on data gathered from airline flight-data recorders, could eventually lead to CAT prediction, enabling pilots to fly around the invisible wind swirls.

The findings provided the first detailed description of hazardous, clear-air turbulence, a series of swirls or vortices of air embedded in upper-level wind streams at altitudes between 35,000 and 40,000 feet. The findings also showed that a jet traveling at 500 mph flew through a single-vortex core in about one second. During that second, the wind would push the plane upward and then down, exerting the weight of an extra G (a measurement of gravity) on passengers. Planes might encounter two to four vortices in succession about four seconds apart, possibly causing injury to flight attendants and passengers without fastened seat belts.

Ames Research Center investigators had found that the strongest vortices occurred at about 25 miles downwind of large thunderstorms or mountain ranges, which caused a swell in the upper-level wind stream, and that wind shears formed vortices at the tropopause, the boundary between the troposphere and the stratosphere. At the tropopause, higher-velocity jet streams traveled just above lower-speed windstreams, the difference in the speeds forming a wind shear. The investigators had verified that wind-shear layers, pushed up over a thunderstorm or a mountain range, acted like ocean waves when forced to rise over an obstacle. The air would form a series of swells that turned to waves that curled like those on the sea. They would continue to curl until they formed complete circles or vortices that whirled at high speeds before disintegrating.

As a result of its research, NASA hoped one day to be able to give pilots better indications of when and where they might encounter the vortices. (NASA announcement, Jan 85, 9)

February 27: In its just-released annual "Aviation Forecasts," which covered FY 85–86, the Federal Aviation Administration (FAA) forecasted healthy growth for the nation's major air carriers in the next dozen years and slow but steady growth for general aviation (private and business flying).

The agency noted that U.S.-certificated air carriers had recorded passenger gains in each of the previous three years and in FY 84 achieved their largest operating profit in history—$2 billion. The FAA said airlines, over the 12-year forecast period, would increase passenger enplanements at better than 4.5% annually, from 336 million in FY 84 to 573 million in FY 96. Commuter airlines had increased passenger boardings 14% in FY 84, a figure expected to more than double during the forecast period to 54.2 million in FY 96.

The FAA expected the general-aviation fleet to increase from 213,300 aircraft in 1984 to 270,500 in 1996, an annual growth rate of 2%.

FAA projections for takeoffs and landings at airports with FAA control towers indicated an increase from 57 to 92 million, as the number of aircraft handled by the agency's en route control facilities would rise from 31.6 to 45.7 million. (FAA Release 8–85)

During March: A panel organized by the National Academy of Engineering in conjunction with the National Research Council (NRC) published its assessment of aviation technology in, "The Competitive Status of the Civil Aviation Manufacturing Industry," the NRC *Newsreview* reported.

U.S. aircraft manufacturers in the past had been particularly successful in translating advanced technology into products suited to the marketplace; however, as competition intensified, the timing of the introduction and the fit of the product to customers' needs had become increasingly important. European countries had tried repeatedly to create a viable air transport manufacturing industry; in 1970 their efforts were realized in the creation of Airbus Industrie, which drew on the resources of many companies in a number of countries. Those foreign companies created a dilemma for U.S. manufacturers, whose product lines were not extensive. Furthermore, U.S. markets were relatively open to competitors, while many foreign markets were closed to American-made products.

The panel foresaw a need for U.S. manufacturers to form international partnerships, especially as the U.S. aircraft industry was often in virtual competition with governments as well as with private commercial companies. And the panel determined that U.S. manufacturers had to be even more sensitive in interpreting the needs of foreign customers.

The panel did conclude that it was possible to further improve reliability of aircraft and air travel, as well as increase efficiency in fuel consumption and operations. Studies cited by the panel indicated that a variety of technological changes together could improve fuel efficiency by as much as 30–50%. Introduction of advanced turboprops or propfans could provide up to 20% additional improvement, and the experimental unducted propfan engine could raise that figure.

In the technology area of advanced structures, the panel viewed the U.S. and Europe as on a par in developing this technology. Although the U.S. led in application experience, Europe threatened the U.S. position. In propulsion technology, the panel saw the U.S. lead as not unassailable; Rolls Royce was the principal foreign competitor, and the U.K. was committed to maintaining

Aviation

a comparative position with the U.S. The panel rated U.S. R&D facilities as the best in the world, European facilities as adequate, and Japanese facilities as handicapping their efforts to benefit from technological developments. (NRC *NewsReport*, Mar/85, 11)

June 17: The Air Transport Association reported that a record 343 million passengers and five million tons of cargo traveled without a single passenger fatality on 5.4 million scheduled jet airline flights in 1984. The industry achieved a 1984 net profit of more than $800 million, after losses for three consecutive years. The year's operating profit was a record $2.2 billion on revenues of $44 billion.

Other facts noted in the association's annual report included lengthened airline passenger trips (average passenger trip length was 887 miles, up from 785 miles 10 years previously), average passenger's cost was 12.1 cents per mile (compared with 11.6 cents in 1983 and 12.3 cents in 1981), airlines in 1984 accounted for more than 88% of the intercity public passenger traffic miles in the U.S. (up from 80% in 1974), and Chicago's O'Hare was the busiest passenger airport in the country, handling more than 45.7 million people (New York's JFK airport handled the most air cargo, 1.3 million tons). (ATA Release No. 38)

June 19: Responding to a suggestion made by President Reagan at a June 19 press conference that the U.S. expand its sky marshal force for use on international flights, Thomas Pyle, speaking for the International Air Transport Association, said security experts found little merit in the suggestion, the *NY Times* reported. "We're not happy about anybody having guns up in the air," he said. "There's a distinct danger to passengers and a bigger danger of damage to the aircraft." Pyle did acknowledge that there might be special occasions, including cases of known threats to particular flights, when marshals would be needed.

Richard Lally, security chief for the association representing U.S. airlines and for the Federal Aviation Administration from 1974 to 1982, agreed with Pyle's assessment. "It is not an answer to the problem and introduces an impression that it is. It provides for a very dangerous situation."

The views expressed by Pyle and Lally reflected the consensus of security experts that the best place to head off hijackings was on the ground. (*NY Times*, June 20/85, A20)

July 26: NASA would once again be one of the chief exhibitors at the 33rd annual Experimental Aircraft Association (EAA) International Fly-in July 26 to August 2 at Oshkosh, Wisconsin, the *Langley Researcher* reported. NASA would broaden its exhibit, "The Shape of Things To Come," which was traditionally only aeronautical, to include a greater representation of its activities in space technology.

Astronautics and Aeronautics, 1985

A 60 × 90-foot tent would house displays of NASA's research and technology development work in aeronautics, the space station, Space Shuttle, and space exploration. Langley Research Center and Ames Research Center would provide aeronautical exhibits, Lewis Research Center would provide aeropropulsion and space exhibits, Goddard Space Flight Center would have its search and rescue satellite van on hand, and Johnson Space Center and Marshall Space Flight Center would exhibit material on space research. Astronauts Robert Gibson and Robert Overmeyer would be EAA guests and would participate in a program the evening of July 29. Two dozen NASA forum speakers would give presentations covering every aeronautical discipline.

Each year for the last several years about 100,000 people and 10,000 privately owned light airplanes were at Oshkosh for the exhibition. Organizers also scheduled the British Airways Concorde to make special fly-bys during the afternoon air shows. (LaRC *Researcher*, July 26/85, 3)

July 26: Secretary of Transportation Elizabeth Hanford Dole announced the award of a $196.9 million contract to IBM to provide new-generation computers that would give controllers extra capacity to handle growing air traffic safely and efficiently in the coming decade. Under the contract, IBM would replace the computer systems in the nation's 20 air route traffic control centers. Contract options for future hardware maintenance and software and technical support, if exercised, could total an additional $235.1 million.

In making the announcement, Dole said that "existing computers are based on outmoded technology . . . The new computer will have the capacity to assume added functions—such as improved conflict detection and resolution—permitting the Federal Aviation Administration (FAA) to make more efficient use of controllers." And it would provide a vehicle for a stable transition to an advanced automation program in the future.

The new computers would have greater storage capacity than the IBM 9020s used since the early 1970s in FAA centers, which control all aircraft operating under instrument flight rules between the nation's airport terminal areas.

The FAA awarded the contract to IBM following a 21-month design competition with the Sperry Corp. The FAA based the selection on a number of cost and performance factors including the results of a "compute off" at the FAA Technical Center near Atlantic City, New Jersey. A key factor in the side-by-side evaluation was the capability of each computer system to run the current 9020 software package with minimum modifications.

Key element in the IBM hardware was the IBM 3080–BX1 model. Each installation would consist of two units with one serving as the primary processor and the other as a backup.

The contract called for IBM to deliver the new computer systems over a one-year period beginning in the summer of 1986 to air route traffic control centers including the FAA Technical Center and the FAA Aeronautical Center where the computer system would be used for training purposes. (FAA Release 32-85)

July 29: U.S., Indian, and Canadian officials were estimating the cost of raising portions of the wreckage of the Air-India Boeing 747 that crashed June 23 into the Atlantic off Ireland, *Aviation Week* reported. Lack of significant data on either the cockpit voice recorder or the flight data recorder, recovered July 10 and July 11, respectively, by a remotely controlled submersible vehicle, stalled the investigation into the cause of the crash until more information became available. Raising portions of the wreckage seemed the only way to accomplish this.

After a British ship initially surveyed the sea bed where the wreckage lay, the Canadian ship John Cabot completed a second sonar survey of that portion of the ocean floor. Officials later said wreckage of the aircraft could be identified only as "lumps."

Officials said they had not determined the cost of raising the wreckage or a significant part of it, but knew it to be high. Although responsibility for the accident investigation legally belonged to India, the U.S. and Canada—both parties to the investigation—might contribute funds to the salvage operation. (*AvWk*, July 15/85, 28; July 29/85, 29)

August 2: In a letter to the editor in the *NY Times*, Saunders Kramer, a Fellow of the American Astronautical Society, commented on the *NY Time's* report of speculation in a London newspaper that debris from a satellite reentering the atmosphere may have struck the Air-India jetliner that crashed in July [see Aviation/Civil Aviation, July 29]. "You dismiss that possibility as 'bizarre,'" Kramer wrote, "Not quite."

Kramer then related a tale of an incident that occurred in the early 1960s, when passengers on a airline flight from Honolulu to Tokyo saw at 30,000 feet a white flash and heard a thud.

"After they landed at Tokyo, inspection revealed a smooth dished-in area (about 18 inches across) on the plane's right wing leading edge between the two starboard engines," Kramer continued. There was no evidence of a bird collision.

When the navigator returned home, he related the tale to Kramer, who checked North American Air Defense Command'a (NORAD) space-tracking data bulletins, which showed reentry of satellite debris at the time and location of the Pan Am flight. "We concluded that the debris had indeed struck the plane," Kramer wrote.

"The propulsion section of the last stage of a satellite booster is frequently massive enough to survive reentry. Had it struck the tail section severely or struck the cockpit from above, it is certainly conceivable, however remote, that destruction of the aircraft could have been instantaneous," Kramer concluded. (*NYT*, Aug 2/85, A24)

August 2: Federal investigation of the Delta Air Line Lockheed L-1011-1 crash August 2 at Dallas/Ft. Worth International Airport showed strong evidence of wind shear and microburst in the aircraft's landing path, *Aviation Week* reported, which raised issues concerning the adequacy of detection and training of pilots to respond to violent weather. The Delta TriStar crashed at 6:05 p.m. CDT on final approach to runway 17L after entering a suddenly developed violent thunderstorm cell that eluded detection by ground-based sensors and was not reported by pilots.

The problems posed by wind shear would force the government in the following weeks to deal with procurement of Doppler radar systems (which the Federal Aviation Administration successfully tested as a wind-shear and microburst detector), regulations requiring simulator training and cockpit resource management instructions to prepare pilots for wind shear, and enhancement of the low-level, wind-shear alert system (LLWAS) to broaden coverage of the detection system.

Rep. Mickey Edwards (R.-Okla.) had called for investigations to go beyond the direct cause of the Delta accident to "basic airline and government operating procedures and priorities" in responding to weather hazards. (*AvWk*, Aug 12/85, 16)

August 26: The investigation into the August 12 crash in a remote mountain area of a Japan Air Lines Boeing 747-100SR after it lost most of its vertical stabilizer and rudders focused on damage in the aft pressure bulkhead of the aircraft, *Aviation Week* reported. Japan's Aircraft Accident Investigation Committee said it was attempting to determine whether the bulkhead damage was caused by a crack or cracks that existed in the bulkhead, by explosive decompression in the cabin, or by the crash impact. The Boeing Co. said inspections as of August 19 of the aft pressure shell of 43 Boeing 747 aircraft showed no evidence of cracks.

In addition to cracks in the bulkhead, lower ends of 5 of 18 triangular panels, or gores, were buckled toward the rear. Boeing Co. had replaced 9 of the 18 panels after the aircraft was involved in a hard landing June 2, 1978.

Further optional or mandated inspections of the wide-body transport fleet stalled the previous week due to lack of information on damage to the JAL aircraft. Information from Japan was sparse, partly because of restrictions placed on access to the wreckage and because of damage done to some of the vital evidence by members of the rescue team. Rescuers had cut the bulkhead into pieces and removed them to clear the way for rescue. Boeing's accident investigation team had not been able to make a close inspection of the bulkhead by the previous week.

One of the few pieces of evidence—damage to the vertical fin—led Boeing to suggest that Boeing 747 operators inspect the fin and rudder structure. And reports that a depressurization occurred prompted the company to suggest inspection of the external and aft portion of the pressure shell structure.

The Federal Aviation Administration declined to take action pending receipt of more information. (AvWk, Aug 26/85, 28)

August 26: At least 54 persons were killed and 83 injured when a Boeing 737-200 operated by British Airtours aborted its takeoff, ran off the runway, and caught fire at Manchester International Airport in northern England, *Aviation Week* reported. Manchester Airport officials said the pilot radioed as he approached rotation speed that he was having trouble with one of his engines. Then the aircraft rolled off the side of the runway causing the fuselage to break. Fire quickly gutted the aircraft.

British Airtours Flight KT 328, en route to the Greek island of Corfu, carried a full load of fuel plus 131 passengers and a crew of six. Most of those who died were in the rear section of the aircraft. Two crew members were among the dead, but both the pilot and copilot survived.

Loss of life was reduced because the Boeing 737 skidded to a halt near the airport's fire station, and all seven of the airport's firefighting vehicles were on the scene within minutes. Firemen were able to rescue many of the passengers, although a secondary explosion during the rescue attempt injured two of the firemen. (AvWk, Aug 26/85, 28)

September 2: Loss of 13 commercial aircraft hulls since January, including four wide-body transports, totaled $318 million and exceeded hull losses in any previous year, *Aviation Week* reported. The wide-body losses included a UTA Boeing 747-300, which burned March 1985 on the ground in Paris, an $85 million loss; an Air-India Boeing 747-200B; a Delta Air Lines Lockheed L-1011-1; and a Japan Airlines Boeing 737-100SR. Observers expected this to result in increased airline insurance premiums, the third round of increases since heavy hull losses in 1982 and 1983 and the onset of large liability awards in U.S. courts.

Insurance officials also predicted the year's losses would narrow capacity in the insurance market, meaning airlines could negotiate for insurance, but it would be harder to get and at higher rates. Payouts from hull losses already incurred had exceeded total premiums paid plus any interest income gained from investment, although interest rates were low and produced comparatively little income.

Premium rates on a world scale averaged a 45% increase in 1984 and would probably rise another 15% in 1985. Also, the levels at which deductible insurance became effective would likely double. Most policies signed in 1984 increased the deductible from $250,000 to $500,000 for narrow-body aircraft and from $600,000 to $1 million for wide-body aircraft. This was part of a long-term trend toward airlines self-insuring for the first few million dollars of liability.

Of greatest concern to insurance officials was the rise of average settlements for passenger fatalities in the U.S., which averaged $450,000 in 1979, $650,000 in 1982, and would likely rise to between $800,000 and $1 million, the officials believed. The average settlement in the rest of the world was approximately $50,000 but varied considerably. (*AvWk*, Sept 2/85, 34)

September 6: Air safety experts, flight attendants, and some members of Congress were complaining that the emergency evacuation tests run by airlines had little resemblance to a real accident, the *Washington Post* reported. Federal Aviation Administration (FAA) regulations required that all jetliners pass an evacuation test that showed a planeful of passengers could be evacuated in 90 seconds. Such tests determined how many seats a plan must have in relations to the number of exits. However, flight attendants typically rehearsed several days for the tests in which there was no smoke, no fire, no elderly participants, or children.

Federal regulations didn't prohibit manufacturers from using their own employees for the tests, although no "passenger" could rehearse for the evacuation. Although regulations required a certain percentage of elderly persons or children for the tests, FAA officials said they often waived the requirement because of concerns that they might be injured.

Rep. James Oberstar (D-Minn.), whose U.S. House public works and transportation subcommittee on investigations and oversight monitored the FAA, said that changes in the FAA's handling of evacuation requirements were long overdue. "We've got . . . no uniformity, just sort of a haphazard approach to rule making and safety, and lives of people are at stake," he said. Rep. Newt Gingrich (R-Ga.) of the same subcommittee called the approach to evacuation taken by the FAA and industry "just totally out of touch with the real world."

The controversy over the adequacy of evacuation requirements had simmered for months since the FAA permitted the Boeing Co. to eliminate two of ten exit doors on its 747 jumbo jet. Boeing officials convinced the FAA to allow elimination of the two over-wing emergency doors on the plane, because federal regulations required two doors for every 110 seats and none of the 747s with sealed doors would carry more than 440 seats.

The FAA's Seattle office, which handled certification of large commercial aircraft, approved the change without a test to see whether passengers actually could evacuate the plane within 90 seconds and instead relied on a mathematical calculation using old evacuation tests. Critics complained that those tests were themselves flawed. (*W Post*, Sept 6/85, A21)

September 9: A mountain climbing team reached the wreckage of an Eastern Airlines Boeing 747 that crashed January 1, 1985, at 19,600 feet into Mt. Illimani in the Bolivian Andes and photographed the site for use in an investigation, *Aviation Week* reported. Judith Kelly, wife of William Kelly, a U.S. Peace Corps official who died in the crash, financed and led the three-person expedition because she was concerned that the accident investigation was lagging.

Photographs of the site showed parts of the tail structure and the fuselage. Kelly also collected cable and honeycomb structures identified by part numbers and turned them over with the photos to the U.S. National Transportation Safety Board (NTSB). Remains of the 29 persons on board were not located.

The Kelly team's ascent was the fourth visit to the site since the crash and prompted questions over the status of the accident investigation. Although Bolivia had primary responsibility to report the accident, government officials said it was unlikely Bolivia would finance an expedition to the site due to lack of funds, but added it would not stand in the way of an investigation by the U.S., which was the country of aircraft registry and manufacture.

Early in January Bolivian Bernardo Guarachi, who was Kelly's guide, reached the crash site and spent several hours there, bringing back pieces of wreckage and luggage tags identifying the flight.

A second Bolivian team in March reached the site and took motion pictures. The film showed the aircraft wreckage, including a tail section that was torn off the plane, and scattered cargo and baggage. The team also found an engine cowling, a piece of the fuselage with the Eastern name on it, and other parts of the tail structure. A team sponsored by the Bolivian Red Cross spent three days in August at the site, but was hindered by new-fallen snow and lack of guidance on where to look for technical evidence.

John Young, the NTSB investigator in charge, said he was discussing with officials in the safety board's accident investigation division whether the safety board should attempt to carry on the investigation and what approach to take. Recovery of the flight data recorder and cockpit voice recorder could aid in determining navigation procedures the crew were taking.

Because of the questions surrounding the crash of the aircraft, which was several miles off course when it struck the mountain 26 miles southeast of La Paz International Airport, the U.S. Air Line Pilots Association was discussing financing an expedition to recover the recorders. (*AvWk*, Sept 9/85, 40)

September 9: Continental Airlines filed a reorganization plan with the U.S. Bankruptcy Court in Houston in which it said creditors would be repaid 100% in a combination of immediate cash and deferred payments with interest, *Aviation Week* reported. Cash payments would total about $121 million at the time of court confirmation of the plan, expected by mid-1986.

Continental had originally filed September 24, 1983, for protection under Chapter 11 or the bankruptcy laws.

"We have submitted a plan that will make whole our obligations to those individuals and companies that stood with us through the worst of times," said Continental Chairman Frank Lorenzo. "It's taken nearly two years to do it, but the pieces of the puzzle are almost all in place."

One missing piece in the plan was provision for union and employee claims. Continental asked the bankruptcy court to estimate these claims, a process that had been used in a number of other companies' bankruptcies.

Continental said its plan to repay its debt in five to ten years had the support of important creditor groups including Continental's banks and the committee representing Continental's unsecured creditors. Included in the plan was an agreement in principle for two groups of banks to provide the airline with $50 million in revolving credit and equipment financing facilities, which the airline said was "clear evidence of the financial community's support and confidence" in the company.

Once the reorganization plan was approved by the court, most creditors were entitled to vote on it. Even if a creditor class did not approve the plan, the bankruptcy court could still confirm it if the court found it "fair and equitable" and that it did not discriminate against the creditor class. (*AvWk*, Sept 9/85, 34)

October 21: The Canadian recovery ship MV Kreuztrum, chartered by the U.S. government, retrieved from the Atlantic Ocean the first piece of wreckage, a section of fuselage skin approximately 15 by 30 feet, of the Air-India Boeing 747 that crashed in June off Ireland, *Aviation Week* reported. Another Canadian recovery ship, the John Cabot equipped with a Scarab 2 underwater search and recovery vehicle, was also at the site. The two ships later recovered three other sections of the aircraft's fuselage and were in the process of lifting a fifth section. Recovery team members said they had confronted no serious problems but that the work was time-consuming.

The Scarab 2 itself lifted a smaller second piece of wreckage; high-capacity winches on the Kreuztrum lifted two other sections. Officials said it was taking 18 to 24 hrs to lift each piece of wreckage.

The ships' crews recovered the larger pieces by using four clamps that the Scarab 2 attached to each corner of a piece. Scarab 2 could take only two clamps down during a dive, and a third dive was necessary to carry down a lifting bridle to attach to the four clamps and a lift line to attach to the bridle. This method made it possible to raise large sections with as little additional damage as possible.

"Since it takes two hours for the Scarab to dive to the 6,700-foot depth, two hours to do its work, and another three hours to swim back to the surface," a recovery official said, "it's going to take some time."

Recovery officials believed wreckage retrieved as of the previous week was from the fuselage both fore and aft of the wing, and they were working to locate its position on the aircraft structure. They believed the fifth piece being recovered was a primary fuselage structure, including ribs and skin.

Recovery officials said they established a two-tier priority system for the order in which they recovered wreckage. They gave top priority to pieces that were likely to confirm or refute primary theories on the cause of the crash. Secondary priority was given to pieces considered likely to contain evidence of what occurred to cause the aircraft to break up in flight. Early efforts focused on fuselage structure.

The officials had located the aft pressure bulkhead of the Air-India 747 and scanned it with TV cameras, but said "there is nothing visible to indicate any common problem" with the August crash of a Japan Air Lines 474. Plans called for raising this bulkhead section toward the end of the recovery operation. (*AvWk*, Oct 21/85, 32)

October 22: Northwest Airlines Inc. completed an agreement today to purchase 10 Boeing 747–400s as part of a $21 billion airplane order, the *Washington Post* reported, making Northwest the first airline to have the long-range, 450-seat 747–400. Northwest president Steven Rothmeier, in a telephone conversation with the *Post*, said, "the 747–400 is really the next logical step in the progression of aircraft for the Pacific . . . It's our belief that this plane will set the economic structure of the Pacific to the end of the century." Northwest also purchased ten more Boeing 757–200s, a high-technology, twin-engine standard-body aircraft.

The 747–400 would be 22% more fuel efficient than current 747s, which it closely resembled; however, it had wings six feet longer that were tipped with "winglets" that bent upward and forward to improve aerodynamic efficiency.

Digital instrument displays and computers would permit the 747–400 to have a two-pilot cockpit instead of the three-person flight crew required for current 747s; and the new plane would have a range of 8,000 statute miles—the longest of any commercial jet—permitting it to fly nonstop from New York to Tokyo, Seoul, or Shanghai.

Boeing would deliver the 747s between December 1988 and 1990; the 757s from 1987 through 1989. (*W Post*, Oct 23/85, F1)

November 7: United Airlines today placed orders with a value of over $3.1 billion for up to 110 737–300s and up to six 747s, the largest order ever placed by an airline, *Aviation Daily* reported. The previous record was a November 12, 1980, order by Delta for 60 757s valued at $3 billion.

United President James Hartigan said that the first two 737s and two 747–200Bs had a June 1988 delivery date and that "we have obtained flexibility in the contracts to change the downline delivery dates of the later aircraft to meet our needs." He added that the orders were made possible because "we reached cost-competitive contracts with all of our work groups. We are now positioned properly for dramatic growth in the competitive marketplace."

All aircraft should be delivered by the end of 1990, giving United a fleet of 478 transports. (A/D, Nov 8/85, 41)

December 13: NASA announced that researchers at its Langley Research Center (LaRC) were developing use of pyrotechnic-activated emergency exit systems that might save lives in an emergency situation aboard commercial transport aircraft.

Although pyrotechnic components aboard commercial aircraft might seem dangerous, Laurence Bement, an aerospace technologist specializing in pyrotechnic-activated aircraft escape systems, pointed out that military aircraft had used such escape systems for more than 20 years and NASA had used them in their manned spaceflight programs as far back as Project Mercury in the 1960s. The emergency egress system he proposed would be more reliable and more effective in aiding the rapid evacuation of airplanes and more cost-effective than existing mechanical and electrical systems. "What we have done is take the best materials and applications from years of pyrotechnic usage and tried to assemble the best escape system, using our past experiences," Bement said.

One U.S. Air Force system, the emergency life-saving instant exit, used pyrotechnic chargers to sever a panel inside the aircraft door. Another, the NASA general aviation egress opening system, created by an explosion an opening in the fuselage without modifying the airframe structure. The pyrotechnic-activated escape systems on U.S. armed forces planes had already saved approximately $10 million by avoiding component replacement costs. However, they added weight and complexity to the aircraft and do not increase the structural efficiency of the airframe.

To improve existing systems, Bement studied the possibility of replacing fuselage skin sections with an explosively severed, composite material panel. Researchers had tested graphite/epoxy and fiberglass composite panels, demonstrating that the graphite/epoxy was the better material. Not only was the graphite/epoxy easier to sever than the original fuselage material, but the composite panel was much lighter and more crashworthy. Once activated, the explosive material severed the panel from the fuselage and jettisoned the panel outward. No debris was projected inward and no sound or over-pressure hazard existed inside the aircraft.

The composite system was more reliable than existing mechanical and electrical systems and would require less maintenance, since the system was expected to last at least 15 years. Bolts would hold the composite panel to the primary structure and it would be a load-carrying component, unlike existing emergency exits.

Bement estimated that using composite panels with the pyrotechnic system could reduce the weight of existing emergency exits by 30 to 50%. If the airframes were designed to include the composite panels, the weight reductions would be far greater, because the light-weight composite panels were capable of carrying loads, meaning the fuselage would not require as much support structure around the emergency door frames.

The explosive in the system was hexanitrostilbene (HVS), an organic compound insensitive to handling, impact, gunfire, and lightning and was unaffected by 50 hours of exposure to temperatures of 350° F. HNS would burn if exposed to a flame but would not explode.

Another advantage of the compound was its explosive power; in a test of the Langley general aviation opening system, less than 0.4 ounce of the compound was sufficient to sever a panel about 30 square inches. (NASA Release 85–171)

December 16: In a recent report to the U.S. House Public Works and Transportation Committee and its subcommittee on aviation, the U.S. General Accounting Office (GAO) said that deregulation of the U.S. air transport system had forced carriers to become more efficient and competitive, lower fares, and improve services for most travelers, *Aviation Week* reported. However, airlines had dropped nonstop service between some cities, and some small communities had lost all scheduled air service. Most communities receiving subsidies were not progressing toward self-sustaining air service and had lost passengers since deregulation began, the report said; and it indicated that without legislative action, many of the small communities with subsidized essential air service would lose all scheduled service when the current subsidy ended in 1988. GAO repeated its recommendation that Congress consider giving the Department of Transportation greater flexibility to increase or decrease subsidies to selected communities.

The report found the airline industry was still in the process of adapting to deregulation with these trends dominant: more airlines offering competing services; declines in average fares; improvements for most passengers in service, availability, and convenience; and industry profitability improved through increased efficiency and declining unit labor and fuel costs.

Travelers between high-traffic, long-distance city-pairs gained the most from deregulation, as fares came closer to cost; and competition fostered greater choice in fares and service. However, the smaller number of passengers flying between many light-traffic, short-distance city-pairs did not benefit from lower fares, the report said.

GAO predicted that the industry would probably remain competitive, although the number of airlines might decrease due to bankruptcies and mergers. And the report said the number of large airports with capacity limitations on service could increase from the current 8 to 61 by the end of the century and that the allocation of limited airport capacity among competing airlines could offset the benefits of regulations. (*AvWk*, Dec 16/85, 33)

December 31: In terms of fatalities, 1985 was the worst year in aviation history, the *Washington Post* reported, as nearly 2,000 people died in 36 accidents. The previous worst year was 1974, when 1,299 people died in 29 accidents.

Among the 1985 aircraft losses were: an Air India jet that disintegrated off the Irish coast, killing all 329 on board; a Delta Air Lines jumbo jetliner that crashed in a violent thunderstorm while approaching the Dallas airport, killing 133; a Japan Air Lines jet on a flight from Tokyo to Osaka that hit a mountain, killing 520; a British Airways charter plane that burst into flames on the runway in Manchester, England, killing 54; a charter plane carrying U.S. military service members that crashed at Gander, Newfoundland, killing all 258 on board; and a twin-engine plane that crashed into the roof of a suburban California shopping mall, killing four persons and injuring 88. In addition, hijackers took over a TWA plane on a flight to Beirut and killed a passenger.

The year's aircraft losses resulted from pilot error and equipment failure, sugar in the fuel line of one plane, and possibly a bomb in the luggage compartment of another. Some blamed the crashes, if only indirectly, on the 1981 firing of the air traffic controllers, on deregulation, and on the enforcement role of federal safety inspectors. However, federal officials said that there was no common thread among the year's major aviation disasters. (*W Post*, Jan 2/86, A3)

Human Powered Aircraft

February 5: Langley Research Center announced that John Langford of the Massachusetts Institute of Technology would speak February 11 on the MIT Monarch, a human-powered aircraft that had won first prize May 11, 1984, in the Kremer World Speed Competition.

An all-volunteer team had designed and built the craft in 88 days during the summer of 1983 and had flown it 29 times before disassembling it. With an improved version, the team had made 35 flights in spring 1984, culminating in the record flight.

Langford would discuss design consideration and construction details of the Monarch, including propulsion and avionics. (LaRC Release 85–5)

Supersonic Transport

February 25: Lewis Research Center (LeRC) researchers believed their engine research, focusing on development of supersonic combustion ramjet (scramjet) technology, would lead to air-breathing engines capable of operating at speeds through Mach 12, *Aviation Week* reported. Testing had already produced net thrust-to-drag ratios of better than one at high Mach numbers. Research included solution of engine aerodynamics, diagnostics and measurement problems, and effects of change to full-scale, flight-weight engine prototypes.

Although the Reagan Administration budget request for NASA FY 86 hypersonic research continued the existing level of effort, NASA believed propulsion technology development would be the pacing factor for hypersonic-flight capability and looked to congressional interest, evidenced by a request for a NASA report on hypersonic research, for increased funding.

Initial scramjet technology applications might lie in missile development. As the potential for higher speeds and longer ranges became attainable, NASA concentrated on basic research for later generations of military cruise missiles rather than applications to aircraft. (*AvWk*, Feb 25/85, 52)

During March: After a detailed review of the history of Concorde development, J.C.D. Baine in his article, "The Concorde Supersonic Airliner—The Struggle for Survival," concluded that the survival of Concordes or their being superseded by second generation supersonic airlines appeared doubtful under the persisting circumstances of international economic, political, and social relationships.

He noted that the confrontation between aviation technology and the environment had brought into focus problems, whether real or emotional, that only advanced aeronautical science and technology would solve. That was especially true with respect to airport noise, sonic booms, and stratospheric pollution caused by emissions from multi-engines that consumed large quantities of petroleum-based fuels. Until those acceptable solutions were found, he concluded, opposition to civilian supersonic aircraft would continue and might further restrict or ban the aircraft's use on national and international routes.

However, he noted that the efforts of Great Britain and France to develop the Concorde should not be dismissed as a waste of human effort and resources. They had contributed to advancements in aeronautical science and technology that, though not financially rewarding, represented a store of scientific knowledge that would be available for future development. (*Aerospace Historian*, Mar/85, 10)

November 13: NASA announced an award of a $400,000 contract to North American Aircraft Operations of Rockwell Internatl. Corp. for preliminary design of a pivoting wing for a supersonic aircraft. The contract covered Phase B of a joint NASA/Navy program to design, develop, and flight test an aeroelastically tailored, pivoting-oblique wing for transonic and supersonic flight evaluation. It was part of a four-phase procurement program to modify the NASA Ames-Dryden F–8 fly-by-wire research aircraft to a supersonic oblique-wing configuration.

For takeoff and landing, the pivoting wing was set in the conventional aircraft flight position. For faster flight speeds, the wing pivoted so that one side was swept forward and the other side swept aft, forming an oblique angle with the aircraft's fuselage. In the oblique wing configuration, an aircraft in high-speed flight encountered less air resistance.

The contract covered preliminary design of the oblique wing, its pivot assembly, and flight control program for the F–8's computers. The company would also define the aircraft's flight envelope and assess the operational capabilities of oblique wings for potential naval applications.

A follow-on contract phase would provide detailed design, fabrication, ground testing, and flight support for a 12-month flight test program of approximately 40 flights. (NASA Release 85–152)

AWARDS

January 30: NASA announced its graphics and visual communications system had won the Presidential Design Award, presented January 30 by President Reagan. The National Endowment for the Arts sponsored the award to recognize excellence in federal design and had selected the NASA project, along with 12 others, from among 91 Federal Design Achievement Award winners.

The Presidential Design Awards jury chairman, architect I.M. Pei, said in his jury report, ". . . Especially noteworthy is the visual-communications system developed by NASA whose posters, publications, and logotype generally maintain a high standard of design, which truly captures the spirit and vitality of the space program." (NASA Release 85–16)

During January: NASA announced it had conferred its Space Act Awards on four Johns Hopkins Applied Physics Laboratory personnel for the invention of a medical device based on space and aeronautics technology. The invention, a programmable implantable medication system (PIMS), was a computerized pump intended for implanting in the human body to dispense medicine automatically to treat disease such as diabetes, cancer, and cardiovascular problems.

A nonhypodermic device introduced medication to a reservoir located in the PIMS device so that its tiny battery-powered pump could send the medicine into the body via a tube in minute doses precisely timed by one of two clocks within the PIMS. A small computer in the PIMS analyzed its performance and operated an alarm if the device was malfunctioning or about to run out of fluid.

NASA technologies employed in the PIMS development were space microcircuitry, titanium welding, and the pump and fluid-handling systems used on the Viking spacecraft and space-program techniques of quality control.

Award recipients were Robert Fischell, Wade Radford, Albert Sadilek, and Arthur Hogrefe. (NASA anno Jan 85, 14)

February 1: The National Society of Professional Engineers had recognized Marshall Space Flight Center's (MSFC) role in the first Spacelab mission as one of the nation's 10 outstanding engineering achievements for 1984, the center announced. Because the 10-day mission occurred so late in 1983, the society had placed it in the 1984 competition. The society also cited MSFC for its management of NASA's role in Spacelab development, which had included the transfer tunnel connecting the Space Shuttle orbiter cabin to the module and other components necessary for successful laboratory operation, for its technical and programmatic monitoring of Spacelab's European design and development activities, and for lending technical expertise and support to the European Space Agency (ESA) in the project.

MSFC's deputy director Thomas Lee had accepted the award on behalf of center employees at the society's winter meeting in Albuquerque, New Mexico. (MSFC Release 85–5)

February 1: The U.S. Navy awarded Dr. Robert Stevenson, an oceanographer with the Office of Naval Research (ONR), its Meritorious Civilian Service Award for his role in the application of space technology to the solution of oceanographic problems, the U.S. Navy announced. The award cited Stevenson as the "father" of space oceanography in the western world, recognizing him as one of the first American scientists to suggest that remote sensing instruments in space could accurately record oceanographic measurements.

As an advisor on space-related programs, Stevenson had worked with the National Council for Marine Resources and Engineering Development, the National Academy of Science, North American Air Defense Command (NORAD), and NASA. Stevenson was a Fellow of the Geological Society of America and author of numerous scientific articles on ocean surface-layers' research, particularly as observed from earth-orbiting satellites. (Dept. of the Navy Release 1–85)

March 5: The Soviet Geophysical Committee of the USSR's Academy of Sciences presented Dr. Alexander Dessler, director of Marshall Space Flight Center's (MSFC) Space Science Laboratory, a bronze medal for his contribution to fulfillment of international geophysical programs, MSFC announced. The committee awarded 1,000 medals, 100 to American scientists, commemorating the 100th anniversary of the first international geophysical program, the First International Polar Year, 1892–1983. Scientists receiving the award made important contributions to international geophysics in such areas as solid earth, oceans, atmospheres, solar-terrestrial relations, and the geophysical programs in Antarctica and space.

Since beginning his scientific career in 1956, Dessler specialized in research in low-temperature physics, atmospheric electricity, and space physics and authored more than 100 technical papers on these subjects. Dessler was a Fellow of the American Geophysical Union and the American Association for the Advancement of Science and served as an officer of the International Association of Geomagnetism and Aeronomy. (MSFC Release 85–11)

March 28: NASA announced it had presented at a ceremony at NASA Headquarters its 1984 inventor of the year award to Dale Kornfeld, Marshall Space Flight Center (MSFC), and John Vanderhoff, Mohammed El–Asser, and Fortunato Micale, all of Lehigh University, Bethlehem, Pennsylvania, for their "process for preparation of large-particle size monodisperse latexes."

The monodisperse latex reactor processor, an experiment flown aboard the Space Shuttle, had produced microspheres in zero gravity in sizes ranging

Awards

from 5 to 30 micrometers. The maximum size particles produced on earth, with the required standards of quality and uniformity, were about 2 to 3 micrometers.

Researchers used microspheres in calibrating sensitive scientific instruments such as microscopes, filters, and particle counters. Medical uses included identification of cancer and glaucoma and the study of the transport of materials inside living organisms. Industrial applications included the production of finely ground products such as paint pigments, inks, toners, explosives, and other powder materials.

NASA took an important step toward making the microspheres commercially available when it presented 15 grams of 10-micrometer particles to the Commerce Department's National Bureau of Standards for certification as "standard reference material." (NASA Release 85-44)

April 5: NASA announced it would award James Harford, executive director of the American Institute for Aeronautics and Astronautics (AIAA), its Public Service Medal at a luncheon ceremony April 10 during AIAA's annual meeting in Washington.

The award cited Harford "for more than 30 years of staff leadership of the nation's principal professional society for the advancement of aerospace."

In 1953 Harford became executive secretary of the American Rocket Society (ARS), than a fledgling organization of about 2,100 missile engineers and scientists. When ARS and the Institute of Aeronautical Sciences merged in 1963 to form the AIAA, Harford became its deputy executive director and a year later its executive director. AIAA, under Harford's leadership, grew to 35,000 members with 66 professional sections and 132 student branches.

Harford was a Fellow of AIAA, the British Interplanetary Society, the American Association for the Advancement of Science and an Associate Fellow of the Royal Aeronautical Society. (NASA Release 85-49)

May 23: In a ceremony today in the East Room of the White House, President Reagan presented the Presidential Medal of Freedom to 12 people including former test pilot Chuck Yeager. "More than you'll ever know," Reagan said, "this world would have been much poorer and a dimmer place without each of you."

In presenting Yeager's award, Reagan said he was "a hero in war and peace," who on October 14, 1947, in a rocket plane named "Glamorous Glennis" after his wife, "became the first human being to travel faster than the speed of sound, and in doing so, showed to the world the real meaning of 'The Right Stuff'." After the ceremony, Yeager gave Glennis Yeager his medal.

In an interview with a *Washington Post* reporter upon his arrival in Washington for the ceremony, Yeagar said, "I was just a lucky kid who caught the right ride. But then I was as naive as could be, living a cloistered life out at Muroc, where the flying was fun and the living was easy.

". . . We didn't know what the word 'macho' meant. We were jes' a bunch of hell raisers . . . It wasn't a case of the right stuff. Just dumb luck.

"When they refer to a pilot 'having the right stuff,' that doesn't mean a rat's ass to me or any other pilot. It's more meaningful to be in the right place at the right time." (*W Post,* May 23/85, D1; May 24/85, B1)

August 7: The Smithsonian Institution announced selection of Robert Gilruth and astronauts Kathleen Sullivan and Bruce McCandless II to receive the newly created National Air and Space Museum Trophy for their achievements in the fields of aerospace science and technology. The Smithsonian on October 4 would present the trophy, created by Washington, D.C., sculptor John Safer, at the Air and Space Museum.

Sullivan, selected by NASA as an astronaut in 1978, was the first U.S. woman to walk in space in October 1984 on Space Shuttle flight 41-G. During that flight, Sullivan and fellow astronaut David Leestma tested an orbital refueling system for use on satellites and the proposed space station.

McCandless, a 1966 astronaut selection, flew for the first time February 1984 aboard Space Shuttle flight 41-B and became the first person to fly in space the manned maneuvering unit that he helped develop.

In 1936 Gilruth joined Langley Memorial Aeronautical Laboratory where he specialized in structures, dynamic loads, and pilotless aircraft. In 1958 NASA named him director of its space task group, the organization responsible for designing, developing, and testing the Mercury spacecraft. From 1961 to 1971, Gilruth was director of the manned spacecraft center in Houston where he was responsible for the Mercury, Gemini, and many Apollo missions. (Smithsonian Institution release, Aug 7/85)

November 11: The Circumnavigators Club in New York City would present Sally Ride, the first American woman astronaut, with its Order of Magellan, the *Washington Post* reported, which was presented to individuals who were dedicated to advancing peace and understanding in all parts of the world and who had circumnavigated the globe. Ride was only the 17th person to receive the award.

Other recipients were Gen. Douglas MacArthur, former President Hoover, Neil Armstrong, Sen. Barry Goldwater, Lowell Thomas, and Thor Heyerdahl. (*W Post,* Nov 11/85, C3)

November 15: NASA's Inventions and Contributions Board awarded the Jet Propulsion Laboratory's (JPL) Dr. Robert Nathan a plaque and a check for $20,000 for his work in planetary image processing, the JPL *Universe* reported. Nathan received the award for his "combined technical contributions to planetary and biomedical image processing and scientific data analysis techniques." Nathan invented digital image processing, a technique in which computers remove noise, correct distortions, and enhance different images such as planetary objects or biomedical images.

Nathan joined JPL in 1959 to manage scientific data analysis for the upcoming Ranger moon missions. The Ranger spacecraft used the best available cameras, but they were subject to distortion and noise contamination. Nathan devised techniques to eliminate extraneous patterns from the images. His processes and the creation of the image processing laboratory (IPL) established JPL as an important center for planetary image processing.

In 1969 Nathan launched JPL into biomedical imaging with a $2 million grant from the National Institutes of Health, which later led to formation of the JPL biomedical image analysis facility.

In 1976 Nathan developed a system at the IPL to handle large amounts of data by means of a large array filter using very-large scale integration (VLSI) system silicon chip technology. The microchip reduced computer time by a factor of 100.

He was currently working on another microchip design that would perform high-quality geometric image manipulations and filter chips that would perform pattern recognition functions. (JPL *Universe,* Nov 15/85, 1)

During November: NASA announced that its Lewis Research Center (LeRC) received four of *Research & Development* magazine's IR-100 awards to honor the 100 most significant new products developed during the past year. A panel of technical judges selected the winners from more than 1,000 entries. The award-winning products were servomechanism for propeller-pitch change, transmit module, communications traffic processor, and rotary power-transfer device, all of which LeRC and private firms under contracts to the center developed jointly.

LeRC researchers Stuart Loewenthal and Bruce Steinetz with General Electric engineers developed the servomechanism for propeller-pitch change, which could accurately control the propeller blade angle of large (10,000 kw) turboprop aircraft propellers over the complete spectrum of flight operating conditions. With certain modifications, the product should be adaptable for applications as aircraft wing pivot mechanisms, tank turret aiming actuators, and other heavy mechanisms, requiring very accurate pointing and tracking control.

The transmit module, developed by LeRC researchers Thomas Kascak, Godfrey Anzic, and Denis Connolly along with Rockwell Internatl. engineers, provided rf signal amplification and phase shifting at a frequency of 20 gigahertz. The device was of monolithic design in which all module functions were contained on a single 6.4 by 4.8 millimeter semiconductor chip with the smallest circuit feature about 40-millionth of an inch. This development made possible the production of weight- and cost-effective phased array antenna systems for future communications satellite systems.

LeRC engineer Russel Jirberg along with electronics engineers from Motorola Corp. developed a communications traffic processor for the next generation of communications satellites. The system provided the capacity and routing flexibility needed to handle the nation's growing demands for telephone and computer message traffic. When used aboard satellites having multiple narrow beam antennas, the device enabled more than 60,000 individual voice messages, video pictures, and computer data to be routed directly to users throughout the U.S. equipped with low-cost ground terminals.

LeRC engineer David Renz in cooperation with Sperry Flight Systems engineers developed a rotary power-transfer device that transferred electrical power through a rotating joint. It had a low loss space-type design capable of transferring high power (hundreds of kilowatts), AC or DC, to frequencies such as 20 kilohertz. One of its outstanding features was loss of only 22 watts per circuit while transferring 100 kilowatts. Renz and his colleagues said the roll-ring assembly device had the potential to be one of the major components of the proposed space station and other spacecraft where large amounts of power are required. (*NASA Activities*, Nov 85, 10)

December 12: President Reagan, at a White House ceremony, presented Distinguished Executive Awards worth $20,000 each to 32 career civil servants for their professional accomplishments, including in many cases, their plans and programs that saved millions of dollars, *USA Today* reported.

Among the recipients were Samuel Keller, NASA's deputy associate administrator for space science and applications; and Andrew Stofan, director of NASA's Lewis Research Center. (*USA Today*, Dec 13/85, 9; LeRC Release 85–84)

COLOMBIA

April 11: Colombia lodged a protest with the U.S. over placement of a private U.S. satellite in geostationary orbit over that country, FBIS, Bogota Emisoras CARACOL Network in Spanish reported, space that Colombia claimed as its own. Colombia's Foreign Minister Augusto Ramirez Ocampo said the satellite was placed there secretly, since the firm owning it had not presented a request to the interested nation in accordance with international legal procedures.

Colombia's United Nations Ambassador Ernesto Rodriguez Medina denounced the situation before the U.N.'s Legal Subcommittee for Outer Space, saying his country planned to place its Satcol satellite in that same position in orbit. (FBIS, Bogota Emisoras CARACOL Network in Spanish, Apr 11/85)

COMSAT

February 1: Comsat Corp. announced unaudited year-end results showing a $29 million decrease in income from continuing operations in 1984 compared to 1983 income, the *Washington Post* reported. Comsat reported consolidated 1984 operating revenue of about $442 million, with consolidated net income of about $51 million and income from continuing operations of about $45 million.

The company attributed $13 million of the decline to a write-off associated with Satellite Television Corp., the direct satellite-to-home TV company, and to the write-down of inventory from Comsat's equipment manufacturing business.

Comsat officials said they would release final figures and per share earnings later in the month (*W Post,* Feb 1/85, F5)

CONFERENCES

January 10: The Jet Propulsion Laboratory (JPL) hosted January 10-11 150 cometary scientists from around the world at a meeting of the Internatl. Halley Watch (IHW), for which JPL was the western hemisphere lead center and University of Erlangen-Nurnberg, W. Germany, the eastern hemisphere lead center, the JPL *Universe* reported. Lead center representatives at the meeting had joined IHW discipline specialists in such fields as astronomy, photometry and polarimetry, radio science, and spectroscopy; representatives of European, Japanese, and Soviet spacecraft missions to Halley; and IHW steering group members. The IHW organized a group of ground-based professional and amateur Halley watchers worldwide to coordinate their observations with those of airborne, earth-orbital, and spacecraft-flyby observations. Nine hundred professional astronomers from 50 countries and some 300 amateur astronomers had signed on as members of the IHW.

JPL expected the IHW meeting would be the last before perihelion, Halley's closest approach to the sun on February 9 at 53.1 million miles. Since its 1948 inbound leg of its 74-year orbit around the sun, Halley would make its closest approaches to earth on November 27 at 57.6 million miles and the following April 11 at 39 million miles.

Astronomers had already observed that Halley's coma (the bright halo of dust and gas surrounding the nucleus) had begun developing and questioned why the comet exhibited this feature at such a great distance from the sun. Early observations had also unexpectedly revealed the comet's brightness was fluctuating.

Amateur observers with small telescopes would be able to see Halley by fall 1985; naked-eye viewing would be possible by March or April 1986. Halley would appear again in 2061. (JPL *Universe,* Jan 25/85, 1)

April 23: "Space and Society—Progress and Promise" was the theme of the twenty-second Space Congress, a nonprofit technical symposium sponsored by the Canaveral Council of Technical Societies, April 23-26 in Cocoa Beach, Florida, *Spaceport News* reported. Speakers discussed new space initiatives, including operation and robotics in space, advanced missions and transportation, Space Station plans and development, and private sector investment and participation in space activities. Lt. Gen. James Abrahamson, director of the Strategic Defense Office, would speak about the Strategic Defense Initiative (SDI).

The traditional "Meet the Astronauts" session would be open to the public, as would be a one-man play entitled *Leviathan 99* by Ray Bradbury, a noted science fiction writer. (*Spaceport News,* Apr 12/86, 8)

DEPARTMENT OF DEFENSE

October 15: Navy Secretary John Lehman said that intercepting U.S. Navy aircraft ordered the Egyptian Boeing 737 Airliner, which Palestinians hijacked and ordered to be flown from Egypt, to land at Sigonella Air Base in Sicily "or else," *Defense Daily* reported. Lehman said one of two Grumman E-2C Hawkey airborne control aircraft, which joined six Grumman F-14 Tomcats and four tankers in the intercept mission, issued the command.

Lehman said four of the F-14s flew a close intercept formation on the 737 to the landing in Sicily; the E-2Cs from the USS Saratoga had loitered overhead until the airliner took off from Egypt with the hijackers.

There "was no deal," Lehman said, and no help from any other country. He noted the operation demonstrated the Navy's readiness, which "doesn't come cheap." (*D/D*, Oct 15/85, 225)

Antisatellite System

January 16: The U.S. Air Force couldn't decide whether the next test of the miniature homing vehicle antisatellite (ASAT) system should be against a target in space, *Aerospace Daily* reported. The Air Force was evaluating data from its November 13 ASAT test, in which the miniature homing vehicle MHV infrared sensor tracked a star, but had said that its next ASAT test would be flown against an instrumented target vehicle. Congressional limitations barred such a test until March 1 and until President Reagan had certified to Congress that the test was needed and wouldn't undercut the potential for negotiations with the USSR.

The previous day the magazine had discussed a House Republican research report recommending raising the limit of three U.S. ASAT tests against objects in space during that fiscal year. (*A/D*, Jan 16/85, 84; Jan 15/85, 73)

May 24: The U.S. Senate voted today, 74 to 9, to approve a proposal by Sen. John Warner (R-Va.) to allow three final-stage antisatellite tests next year as long as President Reagan told Congress the test would not disrupt negotiations aimed at banning antisatellite weapons, the *Washington Post* reported. The Senate previously rejected, 51 to 35, a testing ban proposed by Sen. John Kerry (D-Mass.).

The U.S. antisatellite (ASAT) weapon was a small warhead atop a rocket carried by an F-15 jet to the edge of space, where it was released to pursue its target. Although the U.S. had tested the weapon in stages, the U.S. had not fired the weapon against a target in space. The Pentagon had scheduled that final round of testing to begin within several months.

During the Geneva negotiations on weapons reductions, the USSR sought a halt to ASAT tests, while the Reagan Administration said it was open to such a proposal but had refused to halt ASAT testing as a condition to talks.

In 1984 Congress banned tests of the U.S. system in hopes the pause might help negotiations, which had not then started.

The *Washington Post* earlier had reported that electronic problems would raise the cost of the antisatellite system and stretched out its testing program. The Air Force planned a test, originally scheduled for fall 1984, for late July in which the target would be two metallic balloons attached to an orbiting satellite.

However, the Air Force was cautious about the first test against a target in space, fearing that a well-publicized failure would hurt the space defense program. Sources told the *Post* that it might be impossible for the Air Force to carry out a second test of the weapon against the balloon target in space during the current fiscal year. Originally the Air Force had planned for three tests in 1985. (*W Post*, May 25/85, A4; May 16/85, A27)

July 3: The first test of a U.S. antisatellite weapon against a physical target in space was delayed indefinitely because technical difficulties forced postponement of the target launching, the *Washington Post* reported. The U.S. Air Force cancelled the launching from Wallops Island, Virginia, and sent the targets back to the manufacturer, Avco Systems, for repair. Until the cause of the problem was found, Air Force officials said they could not estimate when the test might be rescheduled.

The *NY Times* reported John Pike, associate director of the Federation of American Scientists, as saying he was told by "usually reliable sources" that the technical problems centered on the radio that would signal the results of the target practice back to earth.

The Air Force had conducted two of the test flights planned for the antisatellite weapon, but aimed at a point in space rather than a physical target. The Air Force had already for apparently technical reasons rescheduled several times in 1985 the latest test, in which the weapon would be fired at one of two six-foot-diameter target balloons in orbit.

The antisatellite weapon was a two-stage rocket with a 13-inch, heat-sensing homing vehicle in its nose. Launched from beneath an F15 fighter, the rocket would intercept and destroy its target on impact.

The U.S. House the previous week approved an amendment to the military programs bill banning tests of antisatellite weapons against targets in space as long as the USSR refrained from similar tests. The Senate voted to permit the testing as long as President Reagan certified that he was trying to negotiate a treaty prohibiting antisatellite weapons. A House-Senate conference committee was scheduled to begin reconciling differences on the military bill the following week. (*W Post*, July 4/85, A4; *NYT*, July 4/85, A9)

August 20: President Reagan informed Congress that the U.S. planned shortly to conduct the first test of an antisatellite weapon (ASAT) against a target in space, which a spokesman said would provide an "incentive" to the USSR to negotiate limits on such weapons, the *Washington Post* reported. Reagan promised the U.S. was bargaining in good faith with the Soviets to limit ASATs, although the White House repeated the administration's opposition to a Soviet proposal for a moratorium on ASAT development and testing.

Officials said they expected the test of an ASAT fired from an F-15 at an obsolete U.S. satellite to take place in September or early October. Technical problems with both the antisatellite weapon, called the miniature homing vehicle, and the intended target, an instrumented balloon to be inflated in space, delayed the first test. Since the original target was still experiencing problems, the test would use an obsolete satellite.

That same day Presidential spokesman Larry Speakes read to reporters a lengthy statement criticizing Soviet advances in antisatellite weapons and technology and said this provided justification for the first U.S. test against a target in space. Speakes added that a moratorium on antisatellite weapons tests would "perpetuate" a Soviet "monopoly" on antisatellite systems. (*W Post,* Aug 21/85, A1)

September 11: Kenneth Adelman, director of the Arms Control and Disarmament Agency, told a U.S. House foreign affairs subcommittee that the U.S. should not change its military plans, such as delaying a test of its new antisatellite weapon, to assist in negotiating an arms control agreement, the *Washington Post* reported. Adelman argued that testing "would not impair" negotiations and "can constitute an incentive to the Soviet Union to reach agreements on a wide range of issues."

Although the Air Force had set September 13 as its first test of the antisatellite weapon against a target in space, congressional sources said the Reagan Administration might step up its test plan because of the negotiations or the November summit meeting. As of mid-July, the Air Force had no plans to test the weapon in September, because it had returned both the weapon and its instrumented target to their manufacturer for repairs expected to take through October. Thus when House-Senate conferees met in July to iron out differences on the 1987 defense spending bill, the Pentagon agreed to a provision that limited the Air Force to three tests against a target in space through the end of FY 86. "They said they would not have a test before the new fiscal year began in October," one conferee recalled.

Pentagon sources said Defense Secretary Caspar Weinberger and President Reagan agreed in early August to test the weapon against an old U.S. satellite "to show resolve."

Critics of the weapon, including four members of Congress, had filed suit in U.S. District Court in Washington seeking to block the test, arguing that Reagan had not met a congressional requirement that he certify he was negotiating "in good faith" to get an agreement with the Soviets limiting such weapons. Oral arguments in the case were set for September 12.

Rep. George Brown Jr. (D-Calif.), one of the plaintiffs in the lawsuit, told the subcommittee the presidential certification was "less than candid" and "circumvented the intent and will of the Congress."

Brown also criticized the weapon, which was two years behind schedule, saying that the Pentagon talked about "how vital to the national security the system is, right up to the day they cancel it." (*W Post*, Sept 12/85, A26)

September 13: An air-launched antisatellite missile (ASAT) hit its target in the Air Force's first ASAT test against a target in space, the Air Force Systems Command *Newsreview* reported. An F–15 from Edwards Air Force Base launched the two-stage ASAT missile 35,000 to 40,000 feet above the Western Test Range at Vandenberg Air Force Base, then two boosters carried the ASAT into space. When the ASAT's infrared sensors locked on to the target, which was about 350 miles over the Pacific, small rocket motors guided the weapon to its kill.

The Air Force in 1979 launched the ASAT's target, an Air Force satellite known as P78–1, which gathered scientific data on the space environment but which had outlived its usefulness.

In 1984 F–15s launched two live-fire tests but not at targets. The first was an ASAT launch to a point in space; the second tested the ASAT sensor's ability to home in on a star's infrared emissions.

Air Force plans called for nine more target tests against specially instrumented balloons and satellites that had completed their missions.

The *Washington Post* later reported that, according to Robert MacQueen, director of the high-altitude observatory at the National Center for Atmospheric Research in Colorado, the Solwind satellite destroyed in the test was providing "very useful data" on solar activity until the moment it was hit. MacQueen said he was surprised and upset at seeing a fruitful experiment being used as a military target and said it was "deplorable" that the Pentagon "had taken a scientifically useful thing and sacrificed it in this way."

The satellite carried seven experiments for the Naval Research Laboratory (NRL) and other government agencies. One NRL experiment used a coronagraph that sent to earth during each of the satellite's orbits, or roughly 15 times a day, images of activity on the sun's surface.

Several months previously, NRL scientists had to draft what one source said "they thought was a routine paper to justify continued operation of their coronagraph." Although the scientists acknowledged problems with the spacecraft system, the source said, they wrote that it should continue.

NRL scientists were told in July that "the satellite would be turned off sometime after August 1, but they weren't told how," the source said. The *Washington Post* reported on September 6 that Solwind was the likely target for the ASAT test because the original target, an instrumented balloon, was plagued with technical problems.

A Pentagon spokesman said the satellite was intended to operate for three years at most and that recent data from the satellite had "marginal value."

The *Washington Post* reported that another Defense Department official said the Solwind research satellite was originally scheduled as a target in the weapon's seventh test, which probably would have come in 1987. The satellite's solar research program was expected to be ending by then, scientists associated with the program said.

Although Secretary of Defense Caspar Weinberger described the satellite as "burned out," an Air Force spokesman said the fact that the satellite was still sending signals back to earth played a key role in its selection as a target. "We had to have an active telemetry system to verify it had been hit," the spokesman said. "A dead satellite would not have given us that." (AFSC *Newsreview,* Sept 27/85, 1; *W Post,* Sept 20/85, A1, Sept 21/85, A9)

December 27: Reps. Les AuCoin (D-Ore.) and Norman Dicks (D-Wash.) sent a letter to Defense Secretary Caspar Weinberger warning that "Congress would not excuse any attempt by the Department of Defense to circumvent" a congressional ban on antisatellite weapon testing, the *Washington Post* reported. The warning came after the *Post* reported that a DOD official discussed plans to continue development of a $4 billion U.S. antisatellite weapon despite a week-old law that banned tests against an object in space as long as the Soviets did not test such a system.

The DOD official said a possibility under study was to fire a test weapon against "a point in space" rather than at two targets put into orbit December 12 [see U.S. Air Force/NASA and DOD, Dec. 12]. The Air Force "won't do anything in direct violation" of the congressional language attached to an omnibus spending bill signed into law by President Reagan, but "we will find a way to go ahead," the DOD official had said.

Although the congressional action stopped testing, it did not halt the overall antisatellite program, for which Congress increased by $15 million a Pentagon request for $150 million in FY 86 to develop the system. The increase, a congressional aide said, was to show the Soviet Union that the U.S. program would be available for testing if Moscow broke its moratorium.

The spending bill also had a special $5 million item for the Air Force "to carry out a research program to develop new and improved verification techniques to monitor compliance with any antisatellite weapon agreement that may be entered into by the U.S and the Soviet Union."

Rep. Les Aspin (D-Wis.), chairman of the House Armed Services Committee, said he hoped to bring arms-control supporters in the House together to discuss antisatellite weapons. The long-term goal of arms control, he said, was "to promote stability." In the antisatellite area, he added, that meant finding a way to make secure the superpowers' high-altitude satellites that monitored nuclear forces and provided early warning of an attack. (*W Post*, Dec 28/85, A5, Dec 25/85, A1)

Budget

March 29: A defense budget frozen to the inflation level could kill the Air Force/McDonnell Douglas C–17 transport program, for which Defense Secretary Caspar Weinberger had approved full-scale development in FY 86, *Defense Daily* reported. In a markup by the Senate Armed Services sea power and force projection subcommittee, funds would be included for the C–17 in the 3% and 4% real growth version of the budget.

The subcommittee also cut the budget request for the Lockheed C–5B, which would have to fill the gap if the C–17 was terminated. Weinberger had said earlier that the C–5B would be one of over 170 aircraft to be deleted if there were a budget freeze, but he had not mentioned the C–17 in his list of threatened programs.

Weinberger had previously approved a request of $453.68 million for the C–17 in FY 86 and $624.8 billion for FY 87. (*D/D*, Mar 29/85, 161)

July 1: Rep. Samuel Stratton (D-N.Y.), chairman of the House Armed Services Committee's procurement subcommittee, said a new bomber force study, done by the Defense Department in response to a U.S. House request, could lead to changes in the planned mix of 100 B–1B bombers and something over 100 Advanced Technology Bombers (ATBs-Stealth), *Defense Daily* reported. The new study was intended to reevaluate and update the conclusions of a 1981 study.

Stratton pointed out that major changes in the strategic environment demanded that the U.S. House take a new look at the air-breathing leg of the strategic triad to determine if those requirements had changed. He emphasized that the subcommittee did not have a prejudged position on the bomber mix question and that the group did not want a rehash of positions taken years before when circumstances were significantly different.

When asked by Rep. Beverly Byron (R-Md.) if the study could change the mix of the bomber force, Stratton responded, "I think that is entirely possible. If the study group finds that a different 'mix' is required, they should clearly state the differences and provide justification for those changes. This study could call for more or fewer of either bomber or for a high aggregate number of both . . . but there must be full debate on the proposed bomber force, because obviously a good deal of money would have to be spent if we are going to continue both those lines into the future." (*D/D*, July 2/85, 9)

Department of Defense

November 7: The Senate Appropriations Committee cut the Air Force FY 86 request for development of the advanced tactical fighter (ATF) by $103 million to $140 million and recommended that a unit cost cap be put on the program, *Defense Daily* reported. The House Appropriations Committee earlier approved $170 million for ATF.

Acting on the recommendation of its defense subcommittee, the Senate committee allocated $100 million for the joint advanced fighter engine development program, $13 million for avionics development, and $27 million for continuation of the competitive air vehicle demonstration and validation "for not less than three contractors." Although the committee endorsed the ATF development program "as a long-term counter to Soviet tactical air improvements," it said it "is concerned that the current program does not provide sufficient time to absorb all of the technology" to be incorporated in the aircraft and that the program "entails unacceptable risks." And the committee pointed out that "these advanced technologies would be very expensive to obtain."

The Air Force estimated R & D costs for the ATF at $11.8 billion in FY 85 dollars and, after considering a figure up to $40 million, assumed the purchase price would be $35 million in FY 85 dollars. However, the committee pointed out that the Air Force "has not identified how it will reduce ATF costs without deleting technical features previously identified as aircraft requirements." Therefore, the committee directed the Air Force to submit an annual report on the future likely cost of ATF, beginning with the FY 87 budget. It recommended that "a 20% limit be placed on the ATF program unit flyaway costs" above F–15 costs and that this limit be made the baseline of the ATF development program.

The ATF would include sustained supersonic dash, internal weapons carriage, blended aerodynamic design including low observable (stealth) technology, and higher thrust (32,000 lb.) engines. It would also have short takeoff and landing (STOL) capabilities, high transonic and supersonic maneuverability, an expanded flight envelope, and improved survivability. (*D/D,* Nov 7/85, 33)

Missiles

January 23: The U.S. Air Force's Aeronautical Systems Division successfully flight-tested a low-cost, state-of-the-art autopilot on an unmanned research vehicle, the Air Force Systems Command's *Newsreview* reported. The division adapted commercially available microprocessor chips in the $50,000, in-house autopilot program to develop flight control-systems technology that would lead to cost reductions in tactical missiles. Researchers made five flight tests of the autopilot since October 1983 on an XBQM–106 unmanned vehicle during which the autopilot provided control augmentation, translating user inputs into vehicle actions.

Current guidance and stabilization packages in missiles and other aerospace systems were costly because they used expensive data processing microchips that had excessive amounts of computational processing capability but no built-in control functions, thus requiring additional electronic interface components to perform the "control" function tasks. The microprocessor chip, in contrast, had sufficient built-in computational and control functions to permit 60% fewer interfacing components and should be appropriate to numerous tactical weapons systems, both surface and air-launched.

The next step in the program would require replacing the wire-wrapped or brassboard model of the autopilot with a refined system on a printed circuit board, which would undergo at least 2 additional flights before the end of FY 85. (AFSC *Newsreview,* Jan 23/85, 8)

March 12: A U.S. House Appropriations subcommittee approved a resolution to release $1.5 billion frozen in Congress the previous year for an additional 21 MX missiles, the *Washington Post* reported. The vote, expected in the defense subcommittee that had supported the MX in the past, was only the first in a series of votes required before funds could be released. But missile opponents acknowledged that the vote was indicative of the uphill battle they faced in trying to defeat the MX in the midst of renewed arms control talks with the Soviets.

The Reagan Administration had lobbied hard for the previous two weeks to win release of the funds, contending that continued funding of the MX was crucial to the arms control talks that had opened in Geneva.

MX missile supporters on the subcommittee argued that the administration should not be denied a weapon that could be a "bargaining chip" in the arms talks. On the other hand, opponents characterized the missile as "the glass jaw in our strategic forces" and "totally irrelevant" to the arms talks.

The resolution would next come before the full committee, which had backed it in the past. Observers agreed the real fight over the missile would occur on the Senate and House floors. Under a complicated arrangement worked out the previous year, representatives of the Republican-controlled Senate and the Democratic-controlled House had agreed that the MX legislation would be sent to the floor of both chambers regardless of what action was taken at the committee level. (*W Post,* Mar 13/85, A4)

May 13: General Electric pleaded guilty to 108 counts of making false claims to the government for work on the Minuteman Mark–12 reentry vehicle systems for January 1, 1980 to April 1983, during which time they had claimed more than $800,00 on nonreimbursable overrun labor costs by altering employees' time cards, *Defense Daily* reported.

Department of Defense

GE faced a maximum penalty of $1.04 million. Because there was not a plea agreement to halt indictment of officials within GE, the investigation would continue in order to find which GE managers were criminally responsible.

A company spokesman said, "We changed our plea because of new information given to us and to the U.S. Attorney's office and the attorney for a former employee. Four years of investigation by GE and various government agencies revealed errors in time card charging in 1980. From the beginning, we offered to reimburse the government for any incorrect charges. However, until now, GE was unable to conclude that any individual has engaged in criminal activity." (*D/D*, May 14/85, 73)

June 28: In May an armament division F–16 from the U.S. Air Force's 3246th Test Wing fired an advanced, medium range air-to-air missile (AMRAAM) over White Sands Missile Range in the first guided launch of its full-scale development program, the Air Force System Command's (AFSC) *Newsreview* reported. Armament division officials said the successful flight verified AMRAAM's interface with the F–16's avionics system and the performance of its active radar guidance capabilities.

AMRAAM is an all-weather, all-aspect radar missile designed to replace the AIM–7 radar missile. AMRAAM is smaller, lighter, and faster than AIM–7 and had an active radar seeker that would permit air crews to "launch and leave" or to execute simultaneous multiple-target attacks during a single intercept.

Pilots could launch AMRAAM from beyond visual range, because it received target information from the launch aircraft's avionics systems. After launch, the aircraft could pass target information to the missile, or the missile could use its own inertial reference unit and microcomputer to guide inertially toward the target. During the terminal phase, the missile's active seeker would take over and guide the missile to the target.

The Air Force and Navy had a contract with Hughes Aircraft Co. for full-scale development of AMRAAM, with Raytheon Co. designated as "follower" contractor to introduce competition during the production phase.

The previous December an AMRAAM separation/control test vehicle flew a preprogrammed course to evaluate the missile's autopilot, control system, aerodynamic characteristics, and safe separation from the aircraft.

The Air Force planned over the next three years more launches from several aircraft including the F–14, F–15, F–16, and F–18. Plans also called for combat aircraft of the United Kingdom and Federal Republic of Germany to carry AMRAAM. (AFSC *Newsreview,* June 28/85, 4)

August 23: Under a 1985 memorandum of agreement, the Eglin Air Force Base's armament division was leading a joint U.S., British, and West German project to determine how a non-nuclear, long-range standoff missile should be developed, the Air Force Systems Command *Newsreview* reported. General Dynamics and Boeing Aerospace Co., which had 15-month contracts to study the possibilities, was working with subcontractors from the other two countries.

The countries could use the conventional, low-altitude subsonic missile, launched from air or ground, primarily against airfields. The immediate goal of the project was to design the system, develop a procurement program and schedule, and identify technological transfers among the three countries.

The point of the agreement was to save money and eliminate duplication of effort in development of a needed weapon system. One country would build the system, but the others would support the effort. While the U.S. completed the procurement package, England provided most of the threat information needed to design the missile. West Germany was taking an active part in all project phases. (AFSC *Newsreview,* Aug 23/85, 3)

NASA and Department of Defense

September 11: The U.S. Air Force announced today that it had selected Air Force Undersecretary Edward "Pete" Aldridge to fly on the first Space Shuttle mission launched from Vandenberg Air Force Base, the *Arizona Daily Star* reported. The Air Force said Aldridge and Air Force Maj. John Brett Watterson would serve as payload specialists with five astronauts named by NASA for the Defense Department's mission set for launch March 20, 1986.

Aldridge, 47, who had been Air Force undersecretary since 1981, said in a statement, "I'm thrilled at the opportunity and thrilled at the prospects that I will be able to apply what I have learned to expanding U.S. efforts in space."

NASA and the U.S. Air Force were building the Vandenberg Space Shuttle launch facility primarily for military missions. Because the Space Shuttle could be launched from there directly south, it could go into a north-south orbit that covered the entire globe, passing over both poles and allowing Space Shuttle crews to observe Soviet military forces anywhere in the world.

Space Shuttles launched from Kennedy Space Center could not go directly north or south because they would fly over inhabited areas during the initial minutes of flight, possibly endangering the population if something went wrong. Those Space Shuttles launched into east-west orbits did not fly any closer than about 2000 miles to the poles. (*Ariz. Daily Star,* Sept 12/85, A3)

Procurement

August 2: McDonnell Douglas Chairman and Chief Executive Officer Sanford McDonnell in a letter to Defense Secretary Caspar Weinberger explained a company policy that instituted "expanded, no-questions-asked"

Department of Defense

refund policies on the sale of spare parts and support equipment to the U.S., foreign countries, and commercial customers worldwide, *Defense Daily* reported.

McDonnell said the government could return "any covered spare part or piece of support equipment if there is any dissatisfaction with its cost—no questions asked."

McDonnell described the refund policy as the most comprehensive in the aerospace industry and said that if the military felt it must retain an item in stock for operational readiness, "their complaint will be addressed without concern for a time limit."

Under the new policy, the military could, if dissatisfied with the price, return within six months of delivery new and unused parts or equipment built by McDonnell Douglas and purchased under prime contracts. The company said the policy applied to prices to the government up to $100,000; at that level the company furnished cost data in advance of establishing price.

Earlier Weinberger directed establishment of a standard industry-wide refund policy based on voluntary refund policies of Boeing Co. and the General Electric Co. The Boeing/General Electric refund programs accepted for credit the return of any spare parts or support equipment that the Pentagon considered to be unreasonably priced. (*D/D,* Aug 2/85, 1)

Satellites

February 18: Sources inside and outside the Reagan Administration said the Department of Defense (DOD) was developing for constant surveillance of all objects in deep space a new generation of navigation, communications, and spy satellites, which would orbit 22,000 miles high and be aided by a nearly completed network of ground stations, the *Washington Post* reported. DOD would harden the new satellites against radiation and laser attacks and give them some small jet engines for maneuvering away from attack. DOD was also determining if the satellites could be armed to defend themselves.

DOD was completing Spacetrack, a worldwide U.S. network of five space-watching facilities, that would provide adequate time to take defensive actions if the USSR launched weapons at U.S. deep-space satellites. DOD's FY 86 budget reportedly would contain $20 million to complete a fully operational Spacetrack sensor system by 1988.

The Soviets had had for 10 years a rudimentary weapon that could knock down low-level U.S. satellites. The Pentagon was beginning tests on its own antisatellite weapon, which would knock down low-level Soviet satellites as they passed over the U.S. (*W Post,* Feb 18/85, A1)

April 26: The Department of Defense (DOD) placed a satellite on the Space Shuttle Discovery [see STS/Military Applications, Jan. 24], which put it in a radically different orbit from many spy satellites, the *NY Times* reported. According to figures made public by the Air Force, DOD put the satellite in a highly elliptical orbit at a low angle above the equator; the majority of U.S. spy satellites were in roughly circular orbits that passed across polar regions so that satellites would spend as much time as possible over the USSR. The satellite's, whose United Nations designation was 1985–10B, highest altitude was 34,670 km (21,543 miles); its low point was 341 km (212 miles). This elliptical path was inclined 28.4° to the equator, and the satellite took 10 hours to complete one revolution. In addition to its elliptical orbits, the satellite could maneuver from its initial orbit into another, perhaps more usual, orbit.

By international treaty, DOD was required to release to the United Nations a technical description of all its satellites' initial orbits. This recording process, which passed through the State Department and included filing the information with the Library of Congress, often took several months after an object was launched. The DOD released the data in early April to the State Department; the *NY Times* then requested and received the information. (*NYT,* Apr 26/85, A19)

June 11: The Rome Air Development Center planned to award three contracts for work in support of the U.S. Defense Advanced Research Projects Agency's (DARPA) proposal to develop a multiple satellite system (MSS) consisting of several hundred low-cost, low-orbiting communications satellites in several orbits, *Defense Daily* reported.

The satellites, expected to cost about $1 million each, would comprise a highly survivable, high-bandwidth global communications system through the combination of high bandwidth packet switches and burst radios with low-cost proliferated, low-orbiting satellites. Each satellite would carry a low-cost burst radio capable of instantaneous data rates on the order of 100 Mbps, along with digital controllers to do the packet switching and a suitable antenna system capable of both omni-directional and directive beams.

The contracts would be for study of burst radio development, low-cost antenna design, and low-cost satellite integration. (*D/D,* June 11/85, 229)

June 28: A B–52 equipped with the Navstar global positioning system (GPS) flew a mission over the North Pole to prove the space-based radio navigation system's ability to navigate a polar mission, the Air Force System Command's (AFSC) *Newsreview* reported. The mission marked first time a preproduction Navstar GPS was used on a polar mission. The flight also tested the system's high-latitude navigation and compatibility with the aircraft's inertial navigation system (INS).

Navstar GPS, when completed in the late 1980s, would use 18 GPS satellites in six orbital planes to give worldwide, three-dimensional position and velocity information to U.S. and allied land, sea, and air forces. To determine the position of an aircraft such as the B-52, an onboard receiver would pick up signals from at least four satellites and measure the time it took each signal to travel from the satellite.

The Navstar GPS on the B-52 aligned the inertial navigation system, giving the crew constant navigation during the 13-hour mission despite the GPS being available only three hours. The GPS and INS kept track of the aircraft's position at all times, even when crossing lines of longitude faster than the onboard screen could update them.

Although testing of the Navstar GPS using the B-52 was almost complete, production integration was just beginning. The production phase of Navstar GPS required bringing the three segments—space, user, and control—to full operational capability. The Air Force expected two-dimensional capability in 1987, three-dimensional capability in 1988. (AFSC *Newsreview,* June 28/85, 7)

August 28: A Titan 34-D booster carrying a classified payload, possibly a reconnaissance satellite, exploded August 28 minutes after launch from Vandenberg Air Force Base, the *NY Times* reported. Because of the classified payload, the Air Force revealed little information about the launch. Booster rockets typically carried an internal destruct package that ground controllers could activate in the event of a malfunction.

The explosion and subsequent crash of the booster started a 20-acre brush fire near the launch pad. An Air Force spokesman said it took firefighters more than five hours to extinguish the blaze.

The Air Force was conducting an investigation to determine the cause of the booster failure. (*NYT,* Aug 31/85, A14)

September 13: New car buyers in 1988 might find a Navstar Global Positioning System (GPS) receiver among their list of options, the Air Force Systems Command *Newsreview* reported.

"It would essentially be a small TV screen that shows your position on a map," said 1st Lt. John Schoenewolf, GPS cargo manager for the Eastern Space and Missile Center. A small cursor, or mobile electronic dot, would pinpoint a car's location to within 300 miles, he added, and map cassettes could be purchased separately.

Before the satellite electronic car map became a reality, along with other GPS applications both military and civilian, the Air Force between 1986 and 1988 would have to launch 28 satellites, the most ambitious launch rate attempted in space-launch history.

"We plan to launch the satellites for the GPS system at a rate of one every seven weeks aboard the shuttle." Schoenewolf said. "We will need a total of

18 to start operating the system, and the target year is 1988 to begin operating."

The East Coast Navstar GPS launch facility opened in August at Cape Canaveral Air Force Station to help make possible the accelerated launch rate. At the facility's ribbon cutting ceremony, Maj. Gen. Donald Henderson said the facility and GPS itself "will revolutionize the way we perform precision navigation."

He noted the satellite system would be the largest constellation ever established and would have an influence on everything that floats, drives, flies, or submerges. (AFSC *Newsreview*, Sept 13/85, 5)

October 11: The Department of Defense (DOD) decided to drop the fee it planned to charge civilian users of the Navstar Global Positioning System, the Air Force System Command *Newsreview* reported. William Taft IV, deputy secretary of defense, said deleting the user fee was necessary to enhance worldwide aviation safety and to avoid charge difficulties. He said the standard positioning service signal, which would be the lower of two accuracy levels, would be broadcast in the clear and available to any properly equipped user. However, Congress could reinstate user fees.

The precise positioning service signal, the higher-accuracy signal, would be encrypted and made available initially to the U.S. and some allied military users. DOD would permit limited civil use if it was shown to be in the national interest, adequate security protection was provided, and comparable accuracy could not be obtained from another source, DOD officials said.

DOD scheduled the Navstar system, a continuous worldwide satellite-based radio navigation system, for operation in the late 1980s. (AFSC *Newsreview*, Oct 11/85, 6)

October 11: American University professor Jeffrey Richelson, testifying at the espionage trial of former Navy intelligence analyst Samuel Morison, said publication in 1984 of three secret KH–11 spy satellite photos in *Jane's Defense Weekly* told the Soviets nothing important that they did not already know, the *Washington Post* reported. Morison, who worked at the Naval Intelligence Support center in Suitland, Maryland, had official approval for his part-time job as U.S. editor of *Jane's Fighting Ships* and in 1984 sent the weekly the three photos in hopes of securing a full-time job there. He was also indicted on charges of keeping in his apartment two classified documents about a May 1984 fire at a Soviet naval ammunition depot.

In his testimony, Richelson said the Soviets already had the KH–11 manual, which they had bought from a CIA officer, as well as earlier published satellite photos to show them how the system worked.

Government witnesses earlier testified that the photos sent to Jane's were potentially valuable to the Soviets in confirming the KH–11's sophisticated workings and in disclosing U.S. targeting interests. Similarly, a Navy intelligence expert testified that the details about the ammunition depot fires, also gleaned from satellite photos, were so precise that it would have been "very damaging" to the U.S. if the documents had been leaked.

Richelson, however, said public sources had provided much detail about the KH–11 and other satellite programs, such as their flight paths over the Soviet Union, their altitude (75 to 155 miles), and the fact that a Titan 3D rocket launched them and another so-called Keyhole satellite, the KH–9. He said it was well known that the KH–11 sent its pictures back to Washington in a matter of seconds via another satellite and that it passed over targets quite frequently.

Another defense witness, John Pike, associate director for space policy at the Federation of American Scientists, testified on what was publicly available about the KH–11, saying DOD would soon replace it with a longer-lasting KH–12. According to Pike, the KH–11 orbited the Soviet Union 11 times a day, had the capacity to take pictures continuously, and had a peripheral vision that could switch from extreme left to extreme right in an instant.

The *Washington Post* six days later reported that Morison was found guilty of espionage and theft and could be sentenced to up to 10 years in prison and fined $10,000 on each of four counts. (*W Post*, Oct 12/85, A9, Oct 18/85, A1)

Strategic Defense Initiative

January 17: The *Wall Street Journal* printed Robert Jastrow's comments on the Union of Concerned Scientists's (UCS) Space Defense Initiative (SDI, or Star Wars) report, in which Jastrow noted that although the UCS had originally stated the program would require 2,400 satellites, they had corrected this to 300, making the program not a seemingly impractical proposal but one readily affordable within fiscal limits on strategic forces.

The UCS had also stated in its report that the neutral particle beam (to destroy Soviet missiles) would require orbiting a 40,000-ton accelerator, an impossible task. However, Jastrow noted the correct weight of the accelerator was 25 tons, quite reasonable for earth orbit, and the UCS had admitted the error.

Also on the SDI program, *Space World* published the Council on Economic Priorities's (CEP) recommendations, in which CEP suggested the Department of Defense reduce emphasis on developing prototype systems until clarification of the technological uncertainties of deploying an effective overall space defense and Congress should significantly slow SDI's growth rate until demonstration of the plan's technical feasibility. (*WSJ*, Jan 17/85, 2–E; *Space World*, Jan 85, 3)

February 20: Michael Burch, assistant secretary of defense for public affairs, said beginning in 1987, two years earlier than planned, two Space Shuttle flights a year would carry experiments for President Reagan's Strategic Defense Initiative (SDI) research program, with initial experiments testing the ability to detect, track, and aim against targets in space, the *Washington Post* reported.

The tracking and targeting tests, developed in a Pentagon research program that antedated the Reagan Administration, would include a mounting device for attaching the telescope-like sight to the Space Shuttle and sensors that could pick up at a distance equal to the width of the U.S. the "signature of objects" such as the booster plume of a missile.

The Department of Defense had canceled a Space Shuttle test of the system's main telescope, though sources said it would continue land-based tests to develop a more capable telescope. Defense Secretary Caspar Weinberger had said the previous year's congressional SDI program budget cuts would delay by one year an integrated-system demonstration. (*W Post*, Feb 20/85, A5)

February 22: In testimony February 21 before the Senate Armed Services Committee, Lt. Gen. James Abramhamson said the High Frontier organization's proposal for a system of non-nuclear space interceptors based on kinetic-kill vehicles was an attractive weapons concept that could be deployed earlier than others, but questioned whether it was the best system, the *Washington Post* reported.

Retired Gen. Daniel Graham, High Frontier's director and former director of the Defense Intelligence Agency, had proposed the satellite-launched system of kinetic-kill vehicles three years previously as a possibly near-term, space-based missile defense. Abramhamson said the High Frontier system was simpler than other systems being considered and might be useful against the current generation of missiles, but said he feared such a system would only drive the Soviets to deployment of more missiles and development of countermeasures. He pointed out a truly effective space-based, missile-defense system had to be tied in with warhead tracking and command and control systems, which he said would be ready in a decade or less.

Abramhamson went on to say it would be the early 1990s before the Strategic Defense Initiative (SDI)-research effort would be far enough along for a decision on which technologies to develop for a deployed system; an effective system would require a whole family of technologies. The *Washington Post* article quoted Abramhamson concluding that he "wouldn't give a figged nickel" for calculations by groups such as the Union of Concerned Scientists that space defense is impossible. (*W Post*, Feb 22/85, 8A)

March 18: Lt. Gen. James Abramhamson, in testimony before the Senate armed service strategic and theater nuclear forces subcommittee, said that bringing down the cost of the Strategic Defense Initiative (SDI) program was a critical factor in proving that SDI was affordable, *Defense Daily* reported. The aim was to have a program that was "practicable and could be implemented," he said.

Abramhamson told the subcommittee that he believed the SDI program would develop a sufficient base of understanding by the early 1990s so that "we could be reasonably confident that decisions could be made [for moving] into the initial portions of a layered defense."

He pointed out that, although a limited defense system using conventional weapons might be possible by 1995, the emphasis on weapons was not the problem, but rather it was putting together the command and control by that time. However, he added that if a national decision were made to have a limited defense system by 1995, this would be possible. (*D/D,* Mar 18/85, 89)

March 29: Robert Cooper, director of the Defense Advanced Research Projects Agency (DARPA), in testimony before the U.S. House Armed Services Committee's research and development subcommittee, said the U.S. had developed a secret graphite coating that could shield spacecraft, satellites, and missiles from Soviet laser beams, the *Washington Times* reported. The new material could be highly effective at a modest weight penalty, perhaps 10% of a spacecraft or missile's payload, and could withstand as much as 100 times more laser energy deposited on it than the typical aerospace material. The CIA estimated that the Soviets could have a space-based laser capable of shooting at U.S. satellites by 1988 and one capable of shooting at targets on the ground possibly by 1990.

The development added a new dimension to the controversy over President Reagan's Strategic Defense Initiative. Critics had argued that it would be difficult if not impossible for a space-based laser to destroy an ascending missile in its early and vulnerable boost phase before its multiple warheads had been launched. Development of the new class of lightweight shielding materials appeared to bolster this argument, increasing the possibility of missiles as well as spacecraft and satellites being made impervious to laser destruction.

When questioned about the Soviets using such a material to thwart a U.S. space-based defense system, Cooper replied, "I don't know what implementation it has for a Soviet space or missile system. I'm not sure they have the technology, or even could achieve it in the near term." (*W Times,* Mar 29/85, 1A)

April 23: Anthony Battista, the U.S. House Armed Services R&D subcommittee senior staff adviser, said on April 3 that a "reasonable" funding effort for the Reagan Administration's Strategic Defense Initiative (SDI) program would be between $2.2 and $3 billion for FY 86, *Defense Daily* reported. Battista noted that the reductions in SDI should be taken from surveillance, acquisition, and targeting, which represented about $1.5 billion of the $3.7 billion requested by the administration and that "the one troublesome part about SDI, from a command and control point of view right now, is that there is not enough emphasis on the hardening of detectors to live in the thermo-nuclear environment. I'm worried about the ability of missiles to live in that environment."

However, in testimony before the Senate defense appropriations subcommittee, Under Secretary of Defense Fred Ikle said on April 23 that a reduction in the FY 86 SDI budget would be extremely harmful to the program and a poor policy decision. SDI "is not some optional experiment, to be continued or curtailed depending on short-term budgetary or arms control considerations," he said. "It is a vital long-term effort to strengthen our ability to prevent nuclear war.

". . . If we slow down the SDI in this early phase, we would not have the answers in the next decade that we need to chart our long-term strategy and arms control policy." The Administration planned to spend $4.9 billion in FY 87 and $16 billion in FY 88–89 on the program. (*D/D,* Apr 3/85, 185, 24/85, 305)

May 4: France turned down President Reagan's invitation to participate in the U.S. Strategic Defense Initiative (SDI) project because of the subordinate role France would have to play and because it was at that time promoting its own French-led European advanced technology/space research project that included some key technologies related to strategic defense, *Defense Daily* reported.

French President Francois Mitterrand told a news conference May 4 at the close of the Bonn economic summit that President Reagan used the term "subcontractors" in reference to Europe's role in the SDI project, which confirmed his view that the U.S. would not treat France and other European countries as equal partners in SDI and that they would not get access to all research results.

"The technology interests me," Mitterrand said, "but the strategic project is interesting only for the future when man becomes a master of space." He urged other European nations to join France in an independent advanced technology R&D program called "Eureka," which would explore civilian uses of space and other advanced technologies including high-power lasers, optics, microelectronics, and high-speed computers. Although these technologies would apply to a number of areas including strategic defense, the primary aim of Eureka would be to "explore space through advanced research in order to master new technologies," Mitterrand said. He noted that Eureka was important for Europe because of the need to "preserve their fund of intelligence, technology, and brains. All this has to be mobilized in a great project that is European."

Department of Defense

According to reports, Mitterrand told Reagan that he was concerned about actual deployment of SDI because it could alter the strategy of mutual assured destruction that had successfully prevented nuclear war. (France had its own small nuclear deterrent, but it was more vulnerable to ballistic missile defense than the large strategic forces of the U.S. and USSR.)

France had agreed to support the European Space Agency's proposal to define a pressurized module called Columbus for the U.S. space station, if all parties could work out adequate arrangements. (*D/D*, May 7/85, 35)

May 10: The Australian minister of defense, Mr. Beazley, confirmed that Australia would help test a new U.S. satellite system called "Teal Ruby" that was associated with the space defense system known as the Strategic Defense Initiative (SDI), FBIS, Melbourne Overseas Service in English reported. Mr. Beazley stressed, however, that Australia's involvement was not directly related to SDI and would end the following year.

He said the U.S. would use Australian defense vessels and aircraft to test the satellite system using infrared rays and that the project had some benefits for Australia over the long term because of that country's surveillance problems. He pointed out that infrared sensing was a good way to maintain surveillance of countries with electronic warfare systems.

However, the Australian Democrats's spokesman, Senator Mason, condemned the decision, saying the research was an essential part of the SDI system and that the Australian government was guilty of hyprocrisy in agreeing to participate in the testing. (FBIS, Melbourne Overseas Service in English, May 10/85)

May 23: The Pentagon's Strategic Defense Initiative (SDI) office said today that the Space Shuttle Discovery would participate in June in the first SDI experiment in orbit, carrying a mirror intended to intercept and reflect a laser beam fired from earth, the *Washington Post* reported. In some missile defense systems, a large orbiting mirror would receive a powerful beam from a ground-based laser and reflect it to destroy an enemy missile, but in the space shuttle experiment, the laser beam would be too weak to harm the spacecraft. A small mirror would reflect the beam back to the ground so engineers could verify their ability to keep the laser pointed at the orbiting mirror.

The laser would travel from an Air Force Base on Maui, Hawaii, and bounce back by a special 8-in.-diameter "retroreflector" that astronauts would place in one of the Space Shuttle's middeck side windows while the spacecraft was over the Pacific.

An SDI spokesman said the laser had successfully tracked during earlier tests an airplane carrying the mirror at an altitude of 30,000 feet. The Space Shuttle flew in orbit at 100 miles and more.

The test, called the high precision tracking experiment, was the first in a series that the SDI office had booked aboard the Space Shuttle. Beginning in

1987, the Pentagon would fly two major SDI experiments each year. (*W Post*, May 24/85, A8)

During May: Brigadier General Robert Rankine, Jr., U.S. Air Force, writing in *Aerospace* about the Strategic Defense Initiative (SDI), said that, "In the long term, we have confidence that SDI will be a crucial means by which both the U.S. and the Soviet Union can safely agree to very deep reductions and eventually even the elimination of ballistic missiles and the nuclear weapons they carry. This does not represent a shift from the basic deterrent strategy of the U.S., but represents a new means for enhancing deterrence.
". . . A ballistic missile defense capability has the potential of increasing deterrence and adding to stability, by increasing substantially the uncertainties in the success of nuclear attack by an enemy, thoroughly confounding his targeting strategy, thus significantly reducing or eliminating the utility of preemptive attack. The system need not be perfect to accomplish this objective, but most meet three important criteria:
—effective against the systems and countermeasures that exist or could be deployed,
—sufficiently survivable that it would not encourage an attack on the system itself . . .
—effective at lower cost than any proliferation or countermeasure attempts to overcome it.
". . . Some of the opponents of the Strategic Defense Initiative," Rankine said, "have argued that the research and technology program currently under way is inconsistent with the ABM treaty and conflicts with arms control in general. Quite to the contrary, the initiative it totally consistent with current U.S. ABM treaty obligations. The initiative contemplates only research and experimentation on a broad range of defense technologies to provide the basis for a decision in the future whether or not to develop systems which would provide an effective ballistic missile defense capability."
Rankine then described the technical scope of SDI and said the program was broken into five major program elements: surveillance, acquisition, tracking and kill, assessment; directed energy weapons; kinetic energy weapons; systems analysis and battle management; and an assortment of other high-priority technologies that did not warrant separate program elements.
Rankine concluded that the goal for SDI had not changed since the President's March 1983 speech proposing the system, when he "challenged all of us in the scientific community to create a means for rendering ballistic missiles impotent and obsolete." (*Aerospace*, Spring 85, 2)

July 14: The first in a series of laser experiments in the Strategic Defense Initiative (SDI) program was only partly successful because laser operators were unable to lock on to a fast, high-flying rocket target, the *NY Times* reported. A laser station on the island of Maui succeeded in focusing a laser beam on the rocket but failed to bounce the beam off an eight-inch reflective shield attached to it, said unidentified sources.

The Pentagon had announced it was preparing a new round of laser tests similar to one conducted in June when a low-power blue-green laser was reflected off the Space Shuttle Discovery. On July 16 the Pentagon issued a statement saying the first of five planned experiments was conducted in the early morning of July 14 and that "it was only partially successful due to technical difficulties not associated with the experiment."

Officials in the SDI office refused to elaborate beyond that saying the Navy's Pacific Missile Range Facility launched a Terrier-Malemute rocket that traveled almost 450 miles into space and that the laser had been fired. (*NYT*, July 16/85, C7)

August 4: In a study important to the U.S. Strategic Defense Initiative (SDI) missile defense program, Space Shuttle crew aboard Challenger on mission 51–F briefly fired the orbiter's rocket engines while scientists in Australia attempted to analyze the effect on the ionosphere, the *Washington Post* reported.

Challenger's engines burned 600 lb. of fuel as the spacecraft passed over an observatory in Tasmania where telescopes measured the effect of the rocket exhaust on the charged gas particles, called plasma, that made up the ionosphere.

A second shorter rocket firing took place over observatories in Massachusetts. (*W Post*, Aug 5/85, A4)

September 6: The Department of Defense announced that a large high-powered chemical laser destroyed at a distance of six-tenths of a mile a large liquid-fuel booster stage from a Titan 2 rocket in an experiment for the Strategic Defense Initiative (SDI) program, the *Arizona Republic* reported. The experiment, conducted at White Sands Missile Range, New Mexico, was significant "because this was the first full-scale test against an object this large," said Mary Pshak, a spokeswoman for the SDI organization. "We had done a lot of subscale testing, but this particular experiment verified our earlier tests," she noted.

Pshak declined to discuss the amount of power used because the information was classified, but said the laser was a large working model of MIRACL (mid-infrared chemical laser), considered the largest weapon-grade device of its type developed in the U.S. However, the SDI research program also was focusing on other types of lasers, some of which were considered even more promising. (*Ariz. Republic,* Sept 14/85, A28)

October 4: Department of Defense scientists the previous week sent concentrated laser beams adjusted for atmospheric distortion from an Air Force facility on Maui, Hawaii, to a retroreflector on a Terrier-Malemute sounding rocket launched from the Navy Pacific Missile Range Facility in Hawaii, the *Los Angeles Times* reported. The test demonstrated for the first time that the U.S. could fire such a beam through the atmosphere and prevent air molecules from distorting and weakening it. The test appeared to be a significant technical achievement because distortion, or "blooming," of a laser as it passed through air could totally block the beam.

To correct beam distortion, engineers developed a flexible mirror that, taking orders from computers, distorted the beam at the start of its journey to compensate, or correct, in advance the dispersing effects of the air molecules.

The sounding rocket reached an altitude of over 350 nautical miles with a flight time of 10 minutes, *Defense Daily* reported. Return of telemetry data from the diagnostic array aboard the rocket confirmed test success.

Defense Secretary Caspar Weinberger announced the test during a speech at a Philadelphia World Affairs Council luncheon for what appeared to be the start of a Reagan Administration campaign to justify continuing its $26-billion Strategic Defense Initiative (SDI) research effort in the face of intensified USSR protests over the program. (*LA Times*, Oct 4/85, A5; *D/D*, Oct 18/85, 251)

October 11: The U.S. Air Force Space Technology Center's Kirtland Contracting Center awarded Aerojet Electro Systems a $17-million contract and Hughes Aircraft a $21-million contract for work on the Strategic Defense Initiative (SDI) program, the Air Force Systems Command *Newsreview* reported.

Under the contracts the companies would develop and build a space surveillance sensor called "the precursor above the horizon sensor," which would look across space using several frequency bands in the long-wave infrared spectrum to track the trajectory of intercontinental ballistic missiles or space systems. (AFSC *Newsreview*, Oct 11/85, 3)

November 18: Britain and the U.S. in the last week of October reached an outline agreement on British participation in the U.S.'s Strategic Defense Initiative (SDI) program that did not guarantee Britain any specific share in SDI research, development, or production contract awards, but did provide assurances on technology transfer acceptable to Michael Heseltine, British Secretary of State for Defense, *Aviation Week* reported. However, *Defense Daily* later reported that Britain decided to delay signing its formal agreement of participation.

The delay apparently stemmed from Britain's political embarassment over loss of the United States Army's Mobile subscriber equipment (MSE) program. Prime Minister Margaret Thatcher and Heseltine had intervened in the MSE competition, although reportedly unaware that the disparity in the bids was beyond the ability of political influence to set right. Also, Britain's Department of Trade and Industry was said to be objecting to certain provisions of the agreement, including property rights to certain technologies developed during Britain's SDI participation. (*AvWk*, Nov 4/85, 26; *D/D*, Nov 18/85, 81)

December 6: Defense Secretary Caspar Weinberger and British Defense Minister Michael Heseltine signed a memorandum of understanding for Britain's participation in the U.S. Strategic Defense Initiative (SDI) program, *Defense Daily* reported. The signing came one day after British Prime Minister Margaret Thatcher had told Parliament that the negotiations over British participation were not finished but that she hoped for a signed agreement before Christmas.

Joining Heseltine at a press conference following the signing, Weinberger said it "emphasizes both the closeness of our alliance and the special nature of our relationship which is a very vital thing" to both countries. "Britain is the leader in many of these technologies and we want very much to have these capabilities placed at the benefit of the program," he said.

Heseltine explained the signing of the agreement one day after Thatcher's statement to Parliament, saying both he and Weinberger had received communiques overnight from their teams in Washington permitting completion of the agreement. Heseltine said the agreement offered a "very significant opportunity for British industry and for British research capability to be associated with a major and exciting program at the frontiers of human capabilities in many of the technologies of tomorrow." (*D/D*, Dec 9/85, 185)

December 28: The Department of Energy's Lawrence Livermore National Laboratory supervised detonation of a hydrogen bomb buried in a box-car sized canister 1,800 feet below the Nevada desert floor in a test of technology for President Reagan's Strategic Defense Initiative (SDI), the *Washington Post* reported. An Energy Department official declined to discuss the test's purpose, but Defense Department and congressional officials said earlier that it was designed to test the concept of harnessing X-rays produced by a nuclear explosion into a laser cannon to destroy Soviet missiles.

The nuclear explosive had a force of 20 to 150 kilotons, the Energy Department spokesman said, registering 5.6 on the Richter scale, according to an official at the National Earthquake Information Center in Boulder, Colorado. It measured 5.3 at the California Institute of Technology in Pasadena.

The Energy Department postponed the test several times because of unfavorable winds and the Christmas holidays. Scientists routinely delayed underground nuclear tests in the Nevada desert when winds were blowing toward the south and west because of the possibility that in an accident a radioactive cloud might drift over a populated area. (*W Post*, Dec 29/85, A5)

EUROPEAN SPACE AGENCY

January 18: Ministers from Belgium, Denmark, France, W. Germany, Ireland, Italy, The Netherlands, Spain, Sweden, Switzerland, the U.K., Austria, Norway, and Canada would meet in Rome January 30 and 31 to consider proposed space projects and set the agenda for European Space Agency (ESA) through the remainder of the century, the JSC *Roundup* reported. Ministers would consider proposals ranging from development of a space station program, possibly in conjunction with the U.S., to eventual pursuit of manned spaceflight.

ESA said it would propose a substantial increase in its activities, particularly scientific and technological programs; developments in earth observations, telecommunications, microgravity processing, and technology; and construction of a new version of Ariane 5. ESA's efforts with Spacelab and Ariane and in a wide range of other activities such as the Giotto probe to intercept Halley's Comet had required the new decisions.

With Ariane 3 operational, ESA said it expected to introduce an upgraded version, Ariane 4, in March 1986, followed by the proposed Ariane 5, which would be capable of generating up to 220,000 lb. of thrust, that could be used to launch European astronauts into orbit. The ministers had to decide whether to authorize development of the large cryogenic engine to power the launcher. (JSC *Roundup,* Jan 18/85, 1; ESA release Jan 17/85)

February 4: Western European cabinet ministers in a two-day meeting in Rome agreed to participate in the 1992 U.S. launch of a permanently manned space station, the European Space Agency (ESA) announced. The ministers also increased the ESA budget and approved funding for Ariane 5, making clear their intent to pursue independent European space activities, but failed to decide on a French-sponsored plan for a small space shuttle (called Hermes).

ESA would base its participation in the U.S. space station on an Italo-German project called Columbus, with an estimated $2-billion project cost. Although W. Germany, Britain, and Italy had made funding commitments, France had not announced its project share.

The ministers' decision ended controversy over European cooperation with the U.S., which had resulted from an earlier U.S./European cooperative effort that ended negatively when a $750 million investment in the 1973 Spacelab produced little significant research or technological spinoffs, the *Washington Post* reported.

ESA also announced its annual expenditures would increase 70% to almost $1.3 billion by 1990. (ESA release Feb 4/85, Jan 31/85; *W Post,* Feb 1/85, A23)

February 4: The European Space Agency (ESA) announced it would hand over on February 5 *MARECS B–2* to the International Maritime Satellite organization (INMARSAT) for commercial operation. The spacecraft, launched November 10 and operational since January 8, would complete INMARSAT's first-generation satellite system, which would offer worldwide service covering three oceans: the Atlantic region with *MARECS–A* positioned at 26° W, the Indian Ocean with *INTELSAT MCS A* at 63° E, and the Pacific Ocean with *MARECS B–2* at 176.5° E.

In preparation for future communications satellite services, ESA in cooperation with INMARSAT was conducting experiments with *MARECS B–2* to design economically operated equipment compact enough for installation in small vehicles. (ESA release Feb 4/85)

February 9: The European launcher Ariane 3 launched two satellites, Brazil's *Brasilsat 1* and the Arab *Arabsat 1*, FBIS AFP in Spanish reported. Frederick D'Allest, French Center for Space Studies director general, noted Ariane had taken off on schedule, down to the exact hour planned a year previously.

Ariane's three sections functioned perfectly, and the 3rd one had reached satellite-ejection speed at an altitude of 210 km; three minutes later *Arabsat 1* separated from the rocket, followed a half minute later by *Brasilsat 1*. The satellites would require several weeks to reach their geostationary orbits: *Arabsat 1* over Zaire, *Brasilsat 1* above central Brazil. (FBIS AFP in Spanish, Feb 9/85)

May 7: The European Space Agency (ESA) launched at 9:15 p.m. EST today from Kourou an Ariane rocket that successfully deployed two telecommunications satellites including a U.S.-built *G-Star 1*, the *Washington Post* reported. A breakdown in timing between the rocket and the ESA's various earth-based satellite tracking stations delayed launch 79 minutes.

ESA officials described the launch as "another success" in the agency's competition with the U.S. Space Shuttle to capture the world market for satellite launches. The Ariane rocket was the 13th fired, including two aborted launches, since 1979. The previous launch February 9 also put two satellites into orbit. Officials said they had nearly 30 orders to launch commercial satellites, compared with about 60 orders for the Space Shuttle.

The Ariane rocket deployed the *G-Star 1* satellite, owned by the GTE Spacenet Corp., and the French *Telecome-1B* 20 minutes after liftoff at about 22,000 miles above the equator. (*W Post*, May 8/85, A26)

May 15: Roger Bonnet, director of the European Space Agency's (ESA) scientific programs, speaking today at the opening of a meeting with the USSR's "Intercosmos" Council for International Cooperation, Soviet Academy of Sciences, said, "We are keenly interested in extending cooperation with the

Soviet Union in space exploration," FBIS, TASS in English reported. "The first results of our joint work and experiments have given us reasons for making the most optimistic forecasts and opened up new opportunities for studying and using space for peaceful purposes."

The program for the three-day meeting included discussion of details of the Soviet "Vega" mission to study Venus and Halley's Comet and review of preparations for a similar West European program called "Giotto," which would use data from the Soviet probe.

Attendees also heard reports on the Soviet "Phobos" project for the comprehensive study of Mars and its satellites and ESA's plans to develop an orbital radio telescope. (FBIS, Tass in English, May 15/85)

June 12: The European Space Agency (ESA) announced that, as a result of decisions made January 30 and 31 at its council meeting in Rome, it had created an earth observation and microgravity directorate, a telecommunications directorate, and, due to the magnitude of ESA's Columbus program, a directorate dedicated to that program.

The ESA Council at its June 11 and 12 meeting made the following nominations to the new director posts: Philip Goldsmith, United Kingdom, director of earth observation and microgravity program; Giorgio Salvator, Italy, director of telecommunications; and Dr. Fredrik Engstrom, Sweden, director of the Columbus program.

In addition, the council appointed Marius Le Fevre, France, to the post of director of ESA's space research and technology establishment. (ESA Release, June 12/85)

July 1: The European Space Agency (ESA) announced that M. Bignier, ESA director of space transportation systems, and Ing. Heise, executive vice president and president of the Space Systems Group of Messerschmidt/Bolkow-Blohm (MBB)/Erno, signed a contract for the development of the European Retrievable Carrier, EURECA, a payload carrier intended for use on the U.S. Space Shuttle. MBB/Erno would serve as EURECA prime contractor with support from some 24 European industrial firms. The approximately $95 million contract called for delivery of the EURECA flight unit to NASA by the end of 1987 for launch from the Space Shuttle in March 1988 and recovery, by the Space Shuttle, six months later.

The first EURECA payload, developed by European national institutes and space agencies, would consist primarily of experiments in the microgravity sciences (life and material sciences) and a limited number of experiments in space science and technology. Work on the payload would proceed in parallel with the development of the carrier, and plans called for delivery of all payload elements to MBB/Erno by spring 1987 for integration onboard EURECA before shipment to the U.S.

The EURECA payload carrier incorporated the more attractive features of Spacelab and in addition provided for relatively long duration flights (up to eight months), higher power and mass capability for the payload, and lower costs compared with Spacelab and many conventional satellites because the system occupied only about 2.5 m of the Space Shuttle orbiter cargo bay.

ESA also envisioned EURECA's eventual use in association with the proposed U.S. space station, with the EURECA operating either in a free-flying mode (co-orbiting with the space station or in polar orbit) or as a mantended, semipermanent payload carrier with modified docking facilities. (ESA release July 1/85)

August 1: The European Space Agency (ESA) announced that following validation tests of the second Ariane launch site (ELA–2) in Kourou, French Guiana, it accepted the new facilities and made them available to the Arianespace company for international use. ELA–2 would provide an enhanced capability for launching the Ariane 2, 3, and 4 and make it possible to reduce the interval between launches to one month and carry out up to ten launches per year at lower operating costs.

The new complex close to the current Ariane launch site had two separate zones, a launcher preparation zone and the launch zone, linked by a one-km rail line. This configuration made possible two simultaneous launch preparations and provided considerable operational flexibility. Arianespace could assemble one launcher in the preparation zone, while another, previously transported on a mobile table to the launch zone, underwent final checkout prior to launch.

The French National Space Centre (CNES) was prime contractor for ELA–2 construction, which took four years to complete. ELA–2 would become Europe's main launch complex; ELA–1 would serve as the backup. Arianespace scheduled the first launch from ELA–2 for December 1985, using an Ariane 3 to launch Brasilsat 2 and G–STAR 2. (ESA release Aug 1/85)

August 2: The European Space Agency announced that it had signed with the Commission of the European Communities an agreement to cooperate on the APOLLO project, which would provide, through the EUTELSAT 1 series of communications satellites developed by ESA and European industry, a high-speed digital information transfer system suitable for long data messages, particularly document facsimiles, transmitted from a small number of information providers (up to 10) to many widely dispersed users.

The agreement provided for development as specified by ESA of prototype equipment by European industry and for the Commission with the assistance of ESA to provide overall coordination of APOLLO, including archives planned for the Office for Official Publications of the European Communities in Luxembourg and for several other large document archives in Europe.

European Space Agency

The commission estimated that APOLLO would help to provide a long-term infrastructure for a market expected to expand greatly over the next five years and to contribute to experience in the design, costing, and operation of small dish earth terminals and other equipment developed for APOLLO. APOLLO could also stimulate the creation of a market for European earth stations and associated equipment.

Trial users transferring digital records, documents, and computer files would assess the system for high-speed and high-quality information. (ESA release Aug 2/85)

September 12: Arianespace officials blew up the European Space Agency's (ESA) Ariane V15 rocket, which carried the third in the European Communications Satellites (ECS–3) series and the U.S. communications satellite Spacenet F3, less than ten minutes after liftoff when the rocket veered off course and began falling, threatening inhabited areas, the *Washington Post* reported. The failure was Ariane's third in 15 launches. French President Francois Mitterrand, on a stopover on his way to French Polynesia, watched the failed launching.

Liftoff at the Kourou (French Guiana) launch site was on schedule at 8:26 p.m. Reports said Ariane was on course during the first minutes of its ascent, but then it suddenly went off course and lost altitude because of a propulsion problem in the third-stage motor.

Ariane's other failures were on May 23, 1980, on its second launch, and on May 9, 1982, on its fifth. (*W Post*, Sept 13/85, A12; ESA release Sept 2/85)

October 24: The European Space Agency (ESA) announced today that during the 71st meeting October 23 and 24 of the Council of the European Space Agency it unanimously approved the accession of Austria and Norway to full membership status, bringing ESA membership to 13 countries.

The agreement between ESA and the governments of Austria and Norway required governmental approval and parliamentary ratification in the two countries with the intention of giving them full membership status on January 1, 1987.

The decision followed a period of close cooperation between ESA and the two countries, both of which had been closely associated with many of ESA's activities over the previous 20 years including the Spacelab, Marecs, and ERS–1 programs. (ESA release Oct 24/85)

December 10: ESA announced it signed three contracts with Arianespace, two for future launches—ECS–4 scheduled for the second quarter of 1986 and Hipparcos scheduled for June 1988—and one for technical assistance for satellite launches on the first flight of Ariane 4.

ECS–4 would replace ECS–3, lost as a result of the launch failure September 12 of Ariane V15. ECS–4 was in production at that time, and ESA accelerated its completion. Arianespace was providing the earliest possible launch slot in accordance with the relaunch conditions in the ECS–3 launch contract.

Hipparcos would provide measurements of the positions, annual proper motions, and parallaxes of some 100,000 stars. From its position in geostationary orbit, the satellite systematically and repeatedly throughout its two-and-a-half-year lifetime would scan the whole sky, providing measurements that would serve as a reference system of unprecedented precision, important in studies of earth's motion, the solar system, and our galaxy, and forming a basis for future ground and space astrometry.

Under the third contract, the first flight of Ariane 4, part of ESA's Ariane 4 development program, would be carried out under ESA's responsibility. The launch, scheduled to take place during the third quarter of 1986, was designed to demonstrate the operational capability of Europe's most powerful launcher. ESA was supplying part of the payload, Meteosat P2, a refurbished spacecraft from the preoperational series, designed to bridge a possible gap between Meteosat 2 and the first of the operational meteorological spacecraft, MOP–1, scheduled for launch in late 1987. Additional payload elements would be Amsat Phase III–C, the second unit of the third generation of the amateur radio "Oscar" series, and a telecommunications satellite yet to be selected. (ESA release Dec 10/85)

FEDERAL AVIATION ADMINISTRATION

February 4: Under pressure from the Federal Communications Commission (FCC) to define its requirements in the internationally authorized 1,544 to 1,660.5 MHz-frequency band and to make allocations in this band for commercial mobile-user communications and surveillance systems, the Federal Aviation Administration (FAA) indicated satellites might play a key role in air navigation, surveillance, and communications in its follow-on generation, air-traffic control system. Therefore the FCC was trying to protect the radio spectrum required for such services, *Aviation Week* reported. In addition, the International Civil Aviation Organization (ICAO) would review at the 1987 World Radio Conference the earlier-assigned band, seeking to assure sufficient spectrum availability for civil aviation operators' future global needs.

The Special Committee 155 of the Radio Technical Commission for Aeronautics said it would report at an April 26 ICAO meeting results of its yearlong study intended to define future spectrum operational requirements for the year 2010 and beyond for all classes of airspace users. The report called for service over both land and water, from the surface to a 70,000-foot altitude, a requirement most easily met by spaceborne systems.

The FAA had considered spaceborne systems for air traffic communications for nearly two decades, but had rejected them as not cost-effective replacements for terrestrial-based facilities (except for transoceanic communications). However, at the urging of the Air Transport Assn. (ATA), the FAA had initiated in the early 1970s a joint program with the European Space Agency (ESA) to explore satellite use for oceanic aircraft communications, resulting in contracts with Comsat Corp. for the service and GE to build a satellite; the FAA had terminated the program when ATA lost interest. More recently, *Aviation Week* said, the FAA had responded cooly to DOD's push for FAA adoption of its Rockwell Navstar Global Positioning System, partly because of Pentagon-imposed signal accuracy limits of 500 miles (1,640 ft.) for civilian users.

Currently, commercial applications to the FCC for authorization to offer domestic communications- and surveillance-satellite services to users, ranging from trucks to aircraft, had raised serious FAA concerns. FAA officials doubted the technical suitability of some of the proposed systems to meet airspace users' demanding requirements.

The government traditionally had owned and operated the surveillance, navigation, and communications facilities the FAA used for air-traffic control. If the FCC, under the Reagan Administration policy of commercializing space activities, should allocate spectrum currently assigned for aviation to nonaviation use, this could preclude a future optimum global spaceborne-aviation system. Therefore, the FAA task force was studying spectrum needs for several types of spaceborne systems, including multifunction satellites, and expected to have recommendations ready for submission to the Future Air Navigation System Committee and the National Telecommunications and Information Agency, which addressed government frequency-allocation issues. (*AvWk*, Feb 4/85, 36)

February 7: The Federal Aviation Administration (FAA) announced rules to permit new two-engine jetliners to fly lengthy remote routes that had required three- or four-engine planes, but each airline would have to prove its aircraft and flight crews met FAA standards. The FAA took the action because of the reliability of modern jet engines and because military and business jets with two engines had flown across the North Atlantic safely for several years. The rules also addressed special provisions for airframe reliability, backup electrical and hydraulic systems, maintenance, cargo-compartment fire protection, and crew training.

Under current rules, a two-engine plane could be no farther from an airport than 60-minute flying time on one engine; the FAA proposal would extend that to two-hours flying time on one engine.

When the rules became final, they would allow Boeing's two-engine 767 jumbo jet to fly the most fuel-efficient North Atlantic routes, which would take flights far from airports in Greenland and Iceland. Boeing had pushed hard for the rule change, the *Washington Post* reported, to expand the sales potential of its 767. (FAA Release 5–85; *W Post*, Feb 7/85, A7)

March 14: Federal Aviation Administration (FAA) Administrator Donald Engen, citing improvements in scheduling practices and air traffic control, told airlines today that on April 1 he would remove minute-by-minute restrictions on the number of flights at airports in Atlanta, Chicago, Denver, and the New York City area, the *Washington Post* reported. He said delays would continue to occur but should be reduced and more manageable.

Before the restrictions, the air traffic system had experienced an unprecedented number of delays, averaging 1,400 daily 15-minute or longer delays nationwide the previous August. Between November 1 and January 31, 1984, average daily delays dropped to 863, a reduction of 46%. Not all delays were attributable to airport congestion or air traffic control problems; the FAA generally blamed about 60% of them on weather.

Also likely to contribute to a decrease in delays was the increasing ability of the FAA to monitor traffic nationwide through its computer system and to redirect planes to less-busy routes. Engen noted that the airlines had indicated they would continue to operate the schedules adopted under the restrictions and would not revert to the bunching of large numbers of operations within short time periods. (*W Post*, Mar 15/85, A16)

June 7: The Federal Aviation Administration (FAA) announced it was using sophisticated electronic equipment to identify pilots who were illegally using air-traffic control radio frequencies to harass nonstriking United Airlines pilots. The FAA had already initiated enforcement action against the pilots responsible for such incidents near Chicago's O'Hare International Airport and the Seattle-Tacoma Airport. The equipment also identified other flight crew members who had engaged in such harrassment, and additional enforcement actions were likely.

The equipment was a tracking device that helped pinpoint the source of otherwise unidentified radio transmissions, of which more than 50, either jamming a radio frequency by pressing a microphone button or verbal abuse over the radio, had occurred since the strike began on May 17.

FAA Administrator Donald Engen said he would not tolerate misuse of the air-traffic control frequencies, which could lead to suspension or revocation of a pilot's license and possibly criminal penalties of up to 10 years imprisonment or a fine of $10,000. (FAA Release 24–85)

August 28: As a result of a Manchester, England, airline accident in which it was believed a wing-mounted engine on the Boeing 737 disintegrated as the plane was attempting to take off [see Aviation/Civil Aviation, Aug. 26], the Federal Aviation Administration (FAA) was trying to decide today how many of about 5,000 Pratt & Whitney jet engines operated by U.S. commercial airlines required inspection for possible cracks, the *Washington Times* reported. Airline officials believed a fire started after the accident when a faulty combustion chamber in the plane's Pratt & Whitney JT8D engine blew apart and a hot piece of the combustion chamber pierced the plane's wing fuel tank.

British inspectors said preliminary investigations into the Manchester accident showed that deterioration in the combustion chamber caused overheating and eventually failure of the part. Investigators said they found "extensive cracking" in six of the engine's nine combustion chambers.

In addition to use on the Boeing 737, Pratt & Whitney sold the same kind of engine for Boeing 727s and McDonald Douglas DC9s. Those planes made up about two-thirds of the 3,000 planes operated by the major U.S. airlines.

The FAA was attempting to arrange the engine inspections without disrupting U.S. airline service and indicated there would be no inspections of newer engines because there was insufficient hours on them. Inspections were made in one of two ways: an isotope inspection, which was similar to x-raying the engine to search for cracks or a visual inspection, which required dismantling the engine to get to the combustion chambers.

An inspection of JT8D engines ordered by the British Civil Aviation Authority turned up serious combustion chamber cracks in five Boeing 737s, and the planes were grounded. (*W Times*, Aug 29/85, 8B)

September 12: The Federal Aviation Administration (FAA) announced it had proposed a regulation requiring that all new transponders installed in aircraft after January 1, 1992, be compatible with the new Mode S, ground-based radar beacon system to give controllers more accurate aircraft position and identification information.

In October 1984, the FAA ordered 137 Mode S ground stations from joint manufacturers Westinghouse/SDC-Burroughs and scheduled the first Mode S for delivery in the spring of 1987 to the FAA Technical Center in Atlantic City, N.J.

In addition to position and identification information, the Mode S had a "selective" address capability, which gave the system its name and provided a channel for automatically transmitting weather and other data between air-traffic control facilities and aircraft in flight.

The FAA already required that airplanes operating above 12,500 feet or in designated airport terminal control areas be equipped with a less sophisticated transponder. When triggered by the sweep of ground radar, this equipment sent back a signal that gave controllers a clean and enhanced target on their radar displays and also told them the aircraft's identity and altitude.

A limitation of the current radar beacon system, which Mode S would replace, was that ground equipment interrogated simultaneously all aircraft in a given area, often resulting in overlapping and garbled signals on radar displays. Mode S eliminated this problem by addressing each aircraft on an individual or selective basis.

The purpose of the proposed rule was to promote early installation of Mode S transponders, although existing transponders would not be obsolete; pilots could continue to fly with that equipment after January 1, 1992, until it needed replacement. (FAA Release 48–85)

FRANCE

May 31: French President Francois Mitterrand opened today the Paris Air Show, stating that the French-led European high-technology program known as Eureka was "off to a good start" after gaining approval from West Germany and other European partners, FBIS Paris AFP in English reported. France and W. Germany had previously disagreed over a U.S. invitation to participate in the Strategic Defense Initiative (SDI) defense research project. West Germany had endorsed the project, but France had rejected it, fearing it would render France's independent nuclear deterrent impotent and give the U.S. an unassailable lead in advanced technology.

In speaking of Eureka, a space research program primarily for civilian uses and the development of ultramodern technology in such fields as lasers and high-speed computers, Mitterrand repeated his contention that the Eureka and SDI projects were not competing for the same goals, but added that Bonn's participation in both would necessitate choices in its budget and assignment of scientists.

He also said France was determined to develop a European fighter plane by the 1990s and that the European Hermes space plane project was of major importance and could tie in with the European Ariane-5 rocket program. (FBIS Paris AFP in English, May 31/85)

Hermes Program

March 11: France's Centre National d'Etudes Spatiales (CNES) space agency was planning to select by midyear a prime contractor for the Hermes manned shuttle and would open the program to participation by other European countries before the end of 1985, *Aviation Week* reported. CNES had established a competition between Aerospatiale and Dassault-Breguet for the Hermes leadership role, and the two companies would make technical presentations in April. After review of the data, CNES would make a selection within several months.

Normally the two companies had their separate niches in the aerospace field, but CNES had forced them into head-to-head competition over Hermes. In the competition, Dassault-Breguet would draw on expertise gained from its Mirage, Jaguar, and Alpha Jet military aircraft and related electronics/avionics systems; Aerospatiale would emphasize its work on military missiles, reentry vehicles/nuclear warheads, the Ariane launcher, communications satellites, and civil aircraft such as the Concorde and Airbus Industrie A300/A310/A320.

CNES Director General Frederic d'Allest said several countries had shown interest in joining the Hermes program following the European Space Agency's (ESA) council meeting at which ministers approved Phase B definition work for the Ariane 5 launcher [see Feb. 4]. Ariane 5 would be capable of orbiting Hermes. France planned to retain a 50% share in the Hermes program and open the rest to participation by other European nations.

The 1985 and 1986 budgets for Hermes totaled $12 million, followed by about $20 million in study work in 1987. France would supply most of this money in 1985, because other European countries wouldn't be involved until at least October.

Hermes would carry two to six crew members and be about half the size of the U.S. Space Shuttle. Its approximate length would be 59 ft. and wingspan, about 33 ft. Payload bay volume would be 377 sq. ft.; diameter of the bay, 9.8 ft. A typical mission would carry 7,700 to 9,900 lb. of useful load to a 248.5-mile circular orbit inclined 0–30°. Mission duration with four to six crew members would be eight days, extending to 30 days with a reduced crew.

Start of full-scale Hermes development in mid-1988 would permit a first flight in 1997, according to CNES scheduling. (AvWk, Mar 11/85, 19)

October 18: Jacques Louis Lions, president of the French National Center for Space Studies, announced today that Aerospatiale and the Dassault-Breguet Aviation Company under the center's auspices would work together over the next several years to build two models for the European space shuttle Hermes, FBIS Hong Kong AFP in English reported. The European Space Agency (ESA) would likely also sponsor the project, and the center would undertake negotiations with European partners who would develop the main Hermes subsystems.

Aerospatiale would lead the project and assemble Hermes components in its Toulouse factories; Dassault would be in charge of aeronautical aspects.

Hermes, about half the size of the U.S. Space Shuttle, would carry two to six astronauts and cargo up to 4.5 tons into low earth orbits. (FBIS, Hong Kong AFP in English, Oct 19/85)

GEOPHYSICS

During April: In response to a request from the Johns Hopkins University's Applied Physics Laboratory to acquire more information on how to predict ocean wave behavior, the Wallops Flight Facility's P–3 aircraft participated October 1984 in NASA's Seasat Imaging Radar B (SIR–B) experiment to monitor ocean waves while, 140 miles above it, the Space Shuttle Challenger produced images of the same waves with its radar.

Scientists knew what caused waves to swell, but found it difficult to predict exactly how they would behave. Researchers hoped they might be able to predict through the use of satellite data the occurrence of dangerous waves before they caused serious damage to ships and coastal towns.

Analysis of previous data from the SIR had indicated that a synthetic aperture radar could observe ocean waves from space, but the lack of information on the actual directional wave spectrum on the ocean at the time of the SIR observations made it difficult to obtain an accurate assessment of the radar's performance.

During Challenger's eight-day flight, the P–3 underflew the spacecraft for five nights off the coast of Chile to obtain sea surface observations. Each night the aircraft flew along several hundred kilometers of the Space Shuttle's imaging radar taking sea-surface observations with its complement of remote sensing instruments. The P–3's airborne oceanographic lidar provided laser elevation profiles of the wave field and the radars (surface contour radar, radar ocean wave spectrometer, and advanced applications flight experiment altimeter) measured significant wave height and provided information on the directional wave spectrum.

Because state of the sea varied over the series of flights from low seas to waves with 6-m height and 400-m wavelengths, the data sets collected should become classic and provide a basis for the quantitative evaluation of synthetic aperture radars in space. (*Inside Wallops*, Apr 85, 1)

October 4: University of Chicago scientists Edward Anders, Wendy Wolbach, and Roy Lewis found evidence that continent-sized firestorms 65 million years ago raged across much of the earth, blackening the skies with soot and possibly triggering a sudden global freeze that wiped out the dinosaurs, the *Washington Post* reported. The event occurred at the time that other scientists said a giant asteroid collided with earth, which may have generated enough heat to start the fires.

It was the asteroid theory that led the University of Chicago researchers to ask whether a nuclear war could trigger a similar freeze, a phenomenon now known as nuclear winter. Their findings suggested that nuclear winter theorists had greatly underestimated the amount of soot that would enter the atmosphere from wildfires.

The researchers found some of the soot, which eventually settled to the ground, in the same geologic layers that five years previously gave evidence of the asteroid impact. The layer, sampled in regions as far apart as Europe and New Zealand, contained an amount of soot, which was pure carbon, equal to about 10% of the carbon currently incorporated into all forms of life on earth.

The original impact theory held that an asteroid, probably at least six miles wide, hit earth, blasting enough rock dust into the atmosphere to darken the sky for weeks. Deprived of sunlight, the ground would have cooled, bringing on a global freeze that exterminated not only the dinosaurs but many other species all over the world. Paleontologists knew that the dinosaurs died out in the most wide-ranging mass extinction the earth had even seen. Scientists working on the asteroid impact theory already had calculated that airborne rock dust alone would have been thick enough to bring on a devastating freeze.

The discovery that vast quantities of soot might also have been in the upper atmosphere indicated that the period of darkness and freezing would have lasted longer, perhaps months, because soot washed out of the atmosphere more slowly than did rock dust, although it would have been no darker or colder than originally thought.

The finding added two more factors that might have contributed to the extinction—concentrations of fire-produced toxic substances, such as carbon monoxide, and the destruction of plants and animals by fire.

The researchers said the discovery was an accident, resulting from examination of ancient sediments while looking for traces of gases that had been part of the asteroid. Instead, they found that the sediments contained about 10,000 times as much carbon as would have been expected.

The layer, sampled in Denmark, Spain, and New Zealand, was the one in which other scientists previously found unusually high concentrations of iridium, an element that is rare on earth but abundant in meteorites and asteroids. The scientists thought that when the asteroid hit the earth, the impact would have generated enough heat to vaporize the asteroid, sending its iridium into the atmosphere where it spread around the world.

Anders said the impact would have scattered white-hot particles of rock dust as far as 800 and possibly 1,200 miles, igniting forest fires over the entire area. "Once started," he noted, "such a fire could spread over an entire continent, and the resulting winds may disperse the soot worldwide."

The impact, other scientists calculated, would have left a crater 85 miles wide and 20 miles deep. Since no such crater had been found, many scientists assumed the asteroid hit the ocean, vaporizing the water before it hit the sea floor. (*W Post*, Oct 4/85, A2)

October 7: NASA announced that its associate administrator for space science and applications, Dr. Burton Edelson, in a speech today at the 36th

Congress of the International Astronautical Federation in Stockholm warned his audience that the earth's environment as was presently known might be in jeopardy and called on the world's scientific community to accelerate its study of the planet. "Resources, once thought to be limitless, are slowly being depleted," Edelson said. "Earth's atmosphere is changing, and some of its life forms are threatened. It is imperative that we, as scientists and engineers, take action now to maintain the quality of life on our planet and improve its biological productivity."

Edelson cited the many accomplishments of planetary science since the beginning of spaceflight 28 years ago. "We have examined most of the planets in the solar system at close range and have performed systematic studies of our closest neighbors, Venus and Mars, through telescopes and more recently, through data and imagery received from planet-orbiting spacecraft," he said. "We still lack synoptic, systematic, and temporal knowledge, predictive skills, and an understanding of the mechanisms underlying earth's global processes," he pointed out.

Calling for an international study effort, Edelson said, "Our sophisticated spacecraft; new air, sea, and spaceborne sensors; and enormous computing capability will enable us to measure, monitor, model, and finally begin to understand the earth as a system. This mission to planet earth could unlock the secrets of life itself . . . and could well prove to be the most important ever undertaken by humankind."

Edelson called particular attention to the problems of water pollution in rivers, lakes, and streams; potential depletion of the ozone layer; growing atmospheric concentrations of carbon dioxide; and a sharp increase in the levels of carbon monoxide, methane, and nitrous oxide. "Economic development over large portions of the earth have significantly changed the patterns of land and water use," he said. "The results have been mixed—while in some cases the benefits have been significant—we have paid a substantial price. We must study the land and learn to use it properly."

To emphasize the importance of his proposal, Edelson said, "The U.S. National Academy of Sciences has reviewed the scientific merit of the global habitability concept and has found it to be both sound and worthy. The Academy is now participating in the broader efforts of the International Council of Scientific Unions in a program called Global Change. NASA and several other government agencies, notably the National Oceanic and Atmospheric Administration and the National Science Foundation, also will participate." Edelson said he envisioned multifaceted investigations with oceanographers, meteorologists, biologists, and foresters studying the land, sea, atmosphere, and the air-sea and solar-terrestrial environments.

"This is truly an international challenge," he concluded, "involving many scientific disciplines. Everyone on earth has a stake in our success." (NASA Release 85–142)

HISTORY

April 7: In a review of . . . *The Heavens and the Earth* by Walter McDougall, a reviewer for the *NY Times Book Review* said that the book operated on four levels: as a narrative history of space activity, a political analysis of what caused Sputnik 1 and what Sputnik 1 caused, an exposition of the contradictions inherent in the Soviet socialist system and the American free-enterprise system, and an essay on the eschatology of what is called the "pursuit of power." Reviewer Alex Roland noted the book was based on a vast published literature complemented by archival research in the U.S. and Europe, interviews with many of the key principals, and recent declassification of important material, including National Security Council policy paper 5814.1, the first official national policy on space.

Roland called the book the "most comprehensive history of space activity written to date, the most thorough analysis of the political and social forces at work. It provides a plausible and compelling interpretation of how and why the space age has developed as it has. And it poses an original and stimulating paradigm for analysis of the post-industrial state in a world of continuing cold war. With this book," Roland concluded, "the history of space activity has come of age." (*NYT Book Review*, Apr 7/85, 1)

April 11: A prime time TV series for the first time would offer a comprehensive look at the history of manned U.S. and USSR space flight, the Ames Research Center's *Astrogram* reported. SPACEFLIGHT, a series of four one-hour programs sponsored by the Public Broadcasting System and the Du Pont Co., would chronicle man's achievements in space from the rocket plane that first broke the sound barrier to the touchdown of the Space Shuttle orbiter Columbia.

The series' executive producer, Blaine Baggett, said his staff interviewed some 40 key participants in space activities—astronauts, people on the ground, scientists, and historians. In addition, the Soviet science attache in Washington provided about 10 hours of film footage. Among those interviewed for the series were Chuck Yaeger, first man to break the sound barrier; Scott Crossfield, first to travel twice the speed of sound and to pilot the X-15; Werner von Braun, who worked on the U.S. rocket program; Alan Shepard and John Glenn, first American in space and first American to orbit the globe, respectively; and Sally Ride, first American woman in space.

From the point of view of the USSR, the series sought answers to such questions as who the chief designer was of the Soviet space program and why the Soviets were first in the early days of space activities.

SPACEFLIGHT also examined future space plans such as space colonies, space stations, and the Strategic Defense Initiative. (ARC *Astrogram*, Apr 11/85, 4)

November 6: Hermann Oberth, considered by some to be the father of spaceflight, watched on TV at Goddard Space Flight Center (GSFC) the landing of the Space Shuttle Challenger on mission 61-A, the *Washington Post* reported. Oberth is the last scientist survivor of the group that transformed theory into modern space exploration. Konstantin Tsiolkovsky, the Russian who worked out some of the early theory of rocket propulsion, died in 1935; Robert Goddard, the American who experimented with early rockets and for whom GSFC is named, died in 1945. But Oberth, who made the most complete analyses of the problems and prospects of human space travel, lived to "see it happen."

In 1932 Oberth wrote a paper in which he not only showed mathematically that it was possible to escape earth's gravity, but also anticipated a host of other aspects of spaceflight not seriously approached for the next 30 years. Oberth described in the book a spaceship's propulsion system and architectural form down to the rocket engine's nozzles; designed spacesuits and methods of eating in weightlessness; conceived of astronauts performing spacewalks; proposed space stations in earth orbit as transfer points for interplanetary travel; considered the problems of weightlessness and motion sickness (which astronauts still faced); proposed that the stations spin slowly to create an artificial gravity; and suggested the use of flying shuttles that could take off like a rocket, visit the space station, and land back on earth like airplanes.

Oberth also worked out the physics of joining two spacecraft in orbit (referred to now as rendezvous and docking), anticipated that photos of earth from space would be useful for studying the ground and forecasting weather, foresaw that telescopes in earth orbit could gather far better astronomical data than those that looked through the atmosphere, and claimed that practical uses for space travel would someday make it a profitable enterprise.

Working unaware of Tsiolkovsky or Goddard, Oberth wrote the paper as an amateur physicist and mathematician while serving as a soldier in the Austro-Hungarian army during World War I. Oberth's commanding officer sent the paper to the War Ministry, where the generals rejected it as obvious fantasy.

After the war, when Oberth was at the University of Heidelberg, he submitted as his dissertation a longer version of the paper, complete with elaborate mathematical formulas proving his ideas. Again it was rejected.

Eventually Oberth paid to have the paper published. "The Rocket Into Interplanetary Space" gained a wide following and attracted Wernher von Braun, who worked as Oberth's assistant and then left to join the German military rocket research program at Peenemunde. Von Braun brought Oberth to work on the V2 rocket, the first major device based on Oberth's ideas. During three years in the 1950s, Oberth joined von Braun in the U.S. where he was developing the Redstone rocket. In 1958 Oberth retired to West Germany.

Oberth had concluded in his book that "the foregoing demonstrates that it is possible, with present day science and technology, to construct vehicles which could attain cosmic speed and that it is probably possible for men to ride in these vehicles," but that it would take more than a decade to realize these possibilities. At GSFC today, Oberth said it had "proven to be much more complicated that I thought." (*W Post*, Nov 7/85, C1)

INDIA

January 23: NASA Administrator James Beggs and U.R. Rao, chairman of the Indian Space Research Organization (ISRO), signed a launch services agreement covering the reimbursable launch of INSAT 1C around mid-1986 from the Space Shuttle, the *Marshall Star* reported. A multipurpose satellite, INSAT 1C would provide communications and meteorological services to India.

In addition, the two agreed that an ISRO scientist/engineer would serve as a payload specialist during that Space Shuttle mission and discussed other possible NASA/ISRO cooperative activities in space applications and space science. (*Marshall Star*, Jan 23/85, 4)

June 15: On the fourth and final day of his U.S. visit, Indian Prime Minister Rajiv Gandhi, accompanied by Vice President Bush, toured the Johnson Space Center's mission control center and then climbed into the commander's seat of a Space Shuttle mockup to dramatize his commitment to high technology, the *Washington Post* reported. Later Gandhi said the spinoffs of space technology "have become part of our daily lives," and that India needed to keep abreast of technological advances in agriculture, communications, meteorology, and prospecting.

Earlier in the week, the *NY Times* reported the U.S. and India were scheduled during Gandhi's visit to announce a joint space effort that would include the launching in 1986 aboard the Space Shuttle of an Indian payload specialist and a satellite that was partly designed to expand the uses of radio and TV in Indian villages.

Gandhi's visit was intended to ease strains with the U.S., reach a series of space and technology agreements, and open a nationwide cultural program, the Festival of India. The previous year the Soviet Union launched an Indian astronaut into space on an eight-day mission. (*W Post*, June 16/85, A12, *NYT*, June 9/85, A3)

INTELSAT

February 11: NASA announced it had scheduled the 8th launch in a series of 10 Intelsat V-type international telecommunications satellites aboard an Atlas Centaur for no earlier than March 7 from Cape Canaveral Air Force Station. NASA had successfully launched six Intelsat V satellites.

The Atlas Centaur-63 was the 2nd of the new stretched version of the vehicle, able to lift from 159 to 227 kg (350 to 500 lb.) more than the previous design for a total of 2,318 kg (5,100 lb.) NASA had extended the first-stage Atlas by about 2.05 m (81 in.) to accommodate more propellant. This would be the 1st Intelsat V–A spacecraft launched by an Atlas/Centaur.

The Intelsat V–A satellites were similar to the Vs except for improvements that increased reliability and boosting communications-carrying capacity by 25% from 12,000 to about 15,000 simultaneous telephone calls and two TV programs. NASA would for several weeks station the satellite over the equator at about the longitude of Paris for in-orbit testing; once operational, the spacecraft would be in geosynchronous orbit over the Atlantic Ocean.

Lewis Research Center (LeRC) managed the Atlas Centaur development and operations, and the Kennedy Space Center (KSC) managed vehicle checkout and launch. (NASA Release 85–22)

February 11: INTELSAT's Assembly of Parties, representatives of INTELSAT member nations, rejected a proposal by 21 nations that all INTELSAT parties and signatories boycott interconnection discussions with private companies planning to launch international satellites and that the U.S. reconsider its favorable policy toward non-INTELSAT satellites, *Aviation Week* reported. INTELSAT signatories, telecommunications organizations that operated ground facilities and represented INTELSAT member nations in organization management, had resolved in April 1983 that diversion of international traffic would threaten INTELSAT's viability.

A resolution adopted by the assembly urged all signatories and parties to "take into account" the boycott and the signatories' economic-viability resolution and to express opinions on separate systems to the U.S. and at the upcoming INTELSAT assembly.

U.S. officials had joined the consensus after objectionable language was deleted, but said naming a specific nation in the resolution could set a bad precedent. As one official noted, "The criticism of the U.S. we expected turned out to be expressions of concern by developing countries that competition could mean increased rates for them."

The assembly asked the Board of Governors to submit recommendations for separate-system coordination to the assembly as soon as possible, as the U.S.'s Federal Communication Commission was processing five applications for private satellite launches. (*AvWk*, Feb 11/85, 29)

March 13: The USSR was expected to sign an unprecedented information exchange agreement with INTELSAT, with a commitment to join the international consortium within two years, the *Washington Post* reported, a tacit admission that its own Eastern Bloc competitor to INTELSAT—Intersputnik—did not meet all the USSR's telecommunications needs. INTELSAT was awaiting V.A. Shamshin, minister of posts and telecommunications in the USSR, to sign the proposed agreement sent the previous week.

The U.S., creator of the 21-year-old INTELSAT system and its biggest single user, had no legal means to prevent INTELSAT from sharing technology with, or granting admission to, the USSR. The Reagan Administration said it would adopt a wait-and-see approach to the proposed agreement, which posed no security danger because the U.S. used INTELSAT solely for commercial purposes.

In the past the Soviets had used the INTELSAT system on a limited basis and never exchanged technical information with the West nor allowed the Eastern Bloc countries access to the system (with the exception of Yugoslavia, which was a full member of the consortium). INTELSAT, however, did not encourage its members to use the Intersputnik system. (*W Post*, Mar 13/85, A1)

August 27: The Soviet Union signed a memorandum of understanding with the International Telecommunications Satellite Organization (INTELSAT) to formalize its relationship with the 109-member nation global satellite consortium, the *Washington Post* reported. However, neither the USSR nor INTELSAT indicated that the understanding would lead to full-fledged membership for the nation that was the largest nonmember user of the satellite network.

The understanding, which came after seven years of negotiations, laid the groundwork for increased use of INTELSAT's network for global transmission of Soviet voice, data, and TV transmissions. If the Soviets increasingly used INTELSAT, experts said, it would be to expand their broadcasting system to reach Third World countries.

In 1984 the USSR broadcast 441 hours of TV programming over INTELSAT's satellites, representing 1% of the consortium's total TV traffic. By contrast, the U.S. used INTELSAT TV transmission capacity over 6,884 hours, 14% of the organization's TV traffic.

INTELSAT had to accept the membership of any nation that belonged to the International Telecommunications Union, which included the Soviet Union. INTELSAT denied using the Soviet Union as a means to deflect competition the organization was facing from entrepreneurs seeking to launch their own satellite systems. (*W Post,* Aug 28/85, G1)

Satellites

March 18: NASA announced that it would launch Intelsat V-A (F-10), first in a series of improved INTELSAT commercial communications satellites, by an Atlas-Centaur (AC-63) from KSC no earlier than March 19, 1985. The Intelsat V-A series had a capacity of 13,500 two-way voice circuits and two TV channels.

Aerospace manufacturers around the world, under the direction of prime contractor Ford Aerospace and Communications Corp., had contributed to the design, development, and manufacture of Intelsat V-A. These contractors and their responsibilities were: Aerospatiale (France)—designed the main member for the spacecraft's modular construction and supplied the main body structure thermal analysis and control; GEC-Marconi (United Kingdom)—produced the 11-GHz beacon transmitters used for earth station antenna tracking; Messerschmitt-Bolkow-Blohm (Federal Republic of Germany)—designed and produced the satellite control subsystem and the solar array; Mitsubishi Electric Corp. (Japan)—contributed the six-GHz and four-GHz earth coverage antennas and manufactured the power control electronics and the telemetry and command digital units; Selenia (Italy)—designed and built the six telemetry, command, and ranging antennas, two 11-GHz beacon antennas, and two 14/11-GHz spot-beam antennas and built the command receiver and telemetry transmitter that combined to form a ranging transponder for determination of spacecraft position in transfer orbit; and Thomson-CSF (France)—built the 10-w, 11-GHz traveling wave tubes.

The Intelsat V-A spacecraft would weigh about 4,402 lb. at separation from the Centaur, including the solid-propellant apogee kick motor (AKM) for circularization in the geosynchronous orbit. The separated spacecraft weight of 4,389 would include 1,963 lb. of AKM expendables and nine lb. of transfer orbit propellants.

NASA would use spin-stabilization during the transfer orbit coast to geosynchronous altitude. After burnout of the AKM, NASA would despin the spacecraft and deploy the antenna and solar array. In this configuration the spacecraft would be about 51 feet wide (measured across the solar panels) and 22 feet high. In orbital operation, the spacecraft would be three-axis stabilized with the body-fixed antenna pointing constantly at the earth and the solar array rotated to point at the sun.

The INTELSAT global satellite system comprised two essential elements: the space segment, consisting of satellites owned by Intelsat, and the ground segment, consisting of earth stations owned by telecommunications entities in the countries in which they were located. The space segment had 16 satellites in synchronous orbit at an altitude of about 25,780 km (22,240 miles). There were 424 communications antennas at 334 earth station sites in 134 countries and territories in the ground segment. The combined system of satellites and ground stations provided more than 800 earth station-to-earth station communications pathways. (NASA MOR M–491–203–85–08 [prelaunch] Mar 18/85)

March 22: NASA announced it had launched on March 22 at 6:58 p.m. EST the *Intelsat V-A (F–10)* by the Atlas/Centaur 63 from Cape Canaveral. Eighth in a series of 10 Intelsat V-type international telecommunications satellites, the spacecraft would undergo several weeks of on-orbit testing before positioning in geosynchronous orbit.

The AC–63, second of the new stretched version of the Atlas/Centaur for which Lewis Research Center had development and operation management responsibility, had an 80-in. extended first stage enabling it to hold more propellants. (NASA MOR M–491–203–85–08 [postlaunch] Apr 29/85; LeRC *News*, Apr 8/85, 2)

June 29: NASA announced today the launch by Atlas/Centaur 64 from KSC of the *Intelsat V-A (F–11)*, second in a series of improved International Telecommunications Satellite Organization (INTELSAT) commercial communications satellites launched by that vehicle. *Intelsat V-A* had a capacity of 13,500 voice circuits compared with *Intelsat V* that had 12,000; *Intelsat IV-A*, 6,000; and *Intelsat IV*, 4,000. All satellites had two TV channels.

Figures collected as a result of the INTELSAT-sponsored global telecommunications traffic conference indicated that an Intelsat IV–A satellite would have insufficient capacity by the early 1980s to cope with the traffic and load on the Atlantic Ocean primary satellite and on the Indian Ocean satellite. Although one solution could have been to orbit another Intelsat IV–A Atlantic Ocean and Indian Ocean satellite, subsequent planning proceeded toward the development of a high-capacity Intelsat V satellite. After an international bidding process, the INTELSAT Board of Governors at its September 1976 meeting awarded a contract for development and manufacture of seven Intelsat V satellites to Ford Aerospace and Communication Corp. as prime contractor and an international team of manufacturers as subcontractors.

Since that time, the board decided to order two additional Intelsat V satellites and to order six higher-capacity Intelsat V-A spacecraft for launch in 1985 and beyond.

Members of the international manufacturing team included Aerospatiale (France), GEC-Marconi (United Kingdom), Messerschmitt-Bolkow-Blohm (Federal Republic of Germany), Mitsubishi Electric Corp. (Japan), Selenia

(Italy), and Thomson-CSF (France). (NASA MOR M–491–203–85–09 [prelaunch] May 27/85, [postlaunch] July 17/85)

September 28: NASA announced that it had launched *INTELSAT V–A* (F–12), 10th in a series of 11 Intelsat V-type satellites to be launched in the 1980 to 1985 period for the International Telecommunications Satellite Organization (INTELSAT), from its facilities at Cape Canaveral Air Force Station aboard an Atlas/Centaur launch vehicle.

INTELSAT assigned the satellite to the Indian Ocean region to replace *Intelsat V* (F–1), which INTELSAT ordered moved to the Pacific Ocean Region where, in combination with *Intelsat V* (F–8), it would replace Intelsat IVA satellites that were running out of stationkeeping propellants.

INTELSAT awarded Ford Aerospace as prime contractor and an international team of manufacturers as subcontractors a contract for development and manufacture of Intelsat V–A satellites. INTELSAT was also considering a number of follow-on satellites with modified and expanded communications capabilities. (NASA Release 85–134; NASA MOR M–491–203–85–10 [prelaunch] Sept 19/85, [postlaunch] Oct 10/85.)

JAPAN

January 12: Japan's first deep-space probe had escaped earth's gravity on its way to a rendezvous with Halley's Comet early in 1986, the *Washington Post* reported. A domestically developed rocket had launched *Sakigake* (Pioneer) January 8 from the Institute of Space and Astronautical Science's launch site at Uchinoura in southwestern Japan.

Dr. Tamiya Nomura, director of the institute's office of project coordination, said the probe was 580,000 miles from earth the morning of January 12 and was "entering a sphere where the sun's gravity is dominant." (*W Post*, Jan 12/85, A17)

August 7: Japan nominated three astronaut candidates, one a woman, to fly in January 1988 aboard the Space Shuttle Columbia, a UPI bulletin in NASA *Current News* said. The one selected to fly would become the first Japanese in space.

Those selected were Takao Doi, a researcher at a NASA center; Mamoru Mori, assistant professor of nucleonics at Hokkaido University; and Chiaki Naito, an assistant at Keio University's Medical School in Tokyo. Japan's National Space Development Agency selected them from among 533 applicants.

The Japanese astronaut would conduct a 12-minute experiment during Columbia's one-week mission. The other two candidates would serve as backups.

The three would take further medical and space simulation tests at Johnson Space Center and then undergo training before selection in May 1987 of the astronaut to fly on the mission. (UPI bulletin in NASA *Current News*, Aug 7/85)

Launch Vehicles

March 13: Japan's highest space policy board, the Space Activities Commission, adopted in a meeting of its four commission members a plan to begin in FY 85 development of a rocket booster capable of putting a two-ton geostationary satellite into orbit, FBIS, KYODO in English reported. The new booster, code named H–II, a two-stage, 240-ton liquid-fueled rocket, 46-m high and 4-m in diameter, would dwarf the country's biggest rocket, N–II, whose capacity was up to 350 kg. Officials expected the project to cost 200 billion yen and take six to seven years to complete.

The commission also agreed to start work on a science satellite and geostationary weather satellite, both intended for 1989 launch, and authorized the design, beginning in FY 85, of a space module that Japan would construct as part of an international space station program planned for early in the 1990s. (KYODO in English, Mar 13/85)

MATERIALS PROCESSING

February 15: NASA announced it had signed a memorandum of understanding for its materials processing in space research program with Grumman Corp. to experiment with directional solidification of gallium arsenide and other semiconductor materials and various metals and alloys. The process would apply to production of semiconductor crystals and magnets for electrical motors, with initial emphasis on semiconductor materials.

Directional solidification would use precisely controlled temperatures to melt and solidify a material, during which the material's crystalline structure would align in a fashion that should virtually eliminate any compound imperfections. Flawless semiconductor crystals would yield a greater quantity and quality of microcircuit chips, leading to higher-speed electronic devices that would consume less power and to greater miniaturization.

The agreement called for an information exchange, with NASA and Grumman Corp. designating personnel for the program. NASA would put up no funds for the program; Grumman expected to spend $6 million in the first three years. (NASA Release 85–23; *W Times*, Feb 13/85, 8B)

NATIONAL AERONAUTICS AND SPACE ADMINISTRATION

During October: The National Aeronautics and Space Administration (NASA) announced that its university programs, primarily involving undergraduate students, combined futuristic space concepts with realistic engineering design challenges. The first program, cosponsored by NASA Headquarters and Ames Research Center (ARC) and Johnson Space Center (JSC), required students to design a spacesuit glove that would allow astronauts maximum flexibility at eight lb.-per-sq-in., the pressure planned for future extravehicular activity. Ten universities responded to a request for proposals, and a team from Kansas State University won the design competition.

The second broader program, entering its second phase of a two-year pilot effort, required universities to adopt NASA advanced space design projects for senior design classes. Each university received a grant and was aligned with a NASA center that provided guidance, data, and lecturers during the academic year and 10-week summer work assignments for three students.

The initial group of centers, schools, and projects were: ARC/University of Wisconsin/manned Mars habitat; ARC/University of Colorado/geosynchronous space station; Langley Research Center; (LaRC)/Virginia Polytechnic Institute/orbital servicing center; LaRC/Massachusetts Institute of Technology/lunar base-manned Mars mission; Marshall Space Flight Center (MSFC)/Georgia Institute of Technology/lunar site preparation; Lewis Research Center (LeRC)/University of Washington/space manufacturing facility; LeRC/University of Michigan/lunar space transportation system; and JSC/University of Texas and Texas A & M/manned Mars mission. (*NASA Activities*, Oct 85, 10)

December 26: NASA announced the highlights of its 1985 activities, which included nine Space Shuttle flights and launches of three Atlas Centaurs, two Scouts, and an Aerobee vehicle, its last flight. In 1985 NASA introduced the last and lightest-weight Space Shuttle orbiter—Atlantis. McDonnell Douglas inaugurated its upper stage booster, the Payload Assist Module D–2.

Fifty two individuals flew aboard Space Shuttles in 1985. Of this group 19 were scientists performing observations, experiments, or investigations; 27 were U.S. military officers, and six were payload specialists from foreign countries (Mexico, Saudi Arabia, France, the Netherlands, and two from West Germany). The 52 individuals' combined time in space was nearly 55 days or over 880 earth orbits, for a combined travel distance of nearly 22 million miles. Six Space Shuttle astronauts spent a total of more than a day in extravehicular activities during the year.

Two Spacelab missions aboard the Space Shuttle and an encounter by the International Cometary Explorer (ICE) with Comet Giacobini-Zinner highlighted the year for NASA's Office of Space Science and Applications.

On April 19, 1985, a little more than a year after President Reagan directed NASA to develop within a decade a permanently manned space station, NASA awarded competitive contracts to eight industry teams for definition and preliminary design (Phase B) of elements of the station. NASA signed a memorandum of understanding with Canada, the European Space Agency, and Japan for cooperation during the definition and preliminary design phase of the program.

NASA's aeronautical research and technology efforts included continuing work on such programs as the NASA/Department of Defense X–29 X-wing research aircraft, the tilt rotor/JVX aircraft, the advanced X–29 aircraft featuring a forward-swept wing, and the mission adaptive wing that could change its curvature.

In the area of space technology, Langley Research Center completed development, assembly, and testing of a 15-meter hoop column deployable structure and antenna system. Other advances in space technology included significant modification to the Space Shuttle orbiter Columbia to measure dynamic and thermodynamic characteristics, selection of the reactor thermoelectric power system concept for design and testing for the space station space reactor power program, and completion of space construction experiments outside the Space Shuttle orbiter.

NASA's advances in space tracking and data systems included transition from ground network station support for low earth-orbiting spacecraft to use of the first Tracking and Data Relay Satellite in geosynchronous orbit. NASA had scheduled TDRS–2 and 3 for launch in January and July 1986, respectively. In February NASA completed consolidation of the Deep Space Network (DSN) and later an upgrade of the network that would enable the DSN to receive significantly more images from Voyager as it encountered Uranus in January 1986.

NASA's Office of Commercial Programs funded and opened the first five of a planned series of centers for commercial development of space, which NASA expected would become self-supporting through research collaboration by industry, academia, and government agencies. (NASA Release 85–177)

Budget

February 4: NASA Administrator James Beggs, during a press conference, reported that President Reagan had requested a NASA budget of just under $7.9 billion, reflecting the President's determination to continue America's space leadership and to achieve the goal of a permanently manned space station.

The budget covered some unforeseen items not addressed by 1985 budget-planning estimates, including the Congressionally mandated development of the Advanced Communications Technology Satellite (ACTS), scheduled for launch in 1989, and acceleration of the advanced turboprop propulsion system for FY 87 flight testing.

The budget contained four major appropriations requests: a total of $2.9 billion for R&D, which included funds for previously approved space science and applications programs, development of ACTS, initiation of the orbiting maneuvering vehicle program, and promotion of commercial use of space; $3.5 billion for space flight, control, and data communications to support Space Shuttle production, operations, and tracking and data acquisition (a decrease of $92 million from FY 85's budget plan); $149 million for facilities construction (down $1 million from FY 85); and $1.3 billion for research and program management. (NASA release Feb. 4/85; NASA press briefing, Feb. 4/85)

February 6: President Reagan's FY 86 budget for NASA sought $7.9 billion—just under what it would take to sustain the agency's programs at current levels, the *Washington Post* reported. At a press conference [see NASA/Budget, Feb. 4], NASA Administrator James Beggs had acknowledged the $230-million budget request for the space station was $50 million less than the agency had sought to keep the project on schedule. (*W Post*, Feb 6/85, A17)

February 28: NASA Administrator James Beggs, in testimony before the Senate science, technology, and space subcommittee, said that projected cost overruns for the Centaur G and G Prime vehicles-development programs could run as high as $50 to $60 million, but that the agency did not think the problems would affect Galileo and Ulysses mission launches planned for the 2nd half of 1986, *Aerospace Daily* reported. Another NASA official testifying earlier had estimated the overrun at $30 to $40 million.

The discrepancy in estimates apparently stemmed from an agreement between NASA and the Air Force calling for a 50–50 sharing of Centaur G program-design and -development costs. The Air Force, however, had set a $150 million ceiling on basic Centaur-development costs, which had not yet been reached. If the overrun surpassed the ceiling, NASA would have to absorb the costs. (*A/D*, Feb 28/85, 1)

March 26: The U.S. House Science and Technology transportation aviation and materials subcommittee completed its markup of NASA's FY 86 authorization, approving a series of changes that resulted in reallocations for several aeronautics programs and elimination of proposed funding for oblique-wing technology, *Aerospace Daily* reported.

NASA requested $522 million for aeronautics and space technology in its FY 86 budget, of which $354 million was for aeronautical research and technology and $168 million for space research and technology. Under the proposed changes, NASA rotorcraft systems technology funding would increase by $4 million over the agency's request for $20.4 million. Hot section engine technology, for which NASA had requested $5.2 million, would increase by $2 million; and the subcommittee restored high-speed aeronautics funding.

The panel approved a $1.4 million decrease for the fluid and thermal physics research and technology program and deletion of the oblique-wing technology, for which NASA had requested $4.7 million, and of funding for altitude wind-tunnel planning.

The panel based its decision to delete oblique-wing technology funding on an advisory committee finding that the planned flight portion of the NASA/Navy oblique-wing program was one whose design solutions would have low payoff to industry. Since the Navy was the only identifiable user, Chairman Dan Glickman (D-Kans.) said, if it was a program worth their while, the service should provide the funding. (A/D, Mar 27/85, 1)

April 1: The U.S. House Science and Technology Committee approved President Reagan's request for a $7.9 billion NASA budget for FY 86, the *Washington Post* reported, but allocated the funds differently than he had requested. By a 32-to-9 vote, the panel decided to retain a subcommittee's recommendation to shift $45 million to spare parts procurement for the Space Shuttle in order to keep the Space Shuttle assembly line open and thus increase chances for a fifth orbiter. The administration requested $2.1 billion in FY 86 for 14 Space Shuttle flights and delivery of the fourth orbiter, Atlantis; it sought no funds for a fifth orbiter. (*W Post*, Apr 1/85, A9)

May 23: NASA Administrator James Beggs told Congress that the new version of the Centaur rocket designed for Space Shuttle-launched planetary missions would cost about $110 million more ($90 million for NASA; $20 million for the Air Force) than expected [see NASA/Budget, Feb. 28], the *Washington Post* reported.

Wider and shorter than the old-model Centaur to fit in the Space Shuttle's cargo bay, the new rocket would launch in 1986 two spacecraft toward Jupiter. An even shorter version starting in 1987 would launch from the Space Shuttle two classified missions for the Air Force.

Beggs told the House Science and Technology subcommittee on space science and applications that "We underestimated the job of integrating the Centaur into the shuttle and General Dynamics underestimated the cost of changing the configuration of the rocket."

The new estimates raised Centaur program costs from $755 million to $865 million. (*W Post*, May 24/85, A16)

NASA

June 27: The U.S. Senate passed by voice vote H.R. 1714, the NASA Authorization Bill for FY 86, NASA's *Legislative Activities Report* noted. As passed, the bill embraced the concept of a budget freeze at the FY 85 level of $7.510 billion; however, it also provided for three specific augmentations that totalled $142 million, accounting for a FY 86 growth of 1.9%.

Among the authorizations for research and development were $200,000,000 for the space station, $477,200,000 for space transportation capability development, and $608,400,000 for physics and astronomy.

Spaceflight, control, and data communications authorizations included $941,500,000 for Space Shuttle production and operational capability; $1,700,100,000 for space transportation operations; and $745,300,000 for space and ground network, communications, and data systems.

Among the authorizations for space transportation facilities were $14,000,000 for construction of an orbiter modification and refurbishment facility at Kennedy Space Center (KSC), $3,600,000 for construction of a thermal protection system facility at KSC, and $6,500,000 for modification of enhanced life support systems testing at Marshall Space Flight Center. (*NASA Legislative Activities Report,* June 27/85, 1)

July 1: By voice vote the U.S. Senate approved a FY 86 NASA authorization of $7.6 billion, an increase of $142 million above the 1985 NASA appropriation and the House-approved authorization, the *Washington Post* reported. The increase would mainly cover the cost of restoring the Reagan Administration's proposed 5% pay cut for federal employees and provide additional funds for booster rockets. The Senate spending ceiling was $234 million below the Administration's request.

The Senate and House authorization bills were substantially different. The Senate approved cuts in specific programs, while the House simply approved an across-the-board freeze for the entire NASA budget. (*W Post,* July 1/85, A13)

Data Tracking and Relay Systems

January 23: NASA announced that the Goddard Space Flight Center (GSFC) would transfer spacecraft tracking and data acquisition operations from its Beltsville Tracking Station to its Wallops Flight Facility by early 1986, consolidating the operations to support balloon, sounding-rocket, and aeronautical-flight research. Diminishing need for ground stations due to the new Tracking and Data Relay Satellite System (TDRSS) necessitated the realignment.

TDRSS would consist of three communications satellites in geosynchronous orbit, providing global coverage by earth-orbiting satellites and replacing the worldwide network of ground stations. NASA had launched the first

TDRSS satellite in April 1983 and would launch the second about February 20. GSFC would phase out or transfer to other agencies most of its tracking stations. In addition to Wallops, however, GSFC would maintain ground stations at Bermuda and Merritt Island, Florida, to support Space Shuttle launches from KSC. The White Sands, New Mexico, station was the ground terminal for orbiting TDRSS satellities (NASA Release 85-11)

February 26: NASA announced it was donating a 26-m antenna located at the Orroral Valley Tracking Station, Australia, which had ceased operations in December 1984, to the Australian University of Tasmania. NASA, which had used the antenna in programs such as Skylab, the Apollo Soyuz test project, and the Space Shuttle, had offered to assist in antenna dismantling and transfer to Hobart, Tasmania.

The University of Tasmania's physics department, one of Australia's major centers for astronomy and astrophysics, would use the antenna as part of its teaching and research activities, including operation with the Australian telescope under construction in New South Wales to improve telescope performance.

Researchers would also use the antenna in conjunction with other instruments for very-long baseline interferometry, a system employing a number of separate antennas to construct a radio telescope with a high-resolution capability, to obtain more accurate measurements of earth's surface. Researchers would also make the antenna available to NASA for its future geodesy, geodynamics, and astronomy projects. (NASA Release 85-28)

May 17: NASA announced that Harris Corp. successfully deployed on the ground a 50-foot antenna system, marking a milestone in NASA's program to demonstrate that large space antenna concepts were feasible. In the test, a hoop-column antenna unfolded, umbrella-style, from a compact package to a combination of thin structural members, quartz filament cords, and gold-plated mesh.

The mesh, serving as a precision reflecting surface stretching across the diameter of the supporting hoop, was shaped like a dish but could be made flat, spherical, or conical, depending on the intended application. The antenna column was a precise telescoping hub, forming the central structure of the antenna, tensing the cords that shaped the antenna surface, and housing the electronic feed mechanisms.

The size of potential large space antennas meant a significant boost in effective radiated power from space and an increased sensitivity to weak signals from the ground or from space. One potential application was in communications, because at that time each earth station had to have a large antenna to receive the weak signals transmitted through small antennas on satellites. Large antennas in space would greatly reduce the size and cost of the antennas required at ground sites. And a few super-antennas placed in high geosynchronous orbit could cover the globe, instead of the great number of smaller satellites otherwise required. Millions of inexpensive home rooftop or land mobile unit antennas could receive satellite signals then picked up only by a few very large ground stations.

NASA believed the 50-foot antenna system was the largest precision antenna designed for space that could be accommodated in existing ground electromagnetic test facilities. However, the ultimate deployable space antennas might have 150- to 300-foot diameters. Studies showed that these larger antennas required space assembly. (NASA Release 85-76)

June 6: NASA announced that its deep space network (DSN), as part of a French-led international tracking network, would track the first of two international balloon experiments carried aboard the Soviet VEGA spacecraft to study beginning June 10 Venus's atmosphere. The VEGA 1 and 2 spacecraft each would drop an instrumented lander and an instrument-laden balloon into the Venusian atmosphere as they approached the planet on their way to a March 1986 rendezvous with Halley's Comet.

After reaching the equatorial regions of Venus's atmosphere, the balloons would float free in the middle, most active, layer of Venus's three-tiered clouds. The flight plan called for the entry packages, consisting of the atmospheric balloon and a lander, to plunge from a 125-km (78 miles) to a 65-km (40 miles) altitude, where a parachute would deploy. The lander would separate from the balloon at 63 km (39 miles) and head for the surface. The balloon would then inflate and carry its instrument package through the atmosphere at an altitude of 55 km (34 miles) for more than two days.

NASA's deep space network and other stations around the world would use a technique called very long baseline interferometry to measure the balloon's velocity, and therefore the wind velocity, with a precision of about 3 km (2 miles) per hour at Venus's about 108 million-km (67 million miles) distance from earth. A Soviet internal network would also track the balloons.

NASA's Jet Propulsion Laboratory was cooperating with the French National Space Agency (CNES), Paris, in the tracking activity. Scientists believed data from the balloons would further their understanding of Venus's complex weather system. (NASA Release 85-87)

Management and Personnel

January 17: NASA announced appointment effective January 13, 1985, of Frank Penaranda, a NASA exceptional service medal recipient serving as special assistant to the associate administrator in the Office of Aeronautics and Space Technology (OART), as deputy assistant administrator for commercial programs, where he would develop and coordinate policies and procedures for implementing commercial programs through NASA's in-house organizations.

Penaranda had joined NASA in April 1969 as a technical program analyst with OART; NASA later named him director of the resources and management systems division in 1974 and director of institutional operations in the Office of Management Operations in Jan. 1978. Penaranda had previously worked for the Defense Atomic Support Agency of the Department of Defense as a nuclear physicist conducting applied research on nuclear radiation transport and shielding.

Penaranda holds an M.S. in physics from Marquette University and a B.S. from Manhattan College. He also graduated from Harvard's Advanced Management Program. (NASA announcement, Jan. 17/85)

April 5: NASA announced that effective today it permanently established the office of NASA productivity programs to direct, initiate, coordinate, monitor, and evaluate agencywide productivity improvement and quality enhancement initiatives. David Braunstein would continue as the office's director and would report to the NASA administrator.

The office was established to ensure NASA's leadership in the development and application of advanced technology and management practices that contributed to significant increases in agency and national productivity.

In making the announcement, NASA Administrator James Beggs said that NASA would provide a participative and challenging environment for all employees, and it would develop a team approach with its contractors to achieve the highest levels of productivity. (NASA announcement, May 15/85)

April 12: NASA announced it honored at a KSC reception 200 NASA and NASA industrial Quality Circle employees for their volunteer efforts in improving quality, safety, and productivity. Jesse Moore, NASA associate administrator for space flight, read a message from George Burns, president of the International Association of Quality Circles, congratulating the employees on their outstanding performance.

NASA selected the honorees under an office of space flight program that recognized first-level supervisory and support personnel for their productivity efforts. The Manned Flight Awareness Panel, made up of NASA and industry personnel, managed the program. NASA had over 120 employee teams with over 1,000 employees voluntarily participating in Quality Circles. Through the program, NASA hoped to foster increased effectiveness through improved efficiency in the space and aeronautics research and development programs. (NASA Release 85-55)

June 12: Approximately 200 NASA management officials and contractors met June 12 and 13 at Marshall Space Flight Center (MSFC) to discuss initiatives to improve quality and productivity in NASA/contractor operations, the *Marshall Star* reported. The first NASA/Hardware Contractors Productivity Conference was at MSFC in April 1984.

Attendees at the 1985 conference heard an interim report on the status of activities implemented as a result of the 1984 meeting and related recommendations. NASA and contractor panels discussed topics such as quality in relationship to productivity; productivity and quality initiatives and incentives; efforts toward implementing participative management techniques; and improvement in specification, preplanning, and measurement.

The conference was part of NASA's continuing effort to provide national leadership in the development and application of advanced technology and management practices. During fiscal year 1984, about 85% of NASA's $7.2 billion budget was placed with contractors. "It's very important," said William Reynolds, associate director for management in MSFC's science and engineering directorate, "that NASA's productivity effort include an improved relationship with the NASA hardware and service/support contractors." (*Marshall Star*, June 12/85, 1)

June 17: NASA announced it appointed Dr. Raymond Colladay associate administrator for aeronautics and space technology to be responsible for the overall management of the agency's aeronautics, space technology, and terrestrial energy programs including the institutional management of Ames Research Center (ARC), Langley Research Center (LaRC), and Lewis Research Center (LeRC). On April 18, 1982, NASA had named Colladay deputy associate administrator for the office of aeronautics and space technology; since April 17, 1985, he had served as acting associate administrator for that office.

Colladay began his career with LeRC in 1969, where he conducted analytical and experimental research in advanced high-temperature gas turbine engines. In 1974 NASA named him head of the turbomachinery fundamentals section and then temporarily assigned him to NASA Headquarters as acting chief of the propulsion branch in the aircraft energy efficiency office. Later he served as assistant manager of the LeRC efficiency engine project, deputy manager of the advanced propulsion systems office, manager for propulsion research and technology, and beginning in 1982 director of the research and technology division.

Colladay received B.S., M.S., and Ph.D. degrees from Michigan State University, East Lansing. He authored or co-authored more than 20 NASA technical reports and journal articles relating to aeronautical and space research. (NASA Release 85-93)

July 11: NASA announced that S. Neil Hosenball, NASA general counsel since 1975, would retire on August 2 to become director of the University of

Colorado's new Center For Space Law and Policy. Before appointment to his current position, Hosenball served as deputy general counsel since October 1967. Earlier he had been assistant general counsel for procurement and had served for four years at Lewis Research Center.

From 1970 to 1979, Hosenball was a member of the U.S. delegation to the United Nations Committee on the Peaceful Uses of Outer Space and headed the U.S. delegation at legal subcommittee and committee sessions.

Hosenball, who received a B.S. degree from the University of Michigan and his LL.B. degree from Harvard Law School, was awarded the NASA Exceptional Service Medal in 1967, the NASA Distinguished Service Medal in 1973, the National Civil Service League Career Service Medal in 1980, and the Presidential Rank of Distinguished Executive in 1983.

The Center for Space and Law Policy, established in November 1984, was one of a very few such institutions in the world and was intended to guide the university in its space research efforts. (NASA Release 85–104)

July 23: NASA announced that John O'Brien, currently NASA deputy general counsel, was appointed effective August 4 NASA general counsel, succeeding S. Neil Hosenball who was retiring.

O'Brien began his career in 1962 with NASA at the Launch Operations Center, later Kennedy Space Center (KSC), during the Mercury Program. He then served as chief counsel of the KSC and assistant general counsel for procurement matters at NASA Headquarters.

He received an A.B. degree from Niagara University and his J.D. degree from Georgetown University. After joining NASA, O'Brien was designated a Princeton Fellow in Public Affairs at the Woodrow Wilson School of Public and International Affairs, Princeton University, and in 1976 received the NASA Exceptional Service Medal. (NASA Release 85–108)

July 23: President Reagan nominated Anthony Calio, who served at NASA for 16 years, to be administrator of the National Oceanic and Atmospheric Administration (NOAA), the *Washington Post* reported. Pending confirmation of the nomination by the U.S. Senate, Calio, who was deputy administrator of NOAA, would replace John Bryne. (*W Post*, July 23/85, A13)

September 12: President Reagan announced today his intention to nominate William Graham to be NASA deputy administrator to succeed Hans Michael Mark, the *Administration of Ronald Reagan* reported.

Graham, with R&D Associates since 1971, had served there as director of computing operations, division manager, corporate program manager, and most recently senior associate. Before joining R&D Associates, he was a member of the technical staff, physics department, at the Rand Corp.; project officer of the Air Force Weapons Laboratory; and a member of the technical staff of the Hughes Aircraft Corp. research laboratory.

Since 1982, Graham had served as acting chairman of the President's General Advisory Committee on Arms Control and Disarmament. He had also been a member of the Defense Nuclear Agency Scientific Advisory Group on Effects, a consultant to the Defense Nuclear Agency, and consultant to the Office of the Secretary of Defense.

Graham received his B.S. degree from the California Institute of Technology and M.S. and Ph.D. degrees from Stanford University. (*Admin. of Ronald Reagan*, Sept 13/85, 1073)

September 20: NASA announced appointment effective October 7, of Richard Barnes to be director of the International Affairs Division. Barnes would replace Kenneth Pedersen, who had served in the job since November 1978 and would spend the following year on sabbatical as research professor at the Georgetown University School of Foreign Service.

Barnes recently had completed four years as NASA European representative based at the American Embassy in Paris, where he was responsible for liaison with the European Space Agency and the national space agencies of Western Europe on cooperative projects and for identification of potential future joint space projects. In 1961 he joined the NASA Office of International Programs where he served in various capacities, including deputy director of international affairs, before his Paris assignment.

Before joining NASA, Barnes was affiliated with the Atomic Industrial Forum, Inc. and served with the Atomic Energy Commission's Division of International Affairs and the Bureau of Ordinance of the Navy Department.

Barnes received his B.A. degree from Dartmouth College and a master of public administration degree from Harvard University. He also graduated from the Industrial College of the Armed Forces and served on commissioned active duty with the U.S. Navy during the Korean War. (NASA Release 85-132)

October 13: NASA announced that, effective October 13, Carroll Dicus, Jr. was appointed chairman of the NASA Board of Contract Appeals and chairman of the Inventions and Contributions Board, succeeding Frederick Lees who retired in September.

From 1968 to 1974 Dicus was an attorney in the chief counsel's office at Goddard Space Flight Center and from March 1974 to October 1980 an administrative judge on the NASA Board of Contract Appeals. He then went to the U.S. Postal Service where he was an associate judicial officer and vice chairman of the Board of Contract Appeals.

Dicus is a graduate of Johns Hopkins University and the University of Baltimore Law School. (NASA anno., Oct 85)

October 15: NASA announced that Edward Frankle was appointed effective October 27 deputy general counsel succeeding John O'Brien.

Astronautics and Aeronautics, 1985

Frankle had been chief counsel of NASA's Goddard Space Flight Center since September 1982 and prior to that associate director, policy development and administrative legal systems, for the Selective Service System, where he was responsible for the development and promulgation of operating regulations for deferment and classification and the selection and training of local and appeal board members across the U.S. He served from 1974 to 1980 as a member of the Office of General Counsel, Department of the Navy.

Frankle received B.S. and M.S. degrees in aerospace engineering from the Catholic University of America and a J.D. degree from Georgetown University School of Law. (NASA anno., Oct 15/85)

October 18: NASA deputy associate administrator for space science and applications Samuel Keller, speaking at a weekly staff meeting, said "We expect 1986 will be the most demanding year we've ever had," the *Washington Post* reported. In January Voyager would encounter the planet Uranus. In March the Space Shuttle would fly a mission dedicated to observing Halley's comet. In May NASA would launch Ulysses, which would fly to Jupiter and use the planet's gravity to "slingshot" itself around the sun, and Galileo, which would orbit Jupiter. In August the Space Shuttle would carry the Hubble Space Telescope into orbit. The three spacecraft cost $2 billion not including launch costs.

Keller's toughest time would occur in May, the *Post* said, when Ulysses and Galileo at the same time would be in the cargo bays of two Space Shuttles on their launch pads. NASA would try May 15 to launch Ulysses and then Galileo four days later. If anything delayed the May 15 launch, NASA had 24 days to get the Space Shuttle into space, a launch window that allowed the two spacecraft to fly to Jupiter using the least amount of fuel.

"I'll be glad when 1986 is over," Keller said. (*W Post*, Oct 18/85, A21)

October 24: NASA announced that an October 29 conference in Washington sponsored by NASA, the Department of Defense, and other government agencies would provide a forum for government executives to discuss the Reagan Administration's program to increase government efficiency and effectiveness.

Following welcoming remarks by Dennis Whitfield, chief of staff to the Secretary of Labor, and David Braunstein, director of NASA's productivity programs, conference participants would be briefed on several perspectives on the President's productivity improvement program. Stephen Scholossberg, deputy under secretary for labor-management relations, Department of Labor, would then provide a labor-management view of quality and productivity.

In the afternoon, NASA astronaut Bruce McCandless, first human to freely maneuver in space without a tether, would speak on the importance of quality and productivity in the space program. John Franke Jr., assistant secretary for administration, Department of Agriculture, would then give case examples of management's commitment to quality; and Commodore John Kirkpatrick, commander, Naval Aviation Logistics Center, would discuss quality improvement through total quality management. (NASA Release 85–146)

November 25: NASA announced that the U.S. Senate confirmed on November 18 Dr. William Graham to be NASA deputy administrator; Graham, who was a founder and executive of R&D Associates, was nominated by President Reagan on September 12.

Graham had served for three years as chairman of the President's General Advisory Committee on Arms Control and Disarmament and previously served as a member of the President-elect's transition team.

Before founding R&D Associates in 1971, Graham spent six years with the Rand Corp. and before that three years active duty at the Air Force Weapons Laboratory, Kirtland Air Force Base, as a project officer directing a group conducting experimental and theoretical research on strategic system survivability.

Graham had been a consultant to the Office of the Secretary of Defense and served on many international and national boards and advisory groups including the National Academy of Science/National Research Council Committee on Undersea Warfare, the Air Force Science Advisory Board Task Force on Manned Strategic System Vulnerability, the U.S.-United Kingdom Joint Working Group on Atomic Weapons, the Defense Nuclear Agency Scientific Advisory Group on Effects, and the Defense Science Board System Vulnerability Task Force and Associated Task Forces.

His memberships in professional and honorary organizations included Tau Beta Pi, the American Institute for Aeronautics and Astronautics, the New York Academy of Science, the Council on Foreign Relations, the board of directors of the Committee on the Present Danger, the Defense Preparedness Association, the Ethics and Public Policy Center, and the American Association for the Advancement of Science.

In 1959 Graham received a B.S. degree in physics from the California Institute of Technology and in 1961 an M.S. degree in engineering science and in 1963 a Ph.D. in electrical engineering, both from Stanford University. (NASA Release 85–155)

December 2: NASA announced the appointment effective November 25 of Joseph Alexander to deputy chief scientist with responsibility for providing assistance to the chief scientist in advising the Administrator and in establishing policy related to scientific aspects of NASA programs and missions.

Alexander joined Goddard Space Flight Center (GSFC) in 1962, where he participated in the establishment of NASA's program in space radio astronomy and conducted studies of the sun, the planets, and the galaxy from earth-orbit, lunar-orbit, and planetary flyby spacecraft. In 1970 he became head of the Galactic Studies Section where he was responsible for scientific studies associated with the Radio Astronomy Explorer satellite program. Following leave as a visiting scientist at the Department of Astro-Geophysics at the University of Colorado, Alexander became head of the Planetary Magnetospheres Branch and directed a research team conducting both experimental and theoretical studies of planetary environments by using instruments on board spacecraft such as the Interplanetary Monitoring Probes, Mariner-10, Magsat, Pioneer-11, and Voyager 1 and 2.

From January 1984 until March 1985, Alexander was a senior policy analyst at the White House Office of Science and Technology Policy where he concentrated on issues related to space science and technology in the civil space program. He then returned to GSFC as associate chief of the laboratory for extraterrestrial physics.

A member of the American Geophysical Union, the U.S. National Committee of the International Union of Radio Science, and the International Astronomical Union, Alexander received in 1960 a B.S. degree and in 1962 an M.A. degree, both in physics, from the College of William and Mary. (NASA anno., Dec 2/85)

December 2: A Federal grand jury today indicted General Dynamics Corp., three of its current executives, and its former executive vice president, James Beggs, currently Administrator of NASA, for allegedly seeking to defraud the Defense Department in connection with producing two prototypes of the DIVAD, an anti-aircraft gun for the Army, *Time* magazine reported.

The seven-count, 33-page indictment stated that between January 1978 and August 1981, while General Dynamics was working on a $41 million program to build the prototypes, it made false statements to the government and was guilty of fraud. The indictment charged the company with illegally billing $7.5 million of its expenses to other government accounts, of which $3.2 million was paid. Along with Beggs, the indictment also named Ralph Hawes, division general manager; David McPherson, program director; and James Hansen, assistant director.

Beggs, who took a leave of absence from NASA, denied the charge, saying, "I have not been involved in any criminal wrongdoing . . . I do not intend to leave, and this is not the first step to a resignation." A General Dynamics spokesman said, "The issue is a highly sophisticated regulatory and accounting matter, which should be resolved in a civil forum, not in a criminal case." (*Time*, Dec 16/85, 46)

The Navy anticipated a series of NROSS satellites with enough launches to keep an operating satellite in orbit unless sensor or other changes made faster launches desirable. NROSS-1's expected lifetime would be three years; an NROSS-2 would depend on needed oceanographic data. The system would help battle group commanders predict ocean conditions for antisubmarine warfare operations, force placement, and plan use of advanced weapon systems. (*A/D*, Jan 10/85, 50)

December 4: NASA Administrator James Beggs today took an indefinite leave of absence from the agency to fight fraud charges against him [see NASA/ Management and Personnel, Dec. 2], the *Washington Post* reported. With White House approval, associate administrator Philip Culbertson became general manager.

Although Beggs had refused to resign, White House spokesman Larry Speakes said that President Reagan, "while reluctantly acceding to his request for a leave of absence," asked Beggs to assist in an "orderly transition of his responsibilities to his colleagues at NASA to facilitate continuity of management at this critically important agency. Mr. Beggs has agreed to do so." Speakes added that the space program "has been revitalized" under Begg's leadership and "this important record must continue."

Putting Culbertson in charge of day-to-day operations put Begg's stamp on the transition, the *Post* said. Culbertson was a 20-year NASA veteran, well-known to the House and Senate committees that dealt with the agency, and responsible for planning the agency's next big project, construction of an $8 billion permanent space station.

William Graham, a former planning analyst at the Rand Corp. and chairman for the previous three years of the White House Advisory Commission on Arms Control and Disarmament, had been with NASA eight working days. Reagan appointed Graham acting administrator, so that Graham could create the job of general manager and name Culbertson to the post.

"Whoever is running the space agency in 1986 has to know where the space station stands, what money it needs and who to talk to to keep it on track," said a congressional aide who dealt with NASA. "That's why Phil Culbertson is getting the job of running the agency day-to-day. He knows the issues."

In his statement, Beggs reiterated that he is innocent of the fraud charges resulting from a Justice Department investigation. "I have concluded there was nothing I did then that I would not do again," Begg's statement said. "I have not been involved in any criminal wrongdoing or, in fact, of wrongdoing of any kind. I am totally confident I will be exonerated." (*W Post*, Dec 5/85, A3)

December 5: Speaking to a standing-room-only crowd at NASA Headquarters and all NASA centers via closed-circuit TV, former administrator James Beggs said today that federal fraud charges against him were "baseless" [see

NASA/Management and Personnel, Dec. 2], that he expected to be cleared, and expressed thanks for an "outpouring" of support, the *Washington Post* reported.

"These charges relate to things that happened in General Dynamics six and seven years ago," Beggs noted and said that he had reviewed the charges and believed, "we acted in an entirely ethical, legal and moral sense."

"The charges, therefore, are baseless . . . They are outrageous, ridiculous and I feel confident that once this is brought to trial that I'll be completely exonerated of the charges."

Beggs also noted that the suits against defense firms were creating a climate that would blight the work of NASA and of the defense community. "The very adversarial relationship that is being created by the suits against the contractors, the very bad kind of statements that are being made in the press, and elsewhere, is going to make our job in the future much more difficult," he said. "Not just here, but in the Defense Department as well."

Earlier, the *NY Times* reported that all top 27 administrators at NASA and field installations had sent a statement to Congress and the White House endorsing Beggs as "an individual with the highest standards of integrity which have earned him the esteem and respect of his colleagues." The statement said he had revitalized the space program and regretted "the untimely interruption of his work," calling him a man of the "highest integrity, totally dedicated to NASA, an extremely able executive." (*W Post*, Dec 6/85, A16; *NYT*, Dec 6/85, A5)

December 5: Having served as NASA deputy administrator for only two weeks, Dr. William Graham was not known well by most NASA employees; but, as acting administrator replacing James Beggs, Graham had a clear opinion of NASA and his opportunity to lead it, the *Washington Post* reported. Graham, who had a Ph.D. in electrical engineering, said in an interview today that "NASA's a marvelous organization . . . For a technologist, this is like dying and going to heaven."

Thus, Graham characterized himself as a professional technologist, which his background would certainly indicate. He had worked as a project officer with the Air Force Weapons Laboratory in New Mexico, was a founder and executive of R&D Associates in California, and had served on numerous advisory panels on nuclear weapons, strategic military policy, and undersea warfare. However, this emphasis in his career on weapons research and military policy aroused some unease at NASA, the *Post* commented. "He's from the other side of the river," one NASA official said, referring to Graham's long association with the Pentagon. But Graham had sought to assure Congressional committees recently that his appointment was not designed to lead to the "militarization of NASA."

Graham's close ties to the Reagan Administration were seen as possibly beneficial to the space agency. And Graham commented that he would follow President Reagan's "very strong and clear space policy."

Some associates noted that even as acting administrator they expected Graham to "take a strong hand" in agency affairs. Graham had already announced that he intended to meet soon with all the agency's top executives and then visit the research and operations center at Cape Canaveral, Houston, Pasadena, and elsewhere. "I'm a hands-on kind of person," he said. "I like to get out and walk the halls and talk to people." (*W Post*, Dec 6/85, A29)

December 6: NASA announced that Acting Administrator William Graham confirmed the appointment for an indefinite period of Philip Culbertson, associate administrator for space station, to the position of NASA general manager. In that capacity he would assist Graham and have specific responsibility for the Offices of Space Science and Applications, Commercial Programs, Space Flight, Aeronautics and Space Technology, Space Station, and Space Tracking and Data Systems.

Culbertson had held his current position since August 1, 1984, coinciding with the establishment of the Office of Space Station. Since November 1981 he had been associate deputy administrator, serving as senior staff advisor to the administrator and the deputy administrator and directing the formulation of policy, strategy, and planning for the space station. From 1979 to 1981 he served as assistant for the Space Transportation System (STS), providing continuous assessment of STS development, acquisition, and operations status and otherwise advising the administrator and deputy administrator on STS matters requiring policy decisions.

His earlier NASA assignments included the positions of deputy associate administrator for STS/technical, assistant administrator for planning and program integration, director of advanced manned missions, and manager of the institutional assessment conducted in 1977. He also served as the NASA representative in the 1979 antisatellite treaty negotiations and in 1976 and 1977 was detailed to the Executive Office for a five-month period as Executive Director of the President's Committee on Science and Technology.

Culbertson received a B.S. in aeronautical engineering from Georgia Institute of Technology and served as a commissioned officer in the U.S. Navy, after which he spent four years as a research associate at the University of Michigan, where he received his M.S. degree in aeronautical engineering. (NASA anno., Dec 6/85)

December 11: NASA's Thomas DeCair, associate administrator for external relations, announced today the appointment of Shirley Green as NASA's director of public affairs. She replaced Frank Johnson Jr., who was appointed assistant associate administrator for external relations (special projects). In her new position, Green would be responsible for planning and directing all NASA activities that provided information to and responded to inquiries from the public and the media.

Green came to NASA with 20 years' experience in communications and management. Since 1981, as deputy and acting press secretary to the Vice President of the United States, she was responsible for planning and coordinating media activities for the Vice President on matters of domestic policy, including the task forces on regulatory relief and drug interdiction. She had accompanied the Vice President to 61 foreign countries, coordinating all media activities.

Green was a former chairman of public affairs for the Texas Federation of Republican Women, press assistant to Congressman Bob Price, and a recipient of the Ten Outstanding Republican Women award in Texas.

She received a bachelor of business administration degree in 1956 from the University of Texas. (NASA Release 85-168)

December 16: Former NASA administrator James Beggs, General Dynamics, and three executives of the company today pleaded not guilty to charges of plotting to hide cost overruns during development of a prototype of the DIVAD antiaircraft gun [see NASA/Management and Personnel, Dec. 2], the *Washington Post* reported. "I plead not guilty to each count," Beggs, a former company executive, told U.S. District Court Judge Ferdinand Fernandez.

Fernandez scheduled a trial to begin April 8, 1986, although prosecutor Randy Bellows said the case was too complex to be heard so quickly. Bellows said the government had 2.7 million documents to review.

Thomas Sullivan, attorney for General Dynamics, told Fernandez that the defendants wanted the trial to begin in the spring because a government suspension of most contracts with the company until the case was resolved had put the defense contractor "in a real serious bind . . . This company could be literally . . . put out of business by this suspension," Sullivan said.

Attorneys estimated the trial could take more than 12 weeks, with Bellows saying he would call at least 70 witnesses. (*W Post*, Dec 17/85, A13)

December 18: NASA announced that H. William Wood, deputy associate administrator (Networks), Office of Space Tracking and Data Systems (OSTDS), was retiring effective early in January after more than 30 years of government service.

Wood had served in his present post since April 1984. Before that he managed the Network Systems program as the division director since 1981. Wood was a research engineer at Langley Research Center when NASA was formed in 1958; and in 1959 he became the group leader with the Tracking and Ground Instrumentation unit for the Mercury Network. His other NASA assignments included associate director, Network Systems Division; NASA senior scientific representative in Australia; and associate director, Operations, in the Network Directorate at Goddard Space Flight Center. Earlier, Wood had served several years in the U.S. Air Force.

Wood received the NASA Outstanding Leadership Medal and two NASA Exceptional Service Medals, and he is a Fellow of the American Astronautical Society. In 1955 Wood received a BSEE degree from North Carolina State University. (NASA anno., Dec 18/85)

December 23: NASA announced that Dr. Charles Kupperman was appointed effective immediately as executive assistant to the deputy administrator.

Kupperman came to NASA after serving as executive director of the President's General Advisory Committee on Arms Control and Disarmament. From 1978 to 1981 he was the research associate and defense analyst for the Committee on the Present Danger, and he also served as a member of President Reagan's 1980 presidential campaign and defense transition team. Kupperman was a consultant to R&D Associates and taught at the School of International Relations, University of Southern California.

Kupperman graduated Phi Beta Kappa with a B.A. degree in political science from Purdue University, received an M.A. degree in political science from the University of British Columbia, and a Ph.D. in international relations from the University of Southern California. He was the author of articles on defense, the Strategic Arms Limitation Talks (SALT), and national security policy for academic and public journals. (NASA Release anno., Dec 23/85)

December 26: NASA announced that it had named effective immediately John Hodge acting associate administrator for space station.

Hodge became director of the Space Station Task Force on its establishment in May 1982 and was appointed deputy director of the Interim Space Station Program Office in April 1984 when the task force completed its work. He directed activities that resulted in the initial concepts definition for the space station, established program management policies, initiated advanced development programs, defined user needs, and organized potential international cooperation. In 1984 NASA appointed Hodge deputy associate administrator for Space Station, coinciding with establishment of the Office of Space Station.

Hodge joined NASA in 1959, after working with Vickers-Armstrong, Ltd. and AVRO Aircraft Ltd., as chief of Flight Control, supervising a team of over 200 technical and administrative personnel and 100 aerospace contractors and serving as flight director for Mercury, Gemini, and Apollo flights.

In 1970 Hodge worked for the Department of Transportation as director of transportation systems concepts at the Transportation Systems Center and later became associate administrator for Policy, Plans and Program Management before rejoining NASA.

Hodge received a B.S. degree in engineering from Northampton Engineering College in London and in 1966 the honorary degree of Doctor of Science from the City University, London. (NASA anno., Dec 26/85)

NASA Installations

Ames Research Center

February 1: NASA announced that effective January 24, Dr. Dale Compton became deputy director of Ames Research Center (ARC), where he had been director for engineering and computer systems since February 1983. Compton joined NASA in June 1957, serving as deputy director of astronautics, then chief of the space science division and manager of the *Infrared Astronomical Satellite (IRAS)*. Recipient of the NASA Outstanding Leadership Medal, Compton earned a B.S. degree in 1957 and an M.S. degree in 1958, both in aeronautical engineering, and a Ph.D. degree in aeronautical and astronautical engineering in 1969. (NASA Release 85–17; NASA anno., Feb 2/85)

During February: Ames Research Center (ARC) announced it had formed the information sciences office to do basic and applied research in artificial intelligence (AI), automation sciences, and space-related computer science technology, with emphasis on applications to the NASA space station.

NASA had designated ARC as the lead center in AI research, and the new office's research would cover AI programming languages, expert-systems development, knowledge representation and information understanding, machine vision and learning, sensor fusion, and optical processing.

Space-related computer-science research would focus on symbolic-processing and data-flow architectures; network design, protocols, and simulation; and optical read/write information-storage technology.

ARC would combine in-house studies, primarily on applied research and development of flight experiments to validate and demonstrate the technologies being developed, with the work of outside research organizations, such as SRI International, Symbolics Inc., Innovative Optics, Inc., Stanford University, University of California at Berkeley, MIT, University of Texas, and University of Michigan, on basic research issues to accomplish project objectives.

ARC would orient research toward the user community, particularly other NASA centers, focusing on both technology development for user applications and help to users during implementation. Space station automation would be the primary application, and ARC had signed agreements with Johnson Space Center and Goddard Space Flight Center to work together on this implementation. ARC would also advise on space station automation through participation on NASA advisory groups such as the space station automation study team, space station automation and robotics panel, advanced-technology automation committee, and NASA headquarters space station data systems steering committee. (ARC Release 85-6)

March 12: NASA announced that personnel at Ames Research Center (ARC) would break ground March 14 for the numerical aerodynamic simulation (NAS) facility building to house the world's most powerful supercomputer system, which was intended to provide a national computational capability that would complement NASA's experimental facilities in maintaining national preeminence in aeronautical research. Researchers would use high-speed supercomputers in the system to solve complex aerodynamic equations describing the fundamental fluid physics and large scale aerodynamic flows associated with aircraft flying in earth's atmosphere. In effect, they would test aircraft configurations by "flying" them in the NAS supercomputer system, reducing both time and costs associated with development of new aircraft.

NAS would also support other research, including computational materials and structures, chemistry, and astrophysics; weather predictions; and genetic engineering. ARC also intended to make the supercomputer network available to remote users in universities, private industry, and other government research agencies via satellite.

Cray Research's Cray 2 supercomputer, with an expected operating speed of 250 million calculations per second when working aerodynamic problems, would be the heart of the initial NAS network when it became operational in mid-1986. NASA intended to incorporate even faster supercomputers as they became commercially available, with the aim of including supercomputers with up to one billion calculations per second in 1988 and 10 billion calculations per second in the 1990s.

In an effort to create a partnership between people and machines that, in itself, would advance computer simulation, NASA had architects Hunt and Co. design the 90,500-sq. ft. concrete building so as to encourage scientists who would use the computers for research to work closely with the people who would operate, maintain, and develop the NAS system. Building contractor was Perini Co. (NASA Release 85-37)

April 2: NASA announced that a group of researchers led by Dr. Lelia Coyne, San Jose University, working at Ames Research Center (ARC) discovered new clay-energy storage and transfer processes that supported a clay origin-of-life theory, which contradicted the theory that organic life arose from an accidental combination of chemicals and energy in a primordial "soup" on the primitive earth.

Basic to the clay origin-of-life theory were observations suggesting that clay, although composed of inorganic material, exhibited life-like characteristics in its ability to select out certain chemicals and to serve as a catalyst for chemical reactions. Some scientists also theorized that clay might be able to perform additional chemical functions basic to life including self-replication, growth, and transfer of chemical information to other chemical systems. They believed that organic chemicals that eventually "learned" to reproduce themselves and create life could have come together in an orderly process that first appeared in the structuring and reproductive processes attributed to clay.

The work at ARC strengthened this theory by finding that clay had the ability to absorb, store, and transfer energy, a necessary process of organic life. This work introduced a new and important aspect to the clay-life theory—that clay minerals were capable of engaging in energetic processes that were necessary conditions if it was to be shown that clay had life-like properties.

The NASA-sponsored investigation grew out of 20 years of research at ARC in studies of the chemical basis for the origin of life. Other scientists working on various aspects of clay research at ARC were Drs. Sherwood Chang, Ted Bunch, James Lawless, Noam Lahav, David White, and Glenn Pollock. (NASA Release 85–48)

September 17: NASA announced that the Ames Research Center (ARC) selected Informatics General Corp. and Technology Development of California for competitive negotiations leading to an approximately $40 million five-year contract to provide computational resources for research in large-scale computational fluid dynamics, aerodynamics design and analysis, computational chemistry, astrophysics, atmospheric modeling, and satellite image processing.

The winning contractor would have to deliver a total system capability including the computational facility, operational support, systems engineering, and management.

The anticipated contract was a follow-on to existing services performed at ARC. (NASA Release 85–130)

October 2: NASA's Ames Research Center (ARC) announced that the world's fastest and most powerful supercomputer, the Cray-2, arrived at ARC to assist researchers in taking major steps toward simulating actual aircraft flight and thus making possible important advances, both in cost savings and performance, in aircraft design. The Cray-2 could perform 250 million continuous calculations per second, which was more than three times faster than the previous generation of supercomputers.

The Cray-2 was the first building block in the creation of NASA's Numerical Aerodynamic Simulation (NAS) program, planned to provide the world's most powerful, large-scale high-speed processor system. In addition to its advantages for aircraft design, NAS represented an important national facility in such research areas as aerothermodynamics, computational chemistry, atmospheric modeling, and other computationally intensive scientific applications.

NASA planned the NAS system as an ongoing project with continuous improvements in speed and memory. Its objectives were to establish and maintain a leading-edge national computation capability to ensure leadership in computational fluid dynamics and related disciplines, provide an integrated processing system capable of a sustained 250 million floating-point, operations-per-second processing rate in 1986 and a one billion rate in 1987, and act as a pathfinder in advanced, large-scale computer systems capability.

The Cray-2 was four feet high and four feet in diameter; its small size was made possible by microminiaturization of the electronic circuits and extremely dense packing of the circuit boards. Since the speed of light was a fundamental limit on computer speed, super fast machines had to be smaller in order to reduce distances that information traveled. Computer scientists indicated they expected to get still more speed in future machines by further reducing size and adding more parallel processors.

A main feature of the Cray-2 was its large random-access memory, which was 16 times larger than previous supercomputers' memories. The Cray-2's memory provided random access from any of the machine's four main processors and any of its high-speed data channels. This meant a user could use all or part of this memory quickly, rather than taking hours to access data for large calculations.

Another main feature of the machine was that its tightly-packed, heat-producing electronic components were immersed for cooling in a colorless odorless, inert fluorocarbon liquid, the first such design in computer history. (ARC Release 85–38)

Goddard Space Flight Center

January 14: NASA announced that Goddard Space Flight Center (GSFC) issued a Request for Proposal (RFP) to industry to develop a space platform providing five years of on-orbit services to NASA payloads while allowing the developer to market it to commercial users, an innovative approach considered a first step toward creating a closer partnership between government and industry in space. Industry would finance, develop, own, and operate the platform with first use planned in late 1988 and would be free to market the platform (totally separate from NASA's plans for a government-developed, permanently manned space station) for materials processing or other manufacturing activities.

NASA stipulated that the platform be capable of providing services for four future projects: the extreme ultraviolet explorer (EUVE), the x-ray timing explorer (XTE), a zero-gravity payload, and an as yet unidentified project.

The Space Shuttle or other launch vehicle would place the commercial platform in orbit. Commercial payloads would reimburse the government for use of the orbiter and facilities. (NASA Release 85–7)

Johnson Space Center

September 12: NASA announced that it had selected Rockwell Shuttle Operations Co. for negotiations leading to award of the $685 million four-year space transportation system operations contract (STSOC) at Johnson Space Center (JSC). Follow-on awards could result in a total contract period of 15 years at a value of about $5.5 billion. JSC would manage the work under a cost-plus-incentive/award-fee contract that included incentive fee on sound cost management and an award fee on the basis of performance. Other team members included Bendix Field Engineering Corp., System Development Corp., Omniplan Corp., RMS Technologies, Inc., and System Management American Corp.

Under the contract, Rockwell would have responsibility for six major STSOC functions: project management; maintenance and operations; sustaining engineering; flight preparation requirements and analysis; flight preparation production; and direct mission operations, testing, and support. Work would be done for such facilities as the mission control center, shuttle mission simulator, shuttle avionics integration laboratory, software production facility, central computing facility, and the mockup and integration laboratory.

The selection represented NASA's consolidation into one contract of work previously performed by 16 firms under 22 contracts.

Other firms submitting proposals were Ford Aerospace and Communications Corp., Space Information Systems Division, Grumman Space Operations Corp., and Lockheed Space Flight Co. (NASA Release 85–128)

October 1: NASA's Johnson Space Center (JSC) announced a NASA source selection board was evaluating proposals from the industry teams of Boeing Aerospace Co. and Wornick Corp.; General Electric, Northrop, and Integrated Systems Analysts; and Hamilton Standard, ILC, and RCA for a single Flight Equipment Processing contract (FEPC) that would consolidate work being done by 15 firms. NASA planned by the end of October to award an initial three-year contract with a two-year priced option extension. The contract would be cost-plus-award-fee for the first six months; cost-plus-incentive-fee, plus an award fee, for the remainder of the term. Estimated contract value at the end of 15 years was $300 million.

The contract would cover processing Space Shuttle flight crew equipment, including operation and maintenance of associated ground systems, primarily at JSC. The winning contractor would be responsible for space suit maintenance and testing before and after flights; failure analysis and repair of spaceship galleys; operation of the space food production facility; preparation and delivery of food for each mission; medications and equipment for medical kits; clothing, personal hygiene and flight kits; tools; radio and TV support; and all film and cameras.

The award would be the second large consolidation contract let in 1985 at JSC; on September 12 Rockwell International won the Space Transportation System Operations Contract to consolidate JSC Space Shuttle operations. (JSC Release 85-038)

December 16: NASA announced that Gerald Griffin, director of NASA's Johnson Space Center (JSC), would leave the agency January 14, 1986, to become president of the Houston Chamber of Commerce. Robert Goetz, deputy director of JSC, would become acting director upon Griffin's departure.

Griffin had been JSC's director since August 1982 and had served with NASA for more than 20 years in a number of key positions at three NASA centers and in Washington, D.C. Griffin served in the U.S. Air Force and worked in the aerospace industry before joining NASA in 1964.

"It was a tough decision to leave NASA," Griffin said, but "I'll depart with the comfort that the NASA team of government, industry, and university people will continue their outstanding job in space activities for this country."

William Graham, acting NASA administrator, said, "As one of NASA's key senior executives, Gerry has had a long distinguished career. He has received many honors and awards in recognition of the contributions he has made both to aeronautics and space. We shall all miss him very much and wish him great success as he moves on to new accomplishments." (NASA Release 85-172)

Kennedy Space Center

August 16: Continuing a project begun in 1984, some of the world's most prominent lightning experts gathered during the summer at Kennedy Space Center (KSC) to study lightning strikes and their effects, the *Spaceport News* reported. KSC attracted scientists with a variety of research interests because it was one of the nation's areas of highest frequency of strikes and the presence of French experts who had mastered the technique of triggering lightning.

Taking part in the summer's project were representatives of NASA, the U.S. Air Force's Wright Aeronautical Lab, the Federal Aviation Administration, the U.S. Naval Laboratory, France's Center for Nuclear Studies, three U.S. universities, and two industry organizations.

Since late May, the 36 lightning investigators had conducted studies at a specially equipped lightning research facility located on the shore of Mosquito Lagoon about nine miles north of the vehicle assembly building. The facility included a launch site for the three-foot rockets used to trigger lightning and two nearby trailers that housed the research teams and their equipment.

From their control center, the French experts triggered lightning events by launching a small rocket that trailed a thin copper wire to an altitude of about 2000 feet. The wire acted as a trigger, attracting a stroke, which vaporized the wire while traveling down an electrical pathway to the ground. Special instruments and cameras near the launch site recorded the strike in detail.

The triggered lightning events allowed researchers to obtain ground measurements that they could use in atmospheric and lightning research. Researchers might eventually use information from the studies to design advanced lightning protection systems for KSC facilities and for greater protection for aircraft such as the Space Shuttle that contained electronic flight control systems.

Other research at the facilities included using a Convair C-580 to obtain measurements on a lightning strike; a study to determine how lightning strokes gained access to power lines; investigation of the Maxwell Current, which developed in a particular pattern before, during, and after thunderstorms; and attempts to obtain three-dimensional photos of lightning strikes.

In addition to hosting the project, KSC was contributing funding to some of the university research. Project managers hoped to attract more participants in the future with the promise of research leading to tangible benefits in lightning protection and prediction. (*Spaceport News*, Aug 16/85, 7)

September 11: Kennedy Space Center (KSC) announced it had extended an existing contract with Planning Research Corp. (PRC) to provide engineering services for the Directorate of Engineering Development at KSC and at Vandenberg Air Force Base (VAFB). The $12,421,841 extension brought the total contract value to $66,187,413 and extended the period of performance from January 1, 1986, through September 30, 1986.

Under the contract, PRC would design ground-support systems for the Space Shuttle/Centaur program, which NASA would use to inject space vehicles into interplanetary trajectories after deployment from the Space Shuttle.

PRC would also provide designs for Space Shuttle launch-support equipment for the Department of Defense (DOD) at VAFB, which would become the second launch and landing facility for the Space Shuttle in the mid-1980s, particularly for launching DOD payloads into polar orbit.

Another PRC task under the contract was to provide KSC with designs for Space Shuttle cargo ground-support equipment. (KSC Release 186–85)

September 13: Kennedy Space Center (KSC) selected Boeing Aerospace Operations, General Electric Co., Grumman Technical Service Inc., Hughes Aircraft Co., Messerschmidt-Bolkow-Blohm (MBB)/Erno (the West German firm that produced Spacelab), and McDonnell Douglas Technical Services to serve as prime assessment teams to observe payload processing activities, the *Spaceport News* reported. KSC made the selection in connection with the initiation of a competitive procurement for a single payload ground operations contract (PGOC) during the Space Transportation System's operational period.

NASA expected to request proposals from the companies near the end of 1985 or the beginning of 1986 and would award a contract during fall 1986.

This contract along with those for shuttle processing and base operations were the last major contracts to streamline operations as the Space Transportation System matured. The PGOC contract would consolidate operations, minimize interfaces, and focus clear responsibility on a single payload contractor at KSC.

Under the new arrangement, NASA would consolidate work then performed by four KSC contractors into a single contractor with prime responsibility for processing of payloads at KSC, primarily for the Space Shuttle. Other contract duties would include sustaining engineering, communications, instrumentation, and telemetry operations, and miscellaneous support. (*Spaceport News*, Sept 13/85, 1)

November 7: Kennedy Space Center (KSC) awarded ASEA Robotics Inc. a $1,143,622 contract to build a robotic development prototype system to be installed in the Robotic Applications Development Laboratory (RADL) and serve as KSC's control center for robotics, *Spaceport News* reported. KSC currently had two labs for robotics research: Lab A was a test facility for training devices, sensor and end-effector development, and small robotic projects; Lab B functioned as a control center made up of complex work cells and "smart systems" (independent control systems integrated into one coordinated system) for large robotic applications.

The objective of the new system was to permit engineers to combine existing and new technology to create advanced independent and semi-independent machines and machines with independent decision-making capabilities or a "computer brain," which could accomplish more on their own without step-by-step instructions from human supervisors. The initial thrust of the new system would be to develop the technology and techniques needed to automatically load and unload hazardous fuels aboard space vehicles and payloads during prelaunch ground operations. Future challenges included application of these refueling techniques to other fuels such as cryogenic propellants.

RADL would have an electrically driven articulated robot composed of a six-axes arm and interchangeable end-effectors or "hands," a computer with user-friendly software, and a vision tracking system. The facility would serve as the focal point for combining robotics hardware, algorithms, software, sensors, and control systems. Leon Davis, a development engineer and electrical lead for the Robotic Control Systems, said of the new system, "This is not a manipulator that is teleoperated by a man, but a robot with a computer brain capable of memorizing the contours and panels of its intended subject and carrying out its functions autonomously." Because NASA would be asking KSC to design processing facilities for new-generation launch vehicles such as the second generation Space Shuttle and the heavy lift launch vehicles, KSC would need these state-of-the-art robotics to provide the most efficient and effective processing methods. Hazardous, time critical, and repetitive Space Shuttle and payload operations all had potential uses for robotics.

A special feature of the system was a "real time" tracking system using "closed-loop" responsive adaptive control commands, feedback, and error signals. "Adaptive control" meant that the robot altered its path automatically in response to sensory feedback from its environment. The KSC robot's servo-controls and software utilities would allow its arm to operate adaptively to control positions and orientation of all six axes, not only the end-effector as found in standard applications.

An electrically driven, servo-controlled robot had several advantages over hydraulic robots, the major one being that smooth motion, speed, and trajectory controls were possible when the robot handled loads up to 200 lb. As Davis said, "We really don't have applications for hydraulic robots. Hydraulic robots can leak, proving a potential hazard to the tile and other sensitive areas of the orbiter."

KSC's robotics research would progress in four phases: phase one was the development of the robotic arm in static work cells; phase two proposed advancing the robotic capabilities to operate in dynamic work cells; phase three would include off-line programming using 3-D graphics and simulated work cells; and phase four entertained the possibility of employing artificial intelligence. Possible future capabilities of robots were three-dimensional scanning; higher order processing; artificial intelligence; sonic, laser, and other ranging systems; and "touch" and mobility systems. (*Spaceport News*, Nov 7/85, 6)

Langley Research Center

February 1: NASA announced that Robert Nunamaker, formerly chief engineer at Ames Research Center (ARC), became Langley Research Center's (LaRC) space director with responsibility for advanced space transportation, planetary entry, space station and large space antenna research, and for LaRC's atmospheric science program. Nunamaker began his NASA career at Lewis Research Center in 1958, where he acquired extensive experience with Project Mercury; he joined ARC in 1963. Nunamaker received a B.S. degree in mechanical engineering in 1957, followed by graduate work at Case Institute of Technology and the University of California at Los Angeles. (NASA Release 85-17; NASA anno., Feb 2/85; LaRC Release 85-8)

July 8: Langley Research Center (LaRC) announced that NASA would sponsor an exhibition July 22 through July 30 at the 1985 National Scout Jamboree at Fort A.P. Hill, Virginia. The jamboree, celebrating scouting's 75th anniversary, was expected to draw more than 35,000 U.S. and international scouts and leaders and many thousands of general public participants.

The NASA exhibit, for which LaRC was serving as lead coordinator, would include information on subjects ranging from current aeronautics research to the Space Shuttle and man's future in space. A 100-seat theater would have video presentations and live programs throughout each day.

Astronaut Frederick Gregory, a former scout and long-time scouting enthusiast, would participate in the jamboree's opening program on July 24 and would talk with scouts in the NASA exhibition area on that day.

NASA would sponsor July 24 through 28 model rocket launch demonstrations and presentations on earning the Space Exploration merit badge. (LaRC Release 85-38)

July 23: NASA announced that in a two-and-one-half-year effort costing $15 million it had lengthened and improved its Aircraft Landing Dynamics Facility at Langley Research Center.

The track, used to test aircraft wheels, tires, and landing gear, used a high-pressure water jet system to propel the test carriage along the rail/track system where researchers conducted experiments under simulated runway conditions. Track testing options included choosing between concrete or asphalt runway surfaces and among a full range of weather-related surface conditions, including dry, damp, or flooded runways or slush- or ice-covered surfaces. Even before the expansion, NASA believed the track to be the only facility in the world capable of testing full-size aircraft landing gear systems under closely controlled conditions simulating takeoffs and landings.

Lengthening the track to more than one-half mile increased the facility's maximum test speed from 120 to 250 mph and made possible, for the first time, simulated landing tests of all modern aircraft and the Space Shuttle. Other facility improvements included a higher capacity water jet system and a newly designed high-speed test carriage. The improved catapult system produced a 1.7-million-lb. thrust on the carriage, resulting in a 17-g (gravity force) acceleration that pushed the carriage in only 400 feet from zero to 250 mph in two seconds, and the new carriage had a 20-by-40-foot test bay to accommodate larger test articles. NASA used ten thousand gallons of water in maximum speed run.

Initial research at the modified facility would study the cornering forces and spin-up characteristics of the Space Shuttle main gear tire, which spun up from zero to landing speeds almost instantly at touchdown. NASA would mount a tire-wheel-brake assembly on the carriage for tests of tire wear and applied loads and would run these first tests on the facility's existing smooth concrete surface. Later runs would require the touchdown area of the track altered to simulate the roughness and grooving of the Space Shuttle runway at Kennedy Space Center. NASA would then paint smooth the relatively rough runway to determine the best KSC runway characteristics to lessen Space Shuttle tire wear.

Later tests would include a variety of aircraft landing hardware tests, averaging about 75 to 100 runs each. NASA estimated the facility could accomplish 300 runs a year with as many as six a day during the height of a test program. One test program, for example, would compare the performance of radial tires, not commonly used on aircraft, to that of conventional bias ply tires. NASA would run these tests in conjunction with a Federal Aviation Administration/NASA program to gather information on runway surface traction.

NASA had scheduled over the next several years test programs that included track tests to develop a data based for the National Tire Modeling Program, an analytical computer model to aid in design of new tires, and a tire failure study. (NASA Release 85-109)

Lewis Research Center

April 11: NASA announced that Lewis Research Center (LeRC) was developing a liquid droplet radiator intended to solve the problem of dissipating heat that builds up inside a spacecraft, allowing a livable temperature for astronauts. Heat removal was currently accomplished through a heavy, bulky, rigid metal heat transfer system, which added considerable weight to the spacecraft.

The LeRC system would use the surface of a liquid coolant to radiate away excess heat and save as much as 90% of the current hardware weight. The system entailed exposing a moving stream of hot droplets, the diameter of a human hair, directly into space, allowing the heat to radiate from the droplets' surface. In the process, a generator would eject the droplets into a collector where they would rejoin to form a liquid; the system would then recirculate and reuse this coolant.

Selection or development of a heat transfer fluid with proper vapor pressure and a sufficiently long life was essential to the project. Also crucial to system feasibility was development of a micromachining capability to produce holes with 0.002-in. diameters for use in the liquid droplet generator.

"This is a very advanced concept," Alden Presler, LeRC program manager, said. "The technology we're developing here at Lewis will result in a very lightweight and compact piece of hardware." (NASA Release 85-53)

April 22: NASA announced Lewis Research Center awarded a $58,291,440 contract to General Dynamics Corp. to modify three Centaur G vehicles, the nation's most powerful upper stage vehicles, to fit into the Space Shuttle's payload bay to launch various large spacecraft including the Galileo and Ulysses missions, both scheduled for spring 1986.

The sole-source, cost-plus-incentive-fee contract for fabrication, assembly, test, checkout, and delivery of the vehicles would begin April 1985 and continue through October 1987. (NASA Release 85-61)

May 2: Lewis Research Center (LeRC) researchers were working on improvements in the National Airspace System (NAS) air-traffic control system to provide more efficient air transportation in the U.S. and to maintain the U.S. position of leadership in civil aviation, the *Lewis News* reported.

A major NAS requirement was accurate upper-air wind information, necessary for efficient movement and control of aircraft. Dating back to 1982, LeRC recognized this need and proposed and developed, in cooperation with the Environmental Research Laboratories of the National Oceanic and Atmospheric Administration (NOAA), an interactive data management and enhancement system (MERIT) to improve significantly the accuracy of upper-wind forecasts in the U.S.

To evaluate the potential of MERIT, the Federal Aviation Agency (FAA) formed an aviation weather task force of a dozen experts from NOAA, NASA, and universities that was led by Dr. John McCarthy of the National Center for Atmospheric Research. Over 18 months at a cost of half a million dollars, the task force would examine all aviation weather forecast products concurrently with MERIT. If the MERIT system could provide improved upper-level wind information (initial tests at NOAA showed a 20 to 30% reduction in wind error compared to the existing system), it was possible that MERIT could become an important contribution to the NAS of the 1990s. (*LeRC News*, May 2/85, 2)

June 3: Lewis Research Center (LeRC) announced it awarded Honeywell Inc., Avionics Division, a $14,900,000 contract for fabrication and delivery of three flight inertial measurement groups (IMG) for the Shuttle/Centaur guidance and navigation system, which maintained a fixed reference orientation for and measured acceleration of the Centaur.

Work under the sole-source firm-fixed-price contract would begin in May 1985 at the contractor's plant in Clearwater, Florida.

NASA was modifying the Centaur, which had launched domestic and military communications satellites and sent spacecraft to investigate planetary systems, to fit into the Space Shuttle. NASA would first use the Shuttle Centaur combination to launch the Galileo and then the Ulysses missions. (LeRC Release 85-40)

August 26: Lewis Research Center (LeRC) announced today that it would open in September the microgravity materials science laboratory (MMSL) to help experimenters make better decisions about what is and is not feasible for science experiments in space. Such experiments are expected to provide knowledge leading to new materials, more efficient use of earth's nonrenewable fuel resources, new medicines, advanced computers and lasers, and better communications.

Providing a capability that existed nowhere else in the world, the new laboratory would offer U.S. scientists and engineers a low-cost, low-risk way to test new ideas for materials science research in space. Initial research plans would support metal and alloy solidification and electronic crystal growth studies. The laboratory should give U.S. companies a competitive advantage in developing better ground-based materials and/or processing of materials through microgravity research.

The laboratory, along with LeRC's assignment to develop the power system for the proposed space station, would put LeRC into the mainstream of the U.S.'s space research efforts. (LeRC Release 85–63)

Marshall Space Flight Center

January 29: NASA announced it had selected Grumman Data System Corp. for negotiations leading to the award of a contract for a high-speed (Class VI) computer system for Marshall Space Flight Center (MSFC). The fixed-price contract would require Grumman to provide hardware, software, documentation, and services for installation and maintenance of a scientific- and engineering-computations system in support of MSFC's programs in thermal, electrical, load, and structural-design characteristics that influence flight-vehicle and payload performance.

The five-year contract, to begin no later than August 1985, would cost about $42 million and provide for a total lease period, with renewal in periods of one to 12 months, not to exceed 60 months. (NASA Release 85–15)

July 1: Marshall Space Flight Center (MSFC) announced that July 1 marked the center's 25th anniversary. MSFC currently managed a budget of about $2 billion annually and employed some 3,300 people, while thousands more worked with numerous aerospace and other high-tech organizations in the Huntsville, Alabama, area to support the center's projects.

During the 25 years, MSFC had moved beyond its initial role as NASA's developer of launch vehicles and propulsion systems to that of a scientific and engineering organization also responsible for spacecraft and scientific experiments. The center was currently responsible for a number of NASA programs including the Space Shuttle's engines, propellant tank, and booster rockets; the Hubble Space Telescope; Spacelab orbiting research laboratories; the Space Shuttle upper stage systems; and a significant portion of the proposed space station. (MSFC Release 85–31)

July 3: Marshall Space Flight Center (MSFC) announced it awarded a three-year, $35,181,700 contract to the Space Transportation Systems Division of Rockwell International for Space Shuttle systems integration work at MSFC, Kennedy Space Center, and Vandenberg Air Force Base. The contract called for provision of management, personnel, equipment, materials, and resources from July 1, 1985 to June 30, 1988. (MSFC Release 85–34)

December 19: Marshall Space Flight Center (MSFC) announced selection of Teledyne Brown Engineering for negotiation leading to the award of a five-year contract to provide operation and maintenance of the propellants and pressurants facilities for the center. Total price for the first three contract years was approximately $8.8 million.

Teledyne Brown would furnish the necessary management, personnel, equipment, and materials required to operate and maintain the gaseous pressurants equipment and distribution systems, liquid propellants storage and transfer systems, fluid transporter services, valve and component refurbishment, and retention of certification status of high-pressure systems, including test stands, wind tunnels, and Department of Transportation certification of fluid transporters. (MSFC Release 85–72)

Wallops Flight Facility

June 29: Goddard Space Flight Center announced the Wallops Flight Facility would celebrate its 40th anniversary on June 29. The events planned were in recognition of the scientific and technological achievements at Wallops, which had conducted rocketborne experiments since July 4, 1945, when researchers first launched a 17-foot Tiamat.

Since that original flight, Wallops had launched approximately 13,000 rocket-propelled research vehicles and conducted thousands of aeronautical and aircraft tests to obtain information on aircraft and spacecraft flight characteristics and to increase knowledge of the upper atmosphere and the near-space environment. This research led to major contributions to the U.S.'s aeronautical and space programs.

The breakthroughs to supersonic flight by aircraft and to hypersonic flight by rocket systems were largely attributed to fundamental aerodynamic research conducted at Wallops by means of aerodynamic models propelled by multiple-stage rocket systems.

Using sounding rockets Wallops personnel developed the technologies for measuring atmospheric structure and the space environment, such as temperature, pressure and density, micrometeorite densities, electric fields, energetic particles, and radiation levels. These measurements formed the basis for developing design criteria for scientific satellites and manned spacecraft.

Wallops's researchers also pursued technology development for the manned space program, such as manned capsule-escape techniques and maximum dynamic pressure tests for the Mercury Program.

Most recently Wallops was known as a center for the NASA suborbital program. Sounding rockets, balloons, and aircraft conducted space science and aeronautical research missions.

The island obtained its name from John Wallops, appointed deputy surveyor of Virginia in the 17th century by Col. Edmund Scarburgh. Wallops received a Crown Patent to the island in 1672. At present, the facility, in addition to the island, included the mainland area in back of the island and the main base (formerly the Chincoteague Naval Air Station) about seven miles northwest.

In addition to an open house, the day's anniversary events would include an Air Force F-15 aerial demonstration and demonstrations in crash/fire/rescue, space sciences, model rocketry, and radio controlled model airplanes. (GSFC Release 85-16)

Office of Inspector General

February 1: NASA announced that June Gibbs Brown, appointed NASA's inspector general June 1981 by President Reagan, would leave March 10 to become vice president of finance and administration and chief financial officer of System Development Corp. Prior to her NASA service, Brown served as inspector general for the Department of Interior and worked for the Bureau of Land Management, Bureau of Reclamation, and had been director of the audit division of the U.S. Navy Finance Center. Brown held bachelor and master degrees in business administration and a Juris Doctor from the University of Denver School of Law. (NASA Release 85-17; NASA anno., Feb 2/85)

November 5: NASA announced it has named effective November 3 Lewis Rinker deputy inspector general, Office of Inspector General (OIG). In that position, Rinker would assist the inspector general in the direction and management of nationwide OIG programs.

Since joining NASA in October 1980, Rinker served in various management positions, the most recent of which was assistant inspector general for technical services and also served as the acting deputy inspector general.

Before his NASA work, Rinker served in managerial positions with the U.S. General Accounting Office and was an industrial engineer and procurement analyst with the Air Force Systems Command.

Rinker received his B.S. degree in engineering from Johns Hopkins University and was doing graduate work at George Washington University. He was a senior member of the American Institute of Industrial Engineers and a member of the Association of Government Accountants and the Association of Federal Investigators. (NASA anno., Nov 5/85)

November 24: NASA announced that Leon Snead was appointed effective today NASA's assistant inspector general for auditing (AIGA), Office of Inspector General (OIG). In this position, Snead would provide advice and support to the inspector general and managerial direction for the functional development, implementation, and supervision of NASA/OIG audit activities.

Before this appointment, Snead was acting assistant inspector general for auditing at the Department of Interior. He began his government career in 1964 with the U.S. Army Audit Agency as an auditor trainee; in 1974 he went on to supervisory auditor and then an audit director position with the Department of Treasury. Later he served as division director, branch chief, and branch manager at the Department of Energy and regional audit manager with the Department of Interior.

Snead graduated in 1963 from Spencerian College, Milwaukee, with a BBA degree in accounting and in 1969 from the University of Baltimore with a JD degree. He was a certified internal auditor (CIA) and a member of the Association of Government Accountants (AGA). (NASA anno, Nov 24/85)

Procurement

February 8: NASA announced it had selected Dr. Milton Silveira, NASA's chief engineer, as its competition advocate responsible for recommending annual goals and objectives, approving organization and personal accountability, assuring training, and reviewing noncompetitive procurement to remove constraints to competition. Silveira named David Austin his principal staff member. A NASA management instruction would designate each center deputy director as the center competition advocate.

NASA competitively awarded 72% of its procurements in FY 84, representing a significant increase over 68.8% in FY 83. (NASA Release 85-21)

April 10: NASA announced selection of RMS Associates for negotiations leading to award of a contract for the operation and maintenance of the NASA Scientific and Technical Information Facility (STIF), which acquired, organized, processed and stored worldwide aerospace information including published articles, papers, books, and reports. STIF is located near the Baltimore/Washington International Airport. Under the contract terms, RMS Associates would acquire and process documents and data approved by NASA for entry into the collection; catalog, abstract, index, announce, and disseminate these materials; maintain a supporting reference service; compile specialized bibliographies; and provide other technical support.

The one-year contract, commencing July 1, 1985, and preceded by a one-month phase in, included provision for two one-year priced option extensions and two one-year unpriced options. RMS Associates estimated the one-year contract value at about $5 million, with a total of approximately $15 million for the first three years. (NASA Release 85–54)

September 30: NASA announced selection of 150 research proposals from small business high-technology firms under a program established by the Small Business Innovation Development Act of 1982 (PL 97–219) to stimulate technological innovation in the private sector, strengthen the role of small businesses in meeting NASA's research and development needs, and contribute to the growth and strength of the nation's economy. NASA selected proposals on the basis of their technical merit from the 1,164 submitted and expected to award contracts in four to eight weeks.

Contractors funded under the program would conduct work in two phases: first, a six-month fixed-price effort to explore the feasibility of the small business-proposed innovations and, if the results warranted, a second phase effort of up to two years in duration to proceed further with the technological innovation. Commercial firms or government agency programs would fund any work beyond the second phase.

The selections resulted from NASA'S third solicitation for Phase I proposals in areas of research and development of interest to the agency. NASA's FY 85 budget included approximately $7.5 million for Phase I, managed by the Office of Commercial Programs, NASA Headquarters. (NASA Release 85–136)

November 27: NASA announced selection of 64 research proposals from high-technology small business firms for negotiation of Phase II contracts in its Small Business Innovation Research program established under Public Law 97–219. The contract awards would total approximately $30 million. NASA selected the Phase II programs, follow-ons to previous Phase I contracts, from 113 proposals submitted by Phase I contractors in the 1984 program. Phase I included 127 awards selected from 919 proposals.

Objectives of the Small Business Innovation Research program were to stimulate technological innovation, strengthen the role of small business in meeting federal agencies' research and development needs, and foster increased private sector commercialization of innovations from federal research and development. The program also encouraged minority and disadvantaged business participation.

In Phase I, NASA awarded six-month fixed-price contracts to explore the feasibility of the proposed innovation and the value of conducting further research and development. Continuation of the activity into Phase II, which could extend up to two years, depended on the merit of the work and agency program plans and needs. Selection might also depend on availability of funds from the private sector and/or government for development or commercialization beyond the scope of the Small Business Innovation Research program. (NASA Release 85–157)

Public Affairs

February 1: NASA announced the following assignments in its public affairs organization would become effective February 15: Kenneth Atchison, assistant news chief and NASA internal communications manager; Debra Rahn, public affairs officer; Leon Perry, public information officer; Sarah Keegan, public information officer; Dwayne Brown, editor, *NASA Activities*; and Barbara Selby, public affairs specialist. (NASA Release 85–17; NASA anno., Feb. 2/85)

July 3: NASA announced that Thomas DeCair was appointed effective July 3 associate administrator for external relations. In this position, DeCair would be responsible for policy level management, direction, and coordination of the agency's relationships with public and private organizations both domestic and international. This included the news media, other federal agencies, state and local governments, industry, and private individuals. He would also serve as the principal advisor to the NASA administrator and other senior officials on matters pertaining to NASA's external relations activities.

DeCair came to NASA from the U.S. Department of Justice where he served since February 1981 as director of public affairs and special assistant to the Attorney General.

DeCair was staff assistant in the White House Press Office and then assistant press secretary to Presidents Nixon and Ford after he graduated from Hope College, Holland, Michigan, where he was elected to Phi Beta Kappa as a junior and received his B.A. degree magna cum laude. (NASA Release 85–102)

Astronautics and Aeronautics, 1985

Review Boards

March 8: NASA announced that, following launch of Atlas/Centaur 62 carrying the seventh Intelsat V payload, a significant leak occurred in the Centaur liquid oxygen (LO_2) tank during Atlas and Centaur separation, resulting in loss of LO_2 and precipitation of a series of anomalous events that compromised vehicle performance and caused loss of the mission.

Centaur main engine start and steady-state operation began for the first burn sequence; however, the engines burned fuel rich as a result of an attempted correction of the LO_2 loss. Also, Centaur's first main engine cutoff had occurred approximately 11 seconds early, as a result of LO_2 mass loss (approximately 1483 lb.) through the tank opening. After first main engine cutoff, the leading LO_2 ullage gasses (HE and GO_2) created a disturbing force on the vehicle, causing it to tumble out of control during the ensuing coast period.

During the coast period, the LO_2 tank pressure vented down to less than the hydrogen tank pressure at 1040 seconds, causing reversal and rupture of the intermediate bulkhead so that the two tank pressures were essentially the same throughout the remainder of the flight.

Although 2nd Centaur engine start was achieved, the engines shut down after less than seven seconds of operation because of inadequate tank pressurization and subsequent engine cavitation. This made it impossible to achieve proper orbit, and NASA terminated the mission.

Following the failure, NASA Headquarters initiated a Flight Review Board composed of representatives from the NASA centers, U.S. Air Force, and INTELSAT. During the investigation, NASA obtained special assistance from Physics International for analytical shock analysis of various blast-shield and tank-pressure configurations, General Dynamics/Ft. Worth Division for shaped charge firing tests and consultations, and Pratt & Whitney Aircraft for special engine testing to resolve LO_2 back flow anomaly and special postflight reconstructions.

The investigation teams' findings indicated that the most probable cause of the failure was due to shock induced loads on the LO_2 tank at high tank pressures causing tank failure. Corrective actions taken to clear AC–63 for flight would include reduced tank pressures to prior levels, increase in the interstage adaptor to blast-shield gap, and a check for ambient flight pressure and leaks on each tank. (NASA MOR M–491–203–84–07 [postlaunch] Mar 8/85)

June 4: NASA announced that an eight-member review team at Goddard Space Flight Center concluded that the failure of the Global Low Orbiting Message Relay Satellite (GLOMR) to deploy from its canister during the April 1985 Space Shuttle 51–B mission was due to the failure of two switches designed to activate bolt cutters on the canister. On June 7 the team would issue a final report along with recommendations for a solution to the problem.

The Northern Utah Satellite (NUSAT) along with GLOMR would be the first satellites successfully deployed from canisters aboard the Space Shuttle. The door of the GLOMR canister opened, indicating normal battery operation, but the bolt cutters would not activate. After a second deployment attempt, NASA decided to return GLOMR to earth for evaluation. (NASA Release 85–85)

June 20: NASA announced that a mishap investigation board had completed its report on a March 8 accident in which the Space Shuttle Discovery's payload bay door was damaged and a technician was injured during prelaunch preparations for mission 51–D at KSC [see Space Transportation System/Launch Schedules, March 8].

In its executive summary, the board reported that "the immediate cause of the mishap was the failure of a master link in one of the two redundant hoist systems that raise and lower the payload bay access platform." The report then noted, "The mishap can be characterized as the culmination of a series of events and conditions that pushed the mechanical components to and beyond their limits."

In the accident, a payload bay access platform used to provide access to the orbiter's cargo bay fell from its stowed position on a rolling bridge crane. Tracing the events and conditions that led to the accident, the board noted that "Operators of the payload bay access platforms customarily stowed the platforms by raising them until the telescoping tubes contacted the (single) upper-limit switch that stopped its upward travel.

"On March 4, a Lockheed Space Operations Co. (LSOC) technician reported a broken upper-limit switch that had caused the telescoping structure to impact the supporting structure. In crane and hoist parlance, this is called 'two-blocking.' The inboard master link failed at this time, and cable overwrap was noted on that portion of the winch. The entire up-down portion of this system was tagged out with a 'Do Not Operate' tag, since only one half of the redundant hoist system remained intact. This tag was placed on the operating controls along with two other similarly appearing tags, both several months old, describing limitations on the operation of the platforms."

The board's summary then noted the platforms were operated at least twice and stowed at least once between March 4 and the March 8 accident. "During the stowing operation(s), given the fact of a broken upper-limit switch and the standard operating procedure, it is an inescapable conclusion that additional two-blocking occurred. This imparted extremely high loads to the master link in the remaining wire rope assembly, fracturing it almost to the point of sufficient separation for the assembly to fall. When the bridge assembly was moved on the morning of March 8, the resulting jolt was enough to complete the break, and the platform assembly fell."

The board recommended that there be a revision of operating procedures and operator training to ensure the upper-limit switches not be used as operational stops, of tagout/lockout procedures to prevent unauthorized use of equipment that had been identified as unsafe, and of the platform preventive maintenance procedures to meet all KSC and Occupational Safety and Health Administration (OSHA) standards; a modification of the design of the payload bay access platform to include the addition of an operational stop and load sensing device in the wire rope system; and a redesign of the telescoping tubes to facilitate the required inspection of critical linkages.

The board estimated damage to Discovery and access platform at $200,000 and noted that the accident delayed rollover to the vehicle assembly building from March 8 until March 23. (NASA Release 85-95)

Technology Transfers

February 8: Los Angeles's Cedars-Sinai Medical Center announced development of a new surgical tool, using JPL's excimer (excited "dimer") laser technology, for one of medicine's most difficult and long-awaited procedures—a means to clean out clogged vascular passages without surgical intervention, the JPL *Universe* reported. The lasers would be a significant alternative to existing treatments—that of angioplasty, in which a balloon catheter squeezed blockage out of the way, or by-pass surgery.

A JPL research team with Cedars-Sinai colleagues had developed the new laser technology, which used glass-magnetic switches in which a xenon chloride-excimer laser produced a uniform beam of energy controlled and pulsed to extremely short periods, to overcome the problems of lack of precision and control and high heat of current medical lasers. The JPL/Cedars/Sinai procedure would allow a fiber-optic catheter, placed nonsurgically in the brachial and femoral arteries and threaded to the site of the obstruction, to transmit the laser's energy to the clogged vessel. The fiber-optic imaging system would enable doctors to view the artery on a TV screen as the laser cleared the material forming the blockage. The speed of the laser's delivery—10 to 200 billionths of a second bursts of ultraviolet light—allowed the laser to cut through arterial plaque with precision and little damage to surrounding tissue.

Although researchers were cautious in predictions of the laser's medical potential, they believed results of experimentation, though not yet done on living human patients, signaled additional future medical applications, ranging from removal of kidney stones to changing the eye shape of patients with myopia. (JPL *Universe*, Feb 8/85, 1)

May 2: NASA announced that research scientist Dr. Billy Wolverton and research chemist Rebecca McDonald at its National Space Technology Laboratory (NSTL), Bay St. Louis, Missouri, had shown that the water hyacinth, when used under controlled conditions, was ideally suited for purifying domestic and certain industrial wastewaters. In addition to the water hyacinth's ability to produce large quantities of fresh water, the researchers determined the plant could be harvested and ground into fertilizers and used to produce biogas and fiber. The water hyacinth also showed promise for partially supplying life-sustaining functions for space travel including oxygen, food, pure water, and waste treatment.

Researchers at NSTL had for 11 years tested the vascular aquatic water hyacinth (the floating species) as an inexpensive method of treating wastewater. The research led to installation of a simple and cost-effective wastewater-treatment system at NSTL and development communities in Florida, Texas, and California.

More recently, NASA developed an advanced natural wastewater-treatment process that combined anaerobic microbial filter technology with the vascular plant wastewater treatment technology to produce an efficient hybrid system that used rooted, cold-tolerant plants such as common reed growing on the surface of a microbial rock filter bed. The filter reed system had advantages over the floating aquatic system because wastewater was exposed to the atmosphere only after treatment and higher chemical concentrations could be tolerated because of the high surface microbial filter. Although NASA developed the system for domestic sewage, the system had shown a potential for chemical waste and drinking water treatment. (NASA Release 85-65)

May 2: In elaborating on a report that appeared in the Jet Propulsion Laboratory's (JPL) *Universe* [see NASA/Technology Transfer, Feb. 8], NASA issued information on a system called the excimer laser, a laser system developed by a team of physicians at Los Angeles's Cedars-Sinai Medical Center and laser scientists at JPL to nonsurgically clean clogged arteries with unprecedented precision. The system may eventually allow patients with arteriosclerosis to avoid coronary bypass surgery.

JPL scientists originally developed the excimer laser to measure gases such as ozone in the earth's atmosphere. Investigations into its application to medicine began a year and a half previously when Cedars-Sinai physicians Warren Grundfest, Frank Litvack, and James Forrester, who were conducting research into the potential of lasers in cardiology, sought a more precise and cooler laser than those currently available for use in medicine.

They found such a laser in the excimer developed by JPL laser researchers Drs. James Laudenslager, Thomas Pacala, Stuart McDermid, and David Rider. Working with the Cedars-Sinai physicians and a fiber optics consultant, Dr.

Tsvi Goldenberg, the JPL team refined the laser for the delicate cardiovascular cleaning procedure devised by the medical researchers. Although the researchers were properly cautious in their predictions of the laser's medical potential, the results of experiments were encouraging.

NASA's office of space science and applications funded development of the excimer laser. (NASA Release 85-66)

May 2: NASA announced that Inorganic Coatings, Inc. was providing interior corrosion protection to the refurbished Statue of Liberty by means of a primer coating known as K-Zinc 531, an aerospace spinoff product developed at Goddard Space Flight Center (GSFC) to protect gantries and other structures at NASA's primary launch site, Kennedy Space Center (KSC).

Because KSC was located on Florida's Atlantic Coast, its launch facilities required greater corrosion protection due to constant exposure to salt spray and fog. GSFC undertook a research program to develop a coating that would not only resist salt corrosion, but also protect KSC launch structures from very hot rocket exhaust and the thermal shock created by rapid temperature changes during a space launch. The GSFC research resulted in an inorganic water-based potassium silicate binder, a compound that provided long-term protection with a single application.

NASA granted in 1981 a license for the coating to Shane Associates, which signed an agreement with Inorganic Coatings to allow it to become the sole manufacturer and sales agent under the Shane license. Inorganic Coatings assigned the trade name K-Zinc 531 to the compound, which was nontoxic, nonflammable, and had no organic emissions. The high ratio silicate formulation bonded to steel in 30 minutes and created a hard ceramic finish with superior adhesion and abrasion resistance. It required no straining before application and could be mixed on site. (NASA Release 85-64)

May 6: NASA announced that space age technology was in use in the Republic of Gabon, Africa, where solar photovoltaic power systems installed in four rural villages would improve public health and education facilities and lighting and sanitary water supplies. NASA's Lewis Research Center (LeRC) managed the project, which was jointly funded by the U.S Department of Energy (DOE) and the Ministry of Energy and Hydraulic Resources of the Republic of Gabon; the Gabon government selected the villages to participate in the program.

The project provided a photovoltaic power system, lights, air circulation fan, and a refrigerator/freezer for vaccine storage to each village dispensary and a power system, lights, and a color TV and video cassette recorder/player to each village school. The villages' sanitary water systems had a newly drilled well, a photovoltaic system, a submersible well pump, a water storage tank, and a water distribution system that piped water to village fountains. Public lighting systems included a power system and sodium vapor lamps mounted on poles. All systems were instrumented to collect data for performance analyses.

Total photovoltaic array capacity for all systems was 12.1 kw; output of the smallest array was 80 watts for street lights, the largest 3 kw for one water pump. Batteries for the solar arrays should last about 10 years before replacement; solar arrays had a life expectancy of at least 20 years.

According to Tony Ratajczak, project manager at LeRC, "The goal of DOE and the Republic of Gabon, for this project, is to investigate and evaluate the economic, social and technical value of photovoltaic power systems in aiding progress and improving the quality of life in Gabon." (NASA Release 85-68)

During November: NASA announced that three physicians from Washington University Medical Center, St. Louis, Misssouri, met the previous month with NASA scientists from National Space Technology Laboratories (NSTL) to discuss using remote sensing techniques as an analytical tool for use in the field of body scanning known as nuclear magnetic resonance or NMR. Dr. Michael Vannier, radiologist and principal investigator for the project, said NMR was already used on a number of organ systems in the brain. However, interest had grown in how magnetic resonance imaging might fit into the scope of breast diagnostic techniques. So Drs. John Gohagan and Ed Spitznagel, also of Washington University, on behalf of the National Cancer Institute's Breast Imaging Project joined Vannier at the discussions.

The meeting resulted in a three-year collaborative effort among the Mallinckrodt Institute of Radiology in St. Louis, NSTL, and the Kennedy Space Center.

Vannier commented that the union with NASA was vital to maximize the potential of NMR scanning. "One of the characteristics of these magnetic resonance scanners is that they provide very specific anatomic information about the location and size of a tumor. However there are other ways to get the same information, one being the use of a mammogram, or breast X-ray."

At NSTL, scientists Doug Rickman and Jim Anderson were responsible for applying the analytical capabilities to make NMR scans a notable advancement over other methods of X-ray and body scanning techniques. They had demonstrated the use of ELAS software to enhance NMR images, thus increasing disease detection accuracy. Vannier explained, "No one in medicine really has the kind of experience with classifying or analyzing these types of images this way. That's why we are using NASA expertise to help us."

By incorporating the ELAS software used in processing satellite imagery into the system, NMR scans could provide not only information on the location and size of a tumor, but also its biological behavior. That is, physicians could determine the status of a mass of tumor directly from the NMR image without having to enter the body surgically.

Close affiliation with pathologists while working on actual cases was important in determining whether there had been misclassification on the image processing machines, Gohagan said. "Our goal is to ultimately get the kind of discrimination necessary to distinguish different anamolous conditions," he explained. For example, blood clots would appear distinctly different from benign tumors, and a marked difference would be evident between benign and malignant tumors.

Vannier concluded that, as a result of the collaboration, an improvement should evolve in the way NMR images are acquired. "I think it will have significant influence on what we do in the future," he said. (*NASA Activities*, Nov 85, 7)

December 5: NASA announced that a NASA scientist and a Stanford University engineer developed the bone stiffness analyzer, an instrument that might aid in treating bone fractures and bone-weaking diseases including osteoporosis, which afflicted millions of the elderly. The two originally developed the instrument, which measured bone mass and stiffness, to help scientists combat bone loss that might occur during long-term spaceflights, particularly on manned space stations or extended space journeys such as trips to Mars.

The instrument was based on a theory, initially demonstrated by Dr. Donald Young, a physiologist at NASA's Ames Research Center, that bone behaved as a structural beam and that well-developed concepts for testing the stiffness and displacement properties of structural beams could be applied to measuring the same properties in arm and leg bones. Dr. Charles Steele, professor of mechanical engineering at Stanford University, adapted the instrument for clinical application.

The analyzer gauged the bone's resistance to a small amount of pressure applied to the forearm or leg bone, while the subject's arm or leg was positioned so the ends were immobile. The instrument's probe, an electromagnetic "shaker" or iron core wrapped with wire, was placed at mid bone and current was run through it, causing the bone to vibrate. A microprocessor then measured the bone's displacement using algorithms stored in its memory, deducing the bone's stiffness and effective mass.

Since the analyzer responded quickly—a test took less than one minute and did not damage the bone or tissue—the analyzer might have wide applications for screening diseases such as osteoporosis, which weakened bones but was usually diagnosed only after a fracture occurred when the disease was well advanced. Although the bone analyzer could not be used on the spine where osteoporosis often first appeared, it could detect the disease long before X-rays, which did not show evidence of change until after at least 20% bone loss. Steele was planning to adapt the device for use on fingers, which also showed early evidence of the bone disease.

Since the analyzer could also monitor a bone's strength as it healed, the device might also aid in the treatment of fractures. The analyzer's quantitative measure of bone strength could replace the inference and guesswork that usually was the basis for a decision on removing a cast from a fractured bone.

After three years of clinical testing on more than 300 subjects, Steele believed the device was at a useful level of precision. To find normal values for bone stiffness, Young and Steele tested participants in the Stanford Invitational Rugby Tournament in 1984. In the coming year, they would conduct more tests on healthy subjects and a data search to find normal "loads," the amount of stresses and strains needed to maintain healthy bones.

After further tests, Young and Steele planned to use the instrument to create an exercise program for maintaining bone strength during the weightless environment associated with extended spaceflight missions when bones tended to atrophy.

Young hoped to develop a program efficient enough to place the necessary stresses on bones through short periods of daily exercise. "It would be great if it could be done in an hour a day," Young commented. He believed a trampoline-like device, with restraints to hold the body, might be effective. Eventually, the analyzer itself might go into space with astronauts, who could test their own bone strength, perhaps determining when they needed to return to earth. (NASA Release 85, 163)

December 23: In order to improve communications among Indonesia's Eastern Island University's member campuses, which were spread over 1,600 miles of ocean, NASA's Lewis Research Center (LeRC) recently supervised installation of a solar power system for a teleconferencing center and earth station in the village of Wawotobi. The earth station provided the communications link among Wawotobi, Djakarta, and an earth communications satellite system and helped make the member campuses' scarce professors, research findings, and library resources more widely available to the rapidly increasing number of students throughout the region.

The teleconferencing center, a classroom large enough for 50 people, was equipped with a speaker system, microphones, a telecopier, an electronic blackboard, and a computer graphics display. The system was interactive and could connect with 11 main campus locations of the Eastern Island University Association and a conference room in the offices of the Director General of Higher Education in Djakarta.

Purpose of the teleconferencing system was to demonstrate the use of satellite communication in developing countries, and the solar power system showed the feasibility and practicality of solar (photovoltaic) power systems for small earth stations in remote areas.

Although diesel electric generators generally powered earth stations in the absence of conventional power lines, the high cost of diesel fuel and the high operation, security, and maintenance costs for diesel generators made use of a photovoltaic system attractive. The total photovoltaic array capacity to power the earth station at Wawotobi was 1.5 kilowatts, and LeRC expected the battery used in conjunction with the array to last about 10 years. The photovoltaic array had a life expectancy of more than 20 years.

The existing satellite system provided national communications services by radio, telephone, and telegraph to the 5000 inhabited islands.

Hughes Aircraft Co. was prime contractor for the photovoltaic power system, and their subcontractor in Indonesia was P. T. Elektrindo Nusantara. LeRC and Hughes would provide operations support for repair, maintenance, and spare parts through June 1986, when NASA would turn over complete responsibility for the system to PERUMTAL, the Indonesian national telecommunications authority.

The project completed the renewable energy systems programs managed by LeRC for the Department of Energy and the U.S. Agency for International Development. Since 1976 LeRC had been responsible for the installation of 71 photovoltaic power systems in 27 countries and the United States. These systems provided power for entire villages for water pumps, grain grinders, lights, vaccine refrigerators, schools, highway signs, forest lookout towers, insect traps, rural dispensaries, and the Wawotobi teleconferencing center/earth station. (NASA Release 85–176)

PEOPLES REPUBLIC OF CHINA

February 17: Li Xue, vice minister of the Peoples Republic of China (PRC) astronautics industry, indicated in London recently that the PRC wanted to compete with the U.S. and West European countries for rocket launches of other countries' satellites, FBIS, Shanghai City Service in Mandarin reported.

He said the PRC had not decided to launch charges or how to market the service, but it was ready to talk to western countries needing launch services. (FBIS, Shanghai City Service in Mandarin, Feb 17/85)

May 30: The Peoples Republic of China (PRC) State Council decided to use an international communications satellite over the Indian Ocean to transmit TV programs to people in outlying areas, FBIS Beijing XINHUA Domestic Service in Chinese reported. Although satellite transmission of TV programs covered areas where 62% of the population lived, reception in some areas was poor. Therefore, the State Council decided to create a special fund to present 50 ground stations to outlying areas and over the next two to three years to expand this ground-station network.

In addition to TV programs, the satellite would transmit newspaper facsimile, educational programs, medical consultations, weather information, and computer linkages. (Beijing XINHUA Domestic Service in Chinese, May 30/85)

June 6: Scientists from China's Ministry of Astronautics announced the previous week plans to improve the design of their existing rockets so they could launch heavier satellites into geostationary orbits, thus challenging the western nations' domination of commercial exploitation of space, the *New Scientist* reported.

China had a rocket that could put 750-kg communications satellites into geostationary orbit, which they used the previous year to launch their first such satellite. Jean Vandenkerckhove of the European Space Agency (ESA) said the rocket made "Chinese launch capability equivalent to Europe's Ariane 3 rocket."

China's rocket, the Long March 3, had two liquid-fuel stages and one cryogenic stage fuelled by liquid hydrogen and oxygen. Chinese engineers said they planned to add extra boosters to allow them to put heavier loads, possibly two satellites, into space. The engineers also spoke of plans to develop a cryogenic upper stage for the rocket that could launch 2.5-ton satellites 35,000 km high into geostationary orbit. That would give China a capability close to that of Ariane 4, Europe's newest rocket. However, Wu Ke Li, in charge of Chinese interest at the Paris Air Show where China for the first time had an exhibit, would not confirm or deny the plans.

Western engineers were impressed by China's existing ability because Long March 3 placed their satellite into correct orbit without flight tests, since China could not afford such tests. Instead, the Chinese relied on elaborate space simulators for checking satellites and launch components.

Vandenkerckhove said China's launch facilities were at a disadvantage compared with Ariane because the compartment for satellites on Long March 3 was smaller than on Ariane 3, measuring 2.3 m rather than 2.9 m. However, Yi Zluo Hyang, one of the Long March 3 designers, said, "Our launch costs will be lower than any other service, partly because we have lower labor costs, but also because we will be supported by the government." During the past year, China had signed cooperative agreements in space science and technology with the U.S., W. Germany, France, and Britian; and they were pursuing an interest in remote-sensing satellites. (*New Scientist*, June 6/85, 4)

July 29: Peoples Republic of China President Li Xiannian today made a three-hour visit to the Jet Propulsion Laboratory, the JPL *Universe* reported. JPL Director Dr. Lew Allen welcomed him, then said, "We are aware of the ongoing discussions between our governments regarding future space cooperation . . . We look forward to cooperative scientific investigations."

Xiannian then viewed JPL's newly revised multi-media presentation, "Welcome to Outer Space," which had been translated into Chinese. After the show, Allen presented the president with a color photograph of Saturn and its moons; Dr. Taylor Wang—the first Chinese-American in space—gave him a photographic collage showing Wang in space during his May Space Shuttle flight and his drop dynamics module equipment.

Xiannian completed his visit with a tour of the Spacecraft Assembly Facility where he heard about the Galileo mission and saw a model of the spacecraft as well as actual pieces of its hardware. (JPL *Universe*, July 85, 1)

August 5: Weather satellite engineers from the Shanghai Institute of Satellite Engineering in the Peoples Republic of China planned to visit the U.S. to seek technological help to complete within the next few years assembly and launch of China's first polar orbit weather satellite, *Aviation Week* reported.

The institute was building a primary and backup spacecraft for China's first weather satellite mission and had started preliminary design of a Chinese geosynchronous orbit weather spacecraft planned for launch in the early 1990s.

The Shanghai Institute of Technical Physics was working on the imaging system the weather satellites would use to take cloud photos and collect other data comparable to that from the U.S. Tiros weather satellite series.

Ground test data from the Chinese-designed tape recorder system and elements of the gyro system for the new polar orbit weather spacecraft, however, had not achieved a high enough reliability standard. The weather satellite manager planned to visit the U.S. in late 1985 or early 1986 to try to procure a U.S.-built tape recorder that could fly in the satellite in tandem with a Chinese-built recorder to improve reliability of the spacecraft.

A U.S. space team visiting the satellite engineering institute said the satellite was relatively large, about 4 feet tall and 4.5 feet wide. With solar arrays deployed, total spacecraft span was about 30 feet.

The spacecraft would weigh 1,500 lb. at launch and be in a 500-mile-high orbit at 99° orbital inclination. The Chinese State Bureau of Meteorology, Beijing, supplied performance specifications and funding for satellite.

The Chinese weather spacecraft, like U.S. weather satellites, would transmit data continuously. Any nation with the receivers used for U.S. weather data could pick up the Chinese data as well. (*AvWk*, Aug 5/85, 79)

October 26: Peoples Republic of China astronautics minister Li Xue said China would put their Long March 2 and 3 rockets on the international market to provide satellite launching services for local and foreign clients, FBIS Beijing XINHUA Hong Kong Service in Chinese reported. The decision resulted from the October 21 launch by a Long March 2 rocket of an earth surveying satellite and an April 8, 1984 launch by a Long March 3 rocket of China's first experimental telecommunications satellite.

Li Xue said the recent Long March 2 launch was its seventh consecutive success, indicating the two-stage liquid-propellant rocket could reliably put a satellite of about two tons into a near-earth orbit from the Jiuquan launching center, in addition the Long March 3 launch from the Xichang launching center of a satellite into geosynchronous orbit showed China had mastered the technology for placing a satellite in that orbit.

He added that in addition to launch services from the two centers, China could provide satellite orbital control and support from the Xian control center, which had an ocean-based survey vessel component. And he pointed out that China could provide preferential prices for foreign customers, would train technicians, and had the capability, through the Chinese People's Insurance Co., to underwrite satellites that were launched.

In conclusion, he said that, after the Long March 2 and 3 rockets were on the international market, they would promote international technical and economic cooperation in the astronautics field and contribute to the maintenance of world peace. (FBIS, XINHUA Kong Hong Service in Chinese, Oct 26/85)

SATELLITES

January 10: Four years after cancellation of the National Oceanic Satellite System (NOSS), the U.S. Navy had begun to develop a simple, low-cost system to monitor ocean conditions routinely from space, *Aerospace Daily* reported. Budgeted at about a third the estimated cost of NOSS, the Navy Remote Ocean Sensing System (NROSS) would have nearly the same capability as NOSS by 1989, lacking particularly the synthetic aperture radar, and would use many of the ocean observation capabilities NASA's Seasat program demonstrated in 1979.

The Pentagon had avoided $591.5 million in FY 1981–88 costs by cancelling NOSS (NASA and NOAA contributions would have made the total over $1 billion). NROSS would use a single-satellite system with no backup; the Navy would resolve later the issue of replenishment satellites. NROSS would also use existing or already planned data-processing equipment at the Navy's Fleet Numerical Oceanography Center and the USAF's Defense Meteorological Satellite Program command and control facilities.

The Navy expected to spend about $200 million from FY 85 development through a launch in 1989, with a NASA contribution of $115 million ($35 million for the satellite program and $80 million in research). The Office of Management and Budget had vetoed use of NOAA's NOAA–D spacecraft as the NROSS bus, so NROSS would use a Block 5D–2 DMSP spacecraft. The Navy had planned an Atlas E launch, but was considering a refurbished Titan II ICBM. Planners rejected a Space Shuttle launch as too expensive.

January 25: The Mutual Broadcasting System announced plans to use its excess satellite space to deliver messages at rates low enough to compete with the postal service, the *Washington Post* reported. The company estimated that telecommunications business from the new Mutual Satellite Services division would generate 50% of its revenue by 1990. Mutual's principal business was the Mutual Radio Network, which owned receiving dishes at more than 700 locations nationwide.

Mutual planned starting in mid-1985 to send audio and data transmissions over a little-used portion of the FM radio spectrum to specially tuned radios in clients' offices. Cassette tape recorders attached to the radios would receive voice communications and send them to high-speed printers. Called MultiComm, the system would allow any organization to create its own private network for sending encrypted audio, data, printed material, or computer software to a number of locations simultaneously. Senders could relay a 300- to 450-word message to 500 receivers for 25 to 30 cents per recipient, compared to a first-class letter at 22 cents. Other electronic mail systems use a wire hookup between the company providing the satellite link and the message recipient; Mutual would provide the first system in which the signal was transmitted by radio. (*W Post*, Jan 25/85, D2)

During January: A six-man Goddard Space Flight Center (GSFC) team at the South Pole on December 14, 1984, had taken turns talking to colleagues in Greenbelt, Maryland, using one of two unprecedented satellite communications links the group had installed over the previous month, the *Goddard News* reported. "The conversation was amazingly audible," said Tony Comberiate, a GSFC communications expert and member of the South Pole satellite data-link project team. "The folks at Goddard sounded like they were just a few feet away."

Although voice communications capabilities from the South Pole using Ham radio had existed for some time, the systems were weather dependent and usually very noisy. The GSFC group used the 17-year-old *Applications Technology Satellite* (*ATS-3*) to establish the two-way voice link. The key breakthrough was installation of the scientific data link, which enabled daily transmittal of information from the Pole, across Antarctica, and to the U.S.

Before system installation, scientific data from the pole was stored during the region's winter months and shipped out by aircraft from November through February. The new system permitted transmission of high-quality scientific data by using three existing polar orbiting satellites (a 4th was scheduled soon). Each satellite passed the pole about 14 times daily and relayed data to McMurdo Sound, which retransmitted it to a geosynchronous satellite, which in turn transmitted it to the U.S. Since the Pole signals were too far below the horizon to be acquired by normal communications satellites, the relay route was necessary. Researchers had previously considered a communications link from the South Pole, but decided it was too costly and impractical, perhaps taking several years and costing an estimated $35 million. The GSFC approach had cost about $250,000 and taken nine months, because it used existing satellites, excess equipment, and had support from several organizations.

Scientists could receive daily transmission of reliable data on global weather patterns, the magnetospheric cusp, upper atmosphere, and glaciological and seismic studies to name a few. The link also could evolve into a data collection network for many of the unmanned observatories (ground-based satellites) scattered throughout Antarctica. (*Goddard News*, Jan 85, 6)

March 7: NASA announced plans to deploy for the first time two small experimental satellites from Get Away Special (GAS) containers mounted in the cargo bay of the Space Shuttle orbiter Challenger during the STS 51-B mission scheduled for launch in late April. Under the GAS program, NASA would deploy for $10,000 each the Global Low Orbiting Message Relay Satellite (GLOMR) and the Northern Utah Satellite (NUSAT) in hopes of establishing an inexpensive way to deploy small satellites during routine Space Shuttle operations.

Clarke Prouty, technical liaison officer for the GAS program at Goddard Space Flight Center (GSFC), said the GAS containers had been upgraded with ejection systems for the 51-B mission and that GSFC had developed a motorized door (full diameter motorized door assembly) for the can similar to the one first flown on the seventh Space Shuttle mission, which would allow the GAS payload to be exposed to space. The door assembly permitted GAS container insulation before and after satellite deployment and provided a means for keeping the satellite in the container in case of malfunction. GSFC had also adapted the spacecraft separation system used in the Delta rocket program for the GAS ejection systems.

The GLOMR satellite, designed and built by Defense Systems Inc. was a data relay, communications spacecraft that was expected to remain in orbit for about a year. The NUSAT, designed, built, and tested by Weber State College, Ogden, Utah, in coordination with the Federal Aviation Administration, was an air traffic control radar system calibrator that would measure antenna patterns for ground-based radars operating in the U.S. and member countries of the International Civil Aviation Organization.

NASA would first launch the NUSAT, then the GLOMR, at the end of Spacelab 3 science activities on the sixth day of the seven-day 51-B mission. Independent user ground stations would operate the satellites following deployment.

The GAS program was available to anyone wishing to fly a small (for containment in 2½- by 5-ft area) scientific research and development experiment. The Space Shuttle had flown 29 GAS containers, including those for materials processing, life sciences, biology, seed and crystal growth, and cosmic radiation. (NASA Release 85-35)

March 25: The Arab League's *Arabsat 1* telecommunications satellite, launched February 8 by an Ariane 3 from Kourou, French Guiana, experienced control problems during initial in-orbit checkout, but was reported stabilized with all systems functioning normally, *Aviation Week* reported. Anomalies in the spacecraft's gyro package caused the problems, detected March 16 by operators at Arabsat's Dihrab, Saudi Arabia, ground station when the satellite was at its final on-station location at 19° E longitude. Investigators were attempting to pinpoint the problem.

An electrostatic discharge may have caused one of the gyro package anomalies. Although the gyros provided pitch, roll, and yaw information, "we never lost control of the satellite," said Michel Duigot, Aerospatiale's Space and Ballistic Systems Division communications satellite system manager. "There is redundancy in the satellite because we have two packages of three gyros each," he pointed out.

The satellite experienced solar array deployment difficulties soon after launch, but the arrays eventually extended successfully. Program officials said the two problems were unrelated.

The solar array deployment problems had resulted in minor modifications to the solar panels on the No. 2 Arabsat, originally scheduled for launch by NASA from KSC in May, but rescheduled for June.

The 22-member Arab Satellite Communications Organization would use the *Arabsat 1* for national and regional communications and TV broadcast requirements. The schedule called for an operational satellite by the end of April, and program officials believed this might still be possible. (*AvWk*, Mar 25.85, 22)

March 29: NASA announced that its first *Applications Technology Satellite* (*ATS-1*), after more than 18 years of service, failed to respond to commands to correct its eastward drift from geostationary position over the Gilbert Islands in the western Pacific. Robert Wales, ATS project operations director at Goddard Space Flight Center (GFSC), said that the ground control station in Hawaii could no longer keep *ATS-1* at its present location and it would likely drift out of useful orbital position during the next six months.

ATS-1, launched in December 1966 with an expected lifespan of three years, most recently had provided a voice and data communications capability to several information networks in the Pacific basin. The pan Pacific education and communications experiments by satellite (PEACESAT) program, the major user of *ATS-1*, would dissolve with the loss of the satellite. Program participants had transmitted educational, health, research, technology, and community services through *ATS-1* to 23 autonomous terminals located in Hawaii, Cook Islands, the Mariana and Caroline Islands, Western and American Samoa, the Marshall Islands, Melanesia, New Zealand, and Australia.

Some of *ATS-1's* notable achievements were the first transmission in 1967 of full-earth, cloud cover pictures from geosynchronous orbit; first transmission in 1967 of real-time TV pictures (Apollo 4 splashdown); two-way communication tests with commercial airliners to determine aircraft orientation effects on satellite communications, a cooperative venture with the Federal Aviation Administration (FAA) and the airlines in 1967 and 1968; link-up between U.S. and USSR scientists during an atmospheric, sea, and ice condition experiment in the Bering Sea in 1971; transmission of electrocardiographs from Hawaii to New Zealand and from Alaska to the University of Washington; and presentation of medical conferences over the PEACESAT network.

The medical world had praised the Alaskan "doctor call" service provided by *ATS-1* as the first innovative approach to rural medicine in the U.S. In the program, Public Health Service physicians could communicate daily through *ATS-1* with trained health aides in the remote Alaskan bush country.

Loss of *ATS-1* would leave one other comparable satellite, *ATS-3*, launched in November 1967, in operation. Positioned in geosynchronous orbit over the Pacific Ocean south of Mexico, *ATS-3* covered the U.S., most of the Atlantic Ocean, and a large part of the eastern Pacific including Hawaii. (NASA Release 85–45)

March 29: Arnold Engineering Development Center (AEDC) qualification tests had shown the Star 30C apogee kick motor, designed to place satellites in orbits around the earth and manufactured by Morton-Thiokol Inc., was ready for production, the Air Force Systems Command (AFSC) *Newsreview* reported.

Sverdrup Technology Inc., operating contractor of AEDC's propulsion test facilities, had conducted qualification checks in one of AEDC's high-altitude rocket development test cells at a simulated altitude of about 100,000 feet. Test objectives were to demonstrate component structural integrity and to determine ballistic performance. Sverdrup temperature conditioned the motors and then test fired them while they spun about their axial centerline to simulate spin stabilization during firing in space.

The Star 30C, designed to be carried by the Space Shuttle, was the propulsion system for the RCA G–STAR satellite. Once in space, it would ignite and carry the satellite to its designated orbit. (AFSC *Newsreview*, Mar 29/85, 6)

April 4: President Reagan announced that March 31 through April 6, 1985, would be National Weather Satellite Week in a proclamation that said, "The United States' weather satellites have tracked the earth's weather since April 1, 1960, and have brought unique benefits to the American people and the world.

"Weather satellites have proven exceptionally valuable in detecting, monitoring, and giving early warning of hurricanes, severe storms, flash floods, and other life-threatening natural hazards, on a local, national, and international basis.

"The international weather satellite search-and-rescue program has saved over three hundred lives since 1982. The achievements of the scientific and aerospace communities in developing weather satellites have contributed significantly to the United States' leadership in satellite technology, international cooperation in space, and an integrated global weather forecasting system . . ."

The proclamation concluded that NASA "has been the world leader in the development of experimental and prototypical weather and environmental satellites. The National Oceanic and Atmospheric Administration has demonstrated outstanding leadership in the management of operational weather and environmental satellite systems and programs." (Admin. of Ronald Reagan, Proclamation 5314, Apr 5/85, 421)

April 22: In assessing the status of the *Syncom IV–3 (Leasat 3)* satellite that was drifting in low-earth orbit after its apogee kick motor failed to push it into proper orbit [see Apr. 18], aerospace engineers said astronauts might be able to salvage the satellite, the *Washington Times* reported. Payload specialists originally believed it would be too dangerous to return the satellite to earth on the Space Shuttle because of the nearly six tons of highly volatile rocket fuel it carried. Jerome Hammack, a spacecraft safety operations expert, said deactivating the satellite's electronic systems would significantly reduce the chance of stray sparks igniting the fuel, most of which would turn to gas upon descent. To ensure that there was no power going to any of the electrical circuits, a spacewalking astronaut would have to cut power circuits running to the satellite's batteries and solar power cells. However, NASA would mount a rescue attempt only if Hughes Communications Inc., Leasat 3's manufacturer, and Hughes' insurance underwriters agreed to a salvage contract. (*W Times*, Apr 22/85, 2A)

May 5: In a commencement address today at the George Washington University School of Public and International Affairs, Charles Wick, director of the U.S. Information Agency, said that the "very survival" of the U.S. may depend on the extent to which its citizens use and master the tools of mass communications, the *NY Times* reported.

"Through the explosion of global satellite communications, a technological 'genie' has been unleashed which will change forever the way that governments communicate ideas and information abroad," he said. However, he noted that ". . . not all countries believe in freedom of ideas. Not all accept the Western tradition that free expression is a basic human right. These opposing forces are attempting to persuade other countries, particularly those in the third world, that the state, not the people, must decide which ideas circulate in their magazines and newspapers and on the TV and radio stations . . .

"Since the 1970s, the Soviet Union has introduced resolutions at Unesco, in the guise of a 'New World Information Order,' to impose this control of the mass media on all nations of the world . . .

"Unless you and I do a better job of explaining to the world how the free flow of information can benefit mankind, other nations will not be persuaded to open their borders to the ideas and opinions of others. This could have damaging, even fatal consequences for a dangerous world, in which cooperation—and survival—depend so heavily on mutual understanding," Wick concluded. (*NYT*, May 7/85, B8)

May 7: The European Space Agency (ESA) and the Department of Commerce's National Oceanic and Atmospheric Administration (NOAA) announced that the U.S. agreed to lend ESA its *GOES-4* standby geostationary weather satellite as a temporary substitute for ESA's *Meteosat-1* spacecraft. *Meteosat-1*, launched in 1977 with a planned 3-year lifetime, had run out of

Satellites

station-keeping fuel and would drift out of view of ESA's ground-station network in July.

NOAA would move *GOES–4*, which had been at 140° W longitude over the Pacific Ocean, westward by 4° a day to bring it by mid-June to its new position at 10° W longitude above the Atlantic Ocean. Because ESA group equipment was not compatible with the *GOES–4* command system, NOAA would operate the satellite for ESA from its satellite guidance facilities at Suitland, Maryland, and Wallops Station, Virginia. ESA's Meteosat ground facilities in Darmstadt would receive data from the spacecraft.

When NOAA's Atlantic-area *GOES-East* Satellite lost its imaging capability in 1984, ESA's *Meteosat–2* provided U.S. weather watchers with weather data, including information on the eastern Atlantic hurricane breeding grounds. (ESA Release, May 7/85)

May 24: NASA announced it was negotiating an agreement with Hughes Communications, Inc. (a wholly owned subsidiary of Hughes Aircraft Co.) to develop jointly plans for a Space Shuttle mission to attempt salvage of the Hughes *Leasat 3* satellite then in orbit [see Space Transportation System/ Missions, Apr. 17].

As a result of Hughes's negotiations with New York- and London-based underwriters insuring *Leasat 3*, Lloyd's and other European underwriters had agreed to proceed with the attempted salvage. Hughes was continuing negotiations with U.S. underwriters.

Although the joint salvage effort would include activities never before attempted, it was based in large part on experience gained by NASA during its April 1984 repair of the Solar Maximum Mission satellite and its retrieval November 1984 of the Palapa B–2 and Westar VI satellites.

Pending an independent review of safety considerations by the Aerospace Safety Advisory Panel, the mission would occur during Space Shuttle flight 51–L scheduled for no earlier than August 24. The salvage plan called for modification to the satellite by two of the Space Shuttle crew during rendezvous with it to permit ground command of the satellite. Ground command would then perform the activation sequence, normally performed by an automatic timer onboard the satellite. Modification made during the rendezvous would bypass all hardware likely to have caused the *Leasat 3* failure. (Although Hughes had identified satellite activation components and circuits as the cause of the failure, the company had not determined, due to limited flight data, a specific cause of the failure.)

In its dormant state, the satellite was experiencing temperatures well below the design and test limits of the liquid- and sold-fuel propellant systems, electronic units, batteries, and all other components. These factors, combined with the complexity of the modifications by the Space Shuttle crew, limited the chances of success.

Hughes Communications, Inc. would contract with NASA for the costs, at that time under assessment, for preparation and execution of the mission. (NASA 85-77)

May 24: NASA had brought under control a $50-million weather satellite, *NOAA-8*, that had tumbled helplessly in space for nearly a year due to a malfunctioning oscillator, the *Washington Times* reported.

The defective oscillator gave out during April, allowing NASA and National Oceanic and Atmospheric Administration (NOAA) scientists to activate a backup oscillator and reprogram the satellite via remote control. NOAA was checking and calibrating the spacecraft's instruments before putting it July 8 into service to transmit daily weather photos and information. (*W Times*, May 24/85, 3A; *W Post*, May 24, 85, A8)

May 29: M. Peter McPherson, head of the Agency for International Development (AID), said a U.S.-financed and -built satellite weather-alert system may have given Bangladesh up to 24 hours notice of a devastating cyclone that killed approximately 10,000 people, the *Washington Times* reported. The advance notice may have saved "a substantial number of lives," along coastal areas of the Bay of Bengal, McPherson said.

McPherson pointed out that the satellite system in Bangladesh, developed in phases since 1978 by NASA, the National Oceanic and Atmospheric Administration, and AID warned of the storm. He said Bangladesh technicians had tracked the cyclone for four days and had the capability of predicting landfall and wind speed within 18 to 24 hr. (*W Times*, May 29/85, 5A)

May 30: NASA announced it had selected Ford Aerospace and Communications Corp., Western Development Laboratories, to negotiate a cost-plus-award-fee contract for the next generation of Geostationary Operational Environmental Satellites (GOES) that would include three spacecraft (GOES-I, J, and K) and a two spacecraft option (GOES-L and M). The contract had a proposed cost of $221 million and would provide for a series of GOES satellite systems, each with a five-year design life, that would contribute to continuation of the National Oceanic and Atmospheric Administration's (NOAA) GOES program in the 1990 to 2000 era. NASA had scheduled a late 1989 launch for the first GOES satellite in this new series.

The contract statement of work covered the satellite bus; imaging, sounding, and space-environment monitor instruments; and flight-support equipment and services to accomplish deployment of the spacecraft into geosynchronous orbit from the Space Shuttle. The contract also included operations ground equipment and support to ensure compatibility of the spacecraft systems with established ground systems operated by NOAA.

Satellites

NASA assigned Goddard Space Flight Center responsibility for project management. (NASA Release 85-80)

May 30: NASA announced that NASA, National Oceanic and Atmospheric Administration (NOAA), and RCA engineers had stabilized the *NOAA-8* satellite after it began tumbling out of control when an oscillator in the spacecraft's altitude control system failed. Engineers were continuing a three-week checkout of the spacecraft to ensure that all calibrations were in order, at which time the satellite would be fully operational.

The satellite carried search and rescue (SARSAT) equipment that would resume operation as part of an international program begun in September 1983 to use satellites to save people in downed airplanes or on ships in distress. During the 11 months *NOAA-8* malfunctioned, three USSR and one U.S. satellite were picking up distress signals and relaying them to ground stations.

The reactivated *NOAA-8* would assume weather data collection that *NOAA-6* had been doing. *NOAA-6* had no search and rescue equipment and its infrared radiometer for taking temperature measurements was not working. (NASA Release 85-81)

July 8: Recently completed agreements between insurance underwriters and Hughes Communications Services acknowledged loss of the disabled *Leasat-3* spacecraft but established a plan under which Hughes and the underwriters would share in communication lease revenues if a plan to repair the satellite during the Space Shuttle 51-I mission was successful, *Aviation Week* reported.

Under the agreement with American and European underwriters who insured *Leasat-3*, Hughes would provide the funds, estimated at about $10 million, for the rescue attempt. If the rescue was successful, Hughes would regain the cost of the salvage mission through initial lease revenues. The agreements recognized a loss of $85 million for the spacecraft and launch costs.

The loss resulted from the failure of the spacecraft on April 13 to activate during an automatic deployment sequence after the spacecraft was ejected from the Space Shuttle orbiter Discovery's cargo bay [see Space Transportation System/Missions, Apr. 17].

Hughes and NASA were developing equipment and procedures to mount the rescue attempt. Johnson Space Center had concluded a critical design review for rescue hardware such as a worksite, stowage assembly, and the capture, handling, grapple, and spinup bars that would be used to stabilize, restrain, and spin the satellite once modifications aimed at restoring ground control were made on orbit. The design review also covered components Hughes was building, such as the remote power unit, spun bypass unit, and a checkout box that would be used to test electronic components on orbit prior to the rescue attempt.

Completion of the critical design review enabled NASA to obtain information on the electronic components Hughes was building and to build rescue equipment for the mission, conduct vacuum chamber testing, and begin training for the salvage mission.

NASA had already conducted initial neutral buoyancy training in the weightless training environment facility and mass simulation tests in preparation for the mission. (*AvWk*, July 8/85, 24)

July 23: In its postlaunch report dated July 23, 1986, NASA announced that it had launched on September 9, 1980, GOES–D (subsequently designated *GOES–4*) from KSC by a Delta 3914 launch vehicle. The GOES (Geostationary Operational Environmental Satellite) satellites provided near-continuous high-quality day and night observations of earth and its environment, including cloud cover; weather; proton, electron, and solar X-ray fluxes; and magnetic fields.

During placement of *GOES–4* and its apogee boost motor (ABM) into transfer orbit, NASA observed lower than expected temperatures for the ABM and decided to fire the motor at second apogee in lieu of the nominal third apogee. Performance of the ABM was nominal and resulted in placement of the spacecraft in desired drift orbit. *GOES–4* arrived September 20, 1980, on-station at the 98° west longitude checkout position. On February 26, 1981, the National Oceanic and Atmospheric Administration (NOAA) moved *GOES–4* to the GOES-west operational position at 135° west longitude, replacing the ailing *GOES–3* spacecraft. NASA modified the ABM thermal design for follow-on spacecraft to avoid recurrence of the low temperature problem.

On May 22, 1981, NASA launched GOES–E (subsequently designated *GOES–5*) from KSC on the Delta 3914. Spacecraft performance during transfer orbit maneuvers was nominal, and NASA fired the ABM on the third apogee of the transfer orbit. *GOES–5* arrived on-station June 5, 1981, at the predetermined checkout position of 85° west longitude. On August 5, 1981, NOAA placed *GOES–5* at 75° west longitude where it became the operational GOES-east satellite.

An additional mission objective for both *GOES–4* and *5* was demonstration and assessment of the temperature and moisture soundings from the VISSR Atmospheric Sounder (VAS); Goddard Space Flight Center initiated a VAS demonstration project to achieve this objective. In 1981 the VAS demonstration project conducted a ground truth field experiment that provided four days of simultaneous, high-density ground-based observations and satellite data. This data showed VAS to be a versatile and valuable instrument with potential applications beyond the severe local storm discipline. Some of

Satellites

these additional areas included hurricane and tropical cyclone research, cloud climatology, and diagnosis of moisture patterns and upper-air circulation. As a result of the VAS demonstration project, NOAA made provision for geosynchronous soundings a requirement for the next generation of GOES satellites.

Under a 1973 basic agreement, NASA had the responsibility under NOAA reimbursable funding to design, engineer, procure, and launch polar and geosynchronous weather satellites to implement the U.S.'s operational meteorological satellite program. After on-orbit checkout, NASA handed over the spacecraft to NOAA for routine operations. (NASA MOR E–612–80–02 [postlaunch] July 23/85, E–612–81–03 [postlaunch] July 23/85)

July 23: Astronauts James van Hoften and Dr. William Fisher said at a Johnson Space Center news conference that they would attempt to activate the Syncom satellite deployed in April from the Space Shuttle, the *Washington Post* reported. Although the satellite was loaded with rocket fuel, the astronauts said they didn't think their mission was any more dangerous than the two other rescue missions by astronauts in the previous two years. Because the satellite was "armed," (in the position in which it was ready to fire its engines), the danger was great if the engines ignited accidentally while the astronauts were working.

However, the astronauts indicated the hardest part of their upcoming mission was its awkwardness, not the danger. They said they would attempt the rescue on the seventh day of the August 24 Space Shuttle 51–I mission with the orbiter Discovery. Van Hoften would stand in foot restraints at the end of the Space Shuttle's 50-foot-long mechanical arm 35 feet below the satellite to affix a capture bar to the side of the satellite, then force the satellite to come to a standstill from its once-a-minute spin.

Fisher, standing in temporary foot restraints fixed to the side of Discovery's cargo bay, would put two plugs on either side of the arming switch to prevent the satellite from accidentally turning itself on.

Then, standing on opposite sides of the satellite, the astronauts would bypass the satellite's electronics so that flight directors on the ground could begin to command it from earth. NASA gave the operation a 50–50 chance of succeeding. (*W Post*, July 23/85, A2)

July 26: NASA announced that the International Maritime Satellite Organization (INMARSAT) selected the Space Shuttle to launch July 1988 and mid-1989 two communications satellites in INMARSAT's second generation series. The spacecraft would enhance INMARSAT's existing maritime satellite network and the communications services the organization provided to 43 member nations including the U.S.

British Aerospace Corp. would build the spacecraft; Hughes Aircraft would supply the spacecraft communications payload. A McDonnell Douglas spin-stabilized upper stage, the PAM-D, would boost the spacecraft from low-earth orbit into a geosynchronous transfer orbit. An apogee kick motor, a smaller rocket motor on the spacecraft, would position the spacecraft in a circular 22,300-nautical-mile geosynchronous orbit. (NASA Release 85-111)

July 31: NASA announced that one of the satellites in the COSPAS/SARSAT system, an international program with the U.S., USSR, Canada, and France serving as major partners, made possible the rescue of a man 140 miles off Cape Henry, Virginia.

On April 17 Jack Boye departed Miami to sail to New York Harbor for the dedication of a memorial to Vietnam veterans of Airborne 173, his wartime squadron. Three days into the sail his radio went out, and a storm caused the sloop's electrical and electronic equipment and engine to fail and the boat to capsize. When the boat rolled over and started filling with water, Boye activated an emergency position-indicating radio beacon that trailed at the stern of the boat.

An orbiting Soviet sea and rescue satellite, one of three USSR and two U.S. satellites equipped with sea and rescue equipment in the COSPAS/SARSAT satellite system, first received the distress signal and beamed it to the Rescue and Command Center at Scott Air Force Base, Illinois, who alerted rescue forces. After an electronic search for the distress signal, a Coast Guard aircraft located the sloop and requested assistance from a naval ship in the vicinity. The USS Detroit sent a whaleboat with the ship's doctor to the sloop.

To date, the COSPAS/SARSAT system had been instrumental in the rescue of more than 400 people. The National Oceanic and Atmospheric Administration managed the U.S. portion of the system; Goddard Space Flight Center was responsible for system research and development. (NASA Release 85-113)

August 2: NASA announced that a Scout vehicle today launched the Navy SOOS-I spacecraft from the Western Space and Missile Center (WSMC) at Vandenberg Air Force Base.

As a result of a 1962 NASA/Department of Defense (DOD) agreement for joint use of the Scout launch vehicle, the U.S. Navy asked NASA to provide Scout launches for the Navy Transit and NOVA programs. The Navy reimbursed NASA for the cost of Scout launch vehicles, WSMC launch services, and mission support requirements as needed.

The Navy and other vessels used the Transit program, an operational navigation system, for worldwide ocean navigation. Prior to the August 1 launch there were six operating satellites in the system. NASA maintained two Scout vehicles on standby status to launch replacement satellites to fill any operational gaps occurring in the system.

Satellites

The Navy SOOS–I mission had two Transit satellites in a stacked configuration. The spacecraft, which weighed 283 lb., had orbital parameters of 670 km, apogee; 554 km, perigee; and 89.83°, inclination. (NASA MOR M–490–606–85–01 [postlaunch] Sept 30, [prelaunch] July28/85)

August 9: A conflict between industrial and developing countries over the use of space would likely be the dominant issue at the five-week World Administrative Radio Conference that would open August 9 in Geneva, the *NY Times* reported. The meeting was intended to negotiate a plan to avert traffic jams in the space used by orbiting communications satellites.

The conflict existed as a result of the desire of industrial countries to gain an advantage from their technical lead with a flexible approach to assigning orbital slots and the desire of some developing countries to reserve orbital slots years before they might be used.

Communications experts around the world said the increased use of low-cost satellites to expedite domestic, regional, and world communications was leading to a virtual traffic jam in the satellite belt, 22,300 miles above the equator. The radio frequencies that gave the clearest signals also were rapidly filling. The U.S. and INTELSAT used most of the orbit space and radio spectrum.

Reagan Administration officials said they would offer a proposal that the U.S. refrain for 10 years from using parts of the radio spectrum opened to satellite use by a 1979 conference. By staying out of the expanded spectrum, U.S. officials said, the U.S. would reassure less-developed nations that there would be radio room for them when their satellite-launching plans matured. The U.S. plan would let a country ask for radio frequencies 15 years in advance against the current five years.

The U.S. plan would also establish a requirement that countries with a satellite in operation made compensatory payments to countries forced to put satellites in less desirable orbits or forced to use less desirable radio frequencies.

The most persuasive aspect of the U.S. proposal was the prospect that between 50 and 100 U.S. satellites were at some stage of planning, development, or scheduled for launch over the next three years.

Less-developed countries, particularly those in the tropical zone, wanted rigid assignments of radio bands, including the lower frequencies then dominated by the developed nations, because the bad weather common in the tropics did not noticeably disturb those frequencies. (*NYT*, Aug 7/85, D1)

August 25: NASA said that the *GOES–6* weather satellite turned its eye toward space for about four and a half hours today before NASA personnel reversed it, the *Washington Post* reported. Located at a fixed point about 22,000 miles above earth, the satellite transmitted weather photos widely used by TV stations and newspapers across the nation.

NASA repositioned the satellite in a reprogramming effort that required transmission of 14 groups of 256 commands each, retiming the revolving satellite so that its instruments would come on when facing the earth.

Since August 1984 when *GOES–5* lost its ability to transmit photos because a light in it failed, *GOES–6* was the lone fixed-point satellite making transmissions. NASA had scheduled for launch in spring 1986 a replacement for *GOES–5*.

If a failure like the one that occurred on August 25 were to last longer, it could have a serious impact on the National Oceanographic and Atmospheric Administration's (NOAA) ability to monitor severe storms. (*W Post*, Aug 29/85, A8)

August 31: During Space Shuttle mission 51–I astronauts James van Hoften and William Fisher spent seven hours and eight minutes, a new record for a spacewalk, in the first stage of repair of the *Leasat 3* satellite, the *Washington Post* reported. During the walk, they retrieved the 15,000-lb. satellite, fastened Discovery's mechanical arm onto it, disarmed its Minuteman rocket motor, and exposed its electronic connections. Later Fisher said he could see no evidence of what caused the satellite to go dead after astronauts deployed it in April. "There is no evidence of debris on anything that would cause a problem. It's clean as a whistle," He said.

Van Hoften and Fisher were outside the Space Shuttle from about 8:00 a.m. to shortly after 3:00 p.m. EDT, breaking by one minute the spacewalk record set April 1984 by van Hoften and George Nelson when they repaired the scientific satellite *Solar Max*.

Although the astronauts accomplished more than half of what was needed in the salvage mission, they were planning a four-hour spacewalk on September 1 to finish the job. They would put a "space blanket" over the engine bell that served as the Minuteman motor's rocket nozzle to trap more of the sun's heat to warm it, because *Leasat–3* had been in the cold of space so long that its rocket motor was too cold to fire correctly. Then the astronauts would install a vertical bar along the 20-foot length of the satellite to serve as a handle for van Hoften to spin the satellite and push it away from Discovery into space.

It would then be seven days before the satellite acquired enough power to thaw its liquid-fuel tanks and almost two months before the solid-fuel rocket motor was warm enough for ground controllers to command it to ignite. (*W Post*, Sept 1/85, A5)

September 1: Astronauts James van Hoften and William Fisher on Space Shuttle mission 51–I today completed repair of the *Leasat 1* satellite, the *Washington Post* reported. Steven Dorfman, president of Hughes Communications Corp., satellite owner, referred to the effort as the "most remarkable salvage mission" tried in space.

During the second day of the salvage mission [see Satellites, Aug. 31], astronaut John Lounge, using the Space Shuttle's mechanical arm, raised the satellite 35 feet above the cargo bay and placed its center of gravity on a straight line with that of the orbiter Discovery. Outside, van Hoften stepped into foot restraints on the end of the arm, and Lounge lifted him to the satellite.

Reaching up, van Hoften placed gloved hands on the bar he had earlier attached to the satellite and pushed down as hard as he could to start the satellite spinning. Three more pushes sent the satellite spinning evenly into space over the equator near the west coast of South America where officials planned to operate it as a Navy radio relay.

Although Hughes was still concerned that very cold temperatures might have permanently damaged the satellite, it did beam to earth its first signals since deployment on April 13. The signals indicated that temperatures in the satellite's outer shell were normal, suggesting that the inner drum holding the two onboard rocket engines and more than 7,500 lb. of rocket fuel might have survived the varying temperatures of space and be ready to carry the satellite into permanent position 22,300 miles above earth. (*W Post*, Sept 2/85, A3)

September 16: Hughes Communications Inc., manufacturer and owner of the *Syncom 4* satellite launched August 27 from the Space Shuttle Discovery during mission 51–I, today declared it a failure due to a faulty cable between the satellite's UHF radio transmission system and its broadcast antenna, the *Washington Post* reported.

Failure of the $85 million satellite, known as *Leasat 4* after its lease to the Navy, pushed satellite insurance losses to $234 million in a single week and to about $570 million for the past 18 months. Satellites insured for $84 million and $65 million were also lost that week when Arianespace blew up a malfunctioning Ariane rocket. (*W Post*, Sept 17/85, A11)

September 23: Hughes Communications faced the possibility of a $10,000 per day fine, up to a total of $5 million, if the Leasat communications system was not fully operational by November 30, *Aviation Week* reported. The potential of a fine resulted from the failure of its *Leasat 4* satellite following deployment from the Space Shuttle Discovery during mission 51–I, reducing significantly the likelihood of three Leasat satellites being operational by the November 30 deadline.

Hughes could avoid the fee if the repaired *Leasat 3* activated in late October; otherwise the firm had the option of launching a ground spare. Telemetry data from *Leasat 3*, repaired during the Space Shuttle 51–I mission, continued to confirm the satellite's functioning. The liquid propulsion system was intact, and the solid propellant perigee kick motor temperatures were rising toward acceptable levels. However, the satellite could explode when its main rocket fired late in September, and Hughes officials gave it no more than a 50–50 chance of reaching geosynchronous orbit.

Leasat 1 and *2*, launched in August and November 1984, respectively, were operating according to specifications, and the Navy was leasing them.

Hughes had to provide the U.S. Navy with four operational satellites by March 31, 1986, to meet the second contract deadline and could at that time again face fines for noncompliance. Company officials estimated December as the earliest date for a *Leasat 5* launch.

Albert Wheelon, Hughes Aircraft Co. senior vice president and president of Hughes's space and communications group, said the *Leasat 4* loss and other recent spacecraft losses were having an impact on the industry. "My reading is that generally the whole insurance market for satellites has just gone belly-up," he commented. "I think it's going to have a major impact on startup companies. I think it's going to be a narrowing, both on the operator and the development side." He added that the status of spacecraft underwriters would have less of an impact on companies like Hughes, because it could provide self-insurance. (*AvWk*, Sept 23/85, 21)

September 26: NASA announced that its *AST–3* communications satellite, located at 105° west longitude, was providing communications support through its control center at Malabar, Florida, to American Red Cross and Pan American World Health Organization rescue and relief efforts following Mexico City's earthquake.

The voice communications link with the outside world was crucial, since the earthquake disrupted all other forms of communications in the city. George Manno, director of media relations for the Red Cross, said "the *ATS–3* is providing us with the most critical communications link. Although ham radio operators have been doing a swell job, they are serving as our backup communications system, while we rely primarily on the ATS as our main communications vehicle."

Immediately after the earthquake occurred, NASA implemented the ATS emergency preemption plan; within 24 hours *ATS–3* was on the air, giving priority to satellite communications traffic for the emergency rescue operations. (NASA Release 85–133)

October 3: The *Syncom II* communications satellite lost one of its 13 channels, described as an important wide-band channel, and efforts were underway to reactivate it, *Defense Daily* reported. The U.S. Navy leased the satellite from Hughes Communications Inc., which was under contract to provide four such Leasats to the Navy.

NASA in September 1984 launched *Syncom I* and *II* from the Space Shuttle and launched in April from the Space Shuttle *Syncom III*, which failed to fire in orbit. The Space Shuttle crew in July repaired *Syncom III*, with a reboost attempt scheduled for October. NASA launched in July from the Space Shuttle *Syncom IV*, but its communications failed in orbit. (*D/D*, Oct 3/85, 174)

October 27: Engineers with Hughes Communications Inc., owner of the *Syncom 3* communications satellite, sent at 11:53 a.m. today from the company's Guam ground station a radio command to fire the marooned satellite's rocket booster, causing the satellite's ICBM-type rocket motor to fire for 64 seconds as planned, the *Washington Times* reported. The undertaking capped the most ambitious space salvage effort ever attempted, the *Times* said, a two-day effort in August by two astronauts aboard the Space Shuttle Discovery (see Satellites Apr 22 and Aug 31). The engineers fired the rocket without knowing whether it would blow up or work properly after months exposed to freezing temperatures.

"We really nailed it," said Albert Wheelon, president of the Hughes space communications group. "We'll have it in synchronous orbit round about the first of November and then we still have 30 days of checking out the radio equipment and all the systems."

Insurance underwriters stood to recoup 85% of the insured cost of the satellite, a boost for an industry that had lost more than $600 million in the past 20 months because of seven satellite failures including *Syncom 3*. (*W Times*, Oct 28/85, 3A)

November 7: The Naval Research Laboratory (NRL), Washington, D.C., announced its scientists devised a precise timing system that used highly efficient and relatively small hydrogen maser clocks that the laboratory expected would enhance timing in the Navstar Global Positioning System (GPS).

GPS, a navigational system based on synchronized clocks (accurate to billionths of a second), would permit users to pinpoint their position within a 30-foot radius by measuring differences in the arrival times of radio signals received from Navstar's satellites. And GPS would disseminate the Naval Observatory's master clock time to fleet units for their command, control, communications, and intelligence.

Maser clocks were considered the most accurate clocks available; however their bulky size and weight (lighter ones weighed over 500 lb.) prevented their use aboard GPS satellites. Also, the massive, tuned cavity where the maser action took place required temperature control and magnetic shielding, which made masers impractical for use in satellites.

The research team at NRL led by Ron Beard reduced the size of maser clocks by a factor of 16 and their weight to 50 lb. And by making the cavity smaller, using different resonant structures, and dielectric (sapphire) loading to produce the effect of a larger cavity, NRL "passive" masers (so called because they did not oscillate on their own) proved to be as accurate as the larger versions. Beard noted that NRL's involvement with masers and the subsequent idea to reduce cavity size came about only after two astro-industrial companies were unsuccessful in meeting GPS's timing demands.

NRL built and tested two experimental maser development models and two advanced development models. An additional NRL project with Hughes Research Laboratories used a Q-multiplier approach to offset the inherently higher signal losses in small cavities.

Research in the NRL maser program included advances in small loaded-cavity properties, beam optics, dissociator reliability, quadropole state selection, thermal/mechanical design, magnetic shielding/design, and vacuum design for space applications.

The NRL-developed technology formed the basis for a $12 million contract awarded to Hughes Aircraft Corp. to produce two types of engineering development models of hydrogen-maser clocks. One clock type was for use in satellites; the other, for GPS ground stations and possibly ships. (NRL Release 58–85R)

November 30: The U.S. Navy today took control of *Syncom 3* from Hughes Communications Inc., certifying the satellite as a complete, though tardy, success, the *Washington Post* reported. Space Shuttle astronauts repaired the $85 million satellite in space [see Satellites, Aug. 31] by bypassing a failed circuit and relaunching it on September 1. Hughes ground controllers then tilted the satellite to face the sun, which, with the help of a thermal blanket installed by the astronauts, warmed the solid fuel.

On October 27, Hughes engineers ignited the satellite's engines to take it up to its intended orbit 22,300 miles above the Pacific Ocean, then Hughes and Navy engineers completed a month of testing.

Following the certification, Hughes president Steven Dorfman commented, "We are meeting all our specifications. Six months later than originally intended, we are going into service."

The successful repair turned around a loss for Hughes and its insurance carrier, which had paid the company $85 million. Dorfman said the insurer would get back $65 million by sharing revenues with Hughes over the satellites's ten-year lifespan. (*W Post*, Dec 1/85, A3)

SEARCH FOR EXTRATERRESTRIAL INTELLIGENCE (SETI)

March 22: Dr. John Billingham, chief of the Extraterrestrial Research Division at Ames Research Center (ARC) and founder of the ARC-based Search for Extraterrestrial Intelligence (SETI), said at a Lewis Research Center (LeRC) ALERT colloquium that study of the earth's part in the 15-billion-year cosmic evolution held the key to searching for life beyond the solar system, the *Lewis News* reported. "By studying how our own planet formed and how basic elements required for life evolved, we can theorize that a similar life-structuring process may be occurring or had occurred elsewhere in the universe," Billingham said.

As an example, he mentioned the collection of microorganisms scientists observed in 3.5-billion-year-old rock formations discovered in the marine environment of Western Australia. Scientists could use that data to study rock-like formations on other planets that might display a similar evolutionary process.

Billingham believed that it was reasonable to assume that life, including intelligent life, existed in the universe because, with 400 million stars in earth's galaxy alone, the natural cosmic process included formation of planets—a rule, not an exception—and the basic stuff of life existed throughout the universe.

Because the Viking program had failed to turn up signs of life on Mars, SETI would focus its sights on life beyond the solar system, Billingham said. To accomplish this would require either a manned interstellar vehicle or radio telescopes. The former option would require, to reach the nearest star, a four-stage spacecraft weighing 34,000 tons, traveling 3/10 the speed of light and using electrical power equal to 500,000 years of earth usage driven by antimatter—hardly a feasible option, he noted. However, researchers could design and build radio telescopes to listen to life from other cosmic sources, he said.

Such a system would make use of the microwave window frequency range, a quiet region of the spectrum that provided the best chance of picking up extraterrestrial signals. To accomplish this, SETI's information system would need to develop the technology using the largest radio telescopes to amplify the signal (1 to 3 GHz) and to build a signal processing system driven by a specially designed SETI computer.

Billingham said the biggest challenge facing the SETI program would be identifying the format, frequency resolution, and time of the signals in order to recognize and eliminate as much interference as possible, thus separating noise from sound manifestations. To determine signals of non-natural origin,

SETI would apply a sequence of logical tests coupled with a huge computer-based storehouse of information that would filter out most of the interference signals.

"A radio telescope placed in low earth orbit will help eliminate some of the interference problems," Billingham said, "and the prospect of a much larger lunar orbiting radio telescope would offer an even more advantageous interference-reducing listening position that would cover a radius of 1,000 light years."

SETI was then preparing its first system radio telescope prototype for testing at Arecibo, Puerto Rico, where a 1,000-ft. radio telescope conducted signal searches. Once fully operational and in orbit, the new telescope would receive signals originating 30 to 40 light years away.

"Either we are alone or not," Billingham concluded, "and either has large implications. And we wonder, indeed if there is intelligent life elsewhere, will it help us understand our own." (*Lewis News*, March 22/85, 2)

SMITHSONIAN INSTITUTION

January 29: Regents of the Smithsonian Institution announced they had cancelled plans to construct a new facility at Washington Dulles International Airport that would have housed NASA's obsolete vehicles, the *Washington Times* reported. The Smithsonian had also expected to display the prototype of the French Concorde and numerous other non-NASA aircraft. However, the regents had decided the facility was a low priority.

"I don't think the only place to store those aircraft and spacecraft is Dulles," said Smithsonian secretary Robert Adams. "I would be concerned to take on another major construction program at this time." He concluded that emphasis would be on completing projects already begun. (*W Times*, Jan 29/85, D3)

April 18: The Smithsonian Institution's Air and Space Museum received approximately $400,000 from the Johnson Wax Co., which was needed to construct a life-sized, remote-controlled model of a pterosaur, a flying dinosaur that existed in various forms for about 145 million years and became extinct about 60 million years ago, the *Washington Post* reported. The Smithsonian had announced the project the previous summer with the caveat that it would need corporate funding. "We always planned to do it," said museum deputy director Don Lopez, "but now we can be sure it will be built on the scale that we wanted."

The Smithsonian asked Paul MacCready, a prize-winning aerodynamicist whose experiments with pedal- and solar-powered aircraft were known around the world and whose Gossamer Condor, a lightweight, pedal-powered plane, hung in the Air and Space Museum, to design and build the model. MacCready and a team of experts in paleontology, bird flight, and aerodynamics were experimenting in California with a small, glider version of the pterosaur. The Smithsonian expected the final version to weigh about 125 lb. and have a wingspan of 36 feet, roughly the size of a four-person airplane. MacCready would construct the model of graphite, carbon, and epoxy fibers.

The challenge, a Smithsonian spokesman said, was to design a computerized "brain" for the model to simulate the slight, constant wing movement that researchers believed the pterosaur used to maintain its balance. (*W Post*, Apr 18/85, D2)

May 8: The Smithsonian Institution announced that "Celestial Images: Astronomical Charts, 1500–1900," a special exhibition of 39 rare depictions of star patterns and planetary systems, would open May 16 at its National Museum of American History.

The exhibition, intended to illustrate a unity of science and art from the Renaissance to the 1900s, would include two woodcuts, the first printed star charts with coordinates and scales from which positions of stars could be plotted, from the workshop of Renaissance artist and mathematician Albrecht Durer (1471–1528). The museum would also display from Lahore, India, a 17th-century Islamic celestial globe made of brass and inlaid with silver stars; the only extant copy of engraved paper gores (1599) for the first celestial globe by Willem Janszoon Blaeu (1571–1638), the first to incorporate the Danish astronomer Tycho Brahe's unexcelled positions for 1,000 stars; and a 17th-century celestial atlas by Andreas Cellarius.

In addition to reflecting artistic styles and conventions of their periods, the works represented differing points of scientific and philosophical debate. Charts that accurately fixed many constellations and thousands of stars also depicted Ptolemy's view of the earth as the center of the universe. Constellations rendered as mythological figures in one work might appear as Biblical figures in another. However, as a result of astronomer's access to stronger and technically improved instruments, the works showed increasing accuracy over the 400 years covered in the exhibition. (Smithsonian Institution Release, May 8/85)

June 21: "The Dream Is Alive," a new IMAX movie that was perhaps the most advanced film of its kind ever made, opened today at the Smithsonian Institution's National Air and Space Museum, the *Washington Post* reported. The $3.6 million production, narrated by Walter Cronkite, was the first ever shot almost completely on location in space and was an informal sequel to "Hail, Columbia," the story of the U.S.'s first Space Shuttle.

Because the film was shot on three of the most dramatic and action-packed Space Shuttle flights of the previous year, it was essentially a video yearbook full of special happenings, the *Washington Times* reported.

The film showed the rescue and repair of the Solar Max satellite, the deployment of the 100-foot-tall solar array that someday could help power the space station, launching of two communications satellites, and the first spacewalk by an American woman.

Among what appeared to be special effects produced in Hollywood were the immensity of Hurricane Josephine as seen from 200 miles overhead and the silence of a satellite drifting noiselessly in the blackness of space.

The astronauts who shot the footage said the IMAX format came as close to capturing the real view and the true feel of spaceflight as any format ever had. IMAX expert Graeme Ferguson, who produced and directed the film, said the next step was to modify an IMAX camera so it could be taken outside the Space Shuttle. (*W Post*, June 22/85, D1; *W Times*, June 20/85, 3B)

July 24: In testimony today before the Senate Committee on Rules and Administration, Smithsonian Institution Secretary Robert McCormick Adams said the Smithsonian opposed a bill introduced by Sen. Barry Goldwater (R-Ariz.) that would authorize funds to construct a museum administered by the National Air and Space Museum for historic airplanes and spacecraft at Dulles International Airport, the *Washington Post* reported. He said the Institution had more immediate legislative priorities, although it would support planning appropriations. In 1976 Goldwater was instrumental in pushing funds through Congress for the National Air and Space Museum.

The bill Goldwater introduced in June called for $42.6 million in government appropriations in FY 89 and beyond, contingent on an equal amount in matching donations from the private sector. Adams questioned whether the Institution should, or could, raise that much money, particularly because the Institution had more pressing fund-raising commitments.

Although Goldwater was on the Smithsonian's Board of Regents, he was pushing the issue now, Terry Emerson, his legislative assistant said, because "its just an interest in priorities." Emerson noted that politicians had a greater sense than did Smithsonian officials of the interest in technology-related museums. "Later they'll come in once they see the overwhelming interest in it," he added.

"Once enabling legislation is enacted," Goldwater said, "I believe numerous private individuals and firms will begin a major fund-raising campaign that will produce enough matching monies to get the first construction going by fiscal year 1989."

The Smithsonian's Board of Regents initially approved in September 1983 the concept for the air museum, citing insufficient space at the National Air and Space Museum and the difficulty of transporting large aircraft and spacecraft to the Mall building. Under the plan, the Federal Aviation Administration would lease 100 acres of its land at Dulles without compensation. Proximity to a runway was essential, because many of the aircraft and spacecraft, such as the Concorde and Space Shuttle, were too large to be transported on roads.

The committee would rule on the proposed legislation in September. (*W Post*, July 25/85, B9)

September 20: Rep. Norman Mineta, (D-Calif.) introduced legislation today for the planning and construction of an expansion of the Smithsonian Institution's National Air and Space Museum at Washington Dulles Airport, the *Congressional Record* reported.

When introducing H.R. 3403, Mineta said, "This legislation is proposed to solve two major problems confronting the Air and Space Museum. First, as my colleagues know, this is perhaps the most popular museum in the world. The public's demand for exhibits and information concerning our aviation history is enormous, and growing. We need the additional facilities to respond to this demand.

"But why Dulles? Why not build the new facility in downtown Washington? The answer is simple: there is no site large enough for the needs envisioned and of paramount importance, there is no other site adjacent to a major airport.

"An airport site is critical because of the second problem we are trying to solve, which is how to display objects of the scale of the space shuttle or other modern aircraft. The museum needs to be at Dulles Airport because that is the only site to which these exhibits can be transported and displayed.

"In particular, this legislation authorizes the transfer to the Smithsonian of land at the airport. This bill authorizes a total of $2.4 million in funds for planning of the new museum, to be appropriated in the 1986 to 1988 period.

"In addition, the bill authorizes the appropriation of $42.6 million in Federal funds beginning in 1989 to finance half of the cost of actually constructing the museum. These Federal funds would not be available unless and until a private fundraising effort has raised an identical amount." (*Cong. Record*, Sept 20/85, E4165)

November 8: NASA announced it would transfer on December 6, 1985, the title to the Space Shuttle orbiter Enterprise to the Smithsonian Institution's National Air and Space Museum. The NASA Shuttle Carrier Aircraft, a modified Boeing 747, on November 16 would deliver Enterprise to Washington Dulles International Airport, where cranes would remove Enterprise from the top of the 747 and lower it to Dulles's tarmac.

Rockwell Internatl., NASA's Space Shuttle prime contractor, constructed Enterprise, which was the first orbiter built. NASA used Enterprise, Space Shuttle orbiter vehicle 101 (OV 101), to test airframe loads, flight dynamics characteristics, and other aspects of the orbiter as it flew through earth's atmosphere. NASA also used Enterprise as a test bed for manufacturing techniques, aerodynamics, and flight control tests, and mating and fit checks for the remainder of the Space Shuttle components—the external tank, solid-fuel rocket boosters, and mobile launch platforms at both Kennedy Space Center and Vandenberg Air Force Base.

On September 17, 1976, Enterprise rolled out from Rockwell's assembly facility at Palmdale, California, and was first flown on February 15, 1977, aboard the modified 747 in taxi tests at Edwards Air Force Base.

There were then five approach and landing tests, commanded and piloted alternately by astronaut teams of Fred Haise, Jr. (commander) and C. Gordon Fullerton (pilot) and Joe Engle (commander) and Richard Truly (pilot).

After ground vibration tests, engineers at Kennedy Space Center used Enterprise for fit checks with the other Space Shuttle components and the mobile launch systems. More recently, engineers used Enterprise for fit checks at the Space Shuttle launch complex at Vandenberg Air Force Base. (NASA Release 85-150)

Smithsonian Institution

November 18: The Space Shuttle Enterprise arrived the afternoon of November 18 at Washington Dulles International Airport after a two-hour delay due to inclement weather, the *Washington Post* reported. NASA had announced earlier that it would transfer title to Enterprise to the Smithsonian Institution's National Air and Space Museum [see Smithsonian Institution, Nov. 8]. NASA delayed Enterprise's takeoff from Kennedy Space Center until an early morning fog burned off so more people along the route could see the craft. Enterprise, atop a Boeing 747, was visible to onlookers in Annapolis and Baltimore before it began a counterclockwise circuit over Washington's Capital Beltway.

Officials at Dulles would store the spacecraft in a temporary shed about two miles from the airport terminal until a permanent facility was constructed. (*W Post*, Nov 19/85, B1)

SPACE SCIENCE AND APPLICATIONS

January 7: NASA announced it signed with Rockwell Internatl. Corp. a memorandum of understanding (MOU) covering Space Shuttle flight assignments for Rockwell's materials-processing laboratory, which flew on STS 41-D and, using the float-zone technique, grew a single indium crystal with a lattice structure originating from a crystal seed. The agreement called for Rockwell to develop an industrial space-processing program under which research institutions and commercial firms would install and operate experiments in the modular laboratory. Rockwell designed its fluids-experiment apparatus (FEA), the first zero-gravity laboratory for basic and product research in low-earth orbit, to fit in the Space Shuttle's mid-deck stowage area where crew could operate and monitor it. Experiments could range from processing applications to liquid chemistry, fluid physics, thermodynamics, crystal growth, and biological-cell culturing.

The laboratory, about the size of a 19-inch TV, could heat, cool, expose to vacuum, and manipulate experiment samples that might be gaseous, liquid, or solid, and could mix or stir, remove, and change samples during a mission. A motion-picture or video camera would record sample behavior and instrument data displays. (NASA Release 85-2)

January 17: NASA successfully launched today an Aerobee liquid-fueled, sounding rocket from White Sands Missile Range, New Mexico, marking the end of that rocket series, which was the oldest continuous rocket-firing program. NASA's Sounding Rocket Division had launched 536 Aerobees with a 94% success rate. The Aerobee's reliability made it the workhorse vehicle for high-altitude studies. Commenting on the program, Maury Dubin, Goddard Space Flight Center (GSFC) physicist, said "virtually everything that's been done in space research can be attributed to sounding-rocket technology. The Aerobee and other sounding-rocket research precipitated the rise of many disciplines, from astronomy to the upper atmosphere."

The Aerobee program began in 1946 when Aerojet Engineering Corp. received a U.S. Navy contract to build the rockets, and the Applied Physics Laboratory (APL) had supplied technical direction; Dr. James Van Allen had been project director. The Aerobee name, the Bumblebees, derived from a combination of Aerojet and APL's series of Navy missiles. NASA had used Aerobees extensively during the Internatl. Geophysical Year (July 1957 to December 1958) to gather information on the atmosphere, cosmic radiation, auroras, and geomagnetism.

The Aerobee rockets consisted of a series of five with each rocket carrying as many as six experiments that included equipment such as cameras, vacuum bottles, mirrors, grids, sensing devices, lenses, and other mechanical units. Most rockets recorded data and simultaneously transmitted it via telemetry to ground stations. The final Aerobee payload tested a new spectrograph that would be flown on NASA's Astro mission to record extreme ultraviolet Dayglow emissions in earth's upper atmosphere.

George Kraft, GSFC flight support section director, said that, although the demise of small liquid-fueled sounding rockets was imminent, "sounding rockets would continue to be an important element in NASA's support of the scientific community. They're virtually the only vehicles that can conduct studies in the 40 km (25 miles) to 200 km (125 miles) high zone of the atmosphere." Balloons could travel only about 50 km (31 miles) high and satellites were ineffective below 200 km (125 miles) high. GSFC's Wallops Flight Facility managed the NASA sounding-rocket program. (GSFC Release 85-6)

March 5: The 16th Lunar and Planetary Science Conference, March 11-16 at Johnson Space Center (JSC), would highlight exploration programs planned by the U.S. and the USSR, including a discussion of major trends in space science and the possible role of the space station in planetary exploration, JSC announced. Speakers at the special session would include Dr. Geoffrey Briggs, director of the Solar System Exploration Division, NASA Headquarters; Dr. V. L. Barsukov, Vernadsky Institute of Sciences; Dr. Eugene Levey, chairman of the Planetary Sciences Department, University of Arizona; and Dr. Ronald Greely, a planetary scientist from Arizona State University.

Other sessions of major significance would be the Florensky Memorial Symposium on Venus, possibly including results from the Soviet Venusian probes, and the Shergotty Consortium, focusing on the debate concerning meteorites that may have originated from Mars's surface.

Planners had scheduled 27 regular session with 315 oral presentations taken from 497 abstracts published in the official conference proceedings. (JSC Release 85-11)

May 11: Dr. George Wetherill, director of the department of terrestrial magnetism at the Carnegie Institution of Washington, said in an interview with the *New York Times* that his report to appear in the May 17 issue of *Science* discussed his hypothesis that the inner planets of the solar system appeared to have been formed when a number of planets, some of them three times larger than Mars, repeatedly collided with one another until only one survived in each of the present planetary orbits.

Wetherill believed the present planets had "lots of brothers and sisters" not much smaller than themselves, which collided to form "trial" planets, the *NY Times* reported. "The four we see today [Mercury, Venus, Earth, and Mars]," he said, "are the survivors. One of the final collisions," he added, "probably increased the spin of earth sufficiently to throw off materials that consolidated to form the moon."

Wetherill derived his hypothesis from a computer simulation of what probably happened after the solar system began to take shape from a rotating cloud of dust and gas about 4.5 billion years ago. The analysis took into consideration various factors affecting the formation process once some 500 bodies, each one-third the size of the moon, had formed in the region around the sun now occupied by the four planets mentioned above. These factors included the gravitational fields around each object, the frequency of near misses that would throw the objects into eccentric orbits, and the collisions that may have canceled the eccentricity.

Collisions would have generated enough heat within the earth to melt most, if not all, of its interior, allowing heavy material to sink and form earth's metallic core.

Wetherill's calculations assumed that not enough gas was present to affect motions of the objects as they sped past one another; many scientists concurred that a violent "wind" blowing out from the sun swept the inner solar system clear of gas.

Each collision would have driven off most of any planetary atmosphere that had begun to accumulate, and since impact histories of Earth and Venus were different, this could account for the differences in their atmospheric abundances of such inert gasses as argon.

Initially, material destined to form the planets was so uniformly spread around the sun that their motions were determined by multiple collisions much like those of molecules in a hot gas. By the time this material had formed into larger bodies, Wetherill said, their mutual gravitational attractions would have become a significant factor. During the ensuing collisions, Wetherill believed, the existing planets had acquired 50% of their present material; after 100 million years the process was 99% completed.

Age determinations of moon rocks indicated that the last great crashes, enough to produce the lunar seas, did not end until 3.8 billion years ago, or several hundred million years after the formative process began. By then, however, the impacting bodies were "quite small," Wetherill said, "about 30 miles in diameter." What remained in the form of asteroids and meteorites was far less than that needed to produce even a small planet.

Jupiter and the large planets beyond it were not in a region swept clear of gas, making their formation histories very different, Wetherill concluded. (*NY Times*, May 11/85, 12)

May 14: NASA announced its National Space Technology Laboratories (NSTL), through its Earth Resources Laboratory, signed a memorandum of understanding (MOU) with the Anthropology Department of the University of Colorado at Boulder to use satellite imaging and remote sensing technology to probe the tropical Andean jungles for archaeological remains in Peru's Rio Abiseo National Park.

Of particular interest in the park were the ancient site of Gran Patajen, the subtropical cloud forest, and the park's diverse ecology. Using sophisticated instrumentation that would allow researchers to "see" through the dense vegetation to locate evidence of past settlements in the now uninhabited region, researchers would combine data from the Landsat earth resources satellite with information gathered by a specially equipped aircraft from NSTL. The researchers might also observe and map interesting geographical features and variations in vegetation.

Thomas Lennon, co-director of the university's Rio Abiseo National Park Research Project, said of the undertaking, "NASA's assistance will enable us to take a good look at the 1,060-sq-mile park from above—immensely easier than trekking through the jungle on foot. When cultural resources are identified through image analysis, we'll be able to check it out on the ground. This approach allows us to make the best use of the limited researchers as well as the time we'll have in the field."

The university, NASA, and a remote sensing team at Peru's National Agrarian University in Lima would work together to extract information from various sensing devices. An early outcome of the project should be accurate up-to-date maps of the uncharted parkland.

The project was one of three archaeological research investigations supported by NASA's remote sensing program administered by NASA's Office of Space Science and Applications, Washington, D.C. In another project with the University of Colorado, NASA was examining volcanic destruction of cities and vegetation in Costa Rica. In the third project, NASA was supporting Richard Leakey of the Leaky Foundation who was searching for evidence of human evolution in Kenya, Africa.

NASA's work on the projects resulted from its interest in expanding the scientific applications of space technology, in these cases to archaeology and anthropology, which could benefit from advancements in remote sensing technology. (NASA Release 85-73)

May 17: NASA announced that seven world class gymnasts would undergo testing May 20-22 at Johnson Space Center (JSC) as part of a continuing study of the space adaptation syndrome that affected about half the astronauts who had flown in space. Researchers wanted to know if gymnasts were less susceptible to the malaise because of their experience of moving and spinning in three dimensions.

The gymnasts underwent baseline testing at JSC in mid-April. A second battery of tests would measure their responses in the laboratory and aboard a jet aircraft that induced brief periods of reduced gravity.

Testing was co-sponsored by the U.S. Gymnastics Federation, of which the participants were members. They were Kathy Johnson, Patty Gerard, Megan Marsden, Krista Canary, Tom Beach, Scott Johnson, and Steve Elliot. (NASA Release 85-75)

Space Science and Applications

During May: Goddard Space Flight Center (GSFC) personnel were testing the interface of the software in the space telescope data capture facility (ST DCF) with other elements of the Hubble Space Telescope (HST) ground system, the GSFC *News* reported. The DCF, which had arrived at GSFC ahead of schedule and on target with program costs, would accept science data from the telescope's five instruments through the NASA communications (NASCOM) system via the *Tracking and Data Relay Satellite (TDRS)* and the NASA ground terminal at White Sands, New Mexico.

William Stallings, head of GSFC's data capture systems section and project manager for ST DCF development, said, "Within 24 hours of receipt of the science data stream from NASCOM, the facility will preprocess the data and forward it to the Space Telescope Science Institute (STSI) for further processing and use by scientists." The DCF had two identical Gould 32–87 computer systems and special hardware to provide the science data processing requirements.

The DCF had previously demonstrated the ability to capture the 1.024 megabit and 4 kilobit per-second data streams, processing the data into user data sets, and transmitting them to the Science Institute at the daily required volume level of three billion bits of science data. Because the space telescope's science instrument data were packetized, future refurbishment of the space telescope with new instruments would require only table updates in the DCF's software. Also, the DCF had automated quality control that should reduce operational costs. (*Goddard News*, May 85, 3)

During July: After preliminary examination of Spartan 1 data, mission manager Dave Shrewsberry said the spacecraft appeared to have performed well during the Space Shuttle 51–G mission. Astronauts deployed the spacecraft on the fourth day of the mission into a free-flying orbit and retrieved it on the sixth day, *Goddard News* reported. Final evaluation required analysis of Spartan 1's tape recordings.

"Although the grapple fixture wasn't pointed in the direction we thought it would be when we were retrieved," Shrewsberry said, "that is not a matter of concern. The running lights were on and the experiment doors closed, indicating that the program we had computed was completed." Early indications also showed that all six of the Get Away Special experiments turned on during the mission. (GSFC *News*, July 85, 1)

August 1: NASA announced that it had completed with the Italian National Research Council's National Space Plan Office (PSN/CNR) selection of U.S. and Italian principal investigators for the cooperative tethered satellite system (TSS), which NASA had scheduled for a first flight in 1988 aboard the Space Shuttle. The system, using the Space Shuttle as a base of operation, would provide a unique reuseable facility remote from the orbiter for space science investigations in the upper atmosphere and in plasma-electrodynamic interactions. 101 The overall objectives of the TSS program were to demonstrate the successful operation of a tethered satellite, perform the initial electrodynamic science and plasma investigations, and develop a reusable facility capable of supporting a broad range of electrodynamic experiments in the ionosphere.

On the first TSS mission, a NASA-supplied deployer mechanism would send the Italian-developed and -built satellite upward from the Space Shuttle orbiter to a distance of as much as 12 miles. The motion of the conducting tether across geomagnetic field lines would generate several thousand volts of energy to provide a broad range of electrical operating conditions. Researchers could also use the overall system to artifically generate and study field-aligned currents and associated plasma effects.

On the second TSS mission, the satellite would be at the end of a nonconducting tether at 62 miles in the earth's atmosphere, an area previously explored only sporadically, mainly with sounding rocket payloads. The tethered satellite would permit investigation of the region on a global scale over several days.

The TSS–3 mission would be a follow-up to the TSS–1 electrodynamics mission, with deployment from either the Space Shuttle or the proposed space station.

A memorandum of understanding signed in 1984 by NASA and PSN/CNR established the TSS program. NASA's Marshall Space flight Center had responsibility for project implementation.

NASA's office of space science and applications selected Peter Banks, Stanford University (Shuttle Tether Electrodynamic Tether System); Robert Estes, Smithsonian Astrophysical Observatory (Investigation of Electromagnetic Emissions by the Tether); Gordon Gullahorn, Smithsonian Astrophysical Observatory (Investigation and Measurement of Dynamic Noise in Tethered Observatory Satellite Systems); Konstantinos Papadopoulos, Science Applications, Inc. (Theory and Modeling of the Tether); and Nobie Stone, Marshall Space Flight Center (Research on Orbital Flight Plasma-Electrodynamics).

PSN/CNR selected Silvio Bergemaschi, Instituto Meccanica, Padova University (A Theoretical and Experimental Investigation of TSS Dynamics); Marino Dobrowolny, Instituto Fisica Spazio (Research on Electrodynamics Tether Effects); and Franco Mariana, University of Rome (Magnetic Field Experiment for the TSS Missions). (NASA Release 85–115)

August 15: A sophomore at Purdue University and a senior at San Francisco's Lowell High School designed experiments to be carried on two 1986 Space Shuttle Flights, the *Washington Post* reported. Kentucky Fried Chicken would sponsor an experiment in which eggs would fly in orbit to determine whether chicken embryos could survive the rigors of weightlessness. The Lawrence Berkeley Laboratory at the University of California would sponsor

Space Science and Applications

an experiment to measure the effects of weightlessness on cell division in yeast.

Greg Delory, 16, said he got the idea for flying baker's yeast from the first Spacelab mission where bread mold was grown in orbit. "The mold was supposed to produce spores every 22 hours, but in space it lost its circadian rhythm and produced spores all the time," he said. Delory won a competition to design a Space Shuttle experiment while attending the U.S. Space Camp under the auspices of Marshall Space Flight Center.

John Vellinger, 20, who won a competition to design a NASA experiment, said, "We hope this will give us some data about the feasibility of raising chickens as a food source in space." (*W Post*, Aug 15/85, A19)

August 23: The Jet Propulsion Laboratory's (JPL) Geobotanical Remote Sensing Group was developing technologies of spectral analysis to acquire a total picture of earth's vegetation—wild and cultivated, trees and plants—the JPL *Universe* reported. The group was adapting techniques developed for space exploration to devise, build, and test spaceborne instruments that would work in conjunction with ground-based monitoring, data-recording, and analysis devices.

The group was employing every available technology—radar, microwave, and infrared analysis—as well as effective procedures for using them. The goal was to develop technology to identify types of vegetation, the area of coverage, and general conditions from field to field, area to area, continent to continent. JPL was working closely with the University of California in the project, particularly the university's Kearney Agricultural Center, to diagnose crop conditions based on factors such as irrigation, insect infestation, and the spread of plant diseases.

In the program, satellites bounced radar off earth's vegetation, in a sense taking its pulse and temperature. At the same time, ground-based studies used, among other techniques, a truck-transported basket like a cherry picker to gather radar as well as infrared measurements similar to those from Landsat and other earth-surveillance satellites. Particularly helpful to the program was JPL's Space Shuttle imaging radar, which was unaffected by darkness or obscuring clouds. The instrument was on an October 1984 Space Shuttle mission and would fly again in 1987.

Data analysis in the program employed several approaches. In one, analysts gave a computer a statistical model and commanded it to survey recorded data to seek out similar conditions throughout the globe, thus providing insights into vegetation identity and condition. Other studies concentrated on the use of physical models to relate spectral data to parameters such as green-leaf area and water status.

Comparison of space-derived and ground-based images and data determined how they differed and how they complimented one another. This permitted monitoring of crop and natural vegetation conditions on a worldwide basis and offered insights into the effects of acid rain and other pollution on crops and forests, in lakes and ponds, and into conditions and changes of arid lands. (JPL *Universe*, Aug 23/85, 1)

During October: The Space Shuttle Discovery on mission 51–G was not the first NASA orbiter to carry retroreflectors for laser ranging in space, the *Goddard News* reported. Apollo astronauts in the late 1960s placed reflectors on the moon, and NASA launched in 1976 *Lageos*, a retroreflector equipped satellite. Laser ranging had become routine in space and was available for a variety of measurement tasks.

At Goddard Space Flight Center (GSFC), Crustal Dynamics Project (CDP) scientists analyzed data received from satellite laser ranging (SLR) to study the movement and deformation of crustal plates that determined earth's shape. The scientists also analyzed data received from lunar laser ranging (LLR) to study polar motion and earth rotation, which led to earth's wobble.

At GSFC there was an international system of cooperating laser networks. All systems, such as the Goddard Laser Tracking Network (GLTN), provided data to the CDP through conventional SLR and lunar laser ranging. The GLTN consisted of a series of transportable and fixed systems throughout the world and other international laser networks also supported the CDP.

Satellite laser ranging had attained major advances since the inception in 1965 of the GLTN. Tracking efficiency had improved from 65% in 1981 to 75% in 1985; instrument accuracy, from 10 cm to 1 cm.

The laser transmitter and sensitive photomultiplier receiver were the most essential equipment in any laser tracking system, because they provided and processed the signals from which the measurements were made. Current laser transmitters shot short-pulsed beams that lasted 200 trillionths of a second, the time it took light to travel about two inches.

Recently, the Federal Republic of Germany asked GSFC scientists to assist in preoperational tests of its new Modular Transportation Laser Ranging System (MTLRS) 1, because GSFC had the Mobile Laser Ranging System (MOBLAS) 7 for a calibration standard of comparison. During the daily tests, the two systems simultaneously tracked the same satellite, then scientists compared data obtained from MTLRS 1 to that from MOBLAS 7. The closer the measurements, the better the calibration; MTLRS 1 measured within 1 cm of the MOBLAS 7.

MTLRS 1 later would participate with other European and U.S. systems to perform satellite laser ranging in West Germany, Italy, Greece, Egypt, Israel, and Turkey. (*Goddard News*, Oct 85, 4)

November 6: Eighteen European governments agreed today to finance a plan, known as Eureka, to increase Europe's technological presence through development of 10 pilot technology projects, the *NY Times* reported. Although

Space Science and Applications

some observers considered Eureka a counterweight to the U.S.'s Strategic Defense Initiative (SDI) program, most of the Eureka projects, ranging from high-powered industrial lasers to a diagnostic kit for sexually transmitted diseases, were under discussion well before French President Francois Mitterrand proposed them in April 1985 as the Eureka project.

The agreement was reached at the same time that several European countries were considering whether to support the Reagan Administration's SDI program. Britain agreed in October to full participation in SDI, France had withheld support, and West Germany said it would decide before the end of the year.

In the Eureka agreement, ministers from the European Economic Community (EEC), joined by Spain, Portugal, Sweden, Norway, Finland, Switzerland, Austria, and Turkey, said in a seven-page "declaration of principles" that the program's aim was to "strengthen the basis for lasting prosperity and employment" by furthering "closer cooperation among enterprises and research institutes in the field of advance technologies."

Widely differing conceptions of how Eureka should work had slowed the plan, but the governments finally found ways around several obstacles, including questions on government versus private financing and a secretariat to oversee the agreement. Emphasizing that funding should come largely from the private sector, the declaration called for "adequate financial commitment by participating enterprises."

The ten pilot Eureka projects and nations to participate approved by the ministers were: production of a standard microcomputer for education and domestic use (Britain, France, Italy); production of a new type of computer chip made of amorphous, or uncrystallized, silicon (France, West Germany); development of a high-speed computer (France, Norway); development of a laser for cutting cloth in the apparel industry (France, Portugal); development of membranes for water filtration that could be used to desalinate sea water (Denmark, France); development of high-power laser systems (West Germany, France, Italy, Britain); development of a system to trace pollutants in European air (West Germany, Austria, Finland, the Netherlands, Norway, and the EEC); development of a European research computer network (West Germany, Austria, Finland, France, the Netherlands, Sweden, Switzerland, and the EEC); development of a diagnosis kit for sexually transmitted diseases (Spain, Britain); and development of advanced optic electronics (France, Italy). (*NYT*, Nov 7/85, D1)

November 27: NASA announced that 1986 would be its most productive year ever in space science activities, including a variety of "space firsts" and several major scientific studies that would be started or continued. To increase the public's knowledge and understanding of NASA's scientific programs, NASA's Office of Space Science and Applications and the Smithsonian

Institution's National Air and Space Museum would work together in a year-long program entitled "1986—A Year for Space Science."

The two organizations planned exhibits, audio-visual presentations, publications, and a lecture series at the Air and Space Museum and several other locations throughout the U.S. In addition, the National Air and Space Museum would carry NASA mission events on TV at designated locations in the museum.

Major space science activities in 1986 would include the Voyager 2 encounter in January with Uranus; the culmination of several scientific investigations of Comet Halley; launch in May from the Space Shuttle of Galileo toward Jupiter; and also in May launch from the Space Shuttle Challenger of the European Space Agency's Ulysses spacecraft to conduct comparative studies of the sun and its heliosphere. In addition, NASA scheduled for launch in late summer the Hubble Space Telescope, the largest telescope to be placed in earth orbit. (NASA Release 85-159)

November 29: NASA announced that its National Space Technology Laboratories, working with the University of Colorado at Boulder, uncovered information using remote sensing techniques suggesting a civilization existed before the Incas in subtropical Peruvian jungles. The university's anthropology department requested NSTL collaboration on the investigation because of the laboratories' expertise in satellite remote sensing and image analysis.

After remote sensing by satellites and aircraft permitted mapping and determining priorities for field investigation sites, Tom Sever, NASA principal investigator, and Tom Lennon, an archaeologist and co-director of the university's Rio Abiseo National Park project, completed a five-day expedition into the jungles of Peru's Rio Abiseo National Park.

Remote sensing had discovered Cerro Central, a site including more than 350 buildings. The previous major point of interest was the ancient site Gran Pajaten, which included only 30 buildings.

"We now know that Pajaten is probably the smallest and least important of the sites," Sever explained. "We are fairly confident that we have approached the very edges of a new civilization, and we believe that the farther in we go, the higher and more complicated the elevation and architecture and civilization will be."

Sever added that the investigation represented in his opinion the best example of a remote sensing application to archaeology and perhaps the only known means by which to obtain his project's objectives. (NASA Release 85-160)

During November: Gene Gilbert, a technician in Goddard Space Flight Center's (GSFC) Laboratory for Oceans was searching for a way to measure rainfall at sea by using an oil rig, *Goddard News* reported, in order to assist in the development of a system called the Tropical Rainfall Measuring Mission

(TRMM). TRMM could take the form of a satellite or space station attachment operational in the mid-1990s to improve upon rainfall sensors that flew aboard Nimbus satellites. But developing better sensing capabilities first required accurate ground truth measurements to test the accuracy of remote observations.

Over the years, meteorologists had developed a good picture of weather over land masses. But they still needed to know what happened over the remaining two-thirds of earth. When putting a collection bucket on a ship, Gilbert explained, different locations on deck gave differing results depending on whether the bucket was sheltered from the wind. In addition, the ship's movement gave angular measurement of the rain, not a true vertical one. And a buoy was not an improvement. "It may look stationary," Gilbert said, "but it really is not." Even the largest ones, which were 50 feet across, tossed in a storm, splashing and spraying and negating results.

"What is needed," Gilbert pointed out, "is a surface that is stationary, vibration free, and splash-proof. I think oil platforms could satisfy all those conditions." GSFC's Microwave Sensors and Data Acquisition Systems Branch in July placed the first rain gauge on a oil producing platform in the Gulf of Mexico off Lafayette, Louisiana. "The very next month a hurricane knocked out our data transmission antenna," Gilbert recalled. "But we repaired the damage easily, and we have been getting good data since then."

The rain gauge was a teeter-totter with a shot-glass sized cup at each end. The cups filled successively with 1/100th of an inch of rain and then dropped the measure back into the ocean. Each count was transmitted to the National Oceanic and Atmospheric Administration's (NOAA) Geostationary Operational Environmental Satellite (GOES) and then back to GSFC via NOAA's data processing facility in Suitland, Maryland.

"The goal is to demonstrate that using a gauge on an oil rig gives us our most accurate ocean rainfall data to date," Gilbert said. "If it does, we may expand to include other oil rigs. Then we will finally have good ground truth data for measuring rainfall over the ocean by satellite, at least for those oceans that include oil rigs." (*Goddard News*, Nov 85/ 2)

SPACE STATION PROGRAM

January 17: The *Washington Post* reported that W. Germany had agreed to contribute $900 million over the next decade to participate in an American-led program to build an $8 billion permanently manned space station for initial operations in 1992, the 500th anniversary of Columbus's discovery of America. The W. Germans would negotiate detailed contracts over the next two years to ensure a fair return on their investment and access to space-based technology, as U.S. restrictions on technology transfer (ostensibly to prevent USSR acquisition of sensitive information, the *W Post* said) had increasingly irked European allies. W. German and Italian companies planned a special laboratory module that would plug into the U.S.-built spacecraft's main structure for experiments in zero gravity and vacuum conditions.

A W. German ministerial report had said cooperation embodied in the space station program "should be welcomed in W. Germany not just for technical and economic reasons, but for political ones, as a transatlantic connecting link." However, the venture had evoked controversy, as some European scientists contended that robots could conduct the work planned for the space station, and others alleged that a $750 million European investment in 1973 to underwrite Spacelab had attained few research benefits.

Heinz Riesenhuber, W. German minister for research and technology, said that Bonn would also contribute nearly $500 million to develop a more powerful and versatile Ariane rocket to achieve some European independence in space travel by the 1990s, although he admitted a major W. German investment in the French space shuttle project "Hermes" was impossible. (*W Post*, Jan 17/85, A1)

January 18: The *Washington Post* reported that Britain would accept a Reagan administration invitation to contribute an expected $300 million to build a permanent space station scheduled for launch in 1992, making Britain along with W. Germany [see Jan. 17] the first noncommunist industrialized nations to participate in the space station program. Geoffrey Pattie, British minister of state for industry and information technology who was in Washington to consult with lawmakers and Reagan Administration officials about scientific cooperation, technology transfer, and telecommunications policies, said Britain would announce the contribution at the European Space Agency's (ESA) Rome meeting [see European Space Agency, Jan. 18].

President Reagan had invited Australia, Canada, Japan, and the U.S.'s Western European allies to join in funding and building the modular space station and expected France, Italy, and Japan to announce soon their intention to participate.

Pattie said Britain wanted access to technologies to design and build space stations as well as the results of research carried out in space laboratories; initial space station plans called for biological and physical sciences labs. While France and Germany were interested in rocketry research, Britain had focused its space expenditures on improving data transmission and expected the space station to enhance these technical capabilities. Pattie did note, however, that Britain feared the U.S. would impose export controls on some of this technology and that technology transfer questions required resolution during two years of space station feasibility studies. He emphasized the U.S. had to fashion "sensible" restrictions "instead of saying no to [exports of] everything, which is counterproductive to U.S. as well as European interests." (*NYT*, Jan 18/85, A1)

February 7: In his "State of the Union" address, President Reagan commented on funding requested for activities in space, the *Washington Post* reported. "We have seen the success of the Space Shuttle. Now we are going to develop a permanently manned space station and new opportunities for free enterprise because, in the next decade, Americans and our friends around the world will be living and working together in space.

"In the zero-gravity of space, we could manufacture in 30 days lifesaving medicines it would take 30 years to make on earth. We can make crystals of exceptional purity to produce supercomputers, creating jobs, technologies, and medical breakthroughs beyond anything we ever dreamed possible." (*W Post*, Feb 7/85, A16)

February 8: The *Lewis News* reported that NASA had completed conceptual designs for its space station and that industry had submitted proposals for definition studies and preliminary designs of various station components for which NASA expected to award 18-month contracts in April. NASA would award contracts in April 1987 for the next phase of the program—the final design, development, launch, and assembly of the station.

Contrary to most large NASA projects, for which an industrial prime contractor coordinated various aspects of a project, NASA would oversee the entire space station program because the station's operational life would extend over a long and indefinite period. Johnson Space Center (JSC) would manage the system engineering and integration function; Lewis Research Center (LeRC) was responsible for the power system.

The space station power system was vital for two major reasons: the quantity of electrical power available to governed station capabilities and, since the power system required large areas of solars cells or mirrors, the size of these areas affected station configuration and operation. Plans called for initial power demands of 25 kw for housekeeping needs of a crew of six and 50 kw for experiments and customer applications. Comparing this power

Space Station Program

system to the largest previously in space—the Skylab manned-mission system, which generated 16 kw—the magnitude of the power station challenge became evident.

LeRC personnel would consider a variety of power system technologies, making selections based on detailed studies of the tradeoffs among options. Studies would examine how the options influenced the performance, risk, and cost of the power system and space station, with consideration given to station growth and satisfaction of later needs. In addition, LeRC had to design a system that would be available on schedule, function reliably for an indefinite life with only on-orbit maintenance, be user-friendly, and remain within cost limits. (*LeRC News*, Feb 8/85, 2)

February 20: Neil Hutchinson, Johnson Space Center's (JSC) space station program manager, at a U.S. House Science and Technology space science and application subcommittee authorization hearing, said that ferrying and assembling a space station would require seven Space Shuttle flights over a period of nine months to a year, *Aerospace Daily* reported, and assembly would rely on automation and robotics, not Space Shuttle crew extravehicular activities. Assembly would begin with a beam of solar panels and a 90-ft. truss structure. Space Shuttle orbiters would dock with the space station structure at different points to assemble the facility, using the orbiter's remote manipulator system (RMS) arm and a remote arm operated on the space station.

The 3rd and 4th Space Shuttle flights would take up habitation modules; the 5th flight would ferry two more sets of solar panels and a 3rd module, possibly for logistics. Hutchinson said current planning called for permanent manning of the station after the 5th flight; the 6th flight would carry the 1st of two planned laboratories with the baseline configuration producing 75 kw of power, which would be completed after the 7th flight.

Phil Culbertson, NASA associate administrator, said a new program plan envisioned a 21-month definition and preliminary design effort beginning in April and extending through January 1987, which reflected the lower than anticipated funding level of $230 million for space station activities in the NASA's FY 86 budget. Of the $230 million, $15 million was for utilization, $82 million for advanced development, $52 million for program management/integration, $7 million for operational readiness, and $74 million for system-definition contracts. *Defense Daily* noted that Culbertson said NASA would have to "stretch" to meet the $8-billion price of the space station and that it would ask its contractors to "stretch with us." (*A/D*, Feb 20/85, 1; *D/D*, Feb 20/85, 227)

March 13: In a commentary in the *Washington Times* in response to a U.S. Office of Technology Assessment (OTA) study that had asserted NASA's plans for a permanent space station were not scientifically, economically, or militarily justified, Robert Melton, assistant professor at Pennsylvania State University, wrote that the study missed the point of NASA's current efforts to define better what should go into the station before proceeding with actual design and construction. He wrote that NASA had entered Phase B, the definition study of the project, in which contractors would provide NASA with thorough analyses of exactly what missions would be carried out in what time schedule and with descriptions of necessary equipment and technology together with a breakdown of costs. Such a detailed study was necessary to avoid premature overemphasis on design and to avert technical problems as the program progressed.

Melton argued that the permanent space station was a logical next step after the Space Shuttle program and could serve as both a laboratory and base of operations for scientific, commercial, and security purposes. The space station could also serve as a permanent base for repairing and routinely servicing satellites, as a depot for permanently, space-based orbital transfer vehicles capable of delivering and retrieving high-orbit payloads, and as a base for building other large structures that would remain in orbit.

Melton noted that the benefits of a permanent space station would make the $8 billion expenditure spread over several years seem a fairly small amount, especially when compared to other items in the federal budget. However, Melton acknowledged that the OTA study proved NASA had to do a better job of defining its space station goals and of impressing upon the public the importance of the project. (*W Times*, Mar 13/85, 2D)

March 14: NASA announced it had selected six industry teams for negotiations leading to 21-month fixed-price contracts for definition and preliminary design (Phase B) of elements for a permanently manned space station. Four NASA centers previously did this work.

The responsible centers and the industry teams selected for negotiations were: Marshall Space Flight Center (MSFC) with Boeing Aerospace Co., Seattle, and Martin Marietta Aerospace, Denver; Goddard Space Flight Center (GSFC) with RCA Astro Electronics, Princeton, New Jersey, and General Electric Co., Space Systems Division, Philadelphia; and Lewis Research Center (LeRC) with Rockwell Internatl., Rocketdyne Division, Canoga Park, California, and TRW Federal Systems Division, Redondo Beach, Calif. In addition, NASA would negotiate with Lockheed Missiles & Space Co., Sunnyvale, California; McDonnell Douglas Astronautics Co., Huntington Beach, California; and Rockwell Internatl., Space Station Systems Division, Downey California, the proposers for work to be performed under the management of Johnson Space Center (JSC). Negotiators would present a report to the NASA Administrator who would then award one or more contracts.

Although negotiations would determine values of the contracts, the September 14, 1984, Request for Proposal (RFP) indicated that the approximate

Space Station Program

value of contracts could be $24 million at MSFC, $27 million at JSC, $10 million at GSFC, and $6 million at LeRC.

In addition to Phase B work, NASA required the contractors to study how those elements of the space station would change depending on whether the station was man-tended rather than permanently manned and to pay particular attention to the recommendations of the NASA advanced technology advisory committee that was identifying automation and robotic technologies that could be used in the space station.

Following completion of the contracts, NASA planned to move into final design and development (Phase C/D) of the space station.

A major objective of the space station program was to involve international partners as builders, users, and operators of the space station. The European Space Agency (ESA), Canada, and Japan had indicated interest in participating. They would provide their own funding and award definition and preliminary design contracts in coordination with NASA activities. NASA, through JSC management, would retain responsibility for overall program definition and for systems engineering and integration throughout the program. (NASA Release 85-38)

March 20: Ames Research Center (ARC) announced that Dr. David Black would become chief scientist for the office of space station at NASA Headquarters, leaving a position he held as research scientist in the theoretical studies branch at ARC since 1972.

In his new job, Black would ensure that the space station would accomodate the needs of the scientists who would use it, advising Phillip Culbertson, associate administrator for space station, about steps to make the space station an accessible research facility for scientists from many disciplines. Black's appointment coincided with selection of contractors who would spend 21 months working out details of space station design [see Mar. 14].

Since other countries intended to cooperate with NASA in the space station project, Black would work with the agencies representing the interests of European and Japanese scientists to coordinate their plans with those of NASA.

Black had been studying scientists' needs for the space station since April 1984, when he had joined the task force on scientific uses of the space station. That committee brought together 30 scientists from various universities and represented the disciplines that were interested in using the space station. Also, Black was serving on a National Academy of Sciences study group called "space sciences, 1995-2015," which was attempting to identify the future directions for research in the space sciences, thus identifying which technologies would be needed to support future research. (ARC Release 85-14)

March 21: Aerospace Daily reported that NASA would submit to Congress that week the space station automation and robotics study, "Advancing Automation and Robotics Technology for the Space Station and for the U.S. Economy," which appropriations legislation enacted in 1984 had directed NASA to prepare by April 1. NASA Administrator James Beggs approved the report compiled by the advanced technology advisory committee and augmented by the work of an automation and robotics panel, SRI International, and aerospace contractors.

The report consisted of two volumes, an executive overview, which outlined major findings and contained proposed goals for automation and robotics applications in relation to the initial space station, and a technical report, which outlined the potential of automation and robotics technologies and would serve a major focus of definition and preliminary design (Phase B) space station contractor efforts.

Under an agreement with the Senate Appropriations Committee the previous April, NASA had funded and managed several studies in automation and robotics that had included industry case studies of advanced automation and robotics. From these the committee determined what should be incorporated in space station initial operational capability and what the design should be so that these elements could be incorporated at a later date.

The firms and functional areas studied were: General Electric Co., space manufacturing concepts; TRW, satellite servicing; Hughes, subsystem and mission ground support; Boeing, man/machine interface; and Martin Marietta, automation technology for assembly of the facility.

Although NASA believed a commitment to the efforts outlined in the committee report would increase space station efficiency and result in significant cost savings, a breakdown of actual costs would not be available until NASA contractors completed the overall plan for implementation of the automation and robotics systems in initial space station studies.

The report did recommend that automation and robotics be a key element of the basis space station program and that the initial space station design take into account evolution and growth in robotics. Examples of proposed goals for automation and robotics applications for the initial space station included a mobile remote manipulator with collision avoidance capability and dexterous manipulator systems that could inspect and exchange orbital replaceable units. (A/D, Mar 21/85, 1)

March 22: The NASA exhibit entitled "Living, Learning, Working in Space" at the 36th Paris Air Show, May 30 through June 9 in Le Bourget, would highlight full-scale elements of a space station, including mockups of working and living quarters, NASA announced.

Audiovisual presentations throughout the 6,900-sq. ft. exhibit area would depict life inside a space station. To give viewers the impression they were looking out into space from a space station module, NASA would suspend in a diarama a large model of the power tower configuration space station along with models of future U.S. space science programs such as the space telescope and Galileo. Exhibits would also showcase NASA's latest aeronautical research. (NASA Release 85-41)

March 22: The Canadian government decided to join the U.S. and other allies in producing an orbiting space station, agreeing in principle to spend up to $600 million over the next 10 years, the *Washington Times* reported.

U.S. space officials had encouraged Canada for more than a year to join the U.S.-led project, which already had the participation of the European Space Agency (ESA) and Japan, together expected to spend about $4 billion.

Canada's science minister Tom Siddon said that Canada's involvement in the space station could produce economic benefits valued at more than $2 billion by the year 2000. (*W Times*, Mar 22/85, 5A)

March 29: Tom Rogers, project director of the Congressional Office of Technology Assessment's (OTA) space station study, said in testimony before the U.S. House space subcommittee that OTA in its controversial November 1984 space station report [see Space Station Program, Mar. 13] never meant to oppose development of a space station but only to suggest that its elements might be altered for more adequate attainment of space goals, *Defense Daily* reported.

"I want to make it clear that OTA remains convinced that there is a strong rationale for some long-term habitable infrastructure in a low earth orbit . . . In other words, OTA is very positive about the principle behind the President's call for a 'space station' and Congress's decision to fund a 'space station' line item in NASA's budget," Rogers said. However, he pointed out that OTA's concerns about the space station centered on its cost and the fact that there was no proper assessment of long-range U.S. space goals, which made it infeasible to determine what elements should make up the station.

"To be more certain that the facilities actually constructed are the ones most likely to lead to optimum space development, they must be constructed with an eye to the nation's long-range civilian space goals and objectives," Rogers testified. "Unfortunately, the only space goals discussed to date are those formulated by the space community itself; there has been little broad-based discussion and agreement on a set of long-range goals for the United States." He recommended particularly that, in parallel with NASA's Phase B station studies, "Congress seek independent studies suggesting goals and specific objectives for the nation's future activities in space, studies conducted primarily by people outside the space community."

Rogers then outlined OTA's alternatives for the space station program. These included using the $3 billion in foreign funding as part of the $8 billion station cost, not as an addition; using private investment to pay for elements of the station; increasing use of the human-tended approach and automation/robotics to reduce station costs; examining alternative station designs costing less than "the canonical $8 billion"; developing the orbital maneuvering vehicle and any reusable orbital transfer vehicle "in close concert with those in the private sector who look forward to using them to provide satellite support services," thus reducing NASA's costs; and having contractors work to performance specifications rather than to detailed engineering specifications for other than "cutting edge" technologies. (*D/D*, Mar 29/85, 162)

April 1: NASA announced it had submitted to the U.S. Congress today a report by its Advanced Technology Advisory Committee (ATAC) that said automation and robotics would be significant elements of the NASA Space Station program and that the initial Space Station design should accommodate evolution and growth in these technologies.

The report, required under Public Law 98–371 that appropriated FY 85 NASA funds, noted that all Space Station elements, including the core station and associated unmanned platforms, vehicles, and ground facilities were candidates for the technologies, necessitating accelerated R&D in automation and robotics. The Automation and Robotics Panel, which assisted the Advisory Committee in a six-month study led by the California Space Institute, said that a desired level of funding for automation and robotics research would be 13% of the total Space Station costs; a minimum acceptable level was 7%. An augmented research program, the report said, would result in improved productivity from the initial station and much greater productivity from later versions of the Space Station.

The report also said successful incorporation of automation and robotics into the Space Station program could lead to deployment of a new generation of flexible and adaptable space systems, which could provide the U.S. with new methods of generating and exploiting space knowledge in commercial activities and thus preserving U.S. leadership in space and industrial systems.

Other report recommendations included: development of criteria for incorporating automation and robotics technology in the Space Station; verification of automated equipment performance, including terrestrial and space demonstration to validate technology for Space Station use; use of automation techniques to enhance NASA's management capability; establishment of measurements to verify inclusion of automation and robotics in the Space Station; development of a program for technology transfer to U.S. industries; and design of satellites and payloads accessible from the Space Station to accommodate service and repair by robots. NASA would incorporate report recommendations into the Phase B definition and preliminary design contracts [see Mar. 14] it was negotiating with industry teams.

The Advisory Committee, made up of personnel from NASA Headquarters and centers, would continue to monitor automation and robotics developments and would report semi-annually to Congress. (NASA Release 85–46)

Space Station Program

April 1: NASA overcame the final obstacle to achieving four-way international cooperation (the U.S., Europe, Japan, and Canada) on the U.S. manned space station when NASA and The European Space Agency (ESA) compromised in a memorandum of understanding (MOU) on wording in three areas: technology transfer, nondiscriminatory access for European users to the Space Station, and station elements that would be candidates for European participation, *Aviation Week* reported. NASA Administrator James Beggs and ESA Director General Reimer Lust resolved at an early March meeting the differences over wording concerning technology transfer and European user access. A compromise formulated by W. Germany and approved in late March by ESA members resolved the issue of which components would be open to European development.

"The final step for Europe is approval of the MOU by ESA's council, which has its next meeting at the end of April," Michel Bignier, ESA director of space transportation systems, said. "I don't foresee any difficulties there because the ESA member delegates have been informed of the negotiations and are aware of the final wording."

Bignier went on to say NASA accepted "certain small nuances" concerning nondiscriminatory access for Europe to the station and that both sides made compromises on the issue of which elements would be included for European cooperation. And he added that ESA "basically accepted NASA's proposed text" on technology transfer for the Phase B memorandum but said the wording would not apply to follow-on agreements covering hardware development. "We felt this was not as critical an issue in Phase B as it will be in the later Phase C/D," Bignier explained.

Philip Culbertson, NASA associate administrator for the Space Station, said international participation in the station was essential to overall U.S. planning for the facility. "The investment of our potential partners from around the world will give the station a significantly greater capability than we in the U.S. could provide alone," Culbertson said. "We believe we will end up with a capable Space Station for all of us to use." (*AvWk*, Apr 1/85, 16)

April 8: I. V. Franklin, manager of future projects at British Aerospace Dynamics Group's Space and Communications Division, said at a recent international space meeting in Italy that British Aerospace had completed preliminary definition of a 50- by 60-foot-long, 24,250-lb. unmanned platform that Europe could develop as part of its Columbus space station program, *Aviation Week* reported. The platform would use a long beam as a common backbone that would contain docking ports for standardized plug-in payload and resources modules. A single Space Shuttle flight would carry the basic platform into orbit, and the company envisioned that astronauts could assemble it in six hours of extravehicular activity.

Franklin said the company would submit the platform proposal to the European Space Agency as one of the work packages in Europe's Columbus program. "The space platform is of particular interest to Europe, following President Reagan's invitation for international participation in the space station program," Franklin continued, "because it represents an element with relatively few direct interfaces to the rest of the program. Although it is not a 'core' component, it would be an important component of the space station complex and [a component] in which there would be substantial European user interest. The platform provides a point for reducing space operation costs at a price that Europe could afford."

In addition to evaluating the platform in low earth orbit in conjunction with the U.S. manned Space Station, British Aerospace also considered the platform in a polar, sun synchronous orbit. The polar-orbiting platform offered possibilities for earth observation and earth resources missions. U.S. officials had encouraged European development of the polar platform to supplement low earth orbit activities performed on board the manned Space Station and its associated free-flying elements.

"The polar platform does not need to work with a manned space station—and indeed, it cannot until a station facility is available in polar orbit," Franklin said. "The platform can be man-tended by a polar orbiting shuttle, although the shuttle's altitude is well below that of the platform."

This problem could be overcome by using the platform's on-board propulsion to deorbit and dock with the Space Shuttle for payload replacement, repair, and maintenance and then return to the operational altitude.

Franklin concluded that the concept of a polar space platform was justified by the level of user interest in both Europe and the U.S. and by the relatively favorable economics when compared with other systems. "The polar platform does not require use of a manned space station, and it therefore is not paced by development of the space station," he noted. (AvWk, Apr 8/85, 60)

April 12: Reiichi Takeuchi, chief of Japan's science and technology agency, said Japan had decided to take part in the U.S. space station beginning with the preliminary design stage, FBIS, KYODO in English reported. He said the Japanese would sign with NASA Administrator James Beggs a memorandum on the decision when Beggs visited Tokyo early in May. The agreement would cover two years of work, at which time, Takeuchi said, the two countries would probably sign a similar document for later stages of the project.

Takeuchi noted he expected the U.S. would ask Japan to design the laboratory room of the Space Station, scheduled for operation in 1992. He estimated the cost at 200 to 300 billion yen. Under the memorandum, the U.S. would provide Japan with technical information on how the Space Station would be used. (FBIS, KYODO in English, Apr 12/85)

April 15: NASA announced it selected McDonnell Douglas Astronautics Co. and Rockwell Internatl.'s Space Station Systems Division for fixed-price contracts for definition and preliminary design (Phase B) of the structural framework and other elements of a permanently manned Space Station. Johnson Space Center (JSC) would manage the 21-month contracts, which had an estimated value of $27 million each. NASA previously announced other industry teams selected for negotiations for definition and preliminary design of other Space Station elements [see March 14].

In addition to work on the Space Station structural framework, the JSC contracts would cover interface between the Space Station and Space Shuttle; mechanisms such as the remote manipulator systems; attitude and thermal control; communications and data management systems; plans for equipping a module with sleeping quarters, wardroom, and galley; and plans for extravehicular activity. (NASA Release 85-56)

April 16: NASA announced that its Administrator, James Beggs, and Canadian Minister of Science and Technology, The Honorable Tom Sidden, signed on April 16 a memorandum of understanding to conduct a cooperative program for detailed definition and preliminary design (Phase B) of a permanently manned Space Station. Under the memorandum, the countries for the next two years would conduct parallel Phase B studies and exchange information on their work. Canada, which had approved $8.8 million for the first year of Phase B, would study a space construction and servicing system, a solar array for a platform or as a potential auxiliary power source for the Space Station, and a remote sensing facility.

At the signing, Administrator Beggs said that "we are pleased Canada has become an international partner in [this] phase of the Space Station's development. We look forward to working with our friends in Canada, Europe, and Japan in building a firm foundation for future cooperation on a permanently manned Space Station to serve the needs of the free world in developing the peaceful uses of space well into the next century."

Separate agreements would cover cooperative efforts during development, operation, and use phases of the Space Station. (NASA Release 85-58)

May 5: The text of a joint declaration made public today at the end of a seven-nation economic conference in Bonn, West Germany, contained a statement concerning space activities, the *NY Times* reported. The text read: "We welcome the positive responses of the member states of the European Space Agency (ESA), Canada, and Japan to the invitation of the president of the United States to cooperate in the United States manned space station program on the basis of a genuine partnership and a fair and appropriate exchange of information, experience, and technologies. Discussions on intergovernmental cooperation in development and utilization of permanently manned space stations will begin promptly. We also welcome the conclusions of the ESA Council on the need for Europe to maintain and expand its autonomous capability in space activity, and on the long-term European space plan and its objectives." (*NYT*, May 5/85, A16)

May 9: NASA announced that NASA Administrator James Beggs and Japanese Minister of State for Science and Technology, Reiichi Takeuchi, signed that day in Tokyo a memorandum of understanding (MOU) for the conduct of a cooperative program concerning detailed definition and preliminary design (Phase B) of a permanently manned Space Station.

Under the MOU, the U.S. and Japan would conduct and coordinate over the next two years parallel Phase B studies and exchange information. Japan would study an experimental module that had pressurized workspace and an exposed workdeck. Cooperation during the space station's development, operations, and use phases would require separate agreements.

In a speech that day to the Federation of Japanese Economic Organizations, Beggs outlined initial uses of the space station and reported on specific projects undertaken by other countries participating in space station development. He then noted, ". . . in negotiating with our potential partners, NASA has emphasized that we view the space station as a potential long-term international partnership—one that should last for decades. Any nation that joins with us in such a full partnership must be prepared to make significant investments in the station, and also be prepared to help operate and use it.

"We expect our partners to continue to shoulder responsibility for owning and maintaining their portions of the facility, while continuing to enjoy the overall benefits our joint efforts will make possible. And we expect their contributions to remain a permanent part of the station's infrastructure."

Beggs then added, "To cement our long-term relationships, the United States will provide partners with assurance on equitable access to all of the space station's facilities. We also will protect their technology and intellectual property and ensure them suitable roles in the station's management and operation." (NASA Release 85-71, NASA Note To Editors, May 8/85)

May 19: NASA announced its Administrator James Beggs and the Director General of the European Space Agency (ESA) would sign June 3 at the ESA pavillion at the Paris Air Show a memorandum of understanding (MOU) for the conduct of a cooperative program for detailed definition and preliminary design (Phase B) of a permanently manned space station [see Space Station Program, Apr. 1].

The cost of the Phase B studies carried out by European industry under ESA management, together with the corresponding technology program, amounted to 80 million accounting units or $64 million at the current rate of exchange. Cooperation during the space station's development, operations, and use phases required separate agreements. (NASA Release 85-78)

Space Station Program

July 12: Marshall Space Flight Center awarded a $139,000 follow-on contract to Martin Marietta Corp. to study a system for reclaiming unused propellant from Space Shuttle external fuel tanks and transferring it to an orbiting space station, *Aerospace Daily* reported. Martin Marietta Corp's New Orleans, Louisiana, manufacturing facility would perform the work on the preliminary design study. The contract followed one let in February 1984 and valued at $250,000 to examine the basic "propellant scavenging" concept.

The system under study would consist of two to four propellant collecting or scavenging tanks in an aft cargo carrier affixed to the rear of the Space Shuttle external tank. After Space Shuttle launch and cutoff of the main engines, residual liquid oxygen and liquid hydrogen would drain from the external tank into the collection tanks aided by a thrust system designed for zero-gravity conditions. Once filled with fuel, the scavenging tank assembly would separate from the aft cargo carrier as a self-contained, remote-controlled vehicle with its own propulsion system.

After the vehicle moved into position beside the space station, technicians aboard the station would send an orbital maneuvering vehicle (OMV) to bring it to the space station fuel depot for transfer of the reclaimed propellant to permanent tanks. When out of fuel, the scavenger vehicle would return to earth in the Space Shuttle cargo bay for reuse on later missions.

The proposed system would be capable of reclaiming and transporting up to 25,000 lb. of propellant per mission, depending on the Space Shuttle payload. Unused fuel on Space Shuttle missions to date had ranged from 9,500 to 28,300 lb. of the total 1.6 million lb. of propellant carried by the external fuel tank on each flight. It was currently estimated that the space station would require some 250,000 lb. of liquid hydrogen and liquid oxygen propellants each year.

Martin Marietta said the proposed system would result in reduced costs for transporting the cryogenic fuels to the space station; the company estimated the system could provide propellants at an average cost of approximately $350 per lb. NASA had studied other methods of scavenging fuel from the Space Shuttle external tank but found them to have a higher cost per lb. or to provide fewer opportunities for transporting the cryogenic fuels to the station. (A/D, July 12/85, 61)

During July: Yvonne Clearwater, in an article in *Psychology Today*, described the work of the Space Human Factors team, an interdisciplinary team of which she was a member, to determine what could make the proposed space station a congenial habitat for living and performing highly sophisticated work. The team—representing psychology, architecture, and engineering—was defining what "habitability" meant in space by looking at the critical relationships among environment, psychological well-being, and performance.

Clearwater said the team would begin by developing architectural and interior-design guidelines based on behavioral research. NASA contractors and development engineers at Johnson Space Center would then use the guidelines in designing and building the space station.

The team would then turn to examining social and psychological issues such as crew selection, training, and support; organization and management; and operations planning. Although recommendations would focus on the physical and social-psychological environments separately, the team had to consider all the factors together in the planning process, because they interacted in the real world, whether on earth or in space.

In contrast to the Soviets, who placed a high priority on the mental well-being of their space crews, NASA originally had focused on technological engineering rather than behavioral science. In addition, the early astronauts had little interest in psychological support or intervention; reportedly the first seven U.S. astronauts announced they would not tolerate psychologists.

Clearwater pointed out, however, that NASA had grown more sophisticated and more sensitive to the fact that body-mind-environment interactions affected the health and performance of people. The formation of the Space Human Factors Office at Ames Research Center reflected NASA's commitment to psychological as well as physical health of space workers.

The challenge was to troubleshoot the proposed space station systems and settings while they were still being developed conceptually by, first, identifying the environmental conditions likely to be significant stressors; second, defining the kinds of psychological, emotional, and behavioral problems these could produce; and finally, showing how such problems might affect work performance.

Clearwater concluded that by defining and meeting human needs in space environments, it might be possible to create more supportive places for living and working on earth. (*Psychology Today*, July 85, 34)

October 25: NASA selected the dual keel reference configuration, a modification of the power tower concept, for the proposed permanently manned space station, the *Spaceport News* reported. It was called the dual keel because of twin vertical booms that would provide the framework for attachment of other structures. The concept's primary attributes were that it would provide better customer accommodations and servicing, increased attachment area, and a more versatile design for growth based on future requirements and changing station roles.

Another change in space station planning was placement of the pressurized modules near the station's center of gravity to increase the amount of space available for experiments, such as crystal growth, that needed a microgravity environment.

NASA selected the dual keel configuration from a variety of layouts devised by teams of contractors working on two-year space station definition and

Space Station Program

preliminary studies. Overall space station shape and module placement were just two of the many significant technical decisions program officials had made since the space station definition period began in April 1985.

NASA would conduct over the next three to five months a series of reviews to narrow design choices so that it could select by early 1986 a baseline structure for the proposed space station. Contractor teams would spend the second half of the study phase doing space station subsystem preliminary designs, leading to start in mid-1987 of the planned development phase. (*Spaceport News*, Oct 25/85, 2)

October 25: NASA earmarked about $3.2 million in FY 86 funding for continuing work on the proposed space station at Kennedy Space Center (KSC) to define processing requirements, evaluate maintenance and resupply activities, and assess facility needs, the *Spaceport News* reported. The funding, an increase over the previous year, would pay for studies to further the Phase B definition and preliminary design effort expected to continue through early 1987. NASA planned for the early 1990s the first launch of a space station element.

KSC studies were intended to identify ground processing options and launch preparation concepts. In addition, KSC contractors were evaluating what facilities were needed for processing the station elements and payloads. KSC was also studying how NASA should approach the ongoing maintenance and resupply activities that would support continuous on-orbit operations.

KSC space station activity in FY 86 was expected to require the equivalent of about 180 NASA and contractor workers. (*Spaceport News*, Oct 25/85, 2)

November 11: European Space Agency (ESA), Japanese, and Canadian officials the previous week expressed concern about NASA's expectations of how the costs of operating the proposed space station should be shared, as fears increased that the U.S. Congress or the Office of Management and Budget might slip the planned 1993 operational date for the $10 to $13 billion facility, *Aviation Week* reported. The officials were meeting with NASA managers to lay the groundwork for critical space station decisions due by December that would determine what type of contribution they would make to the basic $8 billion U.S. investment.

The international space officials were unanimous in their view that the non-U.S. contribution to the station's operational costs should be largely amortized through open access and use of their respective portions of the station system—not through funding transferred to the U.S. As the discussions progressed, the possibility of a space station schedule delay to help cut the U.S. budget deficit in FY 87 became a growing factor.

Sen. Slade Gorton (R-Wash.), chairman of the Senate science subcommittee, and Rep. Don Fuqua (D-Fla.), chairman of the House Science and Technology Committee, both influential space supporters, were cautioning that congressional action on the FY 87 NASA budget could trim space station funding and delay the project's 1993 planned operational date by a year or more.

Fuqua told a meeting of the Washington Space Business Roundtable, a group formed to foster space commercialization, that a slip now, while undesirable and likely to drive up costs later, could be absorbed by the program more easily than one later when the hardware phase was underway. (AvWk, Nov 11/85, 18)

November 15: NASA's Lewis Research Center (LeRC) announced it had awarded a total of $8.7 million in contracts to Sundstrand Corp., Grumman Aerospace Corp., Boeing Aerospace Co., and Harris Corp. for advanced development contracts for definition and preliminary design (Phase B) of the power system for the proposed permanently manned space station.

A major technical issue in Phase B was determination of whether photovoltaic arrays or a solar dynamic (heat engine) system should supply solar-generated power for the space station. Photovoltaic arrays were the accepted system for electricity production in manned and unmanned space missions. However, the space station's electrical power requirements were ten times greater than any mission flown to date and would necessitate arrays of approximately one-half acre for the initial station. Therefore, there was interest in solar dynamic systems because of their higher overall efficiency and relatively smaller size.

In a solar dynamic system, an alternator driven by a turbine in a heat engine cycle produces electricity. Focusing the sun's rays by means of a concentrating mirror into a heat receiver heats the engine gas or liquid. The system operates as a closed-cycle heat engine, and a radiator cools the working fluid and rejects waste heat into space.

Sundstrand Corp., under its $1,010,303 cost-plus-fixed-fee contract, would study the magnitude of possible chemical and thermal degradation in the working fluid of an organic rankine cycle engine. Under its $1,010,000 cost-reimbursement contract, Grumman would study solar dynamic waste heat radiator technology. Boeing Aerospace Co., under its $3,117,059 cost-plus-fixed-fee contract, would study the heat receiver/storage unit and identify and recommend testing required for concept verification. And Harris Corp., under its $3,619,870 cost-sharing contract, would generate conceptual designs for the solar dynamic dish concentrator, as well as identify and test materials, identify and recommend testing required for concept verification, perform engineering designs, fabricate the concentrator, and conduct verification and testing. (LeRC Release 85–77)

SPACE TRANSPORTATION SYSTEMS (STS)

Civilian in Space Program

February 21: NASA and the Council of Chief State School Officers announced that 10,690 teachers applied for a Space Shuttle flight. The Council was responsible for application and screening processes; NASA would choose the teacher to go into space.

The Council's review panel would screen all applications to eliminate those not meeting basic requirements and forward the remaining applications to state-review panels to select two teachers per state by May 1, 1985. A national panel would then review applications of the approximately 118 nominees (two per state plus the District of Columbia, Puerto Rico, Virgin Islands, territories and trusts, and Department of Defense and independent schools) to recommend 10 semifinalists to NASA's spaceflight participant evaluation committee.

All nominees would attend a teacher-workshop and orientation program from June 24 to 28, 1985, in Washington to learn of current developments in the aeronautics and space-education program and to undergo further evaluation and screening; 10 semifinalists would report to JSC for thorough medical examinations, in-depth briefings, and interviews by NASA's evaluation committee.

NASA's administrator and the evaluation committee would select a primary and a backup candidate to undergo training. NASA had not decided on a specific flight opportunity, but was aiming to fly the teacher on a mission in late 1985.

California had the greatest number of applications, 926. (NASA Release 85–26)

May 3: NASA announced that the Council of Chief State School Officers (CCSSO) had that day named 114 elementary and secondary school teacher nominees in the NASA Teacher in Space Project. Selection followed the review of more than 10,000 applications [see Space Transportation System/Civilian in Space Program, Feb. 21].

Dr. William Pierce, CCSSO executive director, said after announcing the names that "It is a great pleasure for the Council to be involved in this historic project. The calibre of the applications from teachers throughout the country has been truly impressive. If their applications are any indication, we can be proud of the quality of teaching that occurs in the classrooms of the elementary and secondary school teachers who applied for this unique educational opportunity."

NASA would host the 114 nominees at a national workshop, June 22–27, in Washington, D.C., during which NASA would discuss current developments in the space program and provide information and training on NASA educational materials available for the classroom. At the workshop a national review panel would interview the applicants to determine the 10 semifinalists. (NASA Release 85-67)

June 14: NASA announced finalists in its Teacher in Space Project would arrive in Washington, D.C., beginning June 22 to attend a national conference sponsored by NASA and the Council of Chief State School Officers (CCSSO) as part of the selection process in the program. Through June 27 the 114 elementary and secondary school teachers would hear from NASA officials and other experts in space science and exploration, including discussions about their responsibilities if chosen to fly aboard the Space Shuttle, and participate in workshops sponsored by NASA's educational affairs division to provide a hands-on learning experience about NASA and its programs. During the week, a national selection panel would interview the finalists.

On June 25 the teachers would attend a reception on Capitol Hill with members of Congress and on June 26 meet with President Reagan. (NASA Release 85-92)

July 1: NASA and the Council of Chief State School Officers announced today the 10 finalists in the NASA Teacher in Space Project. The finalists would travel July 7 to Johnson Space Center for medical examinations and initial spaceflight suitability testing. The NASA Spaceflight Participant Committee would then interview the teachers in Washington, D.C., submitting results of the examinations and interviews to NASA Administrator James Beggs who would select the primary and backup candidate. NASA had tentatively scheduled the teacher for a January 1986 Space Shuttle flight.

Finalists were Kathleen Beres, Kenwood High School, Baltimore, Maryland; Robert Foerster, Cumberland Elementary School, West Lafayette, Indiana; Judith Garcia, Thomas Jefferson School for Science and Technology, Alexandria, Virginia; Peggy Lathlaen, Westwood Elementary School, Friendswood, Texas; David Marquart, Boise High School, Boise, Idaho; Sharon Christa McAuliffe, Concord High School, Concord, New Hampshire; Michael Metcalf, Hazen Union School, Hardwick, Vermont; Richard Methia, New Bedford High School, New Bedford, Massachusetts; Barbara Morgan, McCall-Donnelly Elementary School, McCall, Idaho; and Niki Wenger, Vendervender Junior High School, Parkersburg, West Virginia. (NASA Release 85-99)

July 19: Vice President George Bush announced that Sharon Christa McAuliffe would be the teacher to go into space in January 1986 aboard the Space Shuttle. McAuliffe, a social studies teacher at Concord High School,

Concord, New Hampshire, was the finalist in the NASA Teacher in Space Program that was announced by President Reagan in August 1984. Her backup was Barbara Morgan of McCall-Donnelly Elementary School, McCall, Idaho.

McAuliffe proposed that, while in space, she would gather information for a personal journal, "just as the pioneer travelers of the Conestoga wagon days kept personal journals. My journal would be a trilogy. I would like to begin it at the point of selection through the training program. The second part would cover the actual flight. Part three would cover my thoughts and reactions after my return," McAuliffe said.

Vice President Bush, in his announcement of the selection at a White House ceremony in the Roosevelt Room, where McAuliffe was accompanied by the other nine finalists selected from among 11,416 applicants, said, "We're here today to announce the first private citizen passenger in the history of spaceflight . . . We're honoring all [the teacher applicants] today, and we're doing something else because the finalists here with me and the more than a hundred semifinalists will all in the months ahead serve as a link between NASA and the nation's school system."

McAuliffe, a teacher of 15 years and the mother of two children, would work a year for NASA. She and Morgan would report in September to Johnson Space Center for 114 hours of training over four months.

The *Washington Post* reported that a top NASA official said McAuliffe was an early favorite of the 20-member selection panel. The official said "the judges thought McAuliffe appeared to be a good team player and—vital to her image-making duties—stood out as a good communicator," the *Post* reported. (NASA Release 85-107; Admin of Ronald Reagan, July 19/85, 913; *W Post*, July 22/85, A1)

September 6: NASA today invited Rep. Bill Nelson (D-Fla.), who chaired the U.S. House science and technology subcommittee on space science and applications, to be a congressional passenger on an unspecified flight of the Space Shuttle, the *Washington Post* reported. Nelson, who represented the district in which the Kennedy Space Center launch site was located, later held a news conference at his office in Melbourne, Florida, to announce acceptance of the invitation.

The first congressional Space Shuttle passenger on an April 1985 flight was Sen. Jake Garn (R-Utah), chairman of the committee that oversaw NASA spending.

Although Nelson, unlike Garn, had no flying experience, he was an outspoken supporter of the space program. (*W Post*, Sept 6/85, A8)

September 9: Christa McAuliffe, first teacher as well as private citizen to fly onboard the Space Shuttle, and backup Barbara Morgan arrived September 9 at Johnson Space Center (JSC) to begin their training program, the JSC *Roundup* reported. Reporters, photographers, and TV crews accompanied the two as they arrived at building 100 to receive their identification badges.

Mission commander Francis "Dick" Scobee that afternoon welcomed McAuliffe and Morgan at a meeting with the crew of mission 51-L, scheduled for launch in January 1986. Other 51-L crew members were pilot Michael Smith and mission specialists Ronald McNair, Ellison Onizuka, and Judith Resnik.

JSC personnel the next day measured them for flight suits, helmets, and G-suits. McAuliffe said she would be allowed to select either a jumpsuit or a standard flight suit to take home with her following her flight.

After a food tasting and rating session, in which food lab manager Dr. Charles Bourland asked McAuliffe and Morgan to taste about 40 food and drink items and rate them on a scale of 1 to 9, both teachers commented on their surprise in finding space food so flavorful. They would have several more opportunities for taste tests before McAuliffe selected her flight menu.

Other first-week activities included orientation briefings by management and training officials and familiarization tours of training facilities and mission control.

Asked by a reporter how drastically her life had changed since her selection, McAuliffe said that other than having the opportunity to fly onboard the Space Shuttle and the accompanying publicity she didn't perceive that much difference. She pointed out that as a teacher for a long time she had to gather material and present information to students daily. "As I see it, I'm doing basically the same things, only my audience has changed." (JSC *Roundup*, Sept. 27/85, 7)

October 2: NASA announced that live classroom lessons and scientific demonstrations, which would be broadcast live around the country, would be filmed for use in educational products which were just some of the activities planned by Christa McAuliffe, the finalist in the NASA Teacher in Space Project, for Space Shuttle mission 51-L.

The first live lesson, entitled "The Ultimate Field Trip," would allow students to compare daily life on the Space Shuttle with that on earth. McAuliffe would take viewers on a tour of the orbiter, explaining crew members' roles, showing the location of computers and controls, and explaining experiments being conducted on the mission. She would also demonstrate how daily life in space was different from that on earth in the preparation of food, movement, exercise, personal hygiene, sleep, and the use of leisure time.

The second lesson, called "Where We've Been, Where We're Going," would help the audience understand why people use and explore space by demonstrating the advantages of manufacturing in the microgravity environment, explaining technological advances that evolved from the space program, and projecting the future of humans in space.

Also during the flight, McAuliffe would participate in activities that would be filmed and later used in educational products. Possible activities included demonstrating earth magnetism by photographing and observing the lines of magnetic force in three dimensions in a microgravity environment; demonstrating Newton's first, second, and third laws in a microgravity environment; discussing why products might or might not effervesce in a microgravity environment; encouraging creative works from students that reflected their interpretation of the space program/experience; explaining the use of simple machines/tools and the similarities and differences between their uses in space and on earth; showing the effect of microgravity on plant growth, growth of plants without soil (hydroponics), and capillary action; and demonstrating chromatographic separation of pigments in a microgravity environment.

In addition, McAuliffe would assist mission specialists conducting three Shuttle Student Involvement Project experiments that would fly onboard the Space Shuttle. The experiments dealt with using a semipermeable membrane to direct crystal growth, studying chicken embryo development in space, and the effect of weightlessness on grain formation and strength in metals. (NASA Release 85–139)

October 4: NASA announced today that Rep. Bill Nelson (D-Fla.), chairman of the subcommittee on space science and applications, would fly as a payload specialist aboard Space Shuttle mission 61–C scheduled for launch no earlier than December 20.

NASA said it was willing to schedule flights for the chairmen of its four Congressional appropriations and authorization subcommittees in connection with their NASA oversight duties. Sen. Jake Garn (R-Utah), chairman of the subcommittee on HUD/independent agencies, was a payload specialist the previous April aboard mission 51–D.

NASA assigned Nelson to mission 61–C after reassigning Gregory Jarvis, a Hughes Communications, Inc. payload specialist, from that flight to mission 51–L which was scheduled for launch January 22, 1986. Hughes's decision not to launch its Syncom IV–5 spacecraft on the December mission eliminated the need for a company payload specialist on that flight. Jarvis would conduct experiments in fluid dynamics on the January flight. (NASA Release 85–141)

October 24: NASA announced today that an American journalist would fly on a Space Shuttle mission in late 1986 as the second in a series of communicators to be selected as part of the agency's Space Flight Participant Program. The first candidate selected under this program was Christa McAuliffe, a classroom teacher scheduled to fly in January 1986.

In today's announcement NASA said it would select the journalist, like the teacher candidate, after a nationwide competition conducted by professionals representing a broad spectrum of individuals in the candidates' field. The competition would be limited to full-time working media representatives (U.S. citizens) with five or more years' experience covering or commenting on the news for U.S.-based audio, video, or print media. Demonstrated ability to communicate clearly and effectively to mass audiences in both broadcast and print media would be the basis for evaluation of applications, although it was not necessary for the candidate to have worked professionally in both.

The selection process would first require identification of eight candidates from each of five regions in the U.S.; from the 40 regional nominees, a National Selection Panel would recommend five for final consideration by the NASA Space Flight Participant Evaluation Committee composed of seven senior NASA officials.

The five semifinalists would undergo medical examinations and receive briefings on the spaceflight experience at the Johnson Space Center. Based on results of the physicals and subsequent interviews, the evaluation committee would recommend a primary and backup candidate to the NASA Administrator who would approve the final selection.

The Association of Schools of Journalism and Mass Communication in cooperation with the Association for Education in Journalism and Mass Communication and a Journalism Advisory Committee comprised of representatives from 16 professional journalism organizations including the American Newspaper Publishers Association, Radio Television News Directors Association, American Society of Newspaper Editors, National Association of Broadcasters, Society of Professional Journalists/Sigma Delta Chi, and the National Newspaper Association would administer the competition. (NASA Release 85-147)

November 26: NASA announced that students in classrooms throughout the continental U.S. would have an opportunity to observe various aspects of Space Shuttle mission 51–L and to listen, look, and learn from NASA's Teacher in Space, Christa McAuliffe, during her flight on Challenger. A few students would also be able to question McAuliffe about the mission.

TV viewers with satellite dishes would be able to access the live lessons directly from the RCA satellite Satcom F–2R, Transponder 13. The Public Broadcasting Service, as a result of an agreement with NASA, would carry the live lessons via the satellite Westar IV, Transponder 12. PBS would offer the programs to member stations after requesting that they preempt regular instructional TV or classroom programming to carry the lessons live.

McAuliffe would teach two lessons on the sixth day of the flight [see Space Transportation System/Civilian in Space Program, Oct. 2]; the first at approximately 11:00 a.m. EST, the second at about 1:00 p.m. EST.

Classrooms with access to a satellite dish or cable network that carried NASA-Select would also be able to participate in a "Mission Watch," which covered aspects of the entire Space Shuttle flight from the day before launch through the conclusion of the mission. Barbara Morgan, backup for McAuliffe, would moderate the Mission Watch broadcast. (NASA Release 85–156)

December 6: The Johnson Space Center's (JSC) *Space News Roundup* conducted an interview with Christa McAuliffe, selected by NASA to be the teacher in space, and her backup, Barbara Morgan. When asked to give her impressions of JSC, McAuliffe said, "Big. Very big . . . But that's one of the messages we want to get across. There are, what, 100 astronauts in the program, but thousands of employees here . . . And we've gotten so much information. When I'm 62, I'll finally read the last piece of paper that I bring back from here."

The *Roundup* reporter questioned the two teachers about the national attention they had received, the effect it had on them, and the possibility of moving into different careers after the flight. McAuliffe commented, "Oh, but you're talking to a teacher. I didn't choose my career so I could get monetary rewards. My God, I never would have gone into teaching . . . A year of this is going to be fun and I'm enjoying what I'm doing. I see it as an extraordinary year out of my life . . . We don't see this as a stepping stone to something else. When I go into a radio or TV station, I am looking at everything that is happening and I can't wait to tell my kids what happens in a TV studio, because I have never been in one before."

Morgan then said that one of the funny things that had happened since beginning their training was that, "all of a sudden we are being asked questions as if we are the experts on the Space Shuttle." McAuliffe agreed, noting that she received a phone call from a reporter asking her what she believed was the cause of a problem the previous summer with the orbiter Challenger.

In answer to the question, "What message will you take to people after this experience?" McAuliffe said, "That space is for everybody. It's not just for a few people in science or math, or for a select group of astronauts. That's our new frontier out there, and it's everybody's business to know about space." (JSC *Roundup*, Dec 6/85, 3)

Crews

January 17: NASA announced the assignment of Sen. Jake Garn (R-Utah), chairman of the Senate subcommittee overseeing the NASA budget, as a payload specialist on Space Shuttle mission 51–E, a four-day flight using the orbiter Challenger to deploy the second Tracking and Data Relay Satellite

(TDRS) and the Canadian Telesat I communications satellite. The flight was scheduled for launch February 20, 1985, from KSC. Garn was in preliminary training at Johnson Space Center and would soon begin training with the other members of the crew.

The *NY Times* reported that Garn's activities in space would include research on space sickness and, if he did not become sick in orbit, he might be made ill as part of the experiment.

NASA astronauts assigned to the flight included Karol Bobko, commander; Donald Williams, pilot; mission specialists M. Rhea Seddon, S. David Griggs, and Jeffrey Hoffman; and French payload specialist Patrick Baudry. (NASA Release 85–9; *NYT*, Jan 18/85, A13)

January 29: NASA announced Space Shuttle crews for flights in November and December 1985.

Francis Scobee would command orbiter Atlantis flight 51–L in November to deploy the third NASA Tracking and Data Relay Satellite (TDRS) and relaunch one of the communications satellites retrieved during flight 51–A; Michael Smith would pilot; Judith Resnick, Ellison Onizuka, and Ronald McNair would serve as mission specialists.

Michael Coats would command the orbiter Columbia flight 61–C in December carrying Western Union's Westar 7 and RCA's Satcom KU-2 satellites for launch, 3M Corp.'s Material Sciences Laboratory 3, and the EASE/ACCESS space manufacturing experiment. The pilot would be John Blaha; mission specialists, Anna Fisher, Norman Thagard, and Robert Springer.

NASA also assigned Vance Brand, commander, and S. David Griggs, pilot, for flight 61–D/Spacelab 4 in January 1986, and Jon McBride, commander, and Richard Richards, pilot, for flight 61–E/Astro 1 in March 1986. (NASA Release 85–14)

February 15: NASA announced the astronaut crews for two upcoming Department of Defense (DOD) Space Shuttle missions, including the 1st from Vandenberg AFB.

Robert Crippen (Capt., USN) would command mission 62–A, scheduled for launch no earlier than Jan. 29, 1986, from Vandenberg. Other crew members named were Guy Gardner (Lt. Col., USAF), pilot; and mission specialists Dale Gardner (Commander, USN), Jerry Ross (Maj., USAF), and R. Michael Mullane (Lt. Col., USAF).

Crippen had flown with John Young on the orbiter Columbia's maiden flight in April 1981 and was commander of STS–7, 41–C, and 41–G. Dale Gardner had served as a mission specialist on STS–8 and 51–A; Mullane on 41–D.

Karol Bobko (Col., USAF) would command mission 51–J, scheduled for launch September 1985 from KSC. Ronald Grabe (Lt. Col., USAF) would serve as pilot; David Hilmers (Maj., USMC) and Robert Stewart (Col., USA) as mission specialists.

Bobko had served as pilot on STS-6 and would command 51-E scheduled for launch in March; Stewart had flown as mission specialist on 41-B and had been the 2nd person to fly the manned maneuvering unit on that flight. (NASA Release 85-25)

March 6: NASA announced new crew assignments for the STS 51-D mission set for late March-early April would be Karol Bobko, commander; Donald Williams, pilot; M. Rhea Seddon, Jeffrey Hoffman, and S. David Griggs, mission specialists; and Charles Walker (McDonnell Douglas) and Sen. E. J. "Jake" Garn, payload specialists. NASA would assign the originally announced 51-D crew of commander Daniel Brandenstein, pilot John Creighton, and mission specialists Shannon Lucid, John Fabian, and Steven Nagel to a future mission.

The Centre National d'Etudes Spatiales (CNES) and NASA had agreed to reassign payload specialist Patrick Baudry (France) from the 51-E mission to STS 51-G, as the earliest flight opportunity with adequate middeck experiment stowage capability for the French medical experiments. As a seven-day mission, 51-G would also offer more time for data collection.

For fluid-transfer experiments designed to assist Hughes Aircraft in the refinement of their satellite design activities, NASA would assign a Hughes payload specialist to the 51-I flight to substitute for the lost opportunity on 51-D. Hughes would announce later whether John Konrad or Gregory Jarvis would fly the mission scheduled for early August.

Preservation of the year's flight and crew training schedules had necessitated the changes. Bobko's crew had to fly soon in order to preserve subsequent training schedules for the 51-J dedicated Department of Defense (DOD) mission, the first flight of the Atlantis orbiter.

The reassignment was consistent with NASA's crew selection policy of separating crew from payload flight assignments except in the case of Spacelab and dedicated DOD missions, for which substantial crew/payload interaction was required. (NASA Release 85-34)

April 24: NASA announced it selected Dr. F. Drew Gaffney, an associate professor of medicine and cardiology and director of echocardiology at the University of Texas Health Science Center, Southwestern Medical School, Dallas, and Dr. Robert Phillips, a veterinarian and professor of physiology and nutrition at Colorado State University, to serve as payload specialists for the initial Spacelab Life Sciences (SLS-1) flight. NASA also selected Dr. Millie Hughes-Fulford, an associate professor of biochemistry at the University of California Medical Center, San Francisco, and a medical researcher at the Veterans Administration Medical Center, to serve as a payload specialist for

SLS-2. Dr. Hughes-Fulford was the first woman to serve as a prime payload specialist for a Space Shuttle flight.

The flights were intended to improve significantly knowledge about living beings in the space environment and were a major step in preparing men and women for life aboard the space station scheduled for launch in the early 1990s.

NASA would shortly select for the SLS-2 mission a second payload specialist, who would then train with the three already announced. (NASA Release 85-62)

May 4: NASA announced that Sultan Salman Abdelazize Al-Saud, an Arabsat payload specialist scheduled to fly on the 51-G 7-day Space Shuttle mission in June, would conduct 70mm photography over Saudi Arabia, 35mm photography of a fluids experiment, and would participate in the French posture experiment. Al-Saud's flight was part of a reimbursable agreement with the Arab Satellite Communications Organization covering the launch of the Arabsat 1B communications satellite.

Al-Saud would use the 70mm camera to take pictures on daylight orbital passes over Saudi Arabia and the 35mm camera to document such phenomena as surface tension effects on mixed fluids in the absence of gravity. His other activities would include photography of the new moon in a lunar crescent observation and assisting the French payload specialist as a test subject in the French experiment.

At a May 28 news conference at Johnson Space Center, Al-Saud said his flight was bound to improve diplomatic relations between the U.S. and the Islamic world, the *Washington Post* reported. "You will have 800 million Moslems and 155 million Arabs glued to their TV sets watching an American spaceship carrying an Arab into space," he commented.

NASA selected Al-Saud, nephew of Saudi's King Fahd, from hundreds of Saudi applicants. He had logged more than 1,000 hours in jet aircraft.

Other crew for the mission were Daniel Brandenstein, commander; John Creighton, pilot; Shannon Lucid, John Fabian, and Steven Nagel, mission specialists; and Patrick Baudry, French payload specialist. Backup payload specialist for Arabsat was Abdulmohsen Hamad Al-Bassam. (NASA Release 85-69; *W Post*, May 29/85, A12)

May 31: NASA announced that Frederick Hauck would command Space Shuttle flight 61-F scheduled for no earlier than May 15, 1986, to deploy the Ulysses (International Solar Polar) spacecraft and David Walker would command the Galileo mission 61-G scheduled for no earlier than May 21, 1986. The Galileo spacecraft would explore the environment of Jupiter and its moons.

Hauck first flew as pilot on Space Shuttle flight 7 in June 1983 and was commander in November 1984 of mission 51-A, for which Walker was pilot.

Other 61-F crew members would be Roy Bridges, pilot, and mission specialists David Hilmers and J. Mike Lounge. Other mission 61-G crew members would be pilot Ronald Grabe and mission specialists John Fabian and James van Hoften.

The Ulysses mission would be the first to use the liquid-fueled Centaur upper stage; the Galileo mission would also use the Centaur upper stage. (NASA Release 85-82)

June 17: NASA announced that Loren Shriver, pilot of Space Shuttle mission 51-C, would command mission 61-I scheduled for launch from KSC no earlier than July 15, 1986. The pilot would be Bryan O'Connor, also scheduled to fly as pilot on mission 61-B in November 1985. Mission specialists would be William Fisher, also scheduled as a mission specialist on 51-I in August 1985; Mark Lee, making his initial flight; and Sally Ride, who flew June 1983 on STS-7 as the first U.S. woman in space and then again October 1984 on 41-G. (JSC Release 85-027)

June 19: NASA announced it assigned Robert Cenker, a senior staff engineer at RCA Astro-Electronics, as a payload specialist on Space Shuttle mission 61-C scheduled for launch on December 20, 1985. Cenker would support deployment of the Astro-Electronics-built RCA Satcom Ku-Band-1 communications satellite from orbiter Columbia and perform experiments with an infrared camera developed at RCA's David Sarnoff Research Center and manufactured at Astro-Electronics.

During his 13-year career at RCA, Cenker held a number of engineering positions in the Satcom program that included work on spacecraft design, integration and test scheduling, cost control, and launch site activity planning.

During November NASA would launch the first RCA Satcom Ku-Band satellite, KU-2, on Space Shuttle mission 61-B. Three RCA Satcom Ku-Band satellites would complement the operating Satcom C-Band system in providing distribution of TV services to customers in metropolitan areas. (NASA Release 85-90)

June 19: The crew of Space Shuttle Discovery on mission 51-G designated Steven Nagel as the 100th American to reach space, the *Washington Times* reported. Before that flight, 96 Americans on 48 missions dating back to 1961 had flown in space.

There were five Americans aboard the Space Shuttle, but commander Daniel Brandenstein decided Nagel took the 100th honor because three Americans sat in front of him on Discovery's flight deck. "Shannon Lucid was the first runner-up," Brandenstein said. "She got to space three inches ahead of Steven Nagel." To mark the occasion, which was televised back to earth, Brandenstein presented Nagel with a cake that had "100th" written on it.

"I'm not sure if that's true," Nagel said. "Although I sat three inches behind Shannon, my nose is about three and a half inches longer than hers is." (*W Times*, June 20/85, 3A)

September 19: NASA announced that John Young would command Space Shuttle flight 61–J scheduled for launch August 1986, Vance Brand mission 61–K scheduled for September 1986, and Donald Williams mission 61–I.

Other crew for the 61–J mission, during which astronauts would deploy the Hubble Space Telescope, were pilot Charles Bolden Jr. and previously named mission specialists Kathryn Sullivan, Steven Hawley, and Bruce McCandless.

S. David Griggs would pilot mission 61–K, and other crew would be mission specialists Robert Steward, Owen Garriott, and European Space Agency astronaut Claude Nicollier. NASA previously assigned payload specialists Michael Lampton and Byron Lichtenberg to the mission.

Remaining mission 61–I crew would be pilot Michael Smith and mission specialists James Bagian, Bonnie Dunbar and Manley "Sonny" Carter.

NASA also announced that Norman Thagard would replace John Fabian on mission 61–G scheduled for May 1986 during which crew would deploy the Galileo interplanetary spacecraft. Fabian was leaving NASA; he had not announced his plans. James Buchli would replace Thagard on mission 61–H in June 1986. (NASA Release 85–131)

September 19: Saudi Arabian Prince Sultan bin Salman bin Abdul Aziz Al-Saud, the first Arab in space on the June 1985 Space Shuttle flight 61–G, was in Washington to give interviews, attend receptions, and pay courtesy calls on government officials, including President Reagan, as part of a goodwill tour, the *Washington Post* reported.

The prince commented that his flight, coming as it did during Ramadan, the holiest month in Islam, was a spiritual ascent. "I remember being up in space and I'd read a verse from the Koran and then get up and go to the window . . . It's quite an experience when the sun rises and sets every 45 minutes. First you notice countries. Then you start paying attention to continents. And by the fifth day all you can see is one big blue ball tumbling in front of you."

Dispensing, under the circumstances, with the ritual facing of Mecca—after checking first that it would be acceptable to Moslem scholars—the prince nevertheless prayed five times a day and fasted for part of the flight in observance of Ramadan.

He also commented, "I had a small experience after I came back from space. I always jogged over the same area in Houston, sometimes 15 miles a day, but when I came back and jogged over the same course, I noticed at least 50 percent more things in nature."

When the prince returned to Saudi Arabia after the flight he received a ticker-tape parade. "There were religious old people, women, men, children, helicopters throwing ticker tape, people touching you," he said, "Saudi Arabians have never done this before. Our people are usually not so emotional as to throw ticker tape. But the Saudies will always surprise you." The *Post* also noted that his face was on thousands of T-shirts and he received hundreds of letters a day.

The prince pointed out to the reporter that, "I am a bachelor. I don't drink and I'm a nonsmoker, but I'm not boring. No way am I boring." (*W Post*, Sept 19/85, C1)

October 1: NASA announced today the selection of Dr. Samuel Durrance, associate research scientist in the Department of Physics and Astronomy at Johns Hopkins University, and Dr. Ronald Parise, manager of Advanced Astronomy Programs at Computer Sciences Corp., to serve as payload specialists on Space Shuttle mission 61–E, for an ultraviolet astronomy mission known as Astro–1 scheduled for launch March 6, 1986. Dr. Kenneth Nordsieck, associate professor at Washburn Observatory, University of Wisconsin, would serve as the backup.

The Astro–1 mission would study Halley's Comet and other celestial objects through three ultraviolet astronomical instruments. In addition, a special visible-light, wide-field camera was incorporated into the payload to augment the Halley's Comet studies. The payload specialists, all experienced astronomers, would make decisions during the mission to ensure the best possible scientific return.

The payload specialists were each members of one of three science teams that developed the ultraviolet instruments. Durrance and Parise would operate the ultraviolet instruments; astronaut mission specialists would operate the instrument pointing system (IPS), developed by the European Space Agency (ESA) for precise aiming at celestial targets and first flown in early August 1985 on Spacelab 2, on which the three instruments were mounted.

NASA scheduled the Astro–1 flight to coincide with the Halley encounter missions by ESA, the USSR, and Japan. Dr. Burton Edelson, NASA associate administrator for space science and applications, said, "The opportunities for science synergism between Astro–1 and the armada of Halley encounter spacecraft are significant. The scientific study of Halley's Comet will be an internationally coordinated effort."

In addition to the Halley's observations, the Hopkins Ultraviolet Telescope on Astro–1 would study faint astronomical objects such as quasars, active galactic nuclei, and normal galaxies in the far ultraviolet range; the Ultraviolet Imaging Telescope would record image-intensified photos of faint objects such as hot stars and galaxies in broad ultraviolet wavelengths and with a wide field of view; and the Wisconsin Ultraviolet Photopolarimetry Experiment would study the polarization of hot stars, galactic nuclei, and quasars. The instruments would make a total of 200 to 300 observations during the mission, the first of three in a series scheduled for launch within the next two years. (NASA Release 85-137)

November 8: The Indian Department of Space announced today the selection of two Indian astronaut candidates, one of whom would fly aboard the Space Shuttle carrying India's INSAT-1C satellite, FBIS Delhi Domestic Service in English reported. The candidates were N.C. Bhatt of the Indian Space Research Organization, Bangalore, and P. Radhakrishmam of Vikram Sarabhai Space Center, Trivandrum.

The chairman of India's Space Commission, Professor U. R. Rao, said the commission would select one of them to fly on the mission scheduled for July 1986. (FBIS Delhi Domestic Service in English, Nov 8/85)

November 11: European space program managers were urging the creation of a science astronaut classification to distinguish crew members with scientific backgrounds from other payload specialists who flew on the Space Shuttle, *Aviation Week* reported. European managers were using the science astronaut designation for Ernst Messerschmid, Reinhard Furrer, and Wubbo Ockels, the European crew members who flew aboard Space Shuttle mission 61-A with West Germany's Spacelab D-1 [see Space Transportation System/Missions, Oct. 30]; NASA used its regular payload specialist designation for the three.

"There has to be a difference between some senator or Arab prince and the qualified scientists who fly aboard the shuttle," said Ulf Merbold, crew interface coordinator for the D-1 mission. "We demand that the science astronaut concept be developed so that these proficient crew members can be designated for such a flight," said Hans-Ulrich Steimle, Spacelab D-1 mission manager at the German aerospace research establishment. "Instead, NASA now has become involved in running a travel office for visiting dignitaries, and this is counterproductive when you want to perform a serious science mission."

German officials said they might raise the issue again when planning was finalized for a follow-on Spacelab D-2 mission targeted for 1988. (*AvWk*, Nov 11/85, 24)

December 27: NASA announced that its Administrator selected Dr. Charles Chappell of Marshall Space Flight Center (MSFC) and Dr. Dirk Frimout of the European Space Agency (ESA) to serve as alternate payload specialists for the first Earth Observation Mission (EOM), which would use the Spacelab scheduled for launch on the Space Shuttle orbiter Atlantis in the latter half of 1986.

Chappell and Frimout would serve as backups to flight payload specialists Dr. Byron Lichtenberg and his alternate, Dr. Michael Lampton.

Chappell was chief of the Solar Terrestrial Div., Space Science Laboratory at MSFC and was responsible for directing a research group that studied the physics of the sun-earth environment. He served as mission scientist for the 10-day Spacelab 1 mission, during which 70 investigations were carried out.

Frimout was the senior engineer supporting European researchers who would have experiments on the mission, and he had served as ESA crew coordinator and operations manager for Spacelab 1.

Chappell and Frimout would train with the flight crew and serve as members of the mission management team in the Payload Operations Control Center during the flight. They would communicate directly with the crew on orbit, assist the payload operations team during normal operations, and aid in trouble shooting problems and in changing crew procedures when necessary. They would also advise the mission scientist, Dr. Marsha Torr, of the possible impact of problems and timeline changes.

The EOM flight was the first in a series of Space Shuttle missions primarily dedicated to measuring solar irradiance and the chemical composition of the earth's stratosphere and mesosphere during an 11-year solar cycle. NASA would refly several instruments originally carried on the Spacelab 1 and 3 missions to accomplish these measurements.

The mission would use the short version of the Spacelab module, in which a single Spacelab pallet and special support structure would hold instruments that required exposure to the space environment. The mission would consist of 15 experiments conducted in six disciplines—atmospheric science, solar physics, plasma physics, earth resources, astronomy, and life sciences. The international mission included experiments sponsored by Belgium, France, Japan, Federal Republic of Germany, and the U.S. ESA would provide operations support for the European investigations.

The Investigators Working Group (IWG), which consisted of the principal investigator for each of the mission experiments, recommended alternate payload specialist candidates for selection by NASA's administrator. (NASA Release 85-179)

External Tank

November 27: The U.S. Senate awarded Martin Marietta Corp. the United States Senate Productivity Award for the state of Louisiana for manufacturing improvements and costs savings in production of the Space Shuttle's external tank, *Defense Daily* reported. The Senate cited the company for its innovative across-the-board productivity program that included plant modernization, installation of automated production processes, and development of an employee participation program that led to a number of improvements in the manufacturing process.

"Estimates are that this increased productivity will save NASA and ultimately the American taxpayer more than a billion dollars over the long run," said Louisiana Senators J. Bennett Johnston (D) and Russell Long (D) in a joint statement. The two also cited the "flawless performance" of the tanks in Space Shuttle missions to date. (*D/D*, Nov 27/85, 142)

Launch and Landing Facilities

February 7: NASA and the U.S. Air Force completed for the first time Space Shuttle vehicle stacking at Vandenberg AFB's Space Launch Complex-6, the *Marshall Star* reported. Begun January 12, the stacking consisted of inert solid-fuel rocket-motor segments, an external tank, and the orbiter Enterprise. Unlike facilities at KSC where NASA stacked the Space Shuttle in the vehicle assembly building and rolled it to the launch pad, at Vandenberg NASA and the Air Force stacked the Space Shuttle vehicle on the launch mount (pad).

Workers at Vandenberg rolled a mobile service tower and a shuttle-assembly building to the launch mount, then brought booster segments to the launch complex from the booster facility and lifted them into place by a mobile service-tower crane. This first vehicle stacking at the Vandenberg launch complex was part of facility-verification testing before the first Space Shuttle launch from the complex in early 1986.

Col. Walter Yager, commander of the Air Force shuttle assembly task force, said minor problems had occurred, but "that's why we have facility verification. We want to ensure properly working systems before we start handling actual flight hardware." Fit and function checks, payload operations, and launch-processing simulations would complete facility verification. (*Marshall Star*, Feb 7/85, 1)

April 12: Three quarters of a million gallons of water flooded Space Launch Complex-6 at Vandenberg AFB during a March test to determine whether the system, intended to suppress sound waves from the Space Shuttle during a launch, would operate properly, the Air Force Systems Command (AFSC) *Newsreview* reported. Sound waves, reflected off the launch mount and flame ducts, could seriously damage the orbiter and its payload.

To prevent such damage, nozzles on the mount would begin spraying water into the three pad exhaust ducts seven and a half seconds before the Space Shuttle's main engines ignited. By ignition time, the spray system would reach its full flow rate of 985,000 gallons per minute and maintain that rate for nearly 30 seconds before tapering off after the Space Shuttle left the pad.

Although Vandenberg officials had tested the spray system six times previously, the March test was the first time the nonflying prototype orbiter Enterprise, inert solid-fuel rocket boosters, and an external tank were in place on the mount during a test. Remotely controlled by the launch control center, the splash pattern of the gravity-fed water system resulted in no water splashing up into the main engines, where ice might form before ignition. The system sprayed water exactly where it would be needed at launch—down into the flame ducts and out over the concrete pad surface.

George O'Gorman, site manager for complex-6, said most of the complex's construction work was complete, although some modifications were scheduled before the orbiter Discovery would arrive in September for launch in January 1986.

The Air Force expected to make about four Space Shuttle launches a year by the late 1980s from Vandenberg AFB. The facility could handle as many as ten launches a year. (AFCS *Newsreview*, Apr 12/85, 3)

April 15: The turnaround clock began running when the Space Shuttle Discovery left the KSC launch pad at 8:59 a.m. April 12 and stopped 70 hours and 35 minutes later when NASA declared the orbiter Challenger, configured for launch of the Spacelab 3 mission, "hard down" on the pad, the KSC *Spaceport News* reported.

The tightly coordinated effort began approximately one minute after Discovery's launch. After launch control center's firing room 1 issued the command for automatic washdown of the pad structures, an operational TV scan, ordnance system resistance checks, and statusing fire and leak detectors commenced operation. About 15 minutes after launch, when it was certain the Space Shuttle would not return to the launch site, a convoy of waiting vehicles moved into action.

Workers dumped small cryogenic (hydrogen and oxygen) tanks at the 155-foot level of the fixed service structures (FSS). Two Lockheed and EG&G teams inspected in minutes the hydrogen and oxygen farms, high-pressure hydrogen gas battery, and the gaseous battery beneath the pad, clearing the way for another team to safe and secure the oxygen and hydrogen farms. Following dumping of the cryogenic tanks, two other teams began a level-by-level quick-look inspection of the FSS and rotating service structure, setting the stage for safety and security procedures, including configuring purge line valves and doors, making environmental checks in the mobile launch platform (MLP), and checking operation of elevators. At the same time, others removed ordnance from the external tank vent arm and tail service masts, secured guard rails where necessary, and cleared the pad for entry of general repair crews.

Some damage was anticipated during a launch. In this case, pins had melted off some of the solid-fuel rocket booster holddown posts on the MLP; burnaway (or sacrificial) plates needed repairing; FSS light bulbs had blown at various levels; and some SRB waste residue was present. A 40-foot piece of waterpipe at the 95-foot level of the FSS, part of the Firex system, was bent at a 90° angle.

The KSC integrated control schedule called for removal of the MLP by 7 p.m. the day after launch and for it to be at the doors of the vehicle assembly building (VAB) by midnight. To accomplish this, workers had to mount service platforms, disconnect waterpipes, and move flame deflectors. They also serviced the hydrogen burn pond and its several igniters.

On schedule, the crawler/transporter crew moved into position to begin the return of the MLP to the VAB. At 11:30 p.m. April 13, crawler 1 and MLP 1 reached the entrance of high bay 3; at 12:17 the next day the crew started the journey to the pad with Challenger. The Space Shuttle was "hard down" on the pad at 6:25 a.m. April 15. (KSC *Spaceport News*, Apr 26/85, 4)

May 3: Effective today Lockheed Corp. reassigned six officers of its Lockheed Space Operations Co. at Kennedy Space Center after a six-month corporate investigation of the management practices and work procedures used to process the Space Shuttle for launch, *Aviation Week* reported. E. Douglas Sargent replaced A.R. Schroter as president of Lockheed Space Operations Co. and became program manager of the Space Shuttle processing contract, David Owen replaced W. John Denson as deputy program manager and became executive vice president of the space operations company, and David Dickenson replaced Ronald Petersen as vice president and director of Kennedy Space Center operations.

Although the replaced Lockheed managers had previously received unfavorable management evaluations, the management realignment came as NASA completed a report on its investigation of a March 8 incident in which a work platform fell in the Orbiter Processing Facility, damaging the orbiter Discovery and injuring a workman [see Space Transportation System, Launch Schedules, Mar. 8]. The report would assign a large part of the responsibility for the accident to Lockheed safety and quality control procedures.

However, a NASA official said the platform incident did not precipitate the realignment. He said a Lockheed Corp. investigating team, which had analyzed Lockheed's performance at KSC since the previous October, had made detailed recommendations on how to streamline and improve performance. Also, after receiving a grade of excellent in the award fee evaluation during the transition period when Lockheed took over Space Shuttle processing work previously done by 15 separate contractors, NASA gave Lockheed a low score in the next evaluation period of April 1 to September 30, 1984.

Lockheed's basic space processing contract with NASA ran through September 30, 1986, and there was a priced option period for three more years.

The two priced contracts were worth $2.5 billion; three additional unpriced three-year options could bring the contract life to 15 years and value to more than $6 billion. NASA said it had no plans to recompete the Space Shuttle processing contract. (*AvWk*, May 13/85, 14)

May 15: NASA announced it was discussing with representatives of the Chilean Government possible arrangements for landing support for the Space Shuttle in the event of an emergency during launch from Vandenberg Air Force Base. The discussions were consistent with arrangements made with other countries since 1981 supporting launches from Kennedy Space Center (KSC).

Space Shuttle operational planning provided for emergency landing options for all phases of the flight to protect the lives of the crew and the integrity of the space vehicle. Since the first launch of the Space Shuttle from KSC, NASA had plans for each mission for trans-Atlantic landing sites (TAL) and contingency landing sites (CLS) in addition to the primary landing sites. This was consistent with relevant international agreements relating to the rescue and recovery of astronauts in distress and return of the space vehicle.

Space Shuttle launches, scheduled to begin from Vandenberg in early 1986, required one or more trans-Pacific landing (TPL) and CLS sites in the Pacific Basin. NASA had determined that Mataveri Airfield on Isla de Pascua (Easter Island) could serve as a Space Shuttle TPL/CLS in the event of an emergency.

A Chilean contractor was working on preliminary designs for minor improvements at the airfield, which would be necessary in the event an agreement was reached. Improvements would be in accordance with existing land use statutes and regulations on the island. Any actual improvements were contingent on final agreement with the Government of Chile. (NASA Release 85–74)

June 12: Chilean government member Admiral Jose Toribio Merino said today that he favored accepting the NASA request to build a Space Shuttle emergency landing airfield on Easter Island and that it was essential to build a port for the operation of larger and faster ships, FBIS Santiago Radio Chilena in Spanish reported. "If someone offers to extend the Easter Island airport," Merino said, "we should accept because Chile does not have the money to do it."

The need to supplement the project with the construction of a port for security reasons was a new element added by Merino to the NASA proposal. Merino explained that the operation of an international airport near the sea called for fast ships to act in case a plane went down in the ocean. It was imperative to have a port for the operation of such ships, and Easter Island had only a small bay, he said. "If someone offers to build a port, I would let them," he added.

In referring to criticism of the project because of possible military uses of the airfield, Merino said that those who said that about the project were "ignorant people who were politicians in the past but not now." (Santiago Radio Chilena in Spanish, June 12/85)

June 21: For the first time since the 1975 Apollo-Soyuz project, liquid-hydrogen rocket fuel flowed through the lines at Kennedy Space Center's launch pad 39–B, the *Spaceport News* reported. Nearing its January 1, 1986, deadline for completion of pad B renovations and improvements, NASA was entering final work stages including functional testing for such installations as the new 100-foot-tall flare stacks that would replace the old burn pond for disposal of vented gaseous hydrogen. NASA planned first use of pad B for the 51–L Space Shuttle mission scheduled for launch no earlier than January 22, 1986.

One of the flare stacks, 26 inches in diameter, would vent off and burn gaseous hydrogen from the Space Shuttle's external tank and the orbiter fuel cells service system; the other, 18 inches in diameter, would handle hydrogen from the Centaur upper stage, the facility storage tank, and the mobile launcher platform.

Also new to the complex were two rooms on the rotating service structure, one of which was designed for storage of equipment used in the payload changeout room in order to avoid the necessity for temporary removal of the equipment from the site for each launch. The other was a new clean room/ suit changing room.

Another major new item was the rolling beam that would supply liquid hydrogen to the Centaur upper stage. Plans called for use of the Centaur upper stage during the Galileo and Ulysses missions scheduled for May 1986.

Work began on pad B renovations about seven years previously; total cost would be approximately $150 million. (*Spaceport News,* June 21/85, 4)

July 26: Chilean Foreign Minister Jaime del Valle announced that Chilean and U.S. delegations reached agreement on extending the Easter Island airfield for Space Shuttle emergency landings, FBIS Santiago Radio Chilena in Spanish reported. Del Valle said Chile received favorable answers from the U.S. government to two major questions raised by Chile and that he would recommend the Chilean president approve the project.

Earlier Del Valle had said that a possible agreement with NASA on the extension of Mataveri Airport would not imply the surrendering of Chilean sovereignty over the airfield nor would it constitute the establishment of a foreign military base on Chilean territory. Del Valle was quoted as saying, "On the eve of the beginning of formal negotiations, I want to reaffirm to the country that the agreement proposed by the United States refers exclusively to the emergency landings of space shuttles of the type that are currently in operation. It is appropriate to note that the possible emergency landings at Easter Island include only those isolated flights originating in California and those taking off from Cape Canaveral, which take a different orbit."

Del Valle concluded the earlier statement by saying the "cultural, archeological, and ecological structure of the island will be preserved." (FBIS Santiago Radio Chilena in Spanish, July 2/85, July 26/85)

August 2: Kennedy Space Center (KSC) service and contractor personnel completed within 90 days installation of a launch environment instrumentation system (LEIS) for Space Shuttle launches at Vandenberg Air Force Base, the *Spaceport News* reported. The LEIS, a 400-channel instrumented system, was designed to acquire data on the environment in the vicinity of the launch pad and support facilities during launch.

The system's transducers measured in the launch pad vicinity pressure, acoustics, strain, vibration, and temperature for use in determining baseline data, sources of problems, and possible solutions. Such equipment at KSC, for example, pinpointed engine gas leaks during the first mission of the orbiter Challenger and provided design measurements of the overpressure problem that occurred during the first Space Shuttle launch.

The KSC engineering development directorate, with the assistance of Planning Research Corp., completed design, management, and integration of the system, which Fairchild Weston Systems Inc. built under the direction of KSC for the Air Force. Fairchild began work on the system in March 1984 under a $4,549,256 contract. (*Spaceport News*, Aug 2/85, 7)

August 3: NASA announced that the governments of Chile and the U.S. signed agreements concerning use of Mataveri Airport on Isla de Pascua (Easter Island) as an emergency landing site for the Space Shuttle. The intergovernmental agreements provided the basis for contingency planning for launches of the Space Shuttle from Vandenberg Air Force Base beginning in March 1986. This planning was consistent with arrangements already in place to support launches from the Kennedy Space Center (KSC).

Space Shuttle operational planning had to provide for emergency landing options for all phases of a flight to protect the lives of the crew and the integrity of the space vehicle. The Vandenberg launches would also have a requirement (analogous to those from KSC) for one or more trans-Pacific landing sites in the Pacific Basin. Although the probability of a contingency Space Shuttle landing was extremely remote, identification and preparation of suitable contingency landing sites was consistent with prudent operational planning.

A Chilean contractor had prepared designs, in accordance with existing land use statutes and regulations on the island, for improvements at the air field that were necessary to support a Space Shuttle landing. These improvements included an extension of the existing runway and associated lighting system, construction of a storage building, and enhancement of certain permanent and temporary navigation equipment in the area. The modifications would not affect the historical treasures for which Isla de Pascua was known.

As part of the agreement, the two countries would undertake enhancement of the existing program of scientific cooperation in the exploration and use of outer space for peaceful purposes, thereby building upon a long and successful relationship of technical cooperation in this arena. (NASA Release 85–117)

October 11: The information gathered by several Space Shuttle missions underscored the need for improved weather forecasting at Kennedy Space Center (KSC). NASA along with the U.S. Air Force developed a plan to improve the quality of weather data available at KSC as well as the manner in which it was presented to the forecaster, *Spaceport News* reported.

KSC officials recently held a ribbon cutting ceremony at the Range Control Center on Cape Canaveral Air Force Station to inaugurate the new Cape Canaveral Forecast Facility (CCFF), marking completion of a $3 million three-year effort by NASA, the Air Force, and contractor Pan Am World Services. The facility was the first important tool for improving the efficiency of the weather forecaster during launches and landings of the Space Shuttle.

The facility housed a system called MIDDS, meteorological interactive data display system, which animated and overlaid data from various forecasting tools in color graphic displays. For example, a forecaster could begin with display of a current satellite picture; over that he could place a color enhanced radar display of shower activity. Next he might add a graphic presentation of indications from wind towers around KSC; and over that he could display locations of lightning potential as well as where lightning was striking cloud-to-ground. Motion added to the entire composite picture would illustrate trends in weather activity over a period of time.

The system could overlay maps of upper wind data or barometric pressures collected from weather balloons or sounding rockets. The forecaster could also display figures from the data network of other weather stations around the country or the world.

Although the console was at the Cape Canaveral facility, the new rotating antenna was at Patrick Air Force Base to compensate for the blind spot directly above the radar. A National Oceanic and Atmospheric Administration radar located in Dayton Beach supplemented the Cape radar. A forecaster could call up a current graphic display of any official weather radar in the U.S.

A lightning location and protection (LLP) system would show cloud-to-ground strikes up to 100 miles away, from which antennas were located ten miles north of the vehicle assembly building and at Melbourne Regional and

Orlando International Airports. These antennas supplemented an existing network of 30 area field mills that detected lightning potential.

NASA officials hoped within five years to have at the facility a clear air doppler radar, of which there were only six in the country. It would reduce the number of wind towers needed and improve forecasting efficiency. In the meantime, Space Shuttle weather officer Scott Funk said, "We haven't tapped the full potential of the new MIDDS system. Most weather forecasters have never seen anything like this. But there are still a number of potential capabilities the system has which we will add next year." (*Spaceport News*, Oct 11/85, 7)

November 22: The successful landing test on October 30 of a new nosewheel steering system on the Space Shuttle Challenger on mission 61–A meant that the orbiter Columbia would resume Space Shuttle landings on the concrete runway at Kennedy Space Center (KSC) when it returned following mission 61–C, the *Washington Post* reported. All Space Shuttle landings since April had been at Edwards Air Force Base, where the long desert runways allowed pilots to roll to a stop without excessive brake use.

At KSC crosswinds tended to push the orbiter toward the edges of the narrow runway after the spacecraft touched down, which pilots countered by making "preferential" use of brake assemblies on the landing gear under the Space Shuttle's wings. By braking on one side or the other, pilots kept the orbiter on the runway. But a landing in April shredded three of the orbiter's tires and burned out a brake assembly. NASA built the new nosewheel steering mechanism into the orbiter so the pilot could steer the spacecraft without using his brakes.

Challenger Commander Henry Hartsfield said after the Challenger landing that he was "very pleased" with the assembly, but recommended one more test on the concrete runway at Edwards Air Force Base. (*W Post*, Nov 22/85, A21)

December 9: NASA officials confirmed a schedule calling for the next Space Shuttle orbiter to land at Kennedy Space Center (KSC), *Aviation Week* reported, as a result of the successful hard-surface runway landing December 3 at Edwards Air Force Base by the orbiter Atlantis on mission 61–B. The landing was the first on a concrete surface since the orbiter Discovery damaged its brakes and tires while landing April 19 at KSC.

NASA scheduled the orbiter Columbia on mission 61–C to land at KSC following its planned December 18 launch. Columbia was equipped with the necessary nosewheel steering system upgrades and would use steering instead of brakes as a primary control method for the KSC landing.

During the December 3 landing, control inputs by mission commander Lt. Col. Brewster Shaw, Jr. appeared to cause Atlantis to float slightly just before landing on the 15,000 ft. runway. The orbiter's wheels touched down approximately 2400 feet from the runway threshold, and Shaw applied only light braking during a 17,759-ft. landing roll.

NASA slightly altered the Space Shuttle post-landing processing in an effort to remove Atlantis from the runway as soon as possible to allow the Air Force to reopen it for aircraft use. Normally personnel removed the orbiter's brakes in parallel with other close-out tasks before towing the vehicle back to NASA's Ames-Dryden Space Shuttle processing area at Edwards. They left on the brakes so that Atlantis could be moved to an apron area at the end and off to the side of the runway. H.W. Widick, chief of the Space Shuttle Integration Div., said a preliminary inspection of the brakes did not reveal any damage and that overall condition of the orbiter was good. (*AvWk*, Dec 9/85, 23)

December 13: Kennedy Space Center (KSC) director Dick Smith and Lockheed Space Operations Co. president Doug Sargent today hosted a dedication ceremony for KSC's launch pad B after six years of modification at a cost of $150 million, the *Spaceport News* reported. NASA would use pad B, the second of the two complex 39 launch pads to undergo modification from the Apollo configuration, to support Space Shuttle and Shuttle Centaur upper stage launches. First launch from the modified pad would be Space Shuttle mission 51–L, scheduled for January 22, 1986.

NASA first used pad B in May 1969 for the Apollo 10, third manned Saturn V/Apollo launch. Its last use was for the launch of the Apollo/Soyuz mission in July 1975.

Pad modifications included a new Centaur rolling beam, permanent orbiter weather protection system, TV camera floodlights and a TV camera system, boxcars, upgrading of the payload changeout room, a cost-saving lighting system, variable speed motors for loading liquid oxygen, and hydrogen flare stacks.

Although there were currently minor differences between pads A and B, the work underway at pad A would eventually make the two functionally identical. (*Spaceport News*, Dec 20/85, 1; Kennedy Space Center Release 242–85)

Launch Schedules

January 8: NASA announced it had agreed with the U.S. Air Force to delay the first Space Shuttle launch from Vandenberg AFB, originally scheduled for October 15, 1985, until no earlier than January 29, 1986.

NASA and Air Force officials had conducted in December 1984 an extensive review of Vandenberg Space Shuttle launch-facility readiness, the Department of Defense (DOD) payload, and impact of the orbiter Challenger tile problems on the Space Shuttle program schedule. They had decided to delay launch to maintain the current Space Shuttle manifest and to insure adequate margin in development of the DOD payload for the initial Vandenberg launch.

NASA and the Air Force agreed that the orbiter Discovery would be delivered to Vandenberg in early September 1985 instead of May; NASA would use Discovery for two flights from KSC in mid-1985 to accommodate the impact of tile problems on the schedule and would deliver other Space Shuttle flight hardware (filament-wound cases and external tank) as soon as possible to Vandenberg to provide maximum schedule flexibility for the earliest launch.

The Air Force had corrected problems with pipe welds and cleanliness at the Vandenberg launch site and was conducting ground systems tests with the orbiter Enterprise in preparation for initial checkout. NASA had delivered an external tank and solid-fuel rocket booster skirt ahead of schedule.

NASA was completing the remainder of the 1985 Space Shuttle manifest with commercial and DOD payloads and would release a manifest covering August 1985 through 1989 in the near future.

In reporting the delay, the *Washington Post* noted that Vandenberg was the only base from which Space Shuttles could fly into north-south orbits crossing the poles, an orbit ideal for surveillance and other earth-watching spacecraft because they could fly over the entire globe once every 18 days. (NASA Release 85–4; *W Post*, Jan 9/85, A9)

January 23: NASA postponed for 24 hours the first classified Space Shuttle flight, scheduled for launch the afternoon of January 23, as freezing temperatures threatened to cause severe icing on the external fuel tank, the *Washington Post* reported. Officials said ice chunks could drop off, damaging the Space Shuttle, and also that NASA staff was conducting inspection of fuel and water pipes to determine if they had frozen shut.

The Department of Defense (DOD) had classified nearly all aspects of the mission. The crew consisting of Navy Capt. Thomas Mattingly, AF Lt. Col. Loren Shriver, Marine Lt. Col. James Buchli, AF Maj. Ellison Onizuka, and AF Maj. Gary Payton would not conduct the traditional prelaunch news conference nor hold one after landing. The U.S. Air Force would not disclose how long the crew would stay in orbit or what flight path they would take on the premise it might help the USSR determine mission purpose or interfere with the mission.

Previously the *Washington Post* had reported that Challenger would carry into orbit a sigint (signals intelligence) satellite that could intercept telemetry from Soviet missile tests. The Associated Press later had reported that the satellite could pick up radio, ground-to-space communications, and long-distance telephone calls made by microwave relay. Defense Secretary Caspar Weinberger would not comment on the reports' accuracy.

The USAF had not filed a flight-path plan with the United Nations, and the location of two ships that would recover the solid-fuel rocket boosters was also secret. There would be no public air-to-ground communications during the flight. (*W Post*, Jan 23/85, A3)

January 30: Tile work on the orbiter Challenger had remained the "pacing item" in meeting NASA's next scheduled Space Shuttle mission, 51–E, set for no earlier than February 20, *Aerospace Daily* reported. Randy Stone, 51–E lead flight director, said at a press briefing that NASA would make an assessment at KSC and that he didn't know "what the slip will be or if there is one."

Stone had indicated work on the orbiter originating from adhesion problems with thermal protection tiles was progressing well; however, workers still needed to attach over 300 tiles before NASA could move the orbiter to the Vehicle Assembly Building for mating with the external tank and solid-fuel rocket boosters. If all went well, NASA would launch Challenger February 20 from KSC and land it there at about 9:21 am EST February 24. (*A/D,* Jan 30/85, 1)

February 6: NASA announced it had delayed for at least a week the February 20 Space Shuttle flight that would carry Sen. Jake Garn (R-Utah), because of thermal tiles problems and a backlog of paperwork, the *Washington Post* reported. A chemical reaction in the thermal tiles' adhesive on the orbiter Challenger's underside had caused about 4,000 of the more than 30,000 tiles to loosen during return to earth on the orbiter's last flight. Although NASA had replaced all but 47 of the faulty tiles, it had not certified the repair work. (*W Post,* Feb 6/85, A12)

February 7: NASA announced it had scheduled the 51–E Space Shuttle mission from KSC for no earlier than March 3, 1985, with landing on March 7, 1985, at KSC.

The 7th flight of the orbiter Challenger would include deployment of the 2nd Tracking and Data Relay Satellite (TDRS–B) and Telesat Canada's Anik C1 communications satellite and the French echocardiograph and postural experiments.

Mission 51–E crew would be Karol Bobko, commander; Donald Williams, pilot; mission specialists M. Rhea Seddon, S. David Griggs, and Jeffrey Hoffman; and payload specialists Patrick Baudry and Sen. E. J. "Jake" Garn (R-Utah).

Thermal protection system refurbishment [see Space Transportation System/Launch Schedules, Feb. 6] had necessitated the delay from February 20. NASA had yet to accomplish step measurements, gap filler installation, and tile-bond and quality verifications.

NASA said the current 51–E launch date did not affect the planned March 19 launch date of the Space Shuttle 51–D mission. (NASA Release 85–20)

February 21: NASA announced plans for a March 4, 1985, Space Shuttle 51–E launch from KSC, with a landing at KSC on March 8. NASA based its launch decision on flight readiness-review results. The completion of cargo integration and orbiter-systems testing necessitated additional time in the schedule.

The current 51–E launch date would not change the planned launch of STS mission 51–D, which was planned for no earlier than March 19, to retrieve the Long-Duration Exposure Facility and to launch the Navy Syncom satellite. (NASA Release 85–27)

February 27: NASA announced a new launch date of no earlier than March 7 from KSC for the 51–E Space Shuttle mission with landing at KSC on March 11.

Routine checkout of the Tracking and Data Relay Satellite had disclosed that one cell in a 24-cell flight battery would not accept a charge. Rescheduling the launch would permit battery repair and retesting. The three nickel-cadmium batteries would supply full power to the satellite during orbital-flight, solar-eclipse periods when its solar panels could not provide the required electrical power. Final checkout of the Space Shuttle orbiter systems was progressing satisfactorily.

The 51–E launch date would cause an STS 51–D launch date of no earlier than March 22. (NASA Release 85–29)

February 28: The *Washington Times,* in its report on the Space Shuttle 51–E mission delay, said that NASA Administrator James Beggs had indicated the delay might be a week or two.

In announcing earlier delays, NASA had said they would not alter the planned March 19 launch of the 51–D Space Shuttle mission, *Aerospace Daily* reported. However, in announcing the new 51–E launch date, NASA said the current 51–D mission launch would slip three days. (*W Times,* Feb 27/85, 4A; *A/D,* Feb 28/85, 1)

March 1: NASA announced cancellation of the March 7 Space Shuttle Challenger flight due to problems associated with the Tracking and Data Relay Satellite (TDRS–B). NASA officials determined that, in addition to repairing the previously announced problem with one cell of the TDRS's 24-cell flight battery, it was necessary to remove the TDRS–B from the Challenger cargo bay in order to repair a timing problem that became apparent during testing of the TDRS–1 then in orbit. Under certain operational conditions, the timing circuits could cause errors in the system switching sequences, interrupting user support. Tests run February 27 and 28 at spacecraft contractor facilities confirmed the problem.

Although NASA and its contractors had developed procedures to operate the TDRS–1, these procedures were not acceptable for multiple spacecraft operations, requiring modifications for TDRS–B and subsequent spacecraft. Because the TDRS spacecraft had encryption devices to protect the system from interfering signals, NASA could not provide further technical detail on the problem.

NASA expected a delay of several weeks for modifications, at which time it would remanifest TDRS–B.

NASA would place the Anik–C (Telesat–I) satellite scheduled for the STS 51–E mission on the Discovery STS 51–D mission and delay retrieval of the Long Duration Exposure Facility (LDEF) from the 51–D mission to a future Space Shuttle flight. (NASA Release 85–31)

March 8: An accident at Kennedy Space Center (KSC) today damaged the Space Shuttle orbiter Discovery and injured a technician, forcing delay in the orbiter's next flight originally scheduled for March 29, the *Washington Post* reported. The accident occurred about 8 a.m. when the bucket of a "cherry picker" crane hovering over Discovery fell, hitting a Lockheed technician who was on a work platform and then striking Discovery, which was horizontal on the floor of the orbital processing facility.

The 2,500-lb. bucket broke the technician's leg and injured his shoulder, then fell onto the closed left-hand cargo bay door, which was so thin and made of such light-weight material that it could not be opened on earth without elaborate supports to keep it in a fixed position. The bucket made two holes about three feet apart in the heat protection tiles insulating the door and damaged the door's structure.

A NASA spokesman said an investigating board would determine the cause of the accident, extent of the damage, and impact on the flight schedule. Officials speculated that NASA would have to replace the 2,400-lb. door, which would require shipment of a new door from the manufacturer and as many as three or four days of work to complete installation. (*W Post,* Mar 9/85, A2)

March 13: The *NY Times* reported that NASA officials said the next Space Shuttle mission would not occur until mid-April as a result of the time needed to repair two one-foot-sq. punctures in one of the cargo bay doors of the orbiter Discovery. NASA officials said Discovery would definitely fly the mission, although they had considered substituting Challenger. Rockwell Internatl. was fabricating the replacement material for the punctures and would fly the material to Cape Canaveral that day or the next.

Although NASA could not set a launch date until it had completed repairs, unidentified sources said that, if all went well, launch would be about April 19. (*NYT,* March 13/85, A14)

March 27: NASA announced a launch date of no earlier than April 12, 1985, for the 51–D Space Shuttle mission, with two windows for launch on that date: from 8:04 a.m. to 8:18 a.m. EST and from 8:45 a.m. to 9:00 a.m. EST. The orbiter Discovery would land April 17 at approximately 8:14 a.m. EST at KSC.

During the mission, crew would deploy the Canadian Telesat (Anik C1) and the Hughes Syncom IV (LEASAT) and operate the McDonnell Douglas continuous flow electrophoresis system. (NASA Release 85–43)

During March: NASA announced it had scheduled for no earlier than March 3 from KSC the launch of Space Shuttle mission 51–E, which was the 15th Space Shuttle mission and the orbiter Challenger's 7th flight. The Challenger was scheduled to land on March 7 at the 15,000-foot Space Shuttle landing facility at KSC. NASA previously announced the flight crew [see Feb. 7].

Challenger would carry the heaviest cargo taken into space by a Space Shuttle; total weight of the payload bay and cabin payloads would be nearly 53,400 lb., almost 15,000 lb. heavier than the previous record set on Space Shuttle mission 51–A.

After liftoff and insertion into orbit, the crew would prepare to deploy the Tracking and Data Relay Satellite (TDRS–B)/inertial upper stage (IUS). Following release, the crew would maneuver Challenger to a safe distance and observe the IUS perigee kick motor firing, placing the TDRS in geosynchronous transfer orbit.

Following IUS first-stage burn of about 2 minutes 26 seconds, which would then drop off, the TDRS/IUS second stage stack would coast for six hours on the way to geosynchronous orbit altitude of 22,300 miles. Once there, the IUS second stage would fire for 1 minute 49 seconds to stabilize the TDRS in geosynchronous orbit and the IUS would drop off.

The IUS was an advanced solid-propellant, two-stage booster designed to carry heavyweight payloads into orbits higher than the Space Shuttle could reach. NASA first used the IUS on the STS–6 mission for launch of TDRS–A. Changes to the IUS since then corrected problems that developed during the STS–6 mission. The IUS first stage developed about 46,500 lb. of thrust, the second stage about 18,500 lb.

On the second day, the crew would use the Challenger's orbital maneuvering system (OMS) engines to raise their orbit to above 180 miles for deployment during the 22nd orbit of the 7,347-lb. Telesat–1 (Anik C–1) and its payload assist module (PAM). (NASA Release 85–24)

April 2: NASA Administrator James Beggs, testifying today before the U.S. House HUD and independent agencies appropriations subcommittee on NASA's FY 86 budget request, said that orbiter processing constraints at Kennedy Space Center (KSC) made it difficult for NASA to consider adding Space Shuttle flights to get the year's planned commercial launches back on schedule, *Aerospace Daily* reported.

When asked by Chairman Edward Boland (D-Mass.) if reports were true that NASA was considering launching three commercial satellites on Deltas rather than the Space Shuttle to get the Space Shuttle's manifest back on schedule, Beggs responded that it was an option the agency "had to look at" but one that probably would not be exercised. He said NASA talked to several Space Shuttle customers about flying payloads on Deltas and some were interested. Another option NASA considered was adding another Space Shuttle flight to the manifest, but orbiter processing constraints posed problems. "We just can't push the orbiters through the processing facility quickly enough," he said, although NASA still "may be able to squeeze another launch" into the year's manifest.

He pointed out the hardware for another flight was available and NASA proposed building a third orbiter processing facility high bay and Space Shuttle tile facility at KSC with FY 86 funds. (A/D, Apr 3/85, 1)

April 5: NASA announced a new Space Shuttle manifest for 41 Space Shuttle missions through December 14, 1987 that included nine flights for the remainder of 1985, 15 flights in 1986, and 17 flights in 1987.

Highlights for 1985 included the first Atlantis flight in late September for a Department of Defense mission; a Spacelab 3 flight in late April, Spacelab 2 with extensive European Space Agency participation in July, and another Spacelab flight in October dedicated to W. German scientific investigations. Also during 1985 NASA would deploy communications satellites for Hughes, AT&T, RCA, Mexico, Australia, Canada, and the Arabsat consortium.

In 1986 the first launch of the Space Shuttle from Vandenberg Air Force Base would occur. The first liquid-hydrogen-powered Shuttle Centaur upper stage would be used to deploy Ulysses (formerly the International Solar Polar Mission), and the Galileo mission to Jupiter (also using the Shuttle Centaur) and the Hubble Space Telescope flight would be launched.

NASA said launch dates of missions to deploy the second Tracking and Data Relay Satellite (TDRS–B) and to retrieve the Long Duration Exposure Facility (LDEF–1) were under review. Also, NASA assigned Kathryn Sullivan, Steven Hawley, and Bruce McCandless as mission specialists for the Hubble Space Telescope flight on mission 61–J in August 1986. (NASA Release 85–50)

April 24: NASA announced it selected April 29, 12:00 noon EDT, for launch of STS mission 51–B, the Spacelab 3 flight, and Dryden Flight Research Center/Edwards Air Force Base, California, as the primary end-of-mission landing site, with KSC as an alternate. Landing was set for May 6 at 12:03 EDT.

NASA selected Dryden over KSC for the landing as a result of problems encountered during the STS 51–D mission when the orbiter Discovery's right-hand braking systems locked causing a tire to blow out. Landing conditions had included a crosswind—the first experience at KSC of such conditions—and a higher-than-usual sink rate. A landing at Dryden would provide a greater safety margin for Challenger's tires and brake systems due to the availability of the unrestricted lakebed and the smoother surface.

The decision to land at Dryden for the next flight only would enable engineers to determine what corrective actions were appropriate before returning to the KSC runway for nominal end-of-mission landings. (NASA Release 85-63)

During April: NASA announced it had scheduled for launch on April 12 from KSC the remanifested Space Shuttle 51-D mission with landing on KSC's Space Shuttle runway after the five-day 78-orbit flight.

NASA originally scheduled mission 51-D for March 1 for deployment of NASA's Long Duration Exposure Facility, but cancelled the 51-E Challenger mission and revised the 51-D cargo to include LEASAT-3 and the Canadian communications satellite Anik C-1. Other payloads were the continuous-flow electrophoresis system and the American echocardiograph experiment [see Apr. 12], two middeck student experiments, and two Getaway Special canisters. (NASA Release 85-47)

July 12: Space Shuttle mission 51-F was aborted today at three seconds before liftoff when the orbiter Challenger's onboard computer automatically shut down the main engines, *Spaceport News* reported. Seconds after main engine start, main engine No. 2 lost redundancy to operate the chamber coolant valve. NASA's ground rule restricted launching without an operating backup system in that area of the main propulsion system. Although the valve did assume its proper position after a command from a backup system, the loss of redundancy resulted in a major component failure flag being sent to the orbiter's computers.

Challenger Commander Gordon Fullerton later thanked the Kennedy Space Center team on behalf of the crew, saying the team acted quickly and professionally and that at no time did the crew feel apprehensive. In answering a *Washington Post* reporter's question later, Fullerton said, "We all have mixed emotions here, but we're thankful the system worked the way it should. It was the longest three seconds I've ever experienced."

Launch operations manager Thomas Utsman said that crews would strip parts of the malfunctioning engine and that engineers had identified four different parts that could have failed when they should have been pumping fuel from the hydrogen tank into the engine chamber.

When the liquid-fuel engines ignited, valves that fed fuel into the chambers were fully open so engines could attain speed instantly. Just before liftoff, the valves were partially closed to cool engines slightly and prevent too much fuel from entering the chamber. The valve on main engine No. 2 failed to close partially, signaling the computer to shut down the system.

The first Space Shuttle abort occurred June 26, 1984, when an onboard computer shut down two of Discovery's three engines four seconds before liftoff. That abort triggered an investigation that led to a complete overhaul of Discovery's main engines. NASA had to move Discovery back to the Vehicle Assembly Building where its cargo was removed and placed aboard another orbiter that flew into space two months later.

"We don't think the engine valve itself failed the way a valve failed the [other] time we had a launch abort," Space Shuttle program manager Robert Lindstrom said. "I really believe there's a fair likelihood we'll get off again in another week."

The delay did have an impact on the mission's objectives. The infrared telescope in Challenger's cargo bay could be operated successfully only in a completely dark sky. Had the flight left on schedule, the sky would have been dark for most of the following seven days until a new moon.

NASA had scheduled the next Space Shuttle launch using the orbiter Discovery for August 24. It was not immediately clear how the Challenger's abort would affect that mission. (*Spaceport News*, July 19/85, 1; *W Post*, July 13/85, A3)

July 15: NASA announced it had rescheduled for July 29 launch of Space Shuttle mission 51–F with the orbiter Challenger and delayed by as much as two weeks the previously scheduled September 19 maiden voyage of the orbiter Atlantis that was to carry a secret Pentagon payload into orbit, the *Washington Post* reported.

The July 12 Challenger launch abort forced NASA to design a new schedule that maintained a Discovery blastoff for August 24, delayed another Challenger mission from October 3 to early November, and kept a November 27 Atlantis flight and a December 20 Columbia mission.

Kennedy Space Center (KSC) work crews had removed three parts from Challenger's No. 2 engine that engineers suspected could have triggered computers into ordering all three engines to shut down on the launch pad. One was a valve that failed to close when it should have, a second was a hydraulic actuator that ordered the valve to close, and the third was a controller that sent commands to the actuator.

"We still suspect the actuator, and nothing we've found suggests anything else," KSC spokesman Hugh Harris said. "We are replacing all three parts and continue to analyze and test the parts we remove."

NASA would test fresh parts on the pad, leading to a July 23 flight readiness review. This included electronics and leak checks of parts like pumps, turbines, and valves, not a test-firing of the engines.

The delay meant a July 30 flight readiness firing test for Atlantis would take place September 12, which would delay Atlantis's maiden flight from September 19 to late September at the earliest. (*W Post*, July 16/85, A4)

August 14: NASA cleared today the Space Shuttle's main engines for launch, making possible an August 24 launch of Discovery, the *Washington Post* reported. Laboratory tests and a thorough examination of one of Challenger's engines, which prematurely shut down shortly after a July 29 launch, showed that it was not damaged during launch and that a pair of heat sensors, which mistakenly indicated the engine had overheated, caused the shutdown. NASA had fitted Discovery with new heat sensors that engineers believed were vastly superior to the ones that failed on Challenger.

Dominick Sanchini, executive vice president of Rocketdyne, which built the hydrogen-fueled engines, said, "Our analysis has verified what we felt right along was the probable failure mechanism. We're convinced that the sensors we have now have resolved the problem we saw on the last flight." (*W Post*, Aug 15/85, A8)

August 24: Sudden squalls almost on top of the launch pad early today forced postponement for at least a day of Space Shuttle mission 51–I, the *Washington Post* reported. A series of unforecasted thunderstorms blew in from the mainland directly across the flight path, and dark thunderheads surrounded the launch pad for five miles in almost every direction. NASA aborted the launch five minutes before liftoff.

NASA officials were concerned not only about the Space Shuttle climbing through rain and lightning but also about the crew's ability to see the KSC runway in case they had to make an emergency landing the early minutes of the flight. Rain could damage the orbiter's tiles, and lightning could harm its computers and guidance systems.

NASA rescheduled Discovery's launch for between 7:57 and 8:11 a.m. EDT August 25. Among the factors necessitating that launch window were the position of the satellite planned for salvage and timing of deployment of three communications satellites. If an August 25 launch was impossible, there were launch opportunities the following three days.

NASA had announced earlier that highlights of the eight-day mission included an attempt to repair and salvage the *Leasat/Syncom IV–F3* satellite and deploy it for normal operation, and deployment of the *ASC 1*/PAM–D for the American Satellite Co., the *AUSSAT 1*/PAM–D satellite for the Australian government, and the *Leasat IV–F4* satellite for the U.S. Navy. The physical vapor transport of organic solids (PVTOS) experiment sponsored by 3M Corp. would also fly on the mission.

PVTOS was the second of some 70 experiments the 3M Corp. planned to conduct aboard the Space Shuttle over the next ten years. On the 51–I mission, solid materials would vaporize into a gaseous state to form thick crystalline films on selected substrates of sublimable organics. Researchers at 3M would study crystals produced by PVTOS for their optical properties and other characteristics that might ultimately have important applications to 3M's businesses in electronics, imaging, and health care.

The company's first experiment, flown in November 1984, dissolved materials that led to a solid crystalline product. (NASA Release 85-118; *W Post*, Aug 25/85, A3)

August 25: A computer failure 25 minutes before launch today forced postponement for at least two days of Space Shuttle mission 51-I, the *NY Times* reported. Launch crews planned to replace the malfunctioning computer, inspect the Space Shuttle's fuel plumbing, and try to launch Discovery on August 27 at 7:02 a.m. EDT.

At 7:15 the Space Shuttle crew and Johnson Space Center (JSC) flight controllers almost simultaneously noted warnings that a backup guidance and control computer was registering errors. This came as the backup computer was undergoing a final check to see that its programs agreed with those driving the Space Shuttle's four main computers.

NASA halted the launch countdown while engineers at Kennedy Space Center and JSC examined data retrieved from the malfunctioning computer and compared it with data from one of the regular computers. They concluded that the trouble was an apparent failure in the computer, which could not be remedied in time for a launch.

That the computer worked perfectly in tests before August 25 and worked in tests after the failure, although it showed signs of trouble, puzzled engineers. NASA was about 99% sure that it was a hardware failure, although only an inspection of the computer by JSC engineers would determine the exact nature of the problem.

It was the second postponement in two days for the Discovery and its crew of five, and NASA officials were growing concerned that Discovery might not reach orbit in time for repair of the crippled *Leasat 3* satellite. Any delay beyond August 29 would cancel the repair effort, and the crew would have to confine themselves to deploying three communications satellites in the first three days of the mission. The flight would thus fall short of the eight days then planned.

Arnold Aldrich, manager of the Space Shuttle program at JSC, said NASA would not reschedule beyond August 29 any attempt to repair *Leasat 3*, as that would have too disruptive an effect on other flights scheduled in the next few months.

If it had been only a matter of replacing the computer, the delay would have been a day. However, the need to inspect insulated ducts that carried liquid-hydrogen to the Space Shuttle's three main engines necessitated the two-day postponement. After NASA had pumped the super-cooled fuel into the Space Shuttle system two consecutive mornings and drained it again, there was a chance that the contractions and expansions caused by the alternating freezing cold and Florida heat might have damaged the engines, which in extreme cases could cause them to explode in flight. (*NYT*, Aug 26/85, A11)

November 27: NASA announced that it had agreed with the U.S. Air Force to delay the first Space Shuttle launch from Vandenberg Air Force Base until mid-July 1986. NASA originally scheduled this mission, STS 62–A, for no earlier than March 20, 1986.

Under Secretary of the Air Force Edward "Pete" Aldridge, Jr. said, "There are no major problems at the Vandenberg site. We have repeatedly stated that safety and quality would not be sacrificed for schedule. Our decision reflects our continued commitment to this philosophy."

Aldridge added that, "We have had to make some facility modifications because of what we learned from routine Air Force/NASA operational readiness inspections. We have also added time to the schedule to allow for better preparation and evaluation of the operational systems tests and we have extended the training period for the launch crew of this historic, first West Coast shuttle mission.

"This revised schedule will permit us to complete ongoing modifications, inspections, rework and operational testing with higher confidence than could be permitted with the March 20 schedule," he added. "It also minimizes the potential for conflict with the NASA Ulysses and Galileo planetary missions scheduled in May."

NASA Office of Space Flight associate administrator Jesse Moore concurred, adding, "NASA agrees completely with the Air Force regarding Vandenberg. Our first commitment is to the safety of the crew and the reliability of the vehicle and launch systems. The development of the Vandenberg site is proceeding very smoothly. This readjustment gives us all more time to carry out our commitment to safety and reliability."

The decision to delay meant NASA would deliver the orbiter Discovery to Vandenberg around March 1, 1986, and NASA and Air Force officials would continue to evaluate the STS 62–A schedule in order to establish after January 1 a firm launch date. (NASA Release 85–158)

December 17: NASA said today it postponed the December 18 launch of the Space Shuttle Columbia on mission 61–C for 24 hours because tired workers had fallen behind in countdown tasks, the *Washington Post* reported. "Essentially there were too many tasks to complete and too little time to complete them," NASA spokesman George Diller said. "We felt it was more prudent to delay than to take a chance on making a mistake." Diller added that crews used up more than eight hours of contingency time built into the countdown as they worked through the night December 16 to complete checkout of the engine compartment.

NASA rescheduled the launch for 7 a.m. December 19. (*W Post*, Dec 128/85, A17)

December 19: A hydraulic pump in a booster rocket attached to the Space Shuttle Columbia failed today, halting the launch countdown 14 seconds before liftoff and delaying STS mission 61–C until early in 1986, the *NY Times* reported. The shutdown came four seconds before ignition of the three main engines. NASA had called for a delay of 54 minutes earlier in the morning when clouds moved over the Kennedy Space Center (KSC).

NASA aborted the countdown when a computer on the ground sensed that a pump on one of the Space Shuttle's solid-fuel rocket boosters was spinning at 86,000 rpm, exceeding the turbine's "red line" limits by almost 7,000 rpm. The device, described as a hydraulic power unit, provided power so the nozzle of the solid-fuel rocket booster could pivot slightly in flight, helping to control the direction of the orbiter's ascent. NASA officials indicated there was a backup pump on the booster, but that, if a single pump failed before liftoff, safety precautions called for abort of the countdown.

That particular hydraulic power unit had flown twice before on the resuable solid-fuel rocket boosters, NASA officials said, and similar devices had flown up to four times before requiring replacement. It was the first failure in the Space Shuttle program of that particular part.

NASA would have to repair or replace the failed pump and said it had rescheduled the launching for January 4, 1986, so there would be no interference with the holidays. The delay until January would also allow NASA to avoid paying the Air Force to keep open its Eastern Test Range at Cape Canaveral and paying launch pad workers triple-time pay for working between Christmas and New Year's, the *Washington Post* reported. "The team is obviously disappointed," said Robert Sieck, director of Space Shuttle operations at KSC. "But we're going to isolate the problem and fix it and get on with the launch process."

NASA said the delay would have no impact on the 16 Space Shuttle missions, including 61–C, that it had scheduled for 1986. (*NY Times*, Dec 20/85, B10; *W Post*, Dec 20/85, A2)

Main Engines

February 7: NASA announced award of a $2,374,800 fixed-price contract to Stearns Catalytic Corp. to modify a test stand at the National Space Technology Laboratories (NSTL), Hancock County, Mississippi, to give NSTL a third test position to static-fire individually Space Shuttle main engines. The structural, mechanical, and electrical modifications would begin no later than March 1985 for completion in a year.

Space Shuttle main-engine testing had begun at NSTL in June 1975, using test stands A–1 and A–2 for single-engine testing of flight and nonflight engines. Later NASA used the B–2 test position to certify the main propulsion system in a series of three-engine cluster firings. The additional test position would support projected increases in main-engine test requirements, including increased turbopump production rates. NASA would initially use the new

B-2 position to test new and overhauled engine turbopumps using a testbed engine.

NASA had originally used the B-2 test stand in the 1960s to flight-certify the Apollo/Saturn V space vehicle's first stage. (NASA Release 85-19)

July 30: NASA announced that, based on a review of engine data following the July 29 Space Shuttle mission 51-F launch, Marshall Space Flight Center and Kennedy Space Center engineers believed that failed sensors—not an engine failure—were the probable cause of the main engine No. 1 shut down [see Space Transportation System/Missions, July 29].

The data suggested the high-pressure fuel turbopump on that engine performed normally, because there were no indications the pump was running hot. If the pump were running hot, Johnson Space Center flight engineers, who monitored engine performance during ascent, would have expected to see changes in the position of an oxidizer valve indicating an increased flow to supply the pump with more power. Data showed the valve position did not change. A second data indication of normal pump performance was present in the engine discharge pressures, which were monitored by ground controllers and the main engine controller.

Based on these indications, NASA engineers reviewing the cause of the engine shutdown tentatively concluded that the main engine controller was receiving faulty data from failed sensors. Resistance measurements taken by the engine controller provided an indicator of pump temperatures, because the resistance increased as the temperature increased. The controller logic was able to distinguish, within certain limits, a valid reading from one that indicated the sensor wire, a very thin wire, was broken or had otherwise failed.

The controller recognized the bad reading from the Channel B sensor; the reading from Channel A, however, drifted, indicating a temperature rise above the redline limit. Because of the failure mode in the second sensor, which gave readings above the redline limit but below those that would cause the controller to disregard them as unreasonable and thus indicative of a sensor failure, the engine controller believed there was a possible pump problem and shut down the engine.

A temperature sensor on main engine No. 3 also failed. Because there was a history of these temperature sensors failing on previous Space Shuttle flights, NASA decided to develop a new sensor for use on future Space Shuttle flights beginning with mission 51-I.

Until such time as NASA confirmed this explanation by an examination of the engine and its various systems, NASA would not launch another mission. However, the agency did not anticipate that this would result in a delay in the next mission.

The call "abort to orbit" used to describe the action ordered for the July 29 Space Shuttle flight was the preferred type of Space Shuttle launch "abort" in that it gave the crew, the orbiter, and the ground controllers time to review options and to ascertain that the crew and orbiter were in good condition. In nearly all cases, the abort-to-orbit contingency would allow completion of most of a particular mission's flight requirements.

Once NASA determined that an orbiter was in a safe orbit (minimum would be about 105 nautical miles circular), then the crew and ground flight controllers could begin to assess the impact on planned mission activities. In the case of mission 51–F, the orbit achieved was 106 by 146 nautical miles, 40 miles short of the planned designations; the orbiter was at a velocity of 25,760 feet-per-second, which was 110 feet-per-second less than planned.

NASA determined all abort modes based on the weight of the orbiter, the inclination of the desired orbit, and other characteristics. Each mode had a specific applicable time and duration during the ascent phase, and crew and ground controllers practiced these contingency situations regularly. (NASA Release 85–114)

September 4: NASA announced that Lewis Research Center (LeRC) and Marshall Space Flight Center (MSFC) were pursuing a 10-year program scheduled for completion in 1991 to improve Space Shuttle engine performance, reduce operational costs, and develop advanced engine technology for future space travel.

Although the Space Shuttle propulsion system had performed well for a number of space missions, it incorporated state-of-the-art technology at the time of its development some 15 years before, so NASA believed the engine's operational life and performance levels could be improved. According to Stanley Marsik, LeRC's Space Shuttle main engine technology program manager, "The current engine represents the first generation of reusable engines. In a way, the first reusable space engine can be compared with the Model-T automobile. Now that we have experienced numerous flights with this engine, we have learned how to improve its life and performance. We now are working towards advancing the technologies necessary to support those improvements."

In the age of reusable Space Shuttle engines, demands would increase for higher performance with lighter weight components, resulting in higher operating pressures and temperatures and increased mechanical vibrations and fuel flow turbulence. The system also severely taxed engine components such as turbine blades, bearings, seals, fuel ducting, and combustion chambers.

Engineers were studying new high-temperature, high-strength metal alloys and other advanced materials such as new protective coatings and combustion processes. Researchers using advanced computer modeling techniques studied loads on turbine blades and fuel flow, and examined the use of hydrostatic ball bearings where a film of gas would separate the bearing from metal surfaces, resulting in no metal-to-metal contact.

Engineers were developing new methods of measuring wear and deterioration, so parts would be replaced only when necessary, and studying ways to monitor engine health by means of a series of computer-connected diagnostic sensors on the engine to signal deviations from normal performance and to facilitate more economical on-the-ground maintenance.

The two centers had formed ten disciplinary groups to work with some 130 different technology elements of engine design. Rocketdyne Division of Rockwell Internatl., engine manufacturer, and other NASA contractors and several universities were supporting the program. (NASA Release 85–122)

September 6: Marshall Space Flight Center (MSFC) announced it had scheduled for September 12 at Kennedy Space Center a 20-second on-the-pad test firing of the Space Shuttle orbiter Atlantis's three main engines to verify the flight readiness of Atlantis—the fourth and newest of the Space Shuttle orbiter fleet.

The flight readiness firing in a launch-day environment would exercise the Space Shuttle's main propulsion system and computer programs to demonstrate the proper integration of all elements prior to the first Atlantis launch scheduled for October 3 for mission 51–J. Engineers had previously test fired each engine individually at NASA's National Space Technology Laboratories.

On September 9 a firing simulation would kick off preparations for the test. The one-day test would check out critical flight and ground computer systems that controlled the actual engine firing and give the Space Shuttle launch team a chance to re-review test procedures.

Countdown for the flight readiness firing would begin at 4:00 a.m. September 10, with the clock at the T–43 hour mark, and culminate with the 20-second static firing.

Three days of preparations would precede the firing countdown, including purging the external tank with inert gases, installing special firing instrumentation, and mounting a radiation shield on the deck of the mobile launcher platform to protect the solid-fuel rocket boosters and external tank structures during the test firing. Total flight readiness firing, including pre- and post-test activities, would span six days and require support from personnel and operations from various NASA centers including MSFC and Kennedy and Johnson Space Centers.

Events leading to the September 12 engine firing would be nearly identical to those that preceded an actual Space Shuttle launch, with built-in holds distributed throughout the countdown at the same times they occurred in an actual launch countdown.

An automated ground launch sequencer would control the final nine minutes of the countdown. The sequencer performed the final series of events in a specific order, monitored various measurements of out-of-tolerance conditions, and detected system malfunctions for which it would automatically stop the countdown.

Following the test, technicians would inspect externally and internally the three main engines to verify their readiness for flight. These post-firing activities would take about 11 days. (MSFC Release 85-44)

Management

March 1: Construction was proceeding at Kennedy Space Center (KSC) toward the October completion of the Shuttle Processing Contract (SPC) logistics facility to house approximately 190,000 items of Space Shuttle program stock, including orbiter fuel cells, various electrical components, tires, brakes, windows, nuts, bolts, and washers then stored in three KSC warehouses and four areas of Cape Canaveral Air Force Station, the *Spaceport News* reported. NASA would consolidate seven Lockheed Logistics Directorate departments and its own Launch Support Services Directorate at the facility, with 500 Lockheed and NASA employees working in 75,000 sq. ft. (about a quarter of the total) set aside for office space.

The Lockheed-managed facility would be KSC's first automated logistics warehouse for storage of flight hardware and SPC ground-support equipment. An automated inventory system would interface with the Shuttle Inventory Management System (SIMS) and a manual system. Retrieval procedures would require four roboticized forklifts to find and fetch items, an automatic conveyor, and a manned, three-story high forklift.

As part of the SPC agreement, Lockheed selected Austin Co. as the facility's architectural and engineering firm and builder at a basic design and construction cost of about $16 million. (*Spaceport News,* Mar 1/85, 1)

April 12: National Space Transportation System manager Dr. Glynn Lunney, who was a member of the Space Task Group that inaugurated U.S. manned space flight, announced he would leave NASA in the near future, the Johnson Space Center *Space News Roundup* reported.

Lunney, whose career almost exactly spanned the period of U.S. efforts in space, began his government service in 1958 with the National Advisory Committee for Aeronautics, NASA's predecessor, at Lewis Research Center. He then moved to the Space Task Group and with it to Houston when it became the Manned Spacecraft Center. Later he became head of the mission logic and computer hardware section, chief of the flight dynamics branch, flight director for the Gemini program, and, in July 1968, one year before the first manned landing on the moon, chief of the flight director's office. He then served as special assistant to the manager and then manager of the Apollo spacecraft program.

With the advent of the Space Shuttle program, he became manager of the Space Shuttle payload integration and development program responsible for directing the planning and implementation for all payloads and payload carriers in that program.

Since 1981 he served as manager of the National Space Transportation Systems Program with responsibility for overall systems management and integration of all elements of the program.

Among the awards received by Lunney were the Lawrence Sperry Award, the Arthur S. Fleming Award, the Louis W. Hill Space Transportation Award, the Allan D. Emil Memorial Award, the W. Randolph Lovelance II Award, three NASA Group Achievement Awards, two NASA Distinguished Service Medals, an Outstanding Leadership Medal, Senior Executive Service designations, and an honorary doctorate of laws from the University of Scranton. Lunney was a Fellow of the American Astronautical Society and the American Institute of Aeronautics and Astronautics. (JSC *Space News Roundup*, Apr 12/85, 1)

April 29: NASA and the Defense Department (DOD) with White House concurrence abandoned any consideration of transferring the management of Space Shuttle operations to a commercial venture or quasi-public corporation for the foreseeable future, *Aviation Week* reported. The decision would thwart plans of commercial ventures that had sought the combined Space Shuttle management/marketing role. The Shuttle Operations Strategic Planning Group, a team made up of officials from NASA, DOD, the White House, and private enterprise, said ". . . the suggestion that the space shuttle should or could be transferred, sold or leased to private enterprise for its operation, marketing and future development is simply not a viable alternative for the foreseeable future."

Although some commercial space officials called the group's study a "whitewash," others believed the decision could introduce more Space Shuttle launch and cost stability, a critical issue with a large segment of commercial space ventures that needed Space Shuttle transportation no matter who was managing it.

DOD Space Shuttle military space-flight requirements, including the potential for Strategic Defense Initiative Space Shuttle activity, were key elements in the decision not to take management of the program away from NASA.

As a result of the decision, NASA would have to create a new management structure in order to get the Space Shuttle out of existing development-oriented management. "As the shuttle evolves and the flight rate increases," the strategic planning group said, "organizational realignments within NASA will become necessary to ensure that NASA's R&D and shuttle operations do not detract from one another but are mutually supportive." (*AvWk*, Apr 29/85, 42)

Astronautics and Aeronautics, 1985

July 26: In an administrative realignment reflecting the maturation of the Space Shuttle program, NASA announced the appointment of Arnold Aldrich, a 26-year NASA veteran and head of the Space Shuttle projects office at the Johnson Space Center (JSC), to manager of the National Space Transportation System (NSTS). He would fill the vacancy left when Glynn Lunney retired in April.

The Level II NSTS organization at JSC would assimilate the projects office, consolidating all program elements under Aldrich. With this combined responsibility, Aldrich would take charge of integration of all Space Shuttle program elements including flight software, orbiter, external tank, solid-fuel rocket boosters, main engines, payloads, payload carriers, and Space Shuttle facilities. His responsibilities would also include directing the planning for NSTS operations and for management of orbiter and government furnished equipment projects. NASA named Richard Kohrs, who was acting program manager, and Lt. Col. Thomas Redmond, U.S. Air Force, deputy managers.

In a related move, Thomas Utsman, head of Space Shuttle management and operations at Kennedy Space Center (KSC), would become deputy director of KSC, with NASA dividing shuttle management and operations into two primary organizations: Shuttle engineering and Shuttle operations. The engineering directorate, headed by Horace Lamberth, would expand to include skills necessary for sustaining engineering of the orbiter. The operations directorate, headed by Robert Sieck, would retain all functions necessary to manage day-to-day Space Shuttle processing and its logistical support.

NASA would transfer later in the year from Johnson Space Center to KSC launch support services and orbiter thermal protection system manufacturing contracts, functions closely associated with KSC responsibilities for Space Shuttle maintenance and launch preparation. At the beginning of 1986, KSC would take over logistics responsibility for spare parts refurbishment and procurement and would assume sustaining engineering responsibility for orbiter subsystems. (NASA Release 85–112)

July 29: NASA officials said they might alter Space Shuttle launch countdown procedures to avoid the type of last-second problem that forced postponement of mission 51–F earlier in July [see Space Transportation System, Launch Schedules, July 12], the *NY Times* reported.

Jesse Moore, NASA associate administrator, said in an interview that "We want to minimize the chances of a needless shutdown. There may be a way to soften up some of our criteria. "Before you take off on an airplane the pilot normally tries all his flaps and hydraulics," Moore said. "We don't have a chance to do that on the launch pad right now, and we want to perform those kinds of checks."

Moore indicated new procedures being considered for the countdown included increasing the amount of time allowed for the Space Shuttle's computers to verify the correct functioning of the engine valves and broadening the criteria under which they were meant to operate. Moore added that criteria for safety would not change. The new procedures would add "no risk" to crew members, he said.

The consideration of new procedures likely reflected growing confidence in the Space Shuttle as well as concern about foreign competition for launches. NASA officials said Space Shuttle launches must proceed smoothly and regularly if the program is to be a commercial success and that the current schedule called for one launch a month, eventually going to one every two weeks. (*NYT,* July 29/85, A8)

November 5: NASA announced selection of Boeing Service Internatl., Inc. and United Technologies Corp. for negotiations leading to award of a single contract for flight equipment processing at the Johnson Space Center (JSC). The contract, covering a basic period of three years plus a single priced option for an additional two years, would be for services expected to begin in January 1986. Follow-on awards could ultimately result in a total contract period of 15 years.

JSC would manage the work under a cost-plus-incentive/award-fee contract arrangement that would include incentive fee on sound cost management and an award fee on the basis of performance.

The contractor selected after final negotiations would be responsible for receipt, launch preparation, and postlaunch activities related to the overall processing of crew-related flight equipment required to support the Space Transportation System program. The contractor would process/resupply individual equipment items in preparation for launch and would operate and maintain assigned facilities and support equipment required for successful equipment processing. Equipment covered under the contract included extravehicular mobility units (spacesuits), food and medical systems, communications equipment, display and control equipment, and other miscellaneous items.

The competitive procurement consolidated under one contract work previously performed by 16 firms under 19 contracts. (NASA Release 85–149)

December 5: NASA's Johnson Space Center (JSC) announced signing today with Rockwell Shuttle Operations Co. a cost-plus-incentive/award-fee contract for Space Transportation System operations (STSOC) [see Space Transportation System/Management, Sept. 12]. JSC estimated the first two years of the contract, starting January 1, 1986, to be valued at $378,536,000 and the follow-on two-year extension option from January 1, 1988 through December 31, 1989 at about $374,320,000 for a possible four-year total of $752,846,000.

Rockwell's STSOC tasks would include project management; maintenance and operations of mission control center-Houston, shuttle mission simulator, shuttle avionics integration laboratory, software production facility, and the central computing facility; sustaining engineering; flight preparation requirements and analysis; flight preparation production; and direct mission operations and testing and support for Space Shuttle operations at JSC.

The Rockwell team included Bendix Field Engineering Corp., System Development Corp., Omniplan Corp., RMS Technologies, Inc., and System Management American Corp. (JSC Release 85–051)

December 18: NASA announced that it had selected Boeing Aerospace Operations to receive the Flight Equipment Processing Contract (FEPC) at Johnson Space Center (JSC). Wornick Co. was a Boeing subcontractor. The three-year award with a two-year priced option would begin in January 1986 at an estimated total cost of $76.5 million. The contract also included provisions for two additional unpriced, five-year extensions. JSC would manage the work under a cost-plus incentive/award-fee contract arrangement, which included incentive fee for sound cost management and an award fee based on performance.

Under the contract, Boeing would assume responsibility for receipt, launch preparation, and postlauch activities relating to the overall process of crew-related flight and flight-type equipment required to support the Space Transportation System program. The contractor would process and resupply individual flight equipment items, which supported the flight crew in its daily operation of the orbiter vehicle, and would operate and maintain support equipment required for the successful processing of the flight equipment, which included extravehicular mobility units, food and medical systems, communications equipment, and other miscellaneous items.

In addition to Boeing, NASA conducted final negotiations with Hamilton Standard Management Services, Inc. The selection of Boeing for the contract represented the consolidation into one contract of work currently performed by 16 firms under 19 contracts. (NASA Release 85–174)

Military Applications

January 18: The expected January 23 flight of the orbiter Discovery would be the first of 15 Space Shuttle flights devoted wholly to military purposes, *Science* reported, increasing the visibility of the military role in the program. Discovery would carry a reconnaissance satellite designed to intercept a wide variety of electronic communications, radar signals, and telemetry from intercontinental ballistic missile tests. After several orbits, the Space Shuttle would discharge the satellite from its payload bay and a booster rocket would push it to an altitude of about 35,000 km, where it would "park" over the USSR.

NASA estimated about 30% of Space Shuttle flights over the next 10 years would be military missions, although NASA's civilian budget had paid almost all basic costs of the Space Shuttle and associated launch and servicing equipment (estimated at more than $15 billion). In addition, military needs had dictated that the Space Shuttle have unusually large wings, rugged thermal protection, an unusually large payload bay, and unusually powerful engines; however, the Pentagon had borne none of these added costs. Although some NASA expenditures were for items serving both military and civilian functions, such as improved engines for both the orbiter and a rocket booster and onboard experiments in flight aerodynamics, plasma physics, astronomy, biology, chemistry, and radiation, NASA paid for all these projects without any Department of Defense assistance. (*Science*, Jan 18/85, 276)

January 24: NASA launched today at 2:50 p.m. EST the Space Shuttle Discovery mission 51–C with its five-man military crew to carry out secret military objectives, the *NY Times* reported. NASA, which had delayed launch due to subfreezing temperatures [see STS/Launch Schedules, Jan. 23], did not announce liftoff time until nine minutes before launch. One and a half hours after launch, Houston Mission Control reported the Space Shuttle had reached its intended orbit, the altitude and position of which were secret.

How long the astronauts would remain in orbit was unknown; although most Space Shuttle flights lasted five to eight days, the apparent Discovery mission to launch a single satellite might limit it to three or four days.

Discovery carried a 32,000-lb. propulsion, inertial upper stage capable of boosting a large satellite to 22,300 miles above the equator south of the USSR. This was the first test of the rocket system since a predecessor had misfired on a Space Shuttle mission in April 1983.

Many newspapers had already published reports, based on information in the aerospace trade press and other unclassified documents, that the secret payload was an electronic spy satellite. Both the U.S. and USSR had engaged in electronic eavesdropping from orbit for several years, and the USSR had obtained many of the plans and engineering information for the earlier U.S. electronic reconnaissance satellite Rhyolite. The Carter administration had cancelled an advanced version of Rhyolite, the Argus, but the Reagan administration had revived the project. Most reports described the Discovery's payload as the most advanced and sensitive form of orbital electronic spying.

The Air Force began an investigation to find the officials or contractor employees who had provided information about the secret payload aboard Discovery, the *Washington Post* reported. Inquiries were not aimed at news organizations, Michael Burch, a Department of Defense (DOD) spokesman, had said. When questioned about a photo of an early-warning satellite published on the cover of *Aviation Week*, Burch said the picture should not have appeared. However, the Pentagon later acknowledged that the Air Force had given the photo to the magazine after determining it was unclassified. Burch refused to say what action the government would take if it could identify those who leaked information on the Space Shuttle mission.

Brig. Gen. Richard Abel had previously warned reporters not to speculate about the Space Shuttle's cargo. After the *Washington Post* had reported it was an intelligence satellite intended to eavesdrop on the USSR, Defense Secretary Caspar Weinberger had said the story might have given "aid and comfort to the enemy." Abel later had told a Univ. of Georgia journalism class that little or nothing in the *Post* story was not available from public sources.

Aviation Week's January 21 issue had described the Pentagon's next generation of early-warning surveillance satellites, which would monitor missile and spacecraft launches from Soviet territory and be protected against Soviet laser weapons. (*NYT*, Jan 25/85, A1; *W Post*, Jan 25/85, A3)

January 27: The first military mission of the Space Shuttle ended the afternoon of January 27 when Discovery landed, in excellent condition with only minor damage to a dozen heat protection tiles, at KSC surrounded by the same secrecy in which the mission had begun, the *Washington Times* reported. The five-man crew had blown bosuns' whistles and rung ships' bells in greeting to flight controllers; however, in keeping with the flight's secrecy, the public did not hear the salute.

During the flight, a $50 million rocket booster called IUS, which had failed on its only previous Space Shuttle assignment, had successfully propelled a satellite from the Space Shuttle's low-earth orbit to a listening post 22,300 miles high. NASA would use the IUS on its next flight, scheduled for February 20, to boost a communications satellite to a 22,300-mile high orbit.

Discovery's return was a surprise, as observers had believed it would stay up another day, possibly two. NASA had ordered the ship home early, apparently because weather conditions for a KSC landing had deteriorated.

Sources reported the crew had released a spy satellite during the first 10 hours of the mission. The Air Force said only that the rocket carrying the Space Shuttle's cargo to higher orbit "successfully met its mission objectives" and would not confirm reports that the cargo was an advanced spy satellite, first of a new generation capable of intercepting radio, radio-telephone, and digital communications from ground and space, or the satellite's orbit parameters.

Four private planes had violated the security zone around KSC during preparations for Discovery's landing, the *Washington Post* reported the Air Force as saying, the 8th time in 15 Space Shuttle missions that planes had violated restricted airspace during launch or landing. The Federal Aviation Administration had suspended the flying licenses of some violators for up to 90 days because of dangers posed by intruding pilots. The problem had grown so serious that the Air Force had improved the security zone around KSC. (*W Post*, Jan 28/85, A1, A3, A9; *W Times*, Jan 28/85, 3A, Jan 29/85, 4A)

February 5: In a speech today to the Washington chapter of the National Security Industrial Association, NASA Administrator James Beggs addressed the issue of NASA's role in assisting the Department of Defense (DOD) in national defense. After touching on the media attention given the recent secret launch of a military payload, Beggs said the fuss "stemmed, I believe, from a misapprehension that the nation's civilian space program was about to be taken over by the military. As one news organization put it sternly: 'The line is being crossed.'

"Well, let me set the record straight," he continued. "Nothing could be further from the truth. On the last mission we did what we always try to do. And that is to satisfy our commercial customers, be they civilian or military."

Beggs went on to trace the history of U.S. military aeronautical developments and pointed out that many of the early astronauts were former military pilots and that almost 100 military detailees worked at NASA, mostly at Johnson Space Center (JSC). He also referred to other shared assets, including the KSC launch facilities, and to the aerospace technology that the military employed, practically all of which stemmed from a NASA-developed technology base.

Beggs concluded by saying NASA and DOD's Space Shuttle program cooperative efforts began in 1969 and that DOD had priority on the Space Shuttle manifest for missions of national security. "Our challenge is to continue to work together to secure our future, on earth and in space." (NASA Release 85–18)

March 29: The Teal Ruby experimental space-based aircraft detection system, which had encountered technical problems, would fly on the Space Shuttle around March 1986, *Defense Daily* reported. Defense Advanced Research Projects Agency (DARPA) Director Dr. Robert Cooper said DARPA had run into expected problems with the Teal Ruby flight sensor caused by heat leaks in the cryogenic system. The leaks posed two problems: they prevented the temperature of the instrument from getting low enough to provide necessary high sensitivity and threatened the one-year nominal life of the system. Cooper said the DARPA had found all leaks except one.

Teal Ruby was the first large-scale demonstration of a two-dimensional, staring mosaic infrared detector array and lightweight telescope optics. Its goal was to establish the technology base for future space-based infrared surveillance systems capable of detecting aircraft and other low threshold targets against the earth's clutter background; it would develop a comprehensive radiometric background database and space qualify the advanced technologies embodied in its design. Teal Ruby would be the primary payload of the Air Force space test program satellite P80–1. (*D/D*, Mar 29/85, 167)

June 21: The first test of a Strategic Defense Initiative (SDI) component on June 19 ended in failure when crew members of the Space Shuttle Discovery found themselves 180° out of position to receive an Air Force signal beamed at them from Maui, Hawaii; another beam of laser light was sent on June 21 from the same spot, and it successfully bounced off a reflecting mirror on Discovery, which was flying at 17,500 miles per hour and 230 miles high, the *Washington Post* reported.

It was not clear why Discovery was out of position for the first attempt or why the crew had not realized it in time to correct it. Flight directors at Johnson Space Center took the blame for what the newspaper called easily one of the worst navigating mistakes in more than 20 years of American manned space flight.

After the successful attempt, the Air Force noted the test was not intended to solve any of the most difficult problems of setting up an SDI defense system, nor was it the first laser tracking test. Air Force Lt. Col. Thomas Meyer, manager of the laser test program, said the experiment was more a test to see how the atmosphere distorted the laser beam than a test of the beam's ability to track the Space Shuttle, as atmospheric distortion could limit the usefulness of ground-based lasers against targets in space.

In the June 21 attempt, the laser beam "painted" a blue-green light on the nose of Discovery for at least two-and-one-half minutes, three times longer than the minimum time set by the Air Force. So tight and steady was the beam that it never wavered from the nose of orbiter, which measured 110 feet from nose to tail. "We were able to pulse the beam and change the size of the beam from a fine point a quarter of an inch across to a beam 30 feet across at the point where it tracked the shuttle," Meyer said. "We did everything we wanted to do."

Discovery's crew observed from the port windows the beam shining on them, which they could do without danger to their eyes due to the low power output of the four-watt beacon.

"We demonstrated today that we can track a fast-moving target with a laser on the ground," Meyer said. "Our next step is to perform the same kind of test with rockets fired to an altitude of 260 miles to see if ground-based lasers can stay with them all the way to altitude."

A far more difficult and important goal than tracking objects with lasers was the Pentagon's plan to develop more powerful lasers as weapons. The Defense Department would test a 2-million-watt laser soon at White Sands Missile Range, and it was building a 5-million-watt laser. (*W Post*, June 20/85, A3, June 22/85, A3)

September 5: NASA and the U.S. Air Force announced today that the Space Shuttle Atlantis would take off October 3 on its maiden voyage to carry a secret payload into orbit, the *Washington Post* reported. On board Atlantis, the fourth and final Space Shuttle in NASA's fleet, would be commander Karol

Bobko, pilot Ronald Grabe, mission specialists David Hilmers and Robert Stewart, and payload specialist Air Force Maj. William Pailes.

Pailes, who has a master's degree in computer science and is an accomplished pilot, would be on board to tend the classified cargo. (*W Post*, Sept 6/85, A12)

September 9: Astronauts aboard the Space Shuttle Atlantis, scheduled October 3 for its maiden voyage, would deploy two defense satellite communications system (DSCS–III) satellites to support all three military services and to carry a single channel transponder used to transmit emergency messages from the president to nuclear forces, the *Washington Post* reported the Federation of American Scientists said on September 9.

The group said the classified mission could be determined by anyone including the Soviets, who carefully scrutinized public documents, and that it was revealing the payload because keeping the mission secret "inhibits public awareness and discussion of U.S. military activities in space."

The group indicated it found reference to the satellites in various military publications, the most recent of which was a December 28, 1984, issue of *Defense Daily* that said two DCSCs would "be launched together next year on a single space shuttle mission, apparently on the Atlantis mission . . ." (*W Post*, Sept 10/85, A5)

October 4: The military crew of the Space Shuttle Atlantis on mission 51–J deployed today two communications satellites for Department of Defense, the *NY Times* reported, citing reliable sources. After release from the Space Shuttle, the satellites, known as Defense Satellite Communications System (DSCS III) and weighing a ton each and measuring 38 feet long with solar panels extended, rocket engines boosted them to a point 22,300 miles above the earth.

According to General Electric Co., which built the satellites but would not confirm their presence on the Space Shuttle, both satellites were strengthened to withstand radiation from distant nuclear blasts and had a device to send emergency action messages from the President to the nation's nuclear forces.

The Department of Defense (DOD) scheduled Atlantis to land at Edwards Air Force Base on a date that was classified; NASA and DOD said they would announce the time of landing 24 hours in advance. (*NY Times*, Oct 5/85, A46)

October 6: As five military officers on the Space Shuttle Atlantis orbited earth on the secret 51–J mission, a debate was growing in the U.S. over how much secrecy was necessary in the civilian-run space program and what the military's role in it should be, the *NY Times* reported.

Proponents of an expanded role for the Defense Department (DOD) said the U.S. needed to counter a growing Soviet threat in space. They pointed out that the USSR launched four to five times as many spacecraft a year as the U.S., with the majority of missions devoted to military objectives, and that Soviet military officers had logged years in space, while the U.S. military had logged days.

Critics of an expanded DOD role said the U.S. already had an advantage in military space technology. They said U.S. systems worked better and lasted longer, pointing out the USSR was still trying to perfect a vehicle similar to the Space Shuttle. The critics also argued that the U.S. military's assertions about the Soviet Union were often veiled excuses to try to edge civilians out of the astronaut corps and to classify the most mundane Space Shuttle payloads. This goal of secrecy, they said, was not heightened security but was intended to protect DOD's plans and programs from public scrutiny.

The debate was likely to become more vocal, the *Times* said, as the military expanded its manned space activity when secret military missions would account for 25 to 30% of all Space Shuttle flights in the next decade.

Carl Sagan, professor of astronomy and space sciences at Cornell University, said there was a "fundamental tension" between open scientific inquiry and "the necessarily closed world of military activities . . . As military programs expand," he said, "there's a huge deflection of resources, financial and intellectual, from peaceful uses into the production of weapons."

Daniel Graham, a retired Army lieutenant general who formerly headed the Defense Intelligence Agency, disagreed, saying the military had been at the forefront of U.S. space exploration since its earliest days. "The myth since the Eisenhower Administration is that there's a distinction between military and civilian matters in space," he said. "That's a pretense that a lot of people in NASA would like to believe, that all their activities are sweetness and light," he insisted.

However, even military proponents saw emerging tensions in the manned space program, the *Times* reported. William Gregory, editor of *Aviation Week & Technology* said in an editorial, "The original legislation creating [NASA] specified a civilian space program, separate from the military . . . That line is being crossed now not so much as a formal policy change as out of simple economic necessity. The shuttle needs the military as a customer to spread the system's overhead costs."

Some critics feared that NASA's charter for free and open dissemination of scientific information was eroding as the military's role grew. Scientists ineffectually protested last year when NASA announced that its Defense Department Affairs Division would review and possibly censor images from a large camera and radar carried on a civilian Space Shuttle mission.

"There's a real question of how the decision was made to move the space agency from being open to substantially closed," the *Times* quoted Morton Halperin, director of the Washington office of the American Civil Liberties

Union. "What's at stake is the public's right to participate in the process of making government policy." (*NY Times*, Oct 6/85, A1)

October 7: The Space Shuttle Atlantis on mission 51-J landed just after 10 a.m. today at Edwards Air Force Base, ending a four-day flight operated under secrecy rules imposed by the Department of Defense (DOD), the *Washington Post* reported. "Of course, I can't say anything about our mission," said flight commander Karol Bobko, an Air Force colonel and one of five military men on the mission, "but I can say Atlantis performed superbly its first time in space. I've flown Challenger, Atlantis, and Discovery now, and I think that NASA really has quite a fleet of orbiters and that we have a great national asset here," he added.

NASA ground operations manager Fritz Widick said Atlantis was in "excellent condition" at the end of the 1.7 million-mile flight, although engineers were inspecting an underside area of the left wing to see if tile-insulation damage might have exposed its aluminum mainframe to reentry heat. Such heat might have discolored a region along the portside engine pod under the Space Shuttle's tail.

Widick said Atlantis's brakes came through the 190 mph landing on the lake-bed with only "minor damage" to one of four brake assemblies on the main landing gear.

Bobko appeared to apply minimum braking after landing into a stiff headwind that brought Atlantis to a stop on the center line halfway down the 15,000-foot desert runway.

Fewer than a dozen reporters and about two dozen NASA employees were present for the landing, an event that usually drew hundreds of spectators, the *NY Times* reported. The military gave just 24 hours notice of when the mission would end and had barred members of the public from watching. (*W Post*, Oct 8/85, A3; *NYT*, Oct 8/85, C3)

October 10: The U.S. Air Force announced today the military cargo that would fly aboard the first Space Shuttle flight launched from Vandenberg Air Force Base, citing as its reason public interest in the mission, the *NY Times* reported. The move was considered a major step toward making available more information about the entire flight, including views of the poles as seen during the Space Shuttle's polar orbit.

Air Force Maj. Ronald Rand said the Department of Defense (DOD) approved dissemination of information about the payload because it was experimental and not designed to carry out an operational mission. He said there were two payloads: an assembly of six space physics experiments that would remain in the Space Shuttle's open cargo bay and an experimental airplane-detecting experiment.

"In light of the mounting Congressional and public interest in this historic first manned polar orbit and first west coast shuttle mission, we've made an exception to our policy with the understanding that we'll continue to protect performance data and results of these two experiments," Rand said. "This doesn't mean we've changed our policy of classifying DOD payloads."

Flying aboard Discovery, which was scheduled for launch March 20, 1986, would be Comdr. Robert Crippen of the Navy; pilot Guy Gardner; astronauts Jerry Ross, Dale Gardner, and Richard Mullane; Under Secretary of the Air Force Edward Aldridge; Air Force Maj. John Brett Watterson; and a DOD specialist to tend the cargo. (*NYT*, Oct 11/85, A6)

Missions

March 5: NASA confirmed successful launch on January 24 from KSC of the Space Shuttle with a Department of Defense inertial upper stage deployed and an aggregation of red blood cells experiment for the University of Sydney completed. Spacecraft orbital parameters were classified. The mission had lasted three days, one hour, and 33 minutes, with landing at KSC.

Unusually cold weather, causing icing on the external tank, had delayed launch for one day. (NASA FOR M-989-51-C [post flight] Mar 5/85)

April 12: NASA launched today at 8:59 a.m. from KSC the Space Shuttle Discovery mission 51-D, the 16th launch in the Space Shuttle program and the fourth flight of Discovery, the *NY Times* and *Washington Post* reported. Discovery carried a crew of seven, including Sen. Jake Garn (R-Utah), chairman of the Senate subcommittee responsible for the NASA budget.

The launch followed a last minute repair for a salt water leak on the McDonnell Douglas Corp. experiment, the continuous flow electrophoresis system to produce an unidentified hormone that couldn't be manufactured on earth. Also, launch came 55 minutes later than scheduled and 55 seconds short of postponement because of clouds between 14,000 and 30,000 feet above the launch pad. Astronaut John Young, flying a training plane through the clouds, encountered rain. KSC officials then delayed launch because they did not want to risk getting the spaceship wet, which might cause erosion of critical heat-shielding material. When skies lightened and dried at 8:50, NASA resumed the countdown with no further interruption.

After launch, Discovery reached an elliptical orbit ranging in altitude from 185 to 286 miles, one of the higher courses achieved by the spacecraft.

Discovery's crew besides Sen. Garn were Air Force Col. Karol Bobko, making his second Space Shuttle flight, commander; Navy Comdr. Donald Williams, pilot; Dr. M. Rhea Seddon, Jeffrey Hoffman, and Navy Capt. David Griggs, mission specialists; and McDonnell Douglas Corp. engineer Charles Walker, payload specialist.

The astronauts deployed during early evening of launch day the Canadian *Anik C-1* satellite, which then moved to a higher equatorial orbit. They would deploy on April 13 the Hughes Aircraft Co.'s *Leasat 3*, which the Department of Defense would lease and the Navy operate as part of a military communications network. The remainder of the mission would be devoted to medical experiments and a test of how mechanical toys behaved in space.

Dr. Seddon began the first of the mission's medical tests by taking echocardiograms of three of the crew members. There were no reports on Sen. Garn's medical tests. In an effort to determine the causes and effects of space motion sickness, the Senator wore a waist belt with two stethoscopic microphones to record sounds his stomach and intestines made during digestion. His head and chest were wired to record electrical signals from his brain and heart; other instruments would measure the way his bones grew and shrunk in zero gravity. Also, he would perform exercises aimed at inducing nausea. (*NY Times*, Apr 13/85, A1; *W Post*, Apr 12/85, A6, Apr 13/85, A1)

April 17: NASA terminated today efforts by the crew of the Discovery 51-D mission to revive the U.S. Navy's *Syncom IV-3* (Hughes Corp.'s *Leasat 3*) satellite, the *Washington Post* reported. Troubles with the satellite became apparent the morning of April 13 after astronauts M. Rhea Seddon and Jeffrey Hoffman completed commands that deployed the satellite from the Space Shuttle's cargo bay. The satellite's engine, which would lift it into a permanent geostationary orbit 22,300 miles above earth, failed to fire. A fault in the satellite's timing mechanism (which would direct the satellite through a series of steps to put the satellite into permanent orbit) apparently caused the failure.

Johnson Space Center flight director John Cox said that NASA's best guess was that the "post-deploy sequencer (timer) never was activated. The reason we think so is that the antenna should have popped out about a minute and 20 seconds after deployment, and the satellite should have been put into a 15-rpm spin to stabilize it. Neither of those things happened."

Engineers at Hughes Aircraft Corp., satellite manufacturer, tried to simulate the conditions of the satellite failure in hopes of duplicating what had happened. "One of the astronauts might be able to get close enough to the satellite to move the lever to start the timing mechanism, then get out of there and inside in enough time to be far away from the satellite when the engine fires," said Marvin Mixon, Hughes vice president.

Astronauts maneuvered the Space Shuttle about 45 miles ahead of the satellite in order to take pictures of it; however, they could not draw too close because the satellite's main engine contained 7,370 lb. of solid rocket fuel, its second-stage engine had 4,092 lb. of nitrogen tetroxide and dimethyl hydrazine fuel. The satellite also carried 352 lb. of hydrazine for controlling its position in orbit.

NASA made the decision to extend the flight from five to seven days, and on April 15 the Discovery crew cannibalized plastic covers from flight manuals, parts of a window screen, nylon straps, and other paraphernalia in the spacecraft to fashion flyswatters to attach to Discovery's robot arm for an attempt to fire the satellite's engine. Seddon sewed the plastic covers together to form cones that the astronauts would attach to the arm in order to flip the arming lever.

On April 16 astronauts Jeffrey Hoffman and S. David Griggs went on history's first unrehearsed spacewalk to rig Discovery's mechanical arm with the flyswatter devices. The two spent more than three hours in Discovery's open cargo bay putting the extension "hands" on the 50-foot-long arm.

At about 8:00 a.m. April 17, commander Karol Bobko and pilot David Williams manuevered Discovery to within 30 feet of the satellite. Then Seddon swatted at the four-in.-long power switch. Although she hit it at least twice—once so hard that part of the plastic broke—the lever didn't move.

Following the unsuccessful attempt, ground controllers order the Space Shuttle to move away and abandon the rescue attempt on the off-chance that the contact somehow might have activated the satellite's internal timer, which would cause its rockets to fire.

Underwriters who had insured the satellite said the failure to either recover the satellite or boost it into an operational orbit would be a serious blow to the satellite underwriters business, *Defense Daily* reported. "If we have another major claim," James Barrett, president of International Technology Underwriters, said, "the market would be very seriously injured by a total loss." Hughes said the satellite was insured for between $80 and $85 million, and the space insurance market had already suffered three losses over the past year. (*W Post*, Apr 15/85, A1, Apr 16/85, A1, Apr 17/85, A7, Apr 18/85, A1; *NYT*, Apr 16/85, A1, Apr 18/85, B10; *USA Today*, Apr 16/85, 3A; *W Times*, Apr 18/85, 1A; *D/D*, Apr 16/85, 257)

April 19: The Space Shuttle Discovery's crew on mission 51-D landed the orbiter at 8:54 a.m. EST April 19 at KSC, a return characterized by the worst damage in 16 Space Shuttle missions—a burn-hole in a wing, two failed brakes, two blown tires, and at least 123 broken heat-shield tiles, the Marshall Space Flight Center's *Star* and the *Washington Post* reported. NASA scheduled the mission to last five days, but extended it by two days to give the crew a chance to correct a malfunction in the Hughes *Leasat 3* satellite.

The damage to Discovery could set back preparation for its next mission scheduled for June 12. NASA would have to replace protective tiles, probably damaged when the Space Shuttle lifted off through the rain [see Apr. 12]. "There's little question that we suffered more severe damage on this landing than on any other so far," Space Shuttle launch director Robert Sieck said. Concerning the brakes, Sieck said, "Brake damage we've seen on previous flights. Brake failure is something we've never seen."

NASA had waved off Discovery for a 7:16 a.m. landing because of rain over Cape Canaveral. One orbit later, Discovery headed down to KSC, where NASA told Commander Karol Bobko to land on runway 33 instead of 15 because of sun glare. Bobko landed with no headwind to slow him, touching down at 231 mph, the fastest any Space Shuttle had landed.

As Discovery rolled down the three-mile concrete runway, it encountered a nine-knot crosswind and began veering to the right. When Bobko corrected for the wind by braking one side, the strain apparently caused one of the inboard brake assemblies to lock. Moments later the inboard brake assembly on the other side also locked, causing rubber to burn.

A burn-hole the size of a dinner plate nearly penetrated through aluminum on the Space Shuttle's left wingtip where a landing flap apparently dislodged several protective tiles.

Regarding the *Leasat 3,* a Hughes representative said initial belief that a lever failed to release was in error [see Apr. 18]. Once Discovery had maneuvered to within 10 feet of the satellite, the crew reported the lever had released. The crew successfully snagged the lever twice to ensure it was fully released, and reported the only change in the satellite's status was a slowing from two to one revolution per minute.

The first congressional observer in space, Sen. Jake Garn (R-Utah), had difficulty negotiating the steps down to the runway and required assistance into the van that took the crew to the dispensary for medical examinations. There had been some controversy over his presence on the flight, including a *NY Times* editorial that said, "Having pressured NASA to give him his space trip, Mr. Garn is now indebted to the agency. That's a pity for those whom he represents in supervising NASA's budget and its request to build an $8 billion space station." However, Garn commented that "I'm carrying my own weight. I knew I had to prove myself . . . I've dealt with NASA for 10 years. I'm working on the space station right now, setting policy." (*Marshall Star,* Apr 24/85, 1; *W Post,* Apr 11/85, B1, Apr 20/85, A1; *NYT,* Apr 20/85, 22)

April 29: NASA launched at 12:02 p.m. EDT today from KSC the Space Shuttle Challenger mission 51–B carrying seven astronauts, Spacelab 3, 24 rats, and two squirrel monkeys for the most intensive science mission to date, the *Washington Post* reported. Launched 17 days after Discovery left the same launch pad, the Challenger flight bested the previous "turnaround" record of 34 days.

Challenger carried the oldest crew ever to fly in space. Marine Col. Robert Overmyer, commander, was 49; Frederick Gregory, pilot, and Taylor Wang, payload specialist, were 44; Don Lind, mission specialist, was 54; Lodewijk van den Berg, payload specialist, was 53; William Thornton, mission specialist, was 55; and Norman Thagard, mission specialist, 41. The crew called the flight one by scientists for science. "This mission marks the first time that scientists who designed their own experiments will be executing those experiments in space," said Lind, who had been waiting 16 years for the flight. The five scientists would work in around-the-clock shifts on the 15 experiments inside the European-built Spacelab.

Before liftoff NASA decided not to have the crew launch two Get-Away Special satellites, one a student experiment and the other a Pentago payload, because a nine-volt battery had failed on the last Space Shuttle flight. Ground tests showed that three other batteries from the same batch failed after 18 hours in vacuum. The satellites would return to the ground with the Space Shuttle.

The crew successfully deployed a satellite intended to calibrate air traffic control radars on the ground, but encountered difficulties when trying to deploy a U.S. Navy satellite that would locate drifting weather buoys. The satellite would not move from the cargo bay despite repeated signals to trigger deployment.

Later the crew discovered they had no fresh water from their galley faucet and had to bypass the faucet and reconnect themselves to their water supply using an all-purpose hose carried for such emergencies. Other problems ranged from an overheated hydraulic system power unit to a urine collection device that one crew member said sprayed water "all over the place."

The monkeys and rats appeared to be doing well. Four of the rats had surgical implants in their hearts to record changes in heartbeat and blood flow. The others were testing cages and equipment for future animal experiments. The monkeys were on a shakedown flight to determine how they tolerated living in orbit. If they didn't get nervous or frightened in space, later flights would carry squirrel monkeys with surgical implants to test for space sickness and heart changes. (*W Post,* Apr 28/85, A18, Apr 29/85, A3, Apr 30/85, A7)

May 2: The waste products of two monkeys and two dozen rats continued to float through the cabin of the Space Shuttle Challenger [see Space Transportation System/Missions, Apr. 29], forcing crew members to rearrange their tight schedules and operate in full surgical gear to clean up, the *Washington Post* reported. "Be advised we now have feces in the crew compartment and it isn't much fun, guys," commander Robert Overmyer said to the mission control center in Houston. "How many years did we tell them these cages would never work?"

With TV cameras recording the activity, Overmyer and mission specialists William Thornton, Norman Thagard, and payload specialist Lodewijk van den Berg, wearing surgical smocks, gloves, and masks, floated about the Spacelab cabin using vacuum cleaners to suck the waste out of the cabin air. At one pont Thagard said, "Even the vacuum cleaners aren't enough."

The problems began on the day after liftoff with feeding the animals food and water. One rat had to be hand-fed a gelatin bar to get water into him;

when physician-astronaut Thornton tried to press food bars into the cage, the bars would crumble and a cloud of tiny particles would scatter in the cabin. "I'm not exaggerating," Thornton told mission control, "but there are food particles flooding out of every crack in those cages. I don't see any way we can stop this except if we had a seal over the entire cage." Monkey and rat feces also leaked out of the cages.

The five scientists also could not deploy a French-built camera, intended to survey for 17 hours hot stars in distant galaxies, because an airlock hatch failed to open. And physicist Taylor Wang was unable to run an experiment to test behavior of drops of fluids in weightlessness because circuit breakers continually popped open. Nine of the 15 Spacelab experiments were running even better than hoped, and the crew described the monkeys and rats as "real clean and real happy."

On May 1, however, one of the monkeys got spacesick. "We can tell by the way he's behaving that one of our monkeys is not feeling well," said Dr. Paul Callahan of Ames Research Center. "The other monkey was under the weather his first day in space but he's adapted very well since, which is an almost identical reaction we get from human astronauts." Callahan noted "Monkey No. 1" was not eating and drinking normally, seemed to have a headache, and was generally lethargic and dispirited. "He's just not moving around and the other monkey has begun to do somersaults."

Eleven of the Spacelab experiments were running by that time, although mission specialist George Fichtl said "One is a hit-and-miss kind of thing and three look very doubtful." He added the astronauts had proposed to give up on the French-built wide-field camera and that it was doubtful NASA would extend the mission to give the crew additional time to try to deploy it.

On May 2 Fichtl explained that the rats' and one monkey's vigorous movement, which was "much more than expected," was likely the main reason their waste was escaping the cages. "We designed those cages with an airflow control to keep the waste in the cages," he explained. "Our best guess now is that the animals are so spirited and are enjoying weightlessness so much that they induced turbulence in the cage that's too turbulent to contain the waste."

The problems with the animals cast some doubt over future animal flights on Spacelab, the *Washington Post* continued. NASA had scheduled for 1986 a flight to carry 48 rats and four squirrel monkeys. However, Spacelab mission manager Joseph Cremin insisted that research on animals was crucial to the future of the permanent space station where men and women would have to work in orbit for months at a time. (*W Post*, May 1/85, A3, May 2/85, A11, May 3/85, A3)

May 6: Commander Robert Overmyer landed the Space Shuttle Challenger today on a lake-bed runway at Edwards Air Force Base, after it had caused two sonic booms that triggered burglar alarms and calls to police, the *Washington Post* reported.

Of the 15 experiments flown on Spacelab 3, 12 apparently operated successfully and two at least partially so. Among the successful experiments was photography of the Northern and Southern lights, which sent streaks of brilliant flashes at the extreme latitudes near the north and south polar regions at that time of year. Challenger's crew photographed 18 auroras. Another successful device was a laser spectrometer that for the first time measured from orbit the ozone layer that protected the earth from the sun's ultraviolet light.

Crew members were able to restore two experiments that had initially been inoperable. One was an experiment to study the dynamics of droplets of fluids in weightlessness, the other an instrument that counted and measured cosmic rays striking Challenger in orbit. They were unable to deploy the wide-field camera intended to study the ultraviolet light of hot stars.

The afternoon of the landing NASA personnel removed the monkeys and rats in their cages for a flight to Kennedy Space Center, where researchers would kill and dissect the rats to examine their vital organs under a microscope for changes caused by weightlessness.

Physician/astronaut William Thornton said they had brought back two monkeys that were even friendlier than they were before the flight—"Those primates are part of the crew right now," he commented. During the flight, one of the monkeys experienced motion sickness for several days, requiring Thornton to feed it by hand. On May 4 the *Greensboro News & Record* reported that Thornton said, "Our feeding crisis is over. I wouldn't have believed the effect of a caring human hand."

However, on May 14 Overmyer told reporters at Johnson Space Center that, "NASA has a problem that NASA has to solve if we're going to fly those cages again," the *W Post* reported. "I never dreamed that all that stuff would come out of those cages and escape into our atmosphere" [see May 2]. He added that the experience with the animals "never took away the luster" of spaceflight, a sentiment echoed by pilot Frederick Gregory, an Air Force colonel who became the first black astronaut to take the Space Shuttle's controls. (*W Post*, May 5/85, A4, May 6/85, A3, May 7/85, A1, May 14/85, A3; *Greensboro News & Record*, May 4/85, A4)

June 17: NASA launched today from KSC Space Shuttle Discovery mission 51–G with a crew of seven, including Saudi Prince Sultan Salman A1–Saud and Patrick Baudry from France, for the 18th Space Shuttle flight, the *Washington Post* reported.

Earlier NASA had announced that mission 51–G would carry the automated directional solidification furnace (ADSF), Spartan 1, and the French echocardiograph (FEE) and posture (FPE) experiments. The objective of the ADSF was to provide a technology demonstration of the capability of the equipment to perform directional solidification of a magnetic composite material and to study the role of gravity-driven phenomena in such processing. The apparatus used for the directional solidification experiments was a specially designed furnace that melted a sample contained in a quartz tube one cm in diameter and 35 cm long. A furnace module (a heater with an integral quench block) moved along the four tubes in the experiment so that the manganese-bismuth alloy samples melted and solidified at a controlled rate. Electric resistance provided the heat; the quench block was liquid cooled. In that first flight, all four samples were the same alloy with only the solidification rate changed to allow a better understanding of that aspect of the process.

Spartan 1 was the first of a continuing series of low-cost free flyers designed to extend sounding rocket experiment capabilities. The Spartan 1 objective was to map the X-ray emissions from the Perseus Cluster, the nuclear region of the Milky Way galaxy, and the Scorpius X-2 (SCO-X-2). NASA had flown the Spartan 1 instrument several times on sounding rockets; the resulting information led to the need to obtain greater viewing time of the selected targets, thus permitting more detailed studies with greater resolutions. The crew would deploy the Spartan 1 from the Space Shuttle and retrieve it using the Canadian-built remote manipulator system. After deployment, the Spartan 1 would perform scientific observations for up to 40 hours via an onboard microcomputer controller that commanded all pointing sequences and satellite control. There was no command or telemetry link; when observing sequences were complete, the satellite "safed" all systems and placed itself in a stable attitude for retrieval.

The objectives of the FEE and FPE, a part of a cooperative program with France, was to obtain on-orbit data regarding the response of the cardiovascular and sensorimotor systems to weightlessness. French payload specialist Patrick Baudry would perform the experiments with participation from other crew members including the Arabsat payload specialist. The USSR's July 1982 Salyut mission had also flown the FEE system.

By 4:30 p.m. June 17, Discovery's crew had successfully deployed a Mexican satellite, *Morelos-1*, the first of two identical satellites, which headed for a position 22,300 miles above the equator west of the Galapagos Islands. It would beam TV programs to the most remote regions of Mexico. (NASA Flight Operation Report M-980-51-G [prelaunch] June 5/85, NASA MOR E-420-51-G-16 [prelaunch] June 7/85, MOR E-420-51-G-17 [prelaunch] June 5/85, MOR E-420-51-G-18 [prelaunch] June 13/85, MOR-989-51-G [postflight] July 3/85; *W Post,* June 18/85, A3)

During June: NASA announced that the Space Shuttle mission 51-G would carry three communications satellites, a deployable/retrievable Spartan 1 spacecraft, and six Get Away Special canisters. The orbiter Discovery would fly several mid-deck experiments including one for the Strategic Defense Initiative (SDI) organization and a materials processing furnace.

The Mexican spacecraft, *Morelos–A,* was a version of the Hughes HS 376 satellite, a number of which had already been deployed from the Space Shuttle. *Morelos–A* was the first of two domestic communications satellites that would provide advanced telecommunications to the more remote parts of Mexico.

Arabsat–A, owned by the Arab Satellite Communications Organization and built by Aerospatiale, would provide telecommunications links between the member nations.

A U.S. domestic communications satellite owned by American Telephone and Telegraph (AT&T), *Telestar 3–D* would provide telecommunications services to the continental U.S., Alaska, Hawaii, and Puerto Rico.

All three spacecraft would first spring-eject from the orbiter's cargo bay, then on each a small attached payload assist module would ignite about 45 minutes after deployment to propel the satellites to transfer orbits out as far as 22,300 miles above earth. A second small rocket motor would then fire on each spacecraft to circularize the orbits, placing the spacecraft in geosynchronous orbits.

Three of the six Get Away Special canisters contained West German payloads for study of materials processing in space and behavior of liquid propellants. A fourth canister contained a U.S. Air Force/U.S. Naval Research Laboratory investigation of the ultraviolet radiation environment in space. A fifth canister had a package of nine student experiments in biological and physical sciences. A Goddard Space Flight Center (GSFC) investigation of a developmental heat transfer system was in the sixth canister. A capillary pumped loop (CPL) experiment, it was the first flight of a CPL two-phase thermal control system. The advanced development and flight experiments section of the thermal engineering branch at GSFC developed the CPL (NASA Release 85–83)

July 29: NASA launched from KSC at 5:00 p.m. EDT today Space Shuttle mission 51–F with the orbiter Challenger carrying Spacelab–2. The flight was the eighth for Challenger and the 19th Space Shuttle mission. The flight commander was Col. C. Gordon Fullerton, who served as pilot on the third Space Shuttle mission. Air Force Col. Roy Bridges, Jr. was pilot; mission specialists were Dr. Anthony England, a geophysicist; Dr. F. Story Musgrave, a physician; and Dr. Karl Henize, an astronomer who at the age of 58 was the oldest American to travel in space. Payload specialists were Dr. Loren Acton, a solar physicist at the Lockheed Palo Alto Research Laboratory, and Dr. John David Bartoe, an astrophysicist at the Naval Research Laboratory.

Approximately 5 minutes and 45 seconds after launch onboard computers shut down Challenger's center engine (SSME 1) due to failure of the orbiter's two high-pressure, fuel turbo-pump discharge temperature sensors. Challenger was 70 miles above earth at the time, about 50 miles below the orbit considered safe for a manned spacecraft.

The two remaining main engines ran smoothly and burned about a minute and 10 seconds longer than originally planned, putting the Space Shuttle into a lower-than-scheduled orbit.

The crew then used two auxiliary engines to raise the orbit in steps from an initial 122 miles to 194 miles at 10:30 p.m. the night of launch. The crew also dumped as much as 4,400 lb. of fuel to achieve the higher orbit.

In explaining the problem with SSME 1, NASA associate administrator Jesse Moore said the onboard sensors first indicated that the center engine might be overheating four minutes after liftoff, the *Washington Post* reported. "Computers then ordered the center engine fuel pump to bypass one valve and use another to feed fuel into the combustion chamber," he said. Two minutes later computers again sensed that the fuel pump was overheating and automatically shut down the center engine. The mission control center in Houston also received indications of dangerously high temperature readings on a second engine and instructed the crew to disconnect a backup sensor to prevent the second engine from shutting down before the spacecraft reached orbit.

The crew encountered the next problem when attempting to operate a $60 million telescope-pointing device. The instrument pointing subsystem (IPS) and its four solor telescopes were one of 13 experiments carried on the European-built Spacelab–2. Mission specialist Henize told mission control that the pointing attempts were "rather dismal," the *NY Times* reported.

Lee Briscoe, flight director, elaborated to reporters. "It appears that we are able to find the sun, find the stars, get into what we call a rough track mode," he said. "But we never appear to get into a fine track and actually finish the total tracking."

Since the problem appeared to be with the computer programming that drove the unit, NASA radioed up a new version of the programming. However, it initially failed to correct the problem, and NASA told the crew to try harder. Eventually the IPS operated successfully after the crew inserted a series of software patches developed at Marshall Space Flight Center. NASA extended the mission duration by one day to allow additional collection of engineering and scientific data.

The Spacelab–2 aboard Challenger was the second of its two verification test flights and consisted of an igloo attached to a lead pallet, the IPS mounted on it and a two pallet train behind, with an experiment special support structure. Experiments conducted during the Spacelab–2 mission were in the fields of life sciences, plasma physics, infrared astronomy, high-energy physics, solar physics, atmospheric physics, and technology. Experiments were located on the IPS, the three pallets, the special support structure, the orbiter mid-deck, and one on the ground (NASA FOR M–989–51–F [postflight] Sept 27/85, [prelaunch] July 9/85; NASA MOR M–977–51–F–03 [prelaunch] July 1/85; *W Post,* July 30/85, A1, July 31/85, A3; *NYT,* July 31/85, B4)

August 6: The orbiter Challenger on Space Shuttle mission 51–F landed at 12:46 p.m. PDT today on Edwards Air Force Base's Mojave Desert lake-bed runway, the *Washington Post* reported. During the flight, Challenger circled the earth 127 times, a distance of a little over three million miles.

Technicians stood by at the runway to remove immediately the heat sensors from Challenger's main engines to check a NASA theory that the sensors caused the premature shutdown of the center rocket engine during the July 29 launch. NASA was confident enough that the sensors would be found at fault that it had equipped Discovery's engines with redesigned sensors before moving it August 5 to the launch pad.

During the mission, Challenger became an orbiting astronomical observatory with more than $72 million of telescopes and other instruments to study the sun, stars, and distant galaxies. The *NY Times* reported that Dr. Burton Edelson, NASA associate administrator, said, "We met more than 80 percent of all science objectives of this mission. We're absolutely delighted."

Dr. Eugene Urban, chief mission scientist, added, "Everyone has collected tantalizing new data. It's going to take a long time before this data is analyzed and really fully appreciated. We've made some interesting new observations, and some have been very spectacular."

Jesse Moore, NASA associate administrator, commented that the orbiter appeared to be in good shape after the landing, the eighth for the Challenger. "It was a beautiful landing," he said. "The tiles looked very good. The brakes look excellent. The orbiters continue to be magnificent flying machines." (*W Post,* Aug 7/85, A3; *NYT,* Aug 7/85, A1)

August 27: NASA launched Discovery on Space Shuttle mission 51–I today through the worst weather of the Space Shuttle program, the *NY Times* reported. After two postponements in three days because of bad weather and a computer failure, Discovery barely got off the ground before the darkening clouds moved over the launch pad and the rain descended in torrents.

"We could see large holes in the system 50 or 100 miles across so we prayed for the breaks that would let us thread the needle when the time came," launch director Robert Sieck said later. "We had two things in our favor: There was no lightning in the cloud cover, and most of the rain was south of the launch pad, not right over it."

Launch directors gambled that rain would not be falling on the pad at liftoff time, which was delayed three minutes to make sure the gamble was the right one. As heavy rain fell on the press site three miles to the west, Discovery roared away from the pad through a hole in the clouds and sped into the air 10 minutes before the pad was pelted with rain.

Commanding the flight was Air Force Col. Joe Engle, who commanded the second Space Shuttle test flight in 1981 and on August 26 turned 53. Air Force Lt. Col. Richard Covey was pilot; mission specialists were John Lounge, Dr. James van Hoften, and Dr. William Fisher.

A malfunction encountered two hours into the flight forced the astronauts to deploy a day early one of the three satellites in the orbiter's cargo bay. During a cargo inspection by remotely controlled cameras, a shield that was to protect Australia's *AUSSAT 1* from direct exposure to sunlight stuck in a partly open position. An astronaut then used the Space Shuttle's mechanical arm to nudge the sunshade completely open so they could release the spring-loaded satellite. Any delay of more than six hours could have resulted in radiation damage to the exposed satellite's electronics.

Less than five hours after deploying the Australian satellite, Discovery's crew deployed *ASC 1* for the American Satellite Co. The second deployment marked the first time a Space Shuttle crew deployed two satellites in one day. (*NYT*, Aug 28/85, D18)

August 28: Already far ahead of their work schedule, astronauts aboard Discovery on Space Shuttle mission 51-I began preparing tools they would use in an attempt to revive the U.S. Navy *Leasat 3* satellite, the *Washington Post* reported.

Mission planners had hoped the satellite repairs could be made during an August 30 spacewalk, but they added another spacewalk for August 31 because ground tests showed that the Space Shuttle crippled robot arm could not work fast enough to complete the repair job in one day.

"It appears we are faced with a two-EVA (extravehicular activity) plan," flight director Bill Reeves said. He blamed a failed circuit that prevented Discovery's robot arm from operating in an automatic mode. Reeves noted that ground tests showed that the robot arm, when operated manually, would need about 75% more time to maneuver the *Syncom 3* satellite and that it would take two 6-and-a-half hour spacewalks to complete the repair.

Mission specialists Dr. James van Hoften and Dr. William Fisher spent about four hours August 28 checking out electronic gear they would use to repair the satellite.

Mission specialist John Lounge would work the robot arm from inside the cockpit, while van Hoften and Fisher worked outside on the satellite. The robot arm would play a crucial role, holding van Hoften in place while he worked on the satellite 35 feet above the cargo bay.

Lounge would also use the robot arm to help van Hoften turn the satellite so Fisher, standing in foot restraints inside the bay, could remove a panel and disengage a timing lever, then plug two electrical cables into the panel and, in effect, jump start the satellite. (*W Post*, Aug 29/85, A3)

August 29: Astronauts aboard the Space Shuttle Discovery on mission 51-I took time August 29 to observe Hurricane Elena, the *Washington Post* reported. Commander Joe Engle took photos of the storm, a dense white circular mass that stretched across almost half of the Gulf of Mexico, and noted a second storm in the western part of Gulf.

Among the contingencies NASA faced was the possibility that the storm could strike the Houston area and knock out communications with the Johnson Space Center. A NASA spokesman said they could quickly switch operation to the Goddard Space Flight Center.

The crew also that day launched the *Syncom IV-4* satellite, which was almost identical to the satellite the crew would attempt August 30 to salvage. (*W Post*, Aug 30/85, A3)

September 3: Discovery and its five-man crew on Space Shuttle mission 51-I landed at 9:16 EDT today at Edwards Air Force Base after a seven-day mission [see Aug. 27], the *Washington Post* reported. During the mission, the crew deployed *AUSSAT-1* for the Australian government, *ASC-1* for the American Satellite Corp., and *Leasat 4* for Hughes Communications (reports indicated *Leasat 4* had lost its UHF communications link). In addition, the crew successfully retrieved and repaired the *Leasat 3* originally deployed on STS 51-D.

Later Hughes issued a statement saying "*Leasat 3* is under full control by Hughes' ground command, and telemetry data continued to confirm the good health of the satellite. The liquid propulsion systems are intact and the solid propellant perigee kick motor temperatures appear to be rising gradually toward acceptable levels," the JSC *Roundup* reported.

Jesse Moore, NASA associate administrator for space flight, said after the mission that "I would have to characterize this mission as near to perfect as you can get. It was a perfect mission from the outset, one that shows America's space program at work."

Moore added that he hoped there would be only two more Space Shuttle landings in California before pilots could resume landing in Florida. NASA switched to Edwards Air Force Base landings after brakes failed and a tire blew out in an April landing on Kennedy Space Center's concrete runway. (*W Post*, Sept 4/85, A7; JSC *Roundup*, Sept 13/85, 1)

October 3: NASA launched at 11:15 a.m. today from KSC the Space Shuttle Atlantis, the maiden voyage of the orbiter and the 21st mission in the Space Shuttle program, on mission 51-J, a secret flight for the Department of Defense (DOD), the *NY Times* reported. The secrecy was part of DOD's efforts to make it difficult for the Soviet military to monitor the Space Shuttle takeoff and find out the identity and mission of the payload. Despite these precautions, it was widely reported that a primary goal of the flight was to launch a pair of $100-million military communication satellites. DOD apparently planned to station the satellites, designated DSCS III for Defense Satellite Communications System, 22,300 miles above earth, where they would relay messages to U.S. military forces around the world.

The commander of Atlantis was Air Force Col. Karol Bobko, flying his third Space Shuttle mission; the pilot was Air Force Lt. Col. Ronald Grabe. Other crew members were Maj. David Hilmers of the Marine Corps, Army Lt. Col. Robert Stewart, and Air Force Maj. William Pailes. The first four were members of NASA's astronaut corps; Pailes was an Air Force pilot assigned to the mission to assist in deploying its secret payload.

In accordance with security rules, NASA gave out little information about the flight. There were no space-to-ground communications released to the public and no postlaunch news conference. Nearly five hours after the launch, mission control in Houston said, "On the maiden voyage of Atlantis, the crew is doing well and all systems on board the orbiter are performing satisfactorily." NASA officials said it would issue no other status reports during the mission, which might last a week or more.

Atlantis was the lightest and most advanced of the four orbiters. In its planning and construction, engineers removed minor design flaws found in earlier versions and incorporated the latest in electronic technology and light-weight, high-strength structural materials. NASA strengthened Atlantis mainly so it could withstand more vigorous launchings from the Air Force's facility at Vandenberg Air Force Base. These launches, which would take the Space Shuttle into orbits around the earth's poles, required greater thrust from the Space Shuttle's engines.

Although the mission was devoted to the military, NASA officials said there was a civilian experiment on board to study exposure of the orbiter and its crew to cosmic rays, which were extremely high-energy particles from space that bombarded the earth and the orbiting Space Shuttles. (*NY Times*, Oct 3/85, A18, Oct 4/85, B5)

October 30: NASA launched at noon today from KSC the Space Shuttle Challenger on mission 61–A, a planned seven-day flight carrying Spacelab D–1 (for Deutschland) and the largest Space Shuttle crew ever, the *NY Times* reported. The West German Aerospace Research Establishment paid NASA $65 million to have the 22nd Space Shuttle and ninth Challenger flight devoted fully to the Spacelab experiments.

Americans on board were commander Henry Hartsfield, Jr.; Air Force Lt. Col. Steven Nagel, pilot; and mission specialists Dr. Bonnie Dunbar, a biomedical engineer, Marine Corps Col. James Buchli, and Air Force Col. Guion Bluford, Jr., who had a doctorate in aerospace engineering. The two West Germans in the crew were Dr. Ernst Messerschmid and Mr. Reinhard Furrer, both physicists. The eighth crew member, Dr. Wubbo Ockels, a physicist from the Netherlands, represented the European Space Agency (ESA), which provided several scientific instruments in the Spacelab. The Europeans had prime responsibility for conducting the experiments; Bluford and Dunbar would assist them.

Spacelab D-1, designed, built, and checked out in Europe, was flown to KSC for installation in Challenger's cargo bay. The West German space agency's center in Oberphaffenhofen, near Munich, controlled scientific operations during the flight; ground control for flying the Space Shuttle was in Houston, as usual. Spacelab D-1 carried 76 experiments on racks that lined its walls. German researchers designed most of the experiments, which were for German and other foreign universities, research institutes, and industrial enterprises as well as ESA and NASA; other experiments were from France, Spain, Italy, Britain, the Netherlands, Belgium, Switzerland, and the U.S. All experiments were intended to take advantage of the weightlessness of space to study various processes that were normally affected by gravity. The *Washington Post* reported that West German officials said Spacelab was a major step toward what they hoped would be a larger laboratory module attached to the permanently manned space station that NASA envisioned for the 1990s.

Most experiments were in materials processing. Along two sides of the Spacelab, the experiments were in racks containing small furnaces for heating metals and making alloys that were lighter and stronger than anything produced in full gravity of earth. Other materials processing experiments would study the flow of liquids in weightlessness; yet others would grow very pure and large crystals that could lead to improvements in electronic semiconductors.

In a variety of life sciences experiments, the crew would study the growth of plants (corn and lentils) under microgravity conditions. Other studies were on how embryos of frogs and insects developed in weightlessness. Crew would use tadpoles for tests on how weightlessness might affect the development of vestibular, or balance, organs in vertebrates.

ESA's vestibular sled ran down a pair of rails in Spacelab's center aisle. Experimenters would accelerate and tilt crew members riding in the chair-like device to test how the human balance system in the inner ear responded to weightlessness and a moderate gravity force, because it was thought the inner ear, which relied on gravity to function normally, might be the source of astronauts' motion sickness.

Crew would use instruments in the cargo bay outside the Spacelab for navigation tests, such as synchronization of atomic clocks in space with those on the ground, checking for any discrepancies caused by weightlessness, and precise distance measurements by radio signals between the Space Shuttle and the ground.

To accommodate the crew of eight, NASA installed an extra sleep station, bringing the total to four. Crew would conduct laboratory work in shifts around the clock.

Despite early concern over the condition of the Space Shuttle's steering jets, mission control gave the Challenger the go-ahead for orbital operations. Later the mission control flight director said a valve malfunction prevented the flow of fuel through one line to Challenger's steering jets, but NASA considered an alternate line sufficient for regular operation to maneuver Challenger.

Flight controllers also reported erratic temperature readings on one of the three fuel cells that generated electricity, but with crew supervision the affected unit was operating at full strength. (NASA Release 85–145; NASA MOR M–989–61–A [prelaunch] Oct 28/85; *NY Times*, Oct 31/85, B13; *W Post*, Oct 31/85, A30)

November 2: At the halfway point in Space Shuttle mission 61–A in which the orbiter Challenger carried Spacelab D–1 [see Space Transportation System/Missions, Oct. 30], a slow air leak in one of Spacelab's vacuum chambers developed, but flight controllers in Houston did not consider it serious, the *NY Times* reported. Terry White, a public affairs officer at Mission Control, said the oxygen leak was so slight that, "provided there are no additional fluctuations, we have enough gases to go to the end of the mission."

After encountering some Spacelab equipment malfunctions during the mission's first two days, Challenger's crew indicated they had achieved encouraging success with nearly all of the Spacelab's 76 scientific experiments on the effects of microgravity on metals, biological growth, and human physiology. Mission planners in Houston and scientists at the West German science control center near Munich were studying the possibility of extending the flight by a day to permit more time for experimentation.

The crew reported preliminary results showing that, as expected, seedlings growing in the virtual absence of gravity developed roots that curved in odd directions. "The roots, lacking gravity, do not seem to know which way to go," a crew member commented. In another experiment, the *Bacillus subtilis* bacteria showed a higher rate of growth than anticipated. Mission control scientists said this "may be taken as an indication that bacteria do react differently in space." Most of the studies in processing metal alloys and growing crystals for electronics were still underway, so results were unavailable until completion of the mission.

Earlier in the mission, crew member Dr. Ernst Messerschmid, a West German physicist, reported being "disoriented" when he awoke, a symptom of the space motion sickness that had plagued nearly half of all Space Shuttle astronauts. (*NYT*, Nov 3/85, A27)

November 6: The Space Shuttle Challenger on mission 61–A landed at 9:44 a.m. today at Edwards Air Force Base, the *NY Times* reported, before an estimated audience of 9,000. "The bird fspace flight, said at a post-landing news conference. "The brakes look like it's in good shape," Jesse Moore, NASA's associate administrator good coming in and we are very, very pleased with the results of the nosewheel steering test."

During the landing, Challenger commander Henry Hartsfield tested a new steering mechanism built to reduce tire and brake damage that had occurred during several Space Shuttle landings. An orbiter had not landed at Kennedy Space Center (KSC) since the previous April, when both main landing gears collapsed and one tire blew out. The test called for Hartsfield to engage the new steering mechanism seconds after the nosewheel touched down on the runway so he could steer Challenger 20 feet off course and then bring it back to the centerline. One more successful test later in the year would clear orbiters to resume landing in December at KSC.

Officials in West Germany, which paid NASA $64 million to fly Spacelab D-1 on Challenger, called the mission "highly successful" and predicted that 90 to 95% of all the science studies would be completed. Earlier, a West German official at the science control center near Munich said only one study, a heat diffusion test on a sample of salt, would be unfinished by the time Challenger returned to earth.

Crew members Dr. Bonnie Dunbar, Col. Guion Bluford Jr., Dr. Ernst Messerschmid, Dr. Reinhard Furrer, and Dr. Wubbo Ockels did not attend a postflight press conference, the *Washington Post* reported, but traveled to Dryden Flight Research Center where researchers would run tests on them to determine how well they readapted to gravity after a week in weightlessness. Later the five would go to Kennedy Space Center for more elaborate tests.

The European Space Agency (ESA) announced after the flight that it was particularly satisfied with results from Spacelab D-1, on which ESA had 38% of the payload in terms of critical resources (mass, energy, and crew time). All ESA facilities on Spacelab D-1—the Space Sled, for studies of man's behavior under microgravity conditions; the Biorack, a multi-purpose facility for biological investigations of microgravity and cosmic radiation effects on life forms; and the Fluid Physics Module, for studies of basic fluid phenomena in space—performed perfectly with no operational or technical failures. (NASA FOR M-989-61-A [postflight] Nov 8/85; *NYT*, Nov 7/85, A21; *W Post*, Nov 6/85, A4; ESA release Nov 11/85)

November 26: NASA launched at 7:29 p.m. today from KSC the Space Shuttle Atlantis on mission 61-B to practice methods for building a space station and to launch three satellites, the *Washington Post* reported. People more than 400 miles away in South Carolina and Key West, Florida, saw the launch. There had been only two other previous U.S. night launches: in 1972 when the Apollo 17 crew left for the moon in the middle of the night and in 1983 when the Space Shuttle Challenger departed in early morning.

Nine minutes after launch, mission control reported Atlantis was in a secure orbit about 200 miles high, and launch director Gene Thomas said the countdown was the smoothest yet in the Space Shuttle program.

Crew onboard Atlantis were Lt. Col. Brewster Shaw, commander; Lt. Col. Bryan O'Connor, pilot; mission specialists Dr. Mary Cleave, Maj. Jerry Ross, and Lt. Col. Sherwood Spring; and payload specialists Rudolfo Neri Vela (Mexico's first astronaut) and McDonnell Douglas engineer Charles Walker.

During the mission the crew would launch Mexico's *Morelos B* satellite, Australia's *AUSSAT 2*, and RCA's *Satcom KU–2*. Jerry Ross and Sherwood Spring would also take two long spacewalks to assemble 93 aluminum struts into a 45-foot beam and six 12-foot beams into a 64-lb. inverted pyramid to test techniques that might be used to build the proposed permanently manned space station. Following each spacewalk, the mission specialists would disassemble and stow the beams. Cameras would record every move so that later a computer could reduce the photos to a motion study that engineers would use to determine how construction work could best be performed in the zero gravity of outer space. (NASA Release 85–153; *W Post*, Nov 27/85, A3)

November 27: Astronauts aboard the Space Shuttle Atlantis on mission 61–B today launched satellites for Mexico and Australia, the *Washington Post* reported.

The crew launched Mexico's *Morelos B* in the morning, just hours after liftoff. *Morelos B* was a Hughes 376 satellite, a standard design used by many foreign national and private companies, Kennedy Space Center's *Spaceport News* said. It would provide telephone, TV, and wire services to Mexico through a total of 22 transponders. A PAM–D payload assist module then boosted the satellite to geosynchronous orbit 22,300 miles above earth. The satellite would drift, unused, until it reached it final stationary position in 1989. Mexican officials chose to store *Morelos B* in orbit because launch cost for the satellite could increase to four times the amount that Mexico paid to NASA for today's deployment. Payload specialist Rudolfo Neri Vela, the first Mexican astronaut, was onboard to observe satellite deployment. In June 1985 crew in the orbiter Discovery deployed the first Morelos satellite.

At 8:20 p.m. the Australian satellite *AUSSAT 2* spun out of the Atlantis's cargo bay. "We got a good deploy," said mission specialist Sherwood Spring, who supervised the satellite's spring ejection. "It looked like it might have gone a second early," he added. *AUSSAT 2*, also a Hughes 376 satellite, was the second of three operations satellites for the government-owned Australian National Satellite System. It had eleven 12-watt transponders and four 30-watt transponders to provide domestic communications to Australia's 15-million population. Australia would also use the satellite to improve maritime and air traffic control communications, relay digital data for business purposes, provide standard telephone communications, and direct satellite to home TV broadcasts. A PAM–D would also boost the satellite to geosynchronous orbit. In August 1985 the Discovery crew deployed the first AUSSAT.

Mission commander Brewster Shaw had tested all systems aboard Atlantis and reported that one of four videotape recorders was not working. Mission control gave him permission to take apart an electronic cabinet to look for a wiring flaw, but Shaw said he could not see anything that was obviously wrong. Later, following directions from the ground, Shaw cycled a circuit breaker and reported, "Bingo. That seems to get power to VTR1." (*W Post*, Nov 28/85, A28; *Spaceport News*, Nov 22/85, 4)

November 28: Crew aboard the Space Shuttle Atlantis on mission 61–B celebrated Thanksgiving in space today, saying it was one of the best Thanksgivings they ever had, the *Washington Times* reported. It was the first U.S. manned spaceflight during that holiday since the third Skylab flight in 1973.

The crew's main job for the day was deployment of the 4,144-lb. *SATCOM KU–2* satellite owned and operated by RCA American Communications (RCA Americom). The uninsured $50 million satellite was the last of three KU-brand domestic communications satellites that operated in the 12 to 14 gigahertz range. It had 16 operational transponders and six spares, each transmitting 45 watts of power, more than the 12 to 30 watts used for C-Band transponders.

The payload assist module (PAM) on the satellite was instrumented with radio frequency telemetry that would downlink data to ARIA (advanced range instrumentation aircraft) during burn of the solid-fuel rocket motor, a mission requirement for this first flight of a PAM–D2, the Kennedy Space Center *Spaceport News* explained. The uprated upper stage was identical to the PAM–D, except for size and weight. The spinup would be noticeably slower due to the larger mass and inertial components of the payload. NASA designed PAM–D2 to place up to 4,200 lb. in geosynchronous orbit, compared to the 2,800-lb. PAM–D version.

The crew's dinner menu—chicken consomme, smoked turkey, cranberry sauce, green beans, corn, pasta, butter cookies, and lemonade—sounded traditional, but the turkey was preserved before launch by irradiation, sealed in foil, and had to be heated in a device similar to a toaster oven. Crew added warm water to the consomme, vegetables, and pasta to make them edible. (*W Times*, Nov 29/85, 3A; *Spaceport News*, Nov 22/85, 4)

November 29: Astronauts Jerry Ross, an Air Force Major, and Sherwood Spring, an Army Lieutenant Colonel, aboard the Space Shuttle Atlantis on mission 61–B took 40 minutes to erect a 45-foot-high tower with 93 aluminum struts and 33 joints, the *Washington Post* reported. After breaking down the tower and stowing the parts, they assembled a 400-lb. inverted pyramid out of six 12-foot-long aluminum beams. They built and broke down the pyramid eight times in less time than they were given for six. By the end of their task, they were assembling the pyramid in nine minutes and breaking it down in less than six minutes—three minutes faster than their first assembly and disassembly. By 9:30 p.m. they had done in four hours a job given them to complete in five.

NASA referred to the first activity as ACCESS (assembly concept for construction of erectable space structures), researchers at Marshall Space Flight Center explained, and the second as EASE (experimental assembly of structures in extravehicular activity). Together they were the first flight demonstration of construction of large space structures, so called because they were distinguished by different assembly methods and physical characteristics. The astronauts used no tools in the construction, rather they snapped together prefabricated components to form the EASE/ACCESS structures.

EASE/ACCESS should provide NASA with valuable on-orbit construction experience as well as a comparison of assembly rates and techniques used in space to those used during simulations on the ground and in neutral buoyancy water tank tests. The activity should also help evaluate potential assembly and maintenance concepts and techniques for the proposed space station and identify ways to improve erectable structures to ensure productivity, reliability, and safety.

TV views at NASA Headquarters showed the astronauts trading places twice during their work and appearing in almost complete control of a job that had never been done before in space. Ross had the only complaint, saying his gloved hands were sweating and that he was using too much oxygen.

When the two had finished assembling the six required pyramids, astronaut David Leetsma at mission control told them to take a break. "I'd be willing to do at least one more," Ross replied. "It feels good to do some good hard work." The two then proceeded to begin work again.

Spring once hit his feet against the tower, and Ross by mistake hit a switch with his hand that turned on an outside light, the only "accidents" that occurred during the exercise, although they triggered a warning from Atlantis's commander Brewster Shaw, Jr. "You guys are going great but just remember to be careful. The way that thing shakes up there, it's not going to be easy to chase anything down if it breaks loose," he cautioned. (*W Post*, Nov 30/85, A1; Marshall Space Flight Center Release 85–60)

December 1: Astronauts Maj. Jerry Ross of the Air Force and Lt. Col. Sherwood Spring of the Army today stepped outside the Space Shuttle Atlantis on mission 61–B for the second construction exercise in the weightlessness of space [see Space Transportation System/Missions, Nov. 30], the *NY Times* reported. To TV viewers on earth, the two appeared to manipulate with ease the heavy structures they were building, although both were sweating and breathing heavily after the first 20 minutes of work. Two hours into the spacewalk, they reported that though their hands, feet, and backs were moist with sweat, they were working comfortably.

The purpose of today's spacewalk was the same as previously, to test the efficiency of techniques to build a 45-foot-high aluminum tower. However, the astronauts changed the way they worked and had the assistance of astronaut Dr. Mary Cleave, who was operating the Space Shuttle's 50-foot mechanical arm. She moved the men from place to place as their work required.

Later in the exercise, the *Washington Post* reported, Ross and Spring took turns stringing cable along the four-story metal frame they had constructed, just as future crews might lay cable for a space station. They also removed parts of the tower and replaced them with other parts, simulating a space repair.

NASA officials commented after the first spacewalk on November 30 that the work had proved more taxing than expected, perhaps because the astronauts' gloves, unlike the rest of their spacesuits, were not air-conditioned and could not carry away excess heat. Most of the work had entailed repeated squeezing and twisting with the hands. Ross and Spring said after the exercise that they would not want to repeat it before having a day of rest. They indicated that the activity had left their hands so numb and fatigued that they had trouble fastening the airlock of the Space Shuttle as they reentered it.

NASA officials did note they were favorably surprised by the astronauts' rapid improvement in performance as they learned from successive repetitions of the exercise. (*NYT*, Dec 2/85, B6; *W Post*, Dec 2/85, A3)

December 2: In their last full day aboard the Space Shuttle Atlantis on mission 51–B, the crew concluded an array of experiments and prepared for their scheduled landing at Edwards Air Force Base, the *NY Times* reported. Answering questions radioed from reporters at Johnson Space Flight Center, the astronauts reported progress not only in working out space station construction techniques, but also in developing a gravity-free pharmaceutical factory to be based on the proposed space station.

Payload specialist Charles Walker of the McDonnell Douglas Corp. was in charge of the experiment for development of a space-based factory for purification of a human hormone produced by genetically engineered bacteria. Astronauts on previous flights had tested the basic purification technique, called continuous-flow electrophoresis in a gravity-free environment; the current experiment simulated an actual production run of the hormone, which might prove useful in treating certain forms of anemia.

During the questioning, astronauts Lt. Col. Sherwood Spring and Maj. Jerry Ross acknowledged that they were extremely fatigued after each of their spacewalks and that their hands were stiff and numb after the practice assembly, manipulation, and disassembly of a 45-foot-long truss and a pyramidal module. Nevertheless, the two deemed the exercise a success. "We don't know yet what this will mean for a future space station," Spring commented, "but it's a start. Personally, it was really exhilarating for me out there, face to face with the universe." (*NYT*, Dec 3/85, C3)

December 3: The Space Shuttle Atlantis on mission 61–B landed at 1:33 p.m. Pacific time today on the concrete runway at Edwards Air Force Base before 6,700 spectators, the *NY Times* reported. NASA selected the concrete runway because a week of rain had left scattered puddles on the usually dry lake bed runways.

"Atlantis looks beautiful," Jesse Moore, NASA's associate administrator for space flight, said after the landing. Apart from "a few dings" around the ship's nose and landing gear, he added, there appeared to be "no unusual damage."

It was the second mission for the Atlantis, the newest in NASA's fleet of four orbiters. The crew described all aspects of the flight as highly successful. (*NYT*, Dec 4/85, B6)

Revenues

January 15: During NASA's general management status review, the office of space flight reported on the agency objective to undertake a pricing and marketing strategy that would capture for the STS a dominant percentage of the planned free-world, commercially launched payloads, as measured yearly by payments for launch reservations, through the end of CY 88.

R. Wisniewski, assistant associate administrator, noted that a NASA recommendation for full-cost recovery sent to the White House would likely undergo close scrutiny. He anticipated strong competition as NASA marketed the Space Shuttle capability, with 28 to 36 STS launches of the approximately 58 planned payloads through 1988; Ariane would capture a sizeable portion of the remaining launches. Wisniewski reported $140 million in Space Shuttle receipts in FY 84 and predicted $188 million in Space Shuttle revenue in FY 85. (*GMSR Report*, Jan 15/85, 35)

February 7: NASA Administrator James Beggs told the U.S. House Science and Technology Committee that the agency's launch service's market position had eroded "very severely" and that he had to be "a little bit pessimistic" about NASA's ability to hold a lion's share in the future, *Aerospace Daily* reported. Beggs made the comment in response to a question about the need for a 5th Space Shuttle orbiter.

During the previous 14 months, NASA had won five new commercial launches; The European Space Agency's (ESA) Ariane, five. Beggs said the Europeans were "coming along very aggressively" in using the Ariane 4 and 5 vehicles to improve launch capabilities and that agreement by ministers attending the ESA council meeting [see European Space Agency, Feb. 4] to further Ariane 5 development "was significant." (*A/D*, Feb 7/85, 1)

February 25: NASA Administrator James Beggs criticized the Air Force and NOAA for plans to launch some satellites aboard surplus Titan II missiles rather than using the Space Shuttle, saying it would cost NASA $500 million, the *Washington Post* reported. "You know the old syndrome," he said, "if it wasn't invented here [at the Air Force] it can't be all that good . . . I don't like it. It is not good for our short-term future and makes it that much more difficult for us to get on an even footing in the next five years."

The controversy had begun with an Air Force decision to refurbish and redevelop obsolete Titan II intercontinental ballistic missiles to launch up to 12 Air Force satellites that NASA had assumed would be flying on the Space Shuttle.

When NOAA heard of the Air Force decision, its acting administrator, Anthony Calio (a former NASA official), had contacted the Air Force about launching over the next seven years three advanced weather satellites, called Metsats, into polar orbit. When questioned, Calio had said the Air Force offered a deal that would save NOAA $90 million—it would cost $100 million to redesign the Metsats to fly on the Space Shuttle and $105 million in launch fees versus an Air Force charge of $115 million for the three Titan II launches, resulting in a net saving of $90 million. "This agency has a budget of $1 billion a year so we're talking about 10% of our annual budget. This is strictly a business deal," Calio concluded.

Beggs responded that he always assumed the Air Force and NOAA would use the Space Shuttle, although they had not signed a formal agreement. He said he did not believe NOAA would save that much and that the Air Force would bear the refurbishment cost to redevelop the Titan II, estimated by Beggs at $100 million.

"I believe what we're talking here is a savings to NOAA and a net loss to the U.S. Treasury," Beggs said. "I think this whole scheme should be looked at very carefully by the administration and by Congress." And he mentioned his concern that the Air Force plan could set a dangerous precedent. (*W Post*, Feb 25/85, A1)

May 27: A fare war had broken out in space, *Fortune* reported, and Arianespace, a consortium of European governments and private companies that operated the European-built Ariane rocket, was grabbing communications satellite launches away from the U.S. Space Shuttle by aggressively underpricing it. Although NASA had launched 30 satellites, nine aboard the Space Shuttle, over the previous four years and the Ariane less than 12, the Europeans would become major competitors, *Fortune* predicted.

A Space Shuttle launch cost $25 million; Arianespace charged $24 million. And Arianespace had cut that price by as much as $3 million to lure some U.S. customers. Arianespace officials said its throwaway rockets were more efficient at the single task of launching satellites than was the complex Space Shuttle, which NASA had designed to perform a range of scientific and military missions. "Using the shuttle for launching satellites," Charles Bigot, director of Arianespace, said, "is like flying a supersonic Mirage jet fighter when all you need is a puddle jumper."

American entrepreneurs who wanted to get into the satellite-launching business complained that prices charged by both Arianespace and NASA were ridiculously low as a result of government subsidies. Transpace Carriers Inc., for example, held the rights to buy the Delta rocket from its manufacturer McDonnell Douglas Corp., but couldn't launch a satellite for less than $45 million.

The U.S. Department of Transportation, which wanted to encourage private companies like Transpace Carriers, was encouraging President Reagan to set Space Shuttle launch rates high enough to make the commercial launches profitable. So the future of the launch business, *Fortune* said, would rest with President Reagan's decision on Space Shuttle pricing—and on Arianespace's reaction. If Reagan boosted Space Shuttle launch fees, it would please entrepreneurs anxious to get into the space business. But that would also give the Europeans a bigger opening into the space business than they've had so far. (*Fortune*, May 27/85, 138)

August 1: The White House announced that it had sent Congress a new Space Shuttle pricing policy that entailed, starting October 1, 1988, auctioning the Space Shuttle's cargo bay to commercial and foreign customers at a minimum rate of $74 million for a full bay, the *Washington Times* reported. This meant that owners of three satellites could share a mission and pay NASA a little less than $25 million apiece or about $1 million more than they presently paid for Space Shuttle launches. However, if demand for shuttle space exceeded its availability, the auction system would drive prices upward. Europe's Arianespace SA charged $25 million for a launch.

President Reagan, who earlier had suggested that space should be open to private enterprise and that the Space Shuttle might be turned over to industry, directed NASA under the new policy to establish auction procedures "to ensure maximum return to the government and equitable treatment for all potential launch customers." The President also told NASA to review annually Space Shuttle costs and the effectiveness of its pricing policies and report its findings and recommendations to the president's national security adviser and the director of the Office of Management and Budget.

The new policy represented a victory for NASA and defeat for the Transportation Department and Secretary Elizabeth Hanford Dole, who had argued for a full-bay price no lower than $129 million in order to encourage private industry to get into the launch vehicle business on its own, the *Washington Post* reported.

Two companies wanted to compete with NASA and Arianespace. General Dynamics signed a letter of intent with NASA to use the Atlas-Centaur rocket, and Transpace Carriers Inc. signed a similar letter to use the Delta rocket to launch satellites. Both companies said they could not compete with a Space Shuttle price of less than $40 million to send up a single satellite.

The Transportation Department argued for a higher Space Shuttle launch price to improve private industry's bargaining power. But NASA said a higher charge would send customers to Arianespace, which had booked more than a third of the world's future commercial satellite launch traffic. (*W Times*, Aug 2/85, 10C; *W Post*, Aug 7/85, A3)

September 17: Hughes Communications Inc. said today it would negotiate an approximately $180 commitment with NASA for launch by the Space Shuttle of six communications satellites, including two for a new Japanese network, after receiving proposals from NASA and from Arianespace for its expendable Ariane launch vehicle, *Aerospace Daily* reported. Hughes indicated selection of the Space Shuttle was based solely on cost and was not related to the recent Ariane 3 failure [see European Space Agency, Sept. 12].

Space Shuttle flights scheduled for December 1987 and May 1988 would launch HS–393 Ku-band satellites that Hughes was building for the Japan Communications Satellite Co. (JCSat). Hughes reserved between 1988 and 1990 four slots for placement of HS–393s over the U.S.

Hughes Communications owned 30% of JCSat, a joint venture licensed the previous June by Japan's Ministry of Post and Telecommunications to construct and operate a two-satellite system to serve Japan. The Japanese trading companies C. Itoh and Co. Ltd. and Mitsui and Co. Ltd. owned 40% and 30%, respectively, of the joint venture. (*A/D*, Sept 18/85, 1)

October 18: A controversy between the National Oceanic and Atmospheric Administration (NOAA) and NASA [see Space Transportation System/Revenues, Feb. 25] appeared over when NOAA agreed to launch meteorological satellites (Metsats) into orbit in return for which NASA would give NOAA a discount for carrying the satellites into orbit aboard the Space Shuttle, the *Washington Post* reported.

A deal between NOAA and the Air Force that would have saved NOAA $90 million in launch costs never went through, in large part because NASA and NOAA began working behind the scenes to bury the hatchet, the *Post* said. The previous week NASA Administrator James Beggs sent Anthony Calio, NOAA administrator, a letter outlining a deal governing three Metsat launches starting in 1989.

NASA agreed to share with NOAA the $80 million cost of modifying the Metsat for a Space Shuttle launch, to put a Metsat on a Space Shuttle within three months after NOAA asked for it, and to shave $6.5 million off the $105 million in feed NASA ordinarily would charge NOAA for the three launches. "We gave them a good price," Beggs said. "They're on the shuttle at what essentially is our cost for launching them." (*W Post*, Oct 18/85, A21)

October 29: The U.S. House Appropriations Committee deleted $7.2 million in FY 86 funds requested by the Air Force to initiate a $350 million program to refurbish Titan II ICBMs for use as space boosters, *Defense Daily* reported.

The committee warned that the Titan II launchers could "only have a further impact on the Shuttle and result in a less efficient use of Shuttle facilities" and pointed out that the Air Force had "not been able to demonstrate why the expensive Titan II refurbishment program is critical to the defense of the nation."

The Committee also noted that the National Oceanic and Atmospheric Administration dropped its plans to launch some payloads on the Titan II in favor of launching them on the Space Shuttle [see Space Transportation System/Revenues Feb. 25 and Oct. 18], "bringing into question whether Titan II is cost-effective on a per launch basis," the committee said. It also cited classified reasons for its action. (*D/D*, Oct 29/85, 305)

November 12: Frederic J.P. d'Allest, chairman of Arianespace, told the U.S. Senate science, technology, and space subcommittee the previous week that his firm offered satellite insurance at lower than market rates to those using the Ariane rocket to launch their satellites, the *Washington Times* reported. The insurance covered only the booster phase of a satellite launch.

Arianespace charged between 11 and 13% of the total value of the launch for insuring an Ariane launch from lift off to geostationary orbit, that portion of the flight under Arianespace's control. Satellite operators had to go to commercial insurance markets to obtain coverage for the rest of a satellite's operation.

More than $600 million in losses from the failure of seven satellites in almost two years had resulted in a tightened insurance market that was seriously threatening commercial space projects. Satellite insurance firms had cut back on the amount they would commit to insure launches, while boosting premiums to more than 20% of the total insured value of the launch. The subcommittee before which d'Allest testified was investigating a possible federal role in easing the insurance problem. However, commercial underwriters warned the subcommittee that the federal government should stay out of the insurance business.

"In my view," James Barrett, president of International Technology Underwriters, said, "there is no role for government to play in this commercial market. Intervention can only disturb normal market processes." He noted the solution to the problem was improved reliability of satellite manufacturers and launch agencies. "Insurance . . . cannot compensate for unacceptable levels of reliability," he commented. (*W Times*, Nov 12/85, 7C)

December 12: NASA announced it reached agreement with the Department of Defense (DOD) on a pricing and reimbursement policy for DOD use of the Space Transportation System during the period FY 89 through FY 91. The agreement established an average price of $60 million (in 1982 dollars) for each DOD launch, an average which was based on the estimated cost to fly and an exchange of launch and range support services between DOD and NASA.

The price reflected a fixed-base component of $30 million for each planned flight and an incremental component of $30 million for each actual flight. The annual fixed-based component total—$270 million a year, based on DOD projections of nine equivalent flights—would be paid regardless of the number of actual flights, while the $30 million incremental component would be paid for each actual flight.

By combining a fixed annual charge with a low variable cost per flight, the agreement insured that NASA would be able to cover its funding needs and DOD would continue to rely on the Space Shuttle as DOD's primary space launch vehicle. (NASA Release 85–169)

Scientific Research

January 14: NASA announced that Space Shuttle flight 51–C would carry the first of two experiments to investigate effects of different diseases on red blood cell aggregation and blood viscosity. The experiment was originally scheduled for mission 51–A in November 1984 but withdrawn due to orbiter weight and center-of-gravity considerations. The Department of Defense (DOD) had agreed to add the experiment to 51–C, a dedicated DOD mission.

Called aggregation of red blood cells (ARC), the experiment would determine rate of formation (kinetics) and internal structure and organization (morphology) of red cells and the thickness (viscosity) of whole blood at high- and low-flow rates. Healthy donors and donors with various medical conditions such as heart disease, hypertension, diabetes, and cancer would provide blood samples. Researchers would compare results obtained in microgravity with results from a simultaneous and identical ground-based experiment to determine what effects gravity had on the kinetics and morphology of the blood and, therefore, whether researchers could use information obtained in microgravity to formulate new diagnostic tests or improve existing tests for the benefit of clinical research and medical practice.

Flight hardware, weighing about 165 lb. and installed in three mid-deck lockers in the crew cabin, would consist of a container housing a blood pump/storage subsystem, parallel plate slit capillary viscometer, photo/optical subsystem, thermal control system, pressure transducer, and an electronics equipment package to provide automated control and data acquisition. A crew member would activate the experiment; the electronics package would automatically operate all other procedures. Running time would be about eight hours.

Dr. Leopold Dintenfass of the Kanematsu Institute (Department of Medical Research), Sydney, Australia, had developed the experiment and experiment hardware. Marshall Space Flight Center had responsibility for the flight experiment.

After landing, data from the flight experiments would be delivered to Dintenfass for analysis and comparison with ground-based experiment data. (NASA Release 85–6; NASA MOR E–420–51–C–21 [prelaunch] Jan 23/85)

January 18: NASA announced that the Department of Defense (DOD) had agreed that Space Shuttle mission 51–C, scheduled for launch January 23, 1985, from KSC, would carry the Australian aggregation of red blood cells [see Jan. 14] and shuttle storable fluid management demonstration (SFMD) experiments. NASA had scheduled both for previous missions, but withdrawn them due to orbiter-weight and center-of-gravity considerations.

The SFMD, a joint NASA-Martin Marietta Aerospace/U.S. Air Force experiment, would test how weightless fluids behaved in transit from one tank to another to demonstrate the transfer of fluids planned for servicing and maintaining spacecraft in orbit. Space Shuttle crew would manually operate the experiment, consisting of 13-in.-diameter clear acrylic supply and receiving tanks, with a series of valves in which pressurized air would force fluid to and from transparent tanks, a video tape recorder and 35mm still camera photographing the operation and an accelerometer measuring any motion caused by the orbiter that affected the experiment. (NASA Release 85–10)

February 14: NASA indicated it might remove squirrel monkeys from an April 30 Space Shuttle flight because the species carried a herpes virus known to cause cancer in some subhuman primates, the *NY Times* reported.

Research had shown the virus could not survive in high-order primates, including humans, and NASA officials emphasized it was not one of the forms known to cause a variety of ailments in humans. The agency had spent more than a year trying to assess the risk possibilities to astronauts. (*NYT*, Feb 14/85, B8)

March 22: Adm. James Watkins, speaking to reporters after a ceremony during which astronaut Thomas Mattingly was promoted to commodore and named to head the Navy Space Program Office, said that a Navy oceanographer who had traveled on a Space Shuttle flight the previous fall had brought back "some fantastically important" information that would make it easier for U.S. submarines to hide in the world's oceans, the *Washington Post* reported.

Watkins said the oceanographer, whom he would not name but was known to be Paul Scully-Power, a civilian who worked at the Naval Underwater Systems Center, New London, Connecticut, said the information was vital in trying to understand the ocean depths. When asked if satellite technology might not be hastening the day when submarines could be tracked from space, Watkins acknowledged that technology was opening some doors on submarine tracking. However, he pointed out that the question was whether doors opened on detecting submarines faster than doors opened on learning about the oceans depths. He added that "we're still ahead of the game in the latter category.

"So the ability to track submarines—we don't see that as being a threat to our forces until the turn of the century or later, depending on what kind of breakthroughs we might find at the end of this decade or into the next," Watkins concluded. (*W Post*, Mar 22/85, A10)

March 27: NASA had scratched four squirrel monkeys scheduled to be carried into orbit by the Space Shuttle Challenger in April because NASA found they had a form of herpes that could be transmitted to the astronaut crew, the *Washington Post* reported.

The monkeys' herpes, called *Herpes Samirii*, was not the virus transmitted sexually by humans, but was unique to New World primates whose natural habitat was the rain forests of South America. Researchers suspected the virus of causing cancer in lower mammals such as rats and therefore classified it as potentially cancerous in humans. NASA replaced the monkeys with those from colonies bred to be virus-free.

Four monkeys recruited from the National Institutes of Health and one from Harvard University had been in training since January for their flight on the second mission of Spacelab, scheduled for April 29 at the earliest. Dr. Arnauld Nicogossian, NASA's chief of space medicine, said they were looking for a sixth monkey even though only three might be flown. "One of the monkeys in training is too small and will have to be replaced, and it will be nice to have a backup when the time comes for them to go into orbit," he said.

In space the monkeys' cages would have their own oxygen and food and water supplies in the Spacelab portion of the Space Shuttle's cargo bay, which was shut off from Challenger's cabin by an airlock. The upcoming flight was a shakedown cruise for the monkeys to ascertain that their cages functioned properly and to make sure the monkeys wouldn't get frightened or injured during launch or return to earth. (*Washington Post*, Mar 27/85, A20)

April 12: NASA announced that the American Flight Echocardiograph (AFE), sponsored by the life sciences division of NASA's space science and application office (OSSA) and developed by the medical sciences division at Johnson

Space Center, would fly on STS 51-D, the first of three planned flights of the equipment with follow-on activity determined by data results.

The objective of AFE was to obtain data on in-flight cardiovascular changes during space flight. Most changes seemed to result from the shift of body fluids out of the legs and into the chest and head in weightlessness. The brain apparently detected this shift and interpreted it as an increase in blood volume. Normal responses to this "overload" were an increase in urination and a decrease in thirst, reducing blood volume from an acceptable level.

Other effects of weightlessness included changes in heart size, an increase in the amount of blood pumped by the heart, and a decrease in the resistance to the flow of blood through the arteries. Most of these latter changes were normal responses to weightlessness; however, when combined with the reduced blood volume, they made effective functioning on return to earth difficult.

Determining the important cardiac and vascular changes during adaptation to weightlessness and during the readaptation to gravity after a flight would reveal how the cardiovascular system responded to unusual stresses. This would be valuable in the development of more effective countermeasures to the detrimental effects of spaceflight and in understanding the functioning of the circulation in general.

The AFE equipment used very high frequency sound waves (echocardiography) to obtain in vivo cardiodynamic structural and functional data. A probe held next to the skin sent high-frequency sound waves (ultrasound) through the skin and into the body. It then detected their reflections, or echos, from organ surfaces. The electronic circuitry of the echocardiograph calculated the distance travelled by the ultrasound pulse from the delay between the transmission of the pulse and the detection of its echo.

Dr. M. Rhea Seddon, a co-investigator on the project, had trained to use the AFE to image her own heart from several different angles. Every day in flight she would record her own echocardiogram and those of up to three other crew members as time and circumstances permitted. Dr. Seddon would obtain the first echocardiogram as soon as possible after orbital insertion and midway through and prior to sleep on the first flight day.

During the three flights on AFE, NASA planned to acquire data on about 10 crew members, along with pre- and postflight data. (NASA MOR E-420-51-D-14 [prelaunch] Apr 12/85)

April 12: NASA announced that the Phase Partitioning Experiment (PPE), sponsored by the microgravity science and applications division of the office of space science and applications (OSSA) and developed and managed by Marshall Space Flight Center (MSFC), would fly on STS 51-D in a locker used by payload specialist Sen. Jake Garn, who would take sole responsibility for the operation of the PPE.

Phase partitioning was a selective yet gentle and inexpensive technique for the separation of cells and proteins. It required establishing a two-phase system by adding various polymers to an aqueous solution containing the materials to be separated (two-phase systems were similar to the immiscible liquids oil and water). After establishment of such two-phase polymer systems, the biological materials they contained tended to partition into the different phases.

Theoretically, phase partitioning should separate cells with significantly higher resolution than was obtained in the laboratory. Performing the experiments in orbit should provide a long-term weightless environment that would help control the separation of the phases and obtain better fractionation of biological cells. Although the interaction of the cells and phases could be done in a slower and more controlled manner in weightlessness, there remained the problem of combining the phase emulsion back into the two separated phases without the effects of gravity. The PPE would examine this problem through two methods: natural coalescence and preferential wetting to localize the separate phases.

The PPE equipment was a small hand-held unit weighing 427 g (one lb.). MSFC researchers fabricated the unit from plastic materials with transparent sides so that there were 15 separate small chambers, each with its own stainless steel mixer ball, allowing for simultaneous study of a number of experimental variables. After shaking the unit to achieve an emulsification of the phases in the chambers, photographs over the next 45 minutes would record the progress of the separations. The PPE operator would note transient phenomena not recorded by the camera. Researchers hoped the procedure could be repeated three times during the flight. (NASA MOR E–420–51–D–13 [prelaunch] Apr 12/85)

April 12: NASA announced that a protein crystal growth (PCG) experiment, sponsored by the microgravity science and application division of the office of space science and applications (OSSA), would be flown on STS 51–D in a middeck locker used for the McDonnell Douglas Astronautics Corp.'s continuous flow electrophoresis system (CFES). CFES payload specialist Charles Walker would operate the PCG experiment.

Detailed knowledge of their composition and structure was extremely important to understanding the nature and chemistry of proteins and Douglas's ability to manufacture and/or modify them for medical uses. However, for most complex proteins, it was not possible to grow on earth crystals of sufficient size and quality to allow analyses by X-ray or neutron diffraction techniques. Researchers believed growth of such crystals would be possible in the weightlessness of orbital spaceflight where gravity-driven convection currents were minimized and where the crystals remained suspended during their growth cycle. The PCG experiment hardware would accomplish this.

After reaching orbit, Walker would manipulate the small PCG unit to mix about 30 pairs of solutions. Near the end of the flight, he would prepare the unit for reentry and stow it in the equipment locker. Upon return to earth, NASA would send the unit to MSFC for analyses by the principal investigator. Results of the analyses would determine the schedule for further flights. (NASA MOR E–420–51–D–15 [prelaunch] Apr 12/85)

May 29: Dr. Paul Scully-Power, the first oceanographer to orbit the earth, said today at a Baltimore meeting of the American Geophysical Union, that he had observed previously undetected spiral currents in many parts of the world's oceans while he was aboard the Space Shuttle Challenger, the *NY Times* reported. Oceanographers at the meeting said these eddies introduced an entirely new dimension to studies of ocean dynamics.

Scully-Power had seen the eddies, for example, throughout the Mediterranean and between the east coast of the U.S. and the Gulf Stream; they were typically about 25 miles wide, far smaller than large eddies observed in the Gulf Stream. It was the great prevalence of these swirls, whose origin was yet unexplained, that astounded scientists.

Scully-Power said he did not know how long each eddy survived, but guessed it might be days or weeks. Nor did he know how deeply into the ocean they extended.

At the meeting, Scully-Power displayed one photograph of five such spirals in the eastern Mediterranean, all aligned in a single row. (*NYT*, May 31/85, A15)

June 19: NASA announced that it signed today with Instrument Technology Associates (ITA) an agreement for ITA to develop for use on the Space Shuttle flight hardware consisting of a standardized experiment module (ISEM) payload carrier, a turn key module for use by any commercial firm, government agency, research organization, or educational institution wishing to conduct experiments in a microgravity environment. An organization could purchase or lease all or part of the ISEM.

Under the agreement, NASA would provide ITA with two Space Shuttle flights to obtain data and validate the ISEM design. NASA had scheduled for late 1987 the first ISEM test that would carry a payload for the Bioprocessing and Pharmaceutical Center of the Philadelphia University City Science Center.

ITA would design the ISEM to be compatible with the proposed space station with the idea that initial users of the module would progress to larger and more sophisticated payloads to fly in the future aboard the station. (NASA Release 85–94)

August 1: Astronauts aboard the orbiter Challenger on Space Shuttle mission 51–F released today a six-and-a-half-foot-long drum-shaped satellite, the Plasma Diagnostics Package, from the orbiter's payload bay, shot it with a looping beam from an electron gun, and then retrieved it with the Space Shuttle's 50-foot mechanical arm, the *NY Times* reported.

One purpose of the demonstration was to study the electrical effects the Space Shuttle had as it moved through the electrically charged gases of the ionosphere, effects which occasionally caused the Space Shuttle to emit a soft golden glow. Researchers considered solving that riddle important to the success of future mission, which would carry sensitive instruments that the aura might disturb.

Another goal was to achieve a better understanding of the aurora, known in the northern hemisphere as the northern lights, which was a disturbance in the atmosphere that created bright lights in the night sky over polar regions.

The astronauts reported seeing flashes of light as the beam followed natural magnetic lines to the satellite, the *Washington Post* reported; the newspaper said scientists on the ground called the demonstration "a fabulous success." (*NYT*, Aug 2/85, D14; *W Post*, Aug 2/85, A4)

October 14: Wesley Hymer, a biochemist at Pennsylvania State University and one of 20 researchers who analyzed tissue samples from 24 rats flown aboard the Space Shuttle Challenger in April, said today the rats suffered significant reductions in the release of growth hormones, the *NY Times* reported. The findings could signal a problem for astronauts, in that weightlessness could affect the ability of special cells in their pituitary glands to produce growth hormones that govern development and maintenance of muscle and bone tissue.

Richard Grindeland, a researcher at NASA's Ames Reseach Center, reported the previous month that the same rats lost bone and muscle strength. Grindeland said that they "were limp, like dishrags," and that dissection revealed "very drastic changes" in bone and muscle strength. However, researchers would need more test results before they could link growth hormone reduction to muscle and bone atrophy, Hymer said. But he added, "I think there's probably a good chance that there is a relationship."

After a battery of tests that included implanting normal rats with growth hormone cells from the space rats, Hymer noted a reduction of up to 50% in release of the hormone. "Something is radically changed in those pituitary glands as a result of the space flight," Hymer said. And he added, "I think the surprising thing was these changes occurred so quickly. They happened in seven days in flight."

However, Hymer said researchers did not know if astronauts suffered the same kind of changes. "I can tell you that there are data that show that the blood levels of growth hormones in the astronauts from Skylab were reduced by a significant amount," Hymer said.

Hymer planned to send growth hormone cells, some of which would contain inhibitory and stimulatory chemicals, on a Space Shuttle flight the following September. He said he also would like to see similar experiments on primates. (*NYT*, Oct 15/85, C8)

December 10: Marshall Space Flight Center announced that NASA's first experiments to construct large structures in space were a success [see Space Transportation System/Mission, Nov. 30 and Dec. 2]. EASE (Experimental Assembly of Structures in Extravehicular Activity) and ACCESS (Assembly Concept for Construction of Erectable Space Structures) carried out by astronauts Jerry Ross and Sherwood Spring "went exactly as we had planned, and all experiment and Space Transportation System objectives were met," said Ed Valentine, mission manager for the EASE/ACCESS payload flown on Space Shuttle mission 61–B. Those objectives included gathering data to compare assembly rates and techniques in space to those used during simulations on the ground and in neutral buoyancy water tests, evaluating potential space station assembly and maintenance concepts and techniques, and identifying ways to improve erectable structures to ensure productivity, reliability, and safety.

"Preliminary data obtained from live downlink TV looks very good," Valentine explained. "But we won't be able to make a complete analyses until we see the rest of the data," which was in the form of videotape and film shot while Ross and Spring performed the construction tasks.

Once the data were analyzed and reported, Valentine said, it would give large space structure designers baseline data on the two construction approaches studied. "They can then determine which approach would be appropriate for a particular task," he explained.

On December 11 Valentine said he would meet for a debriefing with the Space Shuttle crew and principal investigators. Then in about three weeks the principal investigators would receive the videotape and film shot during the experiments, and Valentine would publish a final report in about three months. (NASA Release 85–66)

December 18: Marshall Space Flight Center's (MSFC) Materials Science Laboratory would make its first spaceflight on Space Shuttle 61–C, scheduled for launch on December 18, the *Marshall Star* reported. Using a mission peculiar equipment support structure (MPESS) in the payload bay, the Materials Science Laboratory-2 (MSL–2) would provide accommodations for three experiments in materials processing—the MSFC-managed Electromagnetic Levitator (EML) and Automated Directional Solidification Furnace (ADSF) and the Jet Propulsion Laboratory-Managed Three-Axis Acoustic Levitator (3AAL).

The electromagnetic levitator experiment would allow scientists to study the effects of materials flow during solidification of a melted material in the microgravity environment. Six samples would be suspended in the electromagnetic field of a coil and melted by induction heating from the coil's electromagnetic field.

The ADSF consisted of four furnace, or sample, units. The experiment would investigate the melting and solidification process of four different materials for later comparison to samples of the same materials processed on earth.

In the 3AAL experiment, 12 liquid samples would be suspended in sound pressure waves and rotated and oscillated in a low-gravity nitrogen atmosphere. Investigators would study the degree of sphericity attainable and small bubble migration similar to that having to do with the refining of glass.

The standard switch panel in the orbiter aft flight deck would provide activation, deactivation, and status monitoring capability. (*Marshall Star*, Dec 18/85, 2)

During December. NASA scheduled for launch on December 18 the Space Shuttle Columbia, making its first journey into space in two years, on mission 61–C, the 24th flight of the U.S. Space Transportation System. Robert Gibson would command the flight, his second trip into space. Charles Bolden, on his first spaceflight, would pilot; mission specialists would be Franklin Chang-Diaz, a physicist born in Costa Rica and the first American of Hispanic origin to fly in space, and Steven Hawley and George Nelson, both making their second spaceflight; and payload specialists would be Robert Cenker of RCA and U.S. Rep. Bill Nelson of Florida, chairman of the House subcommittee on space science and applications and the second congressional observer to fly in space.

During the mission, the crew would deploy RCA's *Satcom KU–2* communications satellite, the second in a series of three, with its payload assist module D–2 (PAM D–2) upper stage. The crew on mission 61–C in November deployed *Satcom KU–1*.

Also onboard Columbia in the payload bay would be the Materials Science Lab–2 (MSL–2) [see Space Transportation System/Scientific Research, Dec. 18], the first Hitchhiker (HHG–1) payload, the RCA Infrared-Imaging Experiment (IR–IE), and 12 Get Away Special (GAS) experiments in specialized canisters mounted to a GAS beam attached to the payload bay.

Hitchhiker G–1, a more sophisticated version of the GAS concept, consisted of a baseplate on which small payloads were mounted directly or enclosed in canisters. The carrier for the first time provided small payloads with access to the orbiter's 1400-watt power supply and to ground communications.

Hitchhiker G–1 experiments would study experimental capillary pumped loop heat transport systems (sponsored by Goddard Space Flight Center-GSFC), provide film images of the environment around the Space Shuttle (Air

Force Geophysics Laboratory), and study the effect of the Space Shuttle environment on coated mirrors (Perkin-Elmer Corp.).

The GAS Bridge was a beam supporting 12 GAS canisters that were mounted across the orbiter's cargo bay. The beam consolidated the canisters into a single easy-to-load and unload unit. Experiments in the GAS canisters would:

- measure the effect convection had on heat flow in a liquid (General Electric/Penn State University);
- determine the behavior and physiological effects of microgravity on brine shrimp cysts (Booker T. Washington Senior High School, Houston, Texas);
- measure the O and O_2 terrestrial nightflow emissions (National Research Council of Canada);
- measure the dynamics of a vibrating beam in the zero-gravity environment (U.S. Air Force Academy);
- determine how unprimed canvas, prepared linen canvas, and portions of painted canvas reacted to space travel (Vertical Horizons);
- determine the biological effects of neodymium and helium-neon laser light on desiccated human tissue undergoing cosmic radiation bombardment, determine cosmic radiation effects on medications and medical/surgical materials, perform analysis of contingencies that developed due to zero gravity in blood typing, and evaluate laser optical protective eyeware materials that were exposed to cosmic radiation (St. Mary's Hospital, Laser Laboratory, Milwaukee);
- measure galactic and extragalactic contributions to the diffuse ultraviolet background radiation, and develop and demonstrate an advanced Get Away Special carrier system capable of providing data and power services to Space Shuttle attached sounding rocket class instruments (GSFC);
- measure the effect of gravity on particle dispersion of packing materials in HPLC analytical columns (All Tech Associates, Inc.)
- study the solidification of alloys for lead-antimony and an aluminum-copper combination, study the comparative morphology and anatomy of the primary root system of radish seeds, study crystal growth of metallic appearing needle crystals in an aqueous solution of potassium tetracyanoplatinate, and provide information on the "project-explorer-payload-elapsed-time" and the operational status of experiments during flight to all amateur radio stations and short-wave listeners around the world (Alabama Space and Rocket Center);
- expose wild and lab research gypsy moth eggs and engorged female American dog ticks to weightlessness (GSFC/U.S. Department of Agriculture); and
- measure the response of the GAS Bridge to the Space Shuttle environment during liftoff, orbit, and landing (GSFC).

RCA developed the IR–IE infrared camera, and during the mission payload specialist Rober Cenker would supervise its operation to acquire radiometric information that appeared within the field of view of the self-contained optical system. RCA hoped it might photograph storms, volcanic activity, or other natural occurrences.

Middeck payloads would include the Comet Halley Active Monitoring Program (CHAMP) [see Astronomy, Dec. 17], Initial Blood Storage Experiment (IBSE), Protein Crystal Growth (PC 6), and three student experiments.

IBSE, funded by Johnson Space Center and led by the Center for Blood Research, would study blood storage and sedimentation characteristics in microgravity.

Rep. Bill Nelson would assist with an experiment sponsored by the University of Alabama, Marshall Space Flight Center, and the Comprehensive Cancer Center to try to grow crystal proteins in space for cancer research.

There would be three Shuttle Student Involvement Project (SSIP) experiments on board Columbia. In the Measurement of Auxin Level and Starch Grains in Plant Roots experiment, bean plants would be grown and frozen in space. Researchers would analyze the plant roots after the flight for auxin location and concentration and correlation with statolith in the cells. The Air Injection as an Alternative To Honeycombing experiment would investigate the feasibility of producing a high stiffness, density ratio, low-weight casting with an internal framework. And the Study of Paper Fiber Formation in Microgravity experiment would further increase basic papermaking knowledge through a drainage study of paper fiber formation in microgravity and later comparison to this same formation on earth.

NASA scheduled Columbia to land on the sixth day of the flight at KSC, the first landing there since mission 51–D on April 19, 1985. (NASA Release 85–107; NASA FOR M–989–61–C [prelaunch] Dec 16/85; *Spaceport News*, Dec 6/85, 4; *Goddard News*, Dec 85, 1)

Shuttle Orbiter

February 4: NASA Administrator James Beggs, during a press briefing, noted that the fourth Space Shuttle orbiter, Atlantis, would join the fleet in the spring to fly its first mission in 1985 and that, despite being well into the Space Shuttle's operational phase, NASA would continue to improve performance, procure space parts, and enhance reliability.

In support of Presidential efforts to cut government spending, NASA would slow space station definition and development efforts, cancel the advanced composite structures program in aeronautics, and not initiate any new projects in the space science and applications program. (NASA release Feb. 4/85, NASA press briefing, Feb. 4/85)

February 11: After 16 Space Shuttle flights, NASA believed it had solved the spacecraft's commode problems, as reports of the last two missions indicated the improved waste-disposal system performed perfectly, the *Washington Post* reported.

A $12 million-GE design had transformed the commode from a multi-pronged, motor-drive blender, which had fogged the Space Shuttle's air with particles of noxious dust or frozen liquid, to a device using a blast of air from a fan to carry waste into the toilet bowl; by pressing a lever, the user closed the commode top and opened a valve in the side of the Space Shuttle, where the waste was frozen and dried by the cold and vacuum of space. A tube collected urine and directed it to a holding tank for later overboard disposal. (*W Post*, Feb 11/85, A5)

February 22: Members of the U.S. House Science and Technology space science and applications subcommittee questioned Jesse Moore, NASA associate administrator, office of space flight, on the agency's decision not to request funding for a 5th orbiter in its FY 86 budget, *Aerospace Daily* reported. Moore restated the agency's position that a four-orbiter fleet was adequate to meet traffic demand for the next five years or so, pointing out a recent decline in commercial and military traffic.

NASA had requested $2.1 billion for 14 FY 86 Space Shuttle flights; Moore said NASA planned 17 missions in FY 87, 19 in FY 88, and projected 24 per year after that.

Subcommittee chairman Bill Nelson (D-Fla.), who had indicated strong support for a fifth orbiter, questioned Moore on the likelihood of increased launch traffic demand, potential schedule impacts of an accident, and the possibility of stretching out orbiter fabrication over a seven-to-eight-year period.

Moore responded that NASA's program included adequate funding for structural spares procurement and maintenance of a viable base for initiating orbiter production over the next several years, pointing out that any funds for orbiter construction over a seven-to-eight-year period would be an add-on to NASA's runout budget and could be in competition with other items. He had also remarked on increasingly pessimistic market projections, even considering upcoming yearly launches of 18 to 20 communications satellites.

Earlier, NASA Administrator James Beggs, in response to questions on NASA's exploring possibilities of private-sector purchase of a 5th orbiter, had said that no Space Shuttle privitization proposals submitted to NASA would result in a cost savings to the government. (*A/D*, Feb 22/85, 1)

March 6: The Space Shuttle orbiter Challenger made a three-mile, six hour journey from the launch pad back to a hanger as a result of mission cancellation due to problems with the intended payload, a tracking and data relay satellite, the *NY Times* reported. NASA would replace Challenger on the launch pad with the orbiter Discovery on about March 15 and would combine elements of the scrubbed mission with some from Discovery's flight originally scheduled for March 22. Officials had selected Discovery for the mission because they decided it could be modified more easily and quickly than Challenger. (*NYT*, Mar 6/85, B10)

April 9: NASA announced that the Space Shuttle Carrier Aircraft's (SCA) functional check flight was successfully flown today following engine repairs made in accordance with specifications. The SCA would fly the Atlantis orbiter to Kennedy Space Center on April 12, with one refueling stop. (NASA Daily Activities Report, Apr 11/85)

April 14: NASA's newest Space Shuttle orbiter, Atlantis, arrived about 5:00 p.m. April 14 at KSC after a two-day flight from California, the *Marshall Star* reported. KSC workers removed the Atlantis from the 747 Shuttle Carrier Aircraft (SCA) and towed it to the processing hangar, where others would reconfigure it for its maiden flight, mission 51–J, scheduled for September. The SCA made a brief stop at Ellington Air Force Base, Texas, during the journey to KSC. (*Marshall Star*, Apr 17/85, 2)

April 22: NASA officials said damage sustained by Discovery during mission 51–D [see Apr. 19] could endanger its next mission and the upcoming week's flight of Challenger, the *Washington Times* reported. "We may have to do a fair amount of tile work on this one," Space Shuttle chief Jesse Moore said. "Fifty tiles has been our standard up until this last one."

However, NASA engineers were more concerned about a hole on a control flap, called an elevon, on the ship's left wing. Such damage had not occurred previously. The chief Space Shuttle mechanic said the hole "most likely" developed after a heat-shield tile, jarred loose during Discovery's liftoff, allowed temperatures of more than 1,200° F to penetrate the wing flap during reentry.

Also, NASA said it would not approve the Challenger mission or Discovery's next flight, scheduled for June 12, until technicians completed their examination of damage and determined why the brakes locked. (*W Times*, Apr 22/85, 2A)

May 24: In reporting on NASA's concerns over the effects of vibrations and noise on the Space Shuttle's passengers and cargo, the *Washington Post* quoted astronaut Air Force Col. Frederick Gregory saying that after his return from a Space Shuttle mission that "Nothing prepared me for the sensation of those main engines starting and those solids lighting up underneath us. That son-of-a-gun really rattles and rolls when it takes off."

Although NASA maintained that the Space Shuttle was no rougher a ride than the Titan launch vehicle, doubts were creeping into the minds of some people planning the Space Shuttle program, the *Post* said.

Troubles arose on Gregory's flight when astronauts discovered the latch that released the airlock holding the Spacelab wide-field camera was badly bent—possibly because of the rough ride. In another Spacelab incident, all pressure was lost after four days in the laser spectrometer at the heart of an experiment to measure molecules of synthetic chemicals that could be attacking the earth's protective ozone layer.

Space Shuttle managers were quick to defend the spacecraft, the *Post* reported. Columbia, Challenger, and Discovery had flown 17 times in four years without a serious incident. At least two of the three satellites that failed in space did so because of onboard rocket engine difficulties, not because the Space Shuttle was too rough.

Johnson Space Center Director Gerald Griffin said in an interview with the *Post* that "occasional" mistakes had led to mishaps, but he insisted that "we are not repeating our mistakes." He used as an example the Syncom (*Leasat-3*) satellite deployed in April with no power to raise it to a higher orbit. He said that NASA would not have deployed the Syncom without any power if NASA had insisted that Hughes Communications Corp install a device that flashed a warning to the Space Shuttle's cockpit that the satellite had no power.

"We've always insisted our customers adhere to our safety standards to protect the crew and the shuttle, but we've never insisted they do anything else," Griffin said. "We don't want to be 'Big Brother' but we're beginning to think that some of the [the customers] design features are not good for us. I'd like to see more data for the crew in the cockpit." (*W Post*, May 24/85, A1)

May 30: NASA would probably use the desert runways of Edwards Air Force Base for landings of the remaining seven Space shuttle missions in 1985 due to brake damage caused by the grooved tarmac at Kennedy Space Center, the *Washington Post* reported.

"Let's face it, these brakes are right at the edge in a vehicle that lands as heavy and as fast as the shuttle does," said astronaut Daniel Brandenstein, commander of the next Space Shuttle mission. "My guess is we'll be landing for awhile out in the Mojave, where there's no end to the runways and there's almost nothing on either side of the runways to worry about either."

Also, studies suggested that the Space Shuttle nose gear might require rebuilding to provide better steering during landing in a crosswind, a situation more likely to occur at KSC than at Edwards. Cost of such rebuilding could run as high as $3 million for each orbiter. (*W Post*, May 30/85, A19)

June 14: NASA expected the Space Shuttle's nose wheel steering capability, previously used only on an experimental basis, would achieve operational status by the fall, the *Space News Roundup* reported.

NASA had attributed brake damage to the orbiter Discovery following the STS 51–D mission in part to crosswinds at KSC, which required the orbiter's commander to use the main gear braking system for differential steering. Nose wheel steering would allow Space Shuttle commanders to use the brake system for stopping orbiters rather than for braking and steering combined.

A nose wheel steering capability was in the Space Shuttle design, but used only on an experimental basis. NASA had spent the last year upgrading the system to operational status. In addition, NASA instrumented the Challenger to provide better landing condition data and planned to add instrumentation to the landing gear system of the Discovery.

In the 17 Space Shuttle missions to date, NASA had used 68 individual brake systems with 27 receiving no damage, 27 receiving slight damage, and 14 receiving significant damage in which either one or more stators or rotors overheated or suffered other problems.

Until the nose wheel steering capability became operational, NASA intended to choose on a flight by flight basis between Kennedy Space Center and Edwards Air Force Base as the primary end of mission landing site. (JSC *Space News Roundup*, June 14/85, 2)

July 19: After spending 18 months at the factory in Palmsdale, California, Columbia, the flagship of the Space Shuttle fleet, returned for operational duty to Kennedy Space Center (KSC), the JSC *Roundup* reported. Columbia had undergone hundreds of modifications to reconfigure the ship from a developmental flight vehicle to a fully operational orbiter.

Columbia flew the first five Space Shuttle missions, and NASA modified the spacecraft at KSC for the October 1983 STS–9 Spacelab 1 flight. Following its return to KSC and removal of the Spacelab 1 payload, NASA in late January 1984 ferried Columbia to California atop the 747 Shuttle Carrier Aircraft.

Rockwell International performed five major modifications to the orbiter during its stay at the orbiter production plant. Structural modifications to the wings and midfuselage included strengthening of lower wing surfaces and installation of heavier straps across the orbiter's belly.

Rockwell added a heads-up display to the forward flight deck to allow the Space Shuttle commander and pilot to view critical flight information on a see-through panel while they looked through the forward cockpit windows.

Production workers removed the commander and pilot ejection seats, which were installed on Columbia for the initial development Space Shuttle flights, and replaced them with standard seats. Overhead blowout panels were also removed.

Rockwell modified two orbital maneuvering system pods to make them compatible with the other orbiters in the fleet and replaced the thermal protection tiles on the pods with the advanced reusable surface insulation blankets. Production personnel also installed supplemental instrumentation to gather developmental engineering data.

In addition, workers removed and replaced approximately 5000 high-temperature tiles from the orbiter's underside, because engineering analysis indicated several areas where greater heat protection was needed. About half of the wing leading edge panel assemblies was removed and modified to strengthen the supporting structure of the reinforced carbon-carbon panels, an infrared imaging device was installed on the vertical stabilizer to provide a temperature profile of the upper wing surfaces during reentry, and instrumentation was added to the nose cap to provide improved entry air data.

During the trip from California to KSC atop the 747 shuttle transport, rain damaged more than 1,000 of its head shield tiles, the *Washington Post* reported. NASA said that between 200 and 300 of the tiles required replacement. Damaged areas were around the windows and on the forward facing tiles. According to NASA policy, Space Shuttle transports were forbidden to fly through visible rain. However, the shower did not appear on ground radar or on the 747's onboard radar system.

NASA had scheduled Columbia to fly next in December on mission 61–C. (JSC *Roundup*, July 19/85, 1; *W Post*, July 19/85, A17)

July 25: At the third in a series of hearings on assured access to space NASA and Defense Department officials said today before the House Science and Technology space science and applications and Armed Services research and development subcommittees that available data did not support the need for a fifth Space Shuttle orbiter in the early 1990s. But the officials acknowledged that there could be a period around 1992 and beyond in which unforeseen launch demands would strain capability, *Aerospace Daily* reported.

Rep. Bill Nelson (D-Fla.) pointed out that earlier testimony had revealed a launch demand for 26 to 33 Space Shuttle equivalent flights per year in the post-1992 timeframe, and Jesse Moore, NASA associate administrator for spaceflight, acknowledged that a flight capability of 30 per year would require a fifth orbiter. He said that NASA projected the annual launch capability of a four-orbiter fleet in the 1990s timeframe to be 24 flights, 20 from Kennedy Space Center and four from Vandenberg AFB, with the ability to surge temporarily to 28 flights per year.

When asked to factor in potential extended downtime of one orbiter, Moore replied that NASA had not made a projection of lost flights because it would be directly related to the length of the downtime and when it occurred. He added that NASA could probably fly 20 to 24 flights with the three remaining orbiters if one were out of service for periods up to one year; a sustained launch capacity for three orbiters would be in the range of 15 to 20 flights per year, he said.

Edward Aldridge, undersecretary of the Air Force, said that, although the Department of Defense (DOD) was committed to at least eight Space Shuttle missions per year, its current known requirements through the mid-1990s averaged 10 launches per year. Including the one-to-five Space Shuttle-equivalent launches per year projected through 1995 for the Strategic Defense Initiative (SDI) research, he said the total DOD requirement became roughly 11 to 15 Space Shuttle-equivalent launches per year. Those numbers excluded any mission supporting potential deployment of SDI-related systems, he added.

At the same hearing, Gen. Robert Herres, commander of the North American Air Defense Command (NORAD), said he thought the "time had passed" during which the U.S. might have built a fifth orbiter. He pointed out that resources needed for it might better be devoted to new technology and options, "which afford us more flexibility and versatility."

Offering an opposing view, Hans Mark, chancellor of the University of Texas System and former NASA deputy administrator, said that the argument for a fifth orbiter could not be based either on currently projected flight rates or on backup requirements. He said only the proposed space station and SDI requirements could increase space operations to the point where a fifth orbiter was necessary. (A/D, July 26/85, 137)

September 12: NASA announced that Rockwell Internatl. Corp. made significant additions to the Space Shuttle orbiter Columbia to accommodate three research experiments that would measure orbiter aerodynamic and thermodynamic characteristics as it reentered earth's atmosphere. Researchers would use the flight data to develop future generation space transportation systems.

The most obvious change was a cylindrical housing that replaced the fintip atop the vertical tail. The new experiment pod, containing equipment for the shuttle infrared leeside temperature sensing (SILTS) experiment, was approximately 20 in. in diameter and was capped at the leading edge by a spherical dome. SILTS would obtain high-resolution infrared images of the upper (leeside) surfaces of Columbia's port wing and fuselage as the orbiter reentered the atmosphere, providing detailed temperature maps at the surface of the leeside thermal protection materials and indicating the amount of aerodynamic heating of surfaces in flight. Tape recorders would store experiment data.

Infrared cameras mounted inside the dome would view Columbia's left wing and fuselage through two windows protected during launch from debris by plugs that filled the window cavities and that fell away when the experiment began about 400,000 feet above earth at reentry. Injection of gaseous nitrogen into the cavities would cool the cameras during that period.

For the shuttle entry air data system (SEADS) experiment, a new nosecap had 14 penetration assemblies distributed about its surface, each containing a small hole through which local surface air pressure was measured from an altitude of about 56 miles through landing. This would allow precise postflight determination of the orbiter's attitude relative to the oncoming airstream and the density of the atmosphere through which the vehicle was flying.

The shuttle upper atmosphere mass spectrometer (SUMS) would complement the SEADS experiment by providing atmospheric density information at altitudes above 50 miles. SUMS would sample air at Columbia's surface through a small hole, located just aft of the nosecap, to measure the number of molecules of various gas species in order to determine the atmospheric density that, with vehicle motion information, would allow determination of orbiter aerodynamic characteristics at altitudes where the atmosphere was extremely thin. NASA originally developed the SUMS mass spectrometer for the Viking spacecraft that landed in 1976 on Mars and modified it to operate in the orbiter's reentry flight environment.

NASA' Langley Research Center developed the experiments as part of the orbiter experiments program managed by NASA's office of aeronautics and space technology. (NASA Release 85–127)

September 30: NASA announced that, since adverse weather conditions such as rain, mist, or ice could damage Space Shuttle tiles, its Ames-Dryden Flight Research Facility, using an F–104 aircraft as a testbed, would test Space Shuttle thermal protection system tiles for moisture impact damage and to verify techniques to record and measure atmospheric moisture. NASA researchers would then correlate this information with existing Space Shuttle launch criteria and determine the need for further tests.

Researchers would install actual Space Shuttle tiles on the leading edge of a flight test fixture mounted below the F–104's fuselage. Some tiles were unused; others had flown in space on the orbiter Columbia.

During initial flights in the 16-to-18 flight test program, the F–104 would fly at subsonic speeds behind a KC–135 tanker aircraft, which would emit a water spray to create artificial rain. Tanker personnel could control the flow rate, nozzle pressures, and size of the artificial raindrops.

Pilots would fly additional flights at subsonic speeds near Vandenberg Air Force Base, a future Space Shuttle launch site, to observe and measure the effects of mist and low stratus clouds. Pilots later would fly at transonic and supersonic speeds flights at high altitude through ice particles in clouds.

A particle-measurement probe located on the F–104's wing pylon would record moisture particle size, while high-frequency load sensors would measure impact forces. A noseboom on the flight test fixture would record velocity, and test fixture pressure orifices would measure pressure distribution. Video cameras, one pylon-mounted and one mounted on the F–104's lower fuselage looking back at the flight test fixture, would also record data. (NASA Release 85–135)

December 13: Langley Research Center announced that the Space Shuttle Columbia would carry three experiments on mission 61–C developed at the center to measure the spacecraft's aerodynamic and thermodynamic characteristics.

Columbia, the first Space Shuttle orbiter to go into space, was making its first flight in two years after it was pulled from flight service in December 1983 for extensive overhaul, including modification to accommodate the LaRC experiments, which were the Shuttle Entry Air Data System (SEADS), Shuttle Infrared Leeside Temperature Sensing (SILTS) experiment, and Shuttle Upper Atmosphere Mass Spectrometer (SUMS).

SEADS would measure the distribution of air pressure around the orbiter's nosecap during entry to provide precise determination of the orbiter's attitude relative to the oncoming airstream and the density of the atmosphere through which it had flown. Lack of this air data had hindered engineers from determining aerodynamic flight characteristics of the orbiters.

The SUMS experiment complemented the SEADS experiment by gathering atmospheric density information at altitudes above 90 km (56 statute miles). SUMS would sample the gas at Columbia's surface through a small hole located just aft of the nosecap and forward of the nosewheel, with its instrument identifying and measuring the quantities of the various gas species present. Data analysis after the mission would allow determination of atmospheric density.

The SILTS experiment would obtain high-spatial-resolution infrared images of the upper (leeside) surfaces of the orbiter's port wing and fuselage during entry through the atmosphere. These infrared images would provide detailed "maps" of the surface temperatures of leeside thermal protection materials, indicating the amount of aerodynamic heating on the leeside surfaces in flight. (LaRC Release 85–99)

December 22: NASA today rolled the Space Shuttle orbiter Challenger to launch complex 39B two miles from where NASA had installed the orbiter Columbia for mission 61–C, the first time a pair of orbiters had been on the launch pads at the same time, the *Washington Times* reported. Columbia was on Pad 39A until repair of a faulty power unit that aborted a launch attempt December 19 [see Space Transportation System/Launch Schedules, Dec. 19]. NASA had scheduled Challenger on mission 51–L for launch on January 22. (*W Times*, Dec 23/85, 4A)

Solid-fuel Rocket Boosters

January 30: NASA and United Technologies's United Space Boosters Inc.'s Booster Production Co. officials would break ground January 30 for Marshall

Space Flight Center's (MSFC) solid-fuel rocket booster (SRB) assembly and refurbishment facility at KSC, the *Marshall Star* reported. The company had designed and would build the 238,000-sq.-ft. facility, scheduled for completion in 1986, that would employ about 700 people to perform most of the Space Shuttle's SRB refurbishment previously done by the company at various locations at the center and adjacent Cape Canaveral Air Force Station.

Refurbishment, performed under an MSFC contract, would include replacement of insulation on booster components, addition of electronic and guidance systems, and reinstallation of parachutes and ordnance. The refurbished booster components, including forward and aft skirts and frustum, would then undergo computerized checkout in the facility before delivery to NASA for booster stacking.

The new facility would process up to 24 flight sets of solid-fuel rocket boosters yearly and would offer maximum efficiency with extensive use of robotics and computer-controlled production techniques developed in MSFC's Productivity Enhancement Center. (*Marshall Star*, Jan 30/85, 1)

March 20: NASA had completed a series of tests to validate parachute systems to be used with advanced lightweight, solid-fuel rocket boosters that would replace the boosters then in use on the Space Shuttle, Ames Research Center (ARC) announced. The new boosters would be first used during a Space Shuttle launch from Vandenberg AFB no earlier than January 29, 1986.

For the tests, researchers used a vehicle a third the size of a solid-fuel booster with one main parachute and, to simulate a splashdown, drop tested it from the Ames-Dryden Flight Research Facility's (DFRC) B-52 over the China Lake Naval Weapons Center at Ridgecrest. Most drops were made at an altitude of about 40,000 feet with an airspeed of about 230 knots. ARC and DFRC conducted the drop tests in cooperation with Marshall Space Flight Center, which was responsible for the parachute system. (ARC Release 85-1)

May 9: Marshall Space Flight Center's (MSFC) prime contractor for Space Shuttle solid-fuel rocket booster motors, Morton Thiokol's Wasatch Division, successfully static fired today a new lightweight version of that motor at its facility in northern Utah, the *Marshall Star* reported.

The firing was the second and final test of the development version of that motor, which contained major sections made from graphite epoxy materials manufactured by Thiokol's subcontractor, Hercules, Inc. Thiokol would perform a final qualification static test of the flight version of the new filament-wound motor in September.

NASA had scheduled the first use of the filament-wound motor segments for early 1986 during the first Space Shuttle launch from Vandenberg Air Force Base, California.

Lawrence Mulloy, manager of MSFC's solid-fuel rocket booster project, said each of the filament-wound boosters would weigh about 28,000 lb. less than the current steel boosters, making possible an increase in Space Shuttle payload carrying capacity of about 4,600 lbs. (*Marshall Star*, May 15/85, 2)

June 3: A fire of unknown cause today destroyed one of the four buildings used to mix rocket solid fuels for the Space Shuttle at Morton Thiokol's rocket motor assembly plant outside Brigham City, Utah, the *Washington Post* reported. No one was injured in the fire that leveled a 400-sq.-ft. building operated largely by robots at a remote part of the Thiokol plant.

Because three of the four fuel mixing plants at Thiokol were needed at all times to maintain solid-fuel rocket motor production schedules, NASA said it was concerned enough to send an accident review board team from Marshall Space flight Center to investigate. (*W Post*, June 5/85, A5)

June 6: NASA was still evaluating data submitted by four companies in response to its request for responses on the possibility of opening a second source for Space Shuttle solid-fuel rocket motors (SRM), *Aerospace Daily* reported. Those responding were Hercules, Aerojet Strategic Propulsion, United Technologies' Chemical Systems Division, and Atlantic Research Corp. NASA also had a proposal from Morton Thiokol, the current sole-source producer, and would not make a final decision until later in the month.

In April Jesse Moore, NASA associate administrator for spaceflight, told the House appropriations HUD and independent agencies subcommittee that he expected to "have a review of the assessment sometime in the early part of May" and that in the interim NASA would proceed with the next buy of SRMs from Morton Thiokol. Agency officials expected it would take about four years to qualify a second source and "get them up to speed."

NASA Administrator James Beggs would make the final decision on whether to open a second source for SRMs. A NASA official said he thought there was "a good possibility" that Beggs would have the needed information to make "a decision by the end of this month." (*A/D*, June 6/85, 1)

December 26: NASA today released a statement on the proposed action to develop a second source for the Space Shuttle's solid-fuel rocket motor (SRM). Since November 1984, NASA had assessed the desirability of a second source, and four firms—Aerojet Strategic Propulsion Co., Atlantic Research Corp., Hercules, Inc., and Chemical Systems Div. of United Technologies Corp.—had expressed interest in becoming the SRM second source. As a result, NASA believed the development of an SRM second source would serve the national interest.

The statement read in part: ". . . Because of the continued strong interest of the four contractors and our belief that a second source would be in the national interest, it is our intent to provide the contractors the opportunity to respond through a formal request for proposal (RFP).

"Industry must, however, be willing to respond with the full knowledge that NASA will provide no firm guarantee of recovery of their qualification costs. If industry is so willing, and where competition remains available with the incumbent excluded, NASA intends to proceed with establishing a second source.

"Should the winner of a second source competition for SRMs become qualified to produce flight-quality motors in time to support our next production buy, NASA would plan to buy contingent on overall cost considerations, 20 flight sets of motors at the rate of not less than four flight sets a year. In addition, NASA would compete the balance of the buy and subsequent buys to determine the split in production between the incumbent and the qualified second source. A decision to dual source any quantity beyond the planned minimum buy of 20 flight sets would be based on the cost of the dual source in comparison to the cost of contracting the balance with the incumbent (Morton Thiokol, Inc.) only.

"Based on the above, NASA intends to notify the four interested firms by letter in early January 1986 of our willingness to proceed with an RFP and the specific conditions that would apply. Their comments and views would be requested. A positive response would provide the basis for NASA to issue an RFP. Should industry not support the RFP under these conditions, NASA would cancel any initiative of establishing a dual source for the SRM." (NASA Release 85-178)

Student Involvement Program

March 18: Some high school student semifinalists in the Space Shuttle Student Involvement Program, sponsored by NASA and the National Science Teachers Association, would meet at Lewis Research Center (LeRC) March 25 and 26 to present before a panel of LeRC scientists and engineers their proposals for candidate experiments to fly aboard future Space Shuttle missions, the center announced. LeRC personnel would evaluate and suggest improvements to the experiments. The program was a nationwide competition to stimulate the study of science and technology by engaging students in projects to develop actual payload experiments for upcoming Space Shuttle flights.

The 37 students from Ohio, Michigan, Minnesota, Iowa, Missouri, and Wisconsin would hear a presentation by NASA astronaut Robert Springer (Lt. Col. USMC), slated to fly aboard Space Shuttle mission 51-H scheduled for launch in late 1985, and a discussion by Brian Vlcek, a previous year's finalist from Parma, Ohio, on his winning experiment entitled "inducing a geotropic-type reaction in radish roots with chemical stimuli." The meeting would close with a tour of LeRC where the students would see such facilities as the 500-foot deep, zero-gravity facility and a propulsion systems laboratory where researchers tested full-scale aircraft engines under actual flight conditions.

Initially students from 10 geographical regions around the country had proposed over 2,000 biology, chemistry, and astronomy experiments. Interdisciplinary teams of teachers, scientists, and engineers had evaluated the proposals to select each region's semifinalists. As of the Space Shuttle 51–C mission in January 1985, NASA had flown 10 student experiments in space. (LeRC Release 85–17)

April 5: NASA announced that through a cooperative effort with Satellite Communications for Learning Worldwide (SCOLA), it would televise live via satellite to hundreds of high schools, colleges, and universities across the U.S. all the science events on the Space Shuttle Discovery following its launch no earlier than April 12 from KSC. Two days prior to launch, NASA scientists and engineers would conduct a permission videoconference with the schools to acquaint students and teachers with the planned mission and science experiments. Students from selected schools would ask questions of the NASA briefers. NASA's educational affairs office was providing Discovery mission and educational materials to the participating schools.

SCOLA, an association of schools, colleges, and universities based at Creighton University, Omaha, Nebraska, was one of the largest voluntary associations of institutional satellite antenna owners in the U.S. (NASA Release 85–51)

May 7: NASA and the National Science Teachers Association announced continuation for the sixth year of the Space Shuttle Student Involvement Program, which provided an opportunity for secondary school students (grades 9–12) to write proposals for space science experiments that would develop their awareness of space and stimulate interest in science and technology.

During the next year's competition, the association would select up to five students from each of eight geographic regions who, along with their teachers, would receive all-expense-paid trips to a Space Shuttle symposium at a NASA center. Later, one semifinalist selected from each region would attend a national Space Shuttle symposium at Kennedy Space Center, during which the association would award scholarships to the three students submitting the most outstanding proposals. (NASA Release 85–70)

May 30: NASA announced that interdisciplinary teams of teachers, scientists, and engineers selected seven finalists in the fifth national Space Shuttle

Student Involvement Program (SSIP), a joint venture of NASA and the National Science Teachers Association. Judges based selection of finalists' entries upon individual scientific or engineering merit for possible flight aboard the Space Shuttle.

The national winners and their teacher-advisors would attend in August a special Space Shuttle conference at Kennedy Space Center and view the launch of the Discovery Space Shuttle mission 51–L. NASA would present the finalists and their teacher-advisors with orbiter models, and their schools would receive orbiter models for trophy case display.

The sixth year of SSIP competition would open in September 1985. (NASA Release 85–79)

During August: NASA announced that the newest solid-fuel rocket booster retrieval ship, the U.S. Air Force's Independence, arrived at Kennedy Space Center for outfitting with retrieval gear. The Air Force specifically designed the ship for retrieval of expended boosters after Space Shuttle launches from Vandenberg Air Force Base.

Independence would be the mother ship for west coast booster retrievals, towing one of the boosters and providing room for the majority of the crew and the six main parachutes and the bottoms for both boosters. The ship's first booster retrieval would be mission 51–F set for launch no earlier than July 15. The Air Force would lease another smaller ship to tow the other booster and carry the remainder of the 42-member retrieval crew.

Independence, at 199 feet long and with a beam of 40 feet, was larger than its sister retrieval ships, the Liberty Star and the Freedom Star. Those two ships were 176 feet long and had beams of 37 feet. The Independence has both fore and aft water jet thrusters, making it highly maneuverable. The other two ships had only stern thrusters. Halter Marine in Moss Point, Mississippi, built Independence. (*NASA Activities*, Aug 85, 11)

SPACELAB

February 25: The European Space Agency (ESA) issued a review of some material and life sciences experiments aboard the first Spacelab mission November 28 to December 8, 1983.

Of the 33 materials sciences experiments, two were not performed due to hardware malfunctioning, and several experiments did not obtain the full set of experimental runs/operations. The advantage of the microgravity environment for materials processing and fluid physics was the practical absence of thermal (gravity-driven) convention, sedimentation, and hydrostatic pressure. In the microgravity environment, secondary-disturbing factors such as surface tension, capillary forces, and intermolecular forces became dominant.

Some materials sciences experiments focused on protein crystallization, including growth of insoluble crystals by precipitation reaction, nucleation and growth experiments in vapor crystal growth, thermomigration (soret diffusion) of cobalt in liquid tin, floating-zone growth of silicon, investigation of free convection and capillary surfaces in low gravity, and eutectic solidification and formation of fault structures in fibrous and lamellar eutectics.

The Spacelab mission carried nine European life sciences experiments. One investigated functioning in zero gravity of the human vestibular system, an acceleration-sensing system in the inner ear. Others were a mass-discrimination experiment aimed at comparing perception of mass under microgravity conditions and of weight on earth, an experiment aimed at understanding fluid-regulation mechanisms in the low-pressure system of the human body, and observations of the proliferation of lymphocytes during weightlessness. (ESA release Feb 25/85)

April 17: NASA announced it had scheduled the Spacelab 3 mission, a European-developed and NASA-operated space laboratory, for launch on STS 51–B no earlier than April 29, 1985. Spacelab 3 was a microgravity mission with 15 investigations in five scientific disciplines: materials science, life sciences, fluid mechanics, atmospheric science, and astronomy. Two of the investigations—one in materials science, mercury iodide crystal growth [France], and one in astronomy, very wide field camera [France]—had flown aboard Spacelab 1. Scientists in the U.S. sponsored 12 of the investigations, and an ionization measurements investigation was from India.

Important new hardware developments in materials science, fluid dynamics, and life sciences would be on the flight, and researchers had designed the experimental hardware for multiple flight use. NASA had selected two payload specialists with expertise in crystal growth and fluid mechanics for the flight.

The Spacelab 3 configuration consisted of a long tunnel, a long module, and a mission peculiar equipment support structure (MPESS). The module and tunnel would provide a pressurized shirtsleeve environment within which the crew could operate. The MPESS also supported experiments that did not require a habitable crew environment. Common payload support equipment used on Spacelab 3 was the scientific airlock that could extend from the module into space for experiments requiring exposure to the space environment and flight crew hands-on activities.

There were three experiments on materials science: the fluid experiment system (FES), the vapor crystal growth system (VCGS), and the French materials science experiment.

In fluid mechanics, the crew would perform fundamental experiments in the drop dynamics module (DDM) to test theoretical predictions of drop behavor in a near-zero gravity environment. The geophysical fluid flow cell (GFFC) experiment would study fluid motion in a microgravity environment.

The six investigations in life sciences included four in the Ames Research Center's (ARC) life sciences payload system to verify a new facility for housing and studying animals in the space environment, observing the animals' reactions to that environment, and evaluating operations and procedures relative to in-flight animal care; the urine monitoring investigation to monitor crew water intake and to prepare urine samples for postflight analysis; and the autogenic feedback training experiment to test a technique to control space adaptation syndrome.

The four investigations in atmospheric science and astronomy were the ionization states of solar and galactic cosmic ray heavy nuclei that would use a new detector system to determine the composition and intensity of ions emitted toward the earth from the sun and other galactic sources; the French astronomy investigation to make an ultraviolet survey of the celestial sphere in a study of large-scale phenomena such as clouds within our galaxy; the auroral observations experiment to observe and record the visual characteristics of pulsating and flickering auroras; and atmospheric trace molecules spectroscopy to examine on a global scale the composition and variability of the upper atmosphere.

In addition to the Spacelab 3 experiments, NASA had scheduled two Get-Away Specials for the flight—a Northern Utah University satellite (NUSAT) and a global low-orbit message relay satellite (GLOMR). NASA would enclose each in a canister mounted in the orbiter payload bay for deployment near the end of the mission following completion of Spacelab operations. (NASA MOR E-977-51-B-02 [prelaunch] April 17/85)

April 17: Marshall Space Flight Center (MSFC) announced its researchers completed at KSC a series of three intensive tests that verified the compatibility of Spacelab 2 experiments in space. Known collectively as the "mission sequence test," the tests determined for the first time that actual experiment flight hardware for the mission operated in tandem with the Spacelab flight systems. MSFC's Spacelab 2 mission manager Roy Lester said they learned they "could operate this complex system with high efficiency within acceptable Spacelab resource allocations."

MSFC had responsibility for managing the first three Spacelab missions in a series of Spacelab flights that extended over several years. The first mission flew in 1983; the next would fly later in April. "This third and last test was a simulated six-hour slice of the actual on-orbit timeline," Lester said. "It simulated the commanding of experiments as if they were being done both from the payload operations control center (at Johnson Space Center) and from the flight deck of the orbiter."

The period chosen was one that would put maximum stress on the computer, "a worst-case scenario," Lester explained. "Without a situation like that, we wouldn't have gotten the confidence we have now that we're ready to go."

The 13 Spacelab 2 experiments, covering seven scientific disciplines, would rest on three pallets and a special support structure. Unlike the other two Spacelab missions managed by MSFC, there was no habitable laboratory. Spacelab 2 crew would operate the experiments from the aft deck of the orbiter interior. (MSFC Release 85–19)

During April: NASA announced that Space Shuttle flight 51–B/Spacelab 3, an ESA-developed Spacelab carrying 15 experiments [see Spacelab, Apr. 17], was scheduled for launch April 29 from KSC. The mission's main objective was to provide a high-quality microgravity environment for materials processing and fluid experiments.

For the second time in U.S. space history, crew members would perform scientific investigations continuously. Two scientists who developed Spacelab 3 experiments, payload specialists Dr. Lodewijk van den Berg, a materials scientist from EG&G Energy Management Corp., and Dr. Taylor Wang, a fluid physicist from JPL, would conduct onboard research during the mission.

Mission specialist Dr. Don Lind, a high-energy astrophysicist, and Drs. Norman Thagard and William Thornton (both medical doctors making their second Space Shuttle flight) would also do scientific research.

The payload operations control center (POCC) at Johnson Space Center would manage all Spacelab 3 operations. Members of the Marshall Space Flight Center mission management team and the investigator teams that developed Spacelab 3 experiments would monitor, direct, and control experiment operations from the ground control center. The mission control center in the same building as POCC would control the orbiter and basic Spacelab systems. (NASA Release 85–60)

May 1: A C–5A aircraft carrying the German Spacelab D–1, the first payload in the history of U.S. manned spaceflight to be controlled from another country, landed May 1 at the Kennedy Space Center (KSC) Space Shuttle runway, *Spaceport News* reported. The Federal German Aerospace Research

Establishment's (DFVLR) German Space Operations Center in Oberpfaffenhofen near Munich would have responsibility for the 80 Spacelab D-1 (D for Deutschland) experiments. During the welcoming ceremony, (KSC) Director Dick Smith said, "We are looking forward to giving your payload 'tender loving care' and seeing it launched aboard the shuttle and getting it safely returned to you after the mission."

Germany's Parliamentary State Secretary Albert Probst responded, "We are pleased to take part in this event and consider the joint ventures between the Federal Republic of Germany, Europe, and the U.S. in manned space flight to be a good example of international cooperation."

Spacelab D-1, scheduled to fly on the seven-day Space Shuttle orbiter Columbia mission 61-A no earlier than October 16, consisted of a long Spacelab module containing the space sled and a unique support structure. Half of the D-1 experiments were from Germany, the remainder from the U.S. and European countries. Investigations included the vestibular sled, a device driven by an electromotor and traction rope, positioned in the center aisle of the Spacelab module, in which astronauts would serve as test subjects to collect information on human reactions to the equilibrium sensing system; the Biorack for botanical and medical/physiological experiments; the navigation experiment (NAVEX) to test systems performance of clock synchronization and one-way distance measurement; the materials experiment assembly (MEA) that would use Marshall Space Flight Center furnaces and alloy, fluid physics, and crystal growth investigations.

The DFVLR, the result of a merger of three organizations and employing about 3,500 people in five research centers, would also participate in the European Retrieval Carrier (EURECA), a free-flying apparatus of experiments, and project "Columbus," internationally manned space station modules. (*Spaceport News*, May 10/85, 5)

May 21: Marshall Space Flight Center announced that experiment teams for the Spacelab 3 mission had begun analysis of the 250 billion bits of data obtained during the mission, and preliminary results were excellent.

In the area of materials science, two of the three experiment teams had begun to examine crystals grown during Spacelab 3. In one experiment, a mercury iodide crystal the size of a sugar cube grew from a seed crystal in the vapor crystal growth system; via the vapor transport technique, the crystal grew at a carefully controlled rate over a 104-hour period. Researchers would remove the crystal from the glass ampoule within the next few weeks for analysis to determine its quality and properties as an X-ray and gamma-ray detector for applications in scientific research, medicine, and industry.

Two fluid mechanics experiments ended successfully; the geophysical fluid flow cell experiment completed more than 102 hours of experiments designed to provide information for the first time on convention in spherical rotating shells with a radial gravity field. Co-investigator Dr. Fred Leslie said of the experiment, "The films of one of the Jovian (Jupiter) scenarios indicated the formation of a 'Red Spot' similar to that present on Jupiter. This data should be useful to meteorologists and astrophysicists in modelling the large-scale circulations of fluids under the influence of rotation, gravity, and convection."

Three atmospheric and astronomical observations were completed during the Spacelab 3 flight. The atmospheric trace molecules spectroscopy experiment operated for 50 hours providing 19 sequences of more than 150 independent atmospheric spectrally resolved measurements. Researchers would use these to analyze the earth's atmospheric composition chemically and physically for the stratosphere and mesosphere between 10 and 150 km. The information also would provide extremely detailed measurements of the minor and trace components of the atmosphere, which was crucial to understanding the evolution of earth's climate, and would aid in the analysis of pollution causes and effects. The experiment also provided the first high-resolution infrared spectrum of the sun, which indicated some surprising evidence about its molecular constituents.

Four of the six life sciences experiments completed successfully were part of the Ames Research Center's life sciences payload. The research animal holding facilities proved a suitable animal habitat in which the two squirrel monkeys and 24 rats adjusted to spaceflight and demonstrated their suitability for research in orbit. One primate apparently developed symptoms of space adaptation syndrome but recovered in a manner analogous to human experience.

"In addition to the successful results from each of these experiments," said Dr. George Fichtl, Spacelab 3 mission scientist at Marshall Space Flight Center, "the overall success of the mission has proven the importance of the Spacelab concept. During this mission, we have proven that a microgravity environment is an important environment for the growth of crystals, the study of fluid mechanics, and measurement of the effect man is having on this environment. We have also proven the need for man's presence in performing these tasks." (MSFC Release 85-28)

May 24: Spacelab 2 mission manager Roy Lester of Marshall Space Flight Center's (MSFC) Spacelab Payload Project Office said today that the Spacelab 2 payload was declared "ready for installation" into the orbiter Challenger, the *Marshall Star* reported. Spacelab 2 would be the second of two Spacelab missions designed to demonstrate the performance of hardware received from the European Space Agency (ESA).

The Spacelab 1 mission verified the majority of Spacelab systems including the habitable module, tunnel, scientific airlock, and pallets. For the second verification mission, Spacelab would not use a habitable module but only a pallet configuration. The crew would operate experiments from Challenger's aft flight deck.

Spacelab 2 consisted of a pressurized igloo approximately the size of a five-gallon oil drum and three pallets, two of which joined to form the train needed to support several experiments. Also new on the mission was the instrument pointing system for pointing four of the onboard solar experiments. Spacelab would carry 13 experiments in 7 scientific disciplines. Primary objective of the mission was to demonstrate the performance of the igloo, the two-pallet train, and the instrument pointing system.

John Thomas, manager of the MSFC Spacelab Program Office, said NASA had removed Spacelab 3 hardware from Challenger's cargo bay and was reconfiguring the bay for Spacelab 2. "We expect to begin installation of the Spacelab 2 payload on June 3," he noted.

The last major test for the hardware before July 12 scheduled launch would be an end-to-end test planned for June 10, in which NASA would test all systems connecting Challenger, Spacelab 2, and the MSFC-operated payload operations control center (POCC) at Johnson Space Center. Thomas said the other primary effort was to complete training of the payload team. (*Marshall Star*, May 29/85, 1)

August 6: When the Space Shuttle Challenger on mission 51–F landed today, it successfully concluded the third mission of the European-developed space laboratory, Spacelab 2, which NASA used for the first time in its pallet-only configuration, NASA reported. All Spacelab subsystems but one provided excellent data, and NASA verified for the first time two new Spacelab systems, the "Igloo" and the instrument pointing system.

NASA would fly many of the Spacelab 2 instruments on future missions, so the 51–F mission provided opportunities not only for scientific data collection but also for engineering checkouts of the new equipment. Since instruments are brought home, engineers could evaluate and refurbish them to improve their performance or modify them to meet different scientific objectives in response to results from prior flights.

Through Spacelab 2's four ultraviolet and visible light instruments, researchers observed a sun that, although it was in a fairly quiet phase of its activity cycle, displayed sunspots, filaments, granules, spicules, and prominences. These solar features changed over periods as brief as five to ten minutes and from orbit to orbit, day to day.

The solar ultraviolet universal polarimeter (SOUP) started its program late in the mission after an unexplained shutdown and startup. Thereafter, the instrument performed almost perfectly, observing the strength, structure, and evolution of magnetic fields in the solar atmosphere. Scientists were confident that despite abbreviated operations SOUP data would be the best and longest run of solar granulation data ever collected.

Although loss of altitude and propellant during initial Space Shuttle ascent jeopardized part of the plasma depletion experiment, planners managed to schedule four of the eight anticipated burns, two over Millstone Hill, Massachusetts, and one each over Arecibo, Puerto Rico, and Hobart, Australia. Preliminary data indicated that the burns did produce "holes" or troughs of depleted plasma that persisted in the ionosphere for more than an hour. The crew observed resultant airglow after the nighttime burns, and there were reports of visual observations from the ground. The Hobart site reported the reception of low-frequency cosmic radio emissions through the window the burn temporarily opened.

In the properties of superfluid helium in zero-gravity experiment, the cryostat performed up to expectation, maintaining a temperature low enough to keep the helium in the superfluid state. Temperature was readily controlled, and the cryostat recovered well from temperature increases. These findings were important to the use of superfluid helium as a cryogen on future missions.

During the gravity influenced lignification in plants experiment, pine seedlings, oats, and bean sprouts grew in self-contained growth chambers. The crew daily monitored chamber temperatures and photographed the chambers early and late in the mission. The oat and bean seeds germinated in orbit, and sprouts grew to a height of five to six inches as expected. The pine seedlings also showed normal growth. Researchers would analyze the plant tissues to determine whether there was any difference in the production rate of lignin between plants grown in space and in a control group grown in a ground laboratory. (NASA Prelim Spacelab Mission Science Report, Aug 6/85; ESA release Aug 7/85)

September 30: NASA announced that Spacelab 2, which flew onboard the Space Shuttle Challenger that was launched on July 29 from KSC on mission 51–F, completed the second of two planned Design Verification Flights required by the Spacelab Verification Flight Test (VFT) program. Monitoring of mission activities and quick-look analysis of data confirmed that the mission achieved the 13 specific VFT requirements and performed the planned multidiscipline science. Based on these results, NASA judged the Spacelab-2 mission objectives accomplished.

NASA researchers were continuing detailed analysis of all data, and Marshall Space Flight Center (MSFC) would produce additional documents when results were finalized for the Spacelab system, payload integration, and the Spacelab-2 experiments managed by MSFC. Principal investigators would produce separate documentation for their experiments. (NASA MOR M–977–51F–03 [postlaunch] Sept 30/85)

October 30: NASA's Marshall Space Flight Center (MSFC) announced it awarded to McDonnell Douglas Technical Services Co. a 33-month $98,105,079 contract extension for continuing Spacelab integration work, bringing the total value of the Spacelab contract through June 30, 1988, to

$341,496,165. Under the contract, the company would provide system integration, selected flight hardware, software, ground support equipment, and mission integration support for manifested Spacelab missions.

Spacelab, developed by NASA under an international agreement with the European Space Agency, when carried in the cargo bay of the Space Shuttle orbiter converted it into an orbiting scientific research center. MSFC was responsible for monitoring Spacelab design and development activities and for management of U.S. development of selected Spacelab components. (MSFC Release 85–53)

November 18: Many of the experiments carried in Spacelab D–1 aboard the Space Shuttle Challenger on Mission 61–A [see Space Transportation System/Missions, Oct. 30] achieved more than 100% of mission goals, due in part to the continuous cooperation between experiment teams who coped with hardware problems and changing conditions on board the space laboratory, *Aviation Week* reported. Peter Sahm, Spacelab D–1 mission scientist, said this coordination allowed a rescheduling of experiment activity during the final mission days, maximizing the overall scientific return in spite of delays caused by hardware problems.

In one of the repairs, Spacelab D–1 crew members used a saw to cut off a plastic cap on a valve that fed liquid into the fluid physics module's experiment zones. "The valve had an incorrect setting, which perhaps was made before launch," said Berndt Feuerbacher of West Germany's Deutsche Forschungs und Versuchsanstalt fur Luft und Raumfahrt's (DFVLR) Institute for Space Simulation. The DFVLR executed and controlled the mission on behalf of West Germany's BMFT ministry of research and technology. "They tried to reset the valve with a wrench, but they couldn't put the wrench to the nut because there was a plastic cap over it. So they asked for, and received authorization, to cut away the plastic cap, then were able to use the wrench."

Feuerbacher said other repair/troubleshooting work by the Spacelab D–1 crew members included West German payload specialist Dr. Ernst Messerschmid using a vacuum cleaner to collect debris from a heating facility in the Werkstofflabor material science double rack and writing a software patch on the Werkstofflabor to work around a malfunctioning vacuum sensor that incorrectly indicated a lack of vacuum. NASA mission specialist Dr. Bonnie Dunbar also used the vacuum to clean up metallic dust particles from a sample that broke in the Spacelab D–1 Medea's gradient furnace with quenching device.

These corrective actions demonstrated the need for trained crew members in complex spaceflight operations and for positive control from the ground, Wolfgang Finke, head of space program activities at the West German BMFT ministry of research and technology, indicated. "I think it also underscored the good working relationship between the German Space Operations Center here and NASA in the U.S. to overcome problems as they occurred," Finke said. (*AvWk*, Nov 18/85, 55 and 65)

November 20: At the preliminary science review held the previous week at Marshall Space Flight Center (MSFC), Dr. Eugene Urban, Spacelab 2 mission scientist, called "a complete success" the Spacelab 2 mission flown aboard the Space Shuttle Challenger on mission 51-F [see Space Transportation System/Missions, July 29], the *Marshall Star* reported. "We saw that the majority of the experiments collected a large amount of new information that should be tremendously beneficial to the world of science and technology," he commented.

As the lead center for Spacelab 2, MSFC hosted the two-day conference in which numerous science teams shared early data from the mission. The teams represented the science disciplines of solar physics, atmospheric physics, plasma physics, life sciences, technology research, infrared astronomy, and high-energy astrophysics. Among those attending the conference were many of the principal and co-investigators for the 13 experiments, all the payload and mission specialists, some invited members from NASA Headquarters who supported mission work, and a number of MSFC science and engineering team members.

Some results thus far, according to Urban, confirmed what scientists had earlier predicted but could not measure, while other results yielded some surprises. For example, the superfluid helium experiment confirmed theoretical predictions of wave behavior of very thin films of super-cooled helium in microgravity. But the infrared telescope surprised researchers when it indicated that the "Shuttle glow" phenomenon observed on past flights was weak in the short infrared wavelength region, where the researchers expected it to be strong.

The solar physics experiment also gathered much data, according to Urban. He cited in particular the experiment called SOUP, the Solar Optical Universal Polarimeter, which studied the visible surface of the sun. "We were able to obtain long sequences of high-resolution photos of the solar surface," he said. "We now have the means to study the growth and fading of various solar features like sunspots over long periods of time."

An example of the special qualities of Spacelab 2 was the Vehicle Charging and Potential Experiment, VCAP, in which a beam of charged particles from an electron generator passed from the orbiter's payload bay through the ionosphere. "For the first time," Urban pointed out, "we had an interactive experiment with the VCAP in which we could observe the experiment remotely—from the Plasma Diagnostics Package, another experiment we allowed to free-fly and sense environmental conditions. Together the diagnostics package and VCAP found some highly interesting correlations between the man-made electron beam simulations of the ionosphere and naturally occuring auroras. The information will help us understand better how such auroras are formed from beams of charged particles from the sun." (*Marshall Star*, Nov 20/85, 1)

December 24: Marshall Space Flight Center (MSFC) announced preliminary scientific results from Spacelab 3 flown on the Space Shuttle Challenger on mission 51-B launched April 29, 1985. Researchers and participants in the seven-day flight gathered in December at the Center to present their initial findings.

"The mission has made some major contributions in the physical and life sciences," said Dr. George Fichtl of MSFC's Systems Dynamics Laboratory. "We have gained a lot of insight for future Spacelab and space station research. And I think we can now say that space research is becoming routine."

Initial results from the three Spacelab 3 crystal growth experiments were very promising, Ficthl noted. The two triglycine sulfate crystals and the single mercuric iodide crystal grown in the space laboratory were at least as good as the best crystals grown to date on earth. And they might actually be better, although additional testing was necessary to confirm this, Fichtl said. "This is of major significance, because normally we must grow between one and two thousand crystals on earth to get just one crystal that is equal to the quality of those grown on Spacelab 3. Mercuric iodide crystals have application in X-ray detectors, and triglycine sulfate crystals are used in infrared detectors," he said.

In addition, the flight proved the reliability of the methods used to grow the crystals. Triglycine sulfate crystals were grown from a solution as part of an experiment provided by Alabama A&M University. The mercuric iodide crystal grew using a vapor transport process in an experiment provided by EG&G Energy Measurements Inc.

The third crystal growth experiment, provided by France and which also used mercuric iodide, performed as expected and added to researchers knowledge about the process of crystal nucleation, a process difficult to study on earth because of gravity-induced convection.

The Drop Dynamics Module, a fluid physics experiment developed by the Jet Propulsion Laboratory, enabled researchers to do some experiments not possible on earth. Results confirmed some of the theories on how drops behaved when rotated, the first opportunity to test theories posed centuries ago including those by Isaac Newton. The experiment also proved that using sound waves was a viable technique for manipulating liquids in a microgravity environment, which had direct application to containerless materials processing in space. The technique allowed processing of materials without incurring the contaminating effects of the container. The experiment showed that the "acoustic bottle" or "crucible of sound" containing the droplet had much better characteristics than originally theorized.

Drop dynamics experiments confirmed theories on how drops behaved when rotated slowly; however, experiment data showed violation of some basic theories relating to the behavior of drops rotated at higher rates. It was found, for example, that drops tended to transition (change) to a new shape at rotation rates lower than predicted. "All this tends to indicate that some of our theories may need to be modified," Fichtl said.

Researchers for the Geophysical Fluid Flow Cell Experiment reported that all 102 hours of data looked good, confirming existing theories of convection, and were expected to lead to a better understanding of the dynamics of stellar interiors and planetary atmosphere. However, the research team noted that at higher heating rates—such as those that might be found on the sun—the data showed some significant departures from that anticipated. After the mission, researchers developed computer models of some of this new information.

The Atmospheric Trace Molecules Spectroscopy (ATMOS) experiment, sponsored by the Jet Propulsion Laboratory, for the first time simultaneously measured the concentrations of chemical compounds associated with carbon, nitrogen, oxygen, and other chemical cycles in the atmosphere. This would provide better insight into the chemical processes that governed the distribution of minor and trace gases in the atmosphere between 10 and 100 kilometers. The instrument recorded concentrations of gases as low as parts per 100 billion, the first time such sensitive measurements were made, and detected and measured traces of nitrogen pentoxide, a compound previously undetected in the atmosphere.

Results from the Research Animal Holding Facility showed that the apparent case of space adaptation syndrome in one of the monkeys supported the contention that this species would be a good model for future studies of vestibular adaptation in microgravity. Results from studies of the largest group of rats flown aboard a single spacecraft showed their hind muscles became 11 to 36% smaller and the bones were less strong after their flight. (MSFC Release 85-73)

UNION OF SOVIET SOCIALIST REPUBLICS

Missiles

February 6: Finnish army divers had recovered the main section of a stray Soviet rocket that had crashed December 28, 1984, through a frozen lake, and military officials said it was not a cruise missile, the *Washington Post* reported. A helicopter recovered the missile's main frame and engine after divers had brought up the nose cone and other debris. Finnish authorities said the debris showed it was "an old-type missile dating from 1971 or 72 and without military capacity."

U.S. officials, contradicting an earlier statement by Defense Secretary Caspar Weinberger that it was an "air cruise missile," agreed with the Finnish that the missile was an older-generation drone and unarmed.

The Soviet Embassy in Helsinki, which said the missile had gone off course while in use for target practice over the Barents Sea, had requested return of the remnants.

The Finnish later announced they would return the fragments, but asked Moscow to pay for the recovery. (*W Post*, Feb 3/85, A17, Feb 6/85, A15)

April 1: The Department of Defense's 1985 edition of *Soviet Military Power* said the USSR was continuing in 1984 deployment of new nuclear and conventional military weapons systems, including test firings of fifth-generation ICBMs, launchings of new classes of submarines, and installing new strategic bombers, the *Washington Times* reported.

The book detailed several significant developments by the Soviet military, the article said, including continued test firings of the SS–24 and SS–25 intercontinental ballistic missiles, launch of two units of new Delta IV-class of strategic ballistic missile submarines to be fitted with the SS–NX–23 submarine-launched ballistic missile (SLBM) then being flight tested, and completion of sea trials of a third 25,000-ton Typhoon-class strategic ballistic missile submarine to join the two Typhoon units already operational and fitted with 20 SS–N–20 SLBMs.

The book also said the Soviets had developed a heavy-lift launch vehicle capable of putting 150-ton payloads into orbit as part of an "extremely high priority" military-related space program. The Soviet version of the U.S. Space Shuttle and manned Space Station would benefit from this heavy booster. Major emphasis in their space program was on long-duration manned missions for military research.

The USSR continued research on ground- and space-based, antisatellite high-energy lasers. "The Soviets currently have the world's only deployed antisatellite weapons system that can attack satellites in near-earth orbit," the book said.

The Soviets also continued upgrading detection and tracking systems for ballistic missile defense and development of new early-warning and air-surveillance radars. Their work on a new, large phased-array radar at Krasnoyarsk violated the ABM treaty, the book stated, and "in addition, the Soviets are actively engaged in extensive research on advanced defenses against ballistic missiles." (*W Times*, Apr 1/85, 1A)

September 3: Soviet leader Mikhail Gorbachev today told a U.S. Senate delegation that the USSR would make "radical proposals" to reduce strategic and intermediate-range offensive nuclear arms one day after the U.S. agreed to prohibit the militarization of space, the *Washington Post* reported. Gorbachev told the eight visiting senators, headed by Sen. Robert Byrd (D-W. Va.), that the Soviet Union opposed research on military space defense programs, such as President Reagan's Strategic Defense Initiative (SDI), that went beyond what was done in laboratories.

Clarifying the Soviets' position, Gorbachev said that any research outside of a laboratory was considered verifiable and subject to limits defined in the Antiballistic Missile Treaty ratified in 1972 by both countries.

During the discussion on SDI research, according to notes taken by Sen. John Warner (R-Va.), Gorbachev said, "You can't verify what's going on in the brain . . . and that's what we refer to as fundamental or basic research.

"But as soon as you go beyond the laboratory, to mock-ups, models, contracts with defense contractors, here surely verification can be done.

"We want a ban on that phase of research that approaches design and manufacture," Warner said his notes concluded. (*W Post*, Sept 4/85, A1)

Satellites

February 1: The USSR today boosted the nuclear-reactor core of its *Cosmos 1607* radar ocean-surveillance satellite (RORSAT) to a higher orbit, marking the end of its 92-day mission, *Aerospace Daily* reported.

Some observers linked the maneuver to the failure the week before of *Cosmos 1625*, a solar-powered electronic ocean reconnaissance satellite (EORSAT). They believed that EORSATs and RORSATs worked together to monitor movements of potentially hostile ships and that, upon the failure of *Cosmos 1625*, the Soviets decided *Cosmos 1607* was not worth maintaining.

Others said the Soviets terminated the *Cosmos 1607* mission because it had operated successfully for three months, and were therefore uncertain of the satellite's reliability if kept in operation any longer. (The Soviets had problems with RORSATs in the past—parts of the *Cosmos 954* reactor came down in Canada in 1978, and the nuclear power-pack components of *Cosmos 1402* reentered over the Indian Ocean and South Atlantic in early 1983.

Cosmos 1266 lasted only about a week in 1981, and in that same year *Cosmos 1299* only a little longer. In 1982 *Cosmos 1412* was operational for only 40 days.)

Cosmos 1607 was the second Soviet RORSAT of 1984; they safely boosted the power pack of *Cosmos 1579*, launched June 28, to a higher orbit on September 28. The Soviets boosted the power packs to high orbit so their decay would occur after half-life of the nuclear elements. (*A/D*, Feb 11/85, 1)

February 7: The USSR had launched *Meteor-2*, a meteorological satellite that would obtain global pictures of cloud cover and the surface below in visible and infrared frequencies in both the recorded and direct transmission mode and would observe penetrating-radiation flow in near-earth space, FBIS Tass International Service in Russian reported.

The satellite also carried an earth-orientation system, equipment to automatically align solar panels to the sun, a radiotelemetry system for satellite monitoring, and a radio complex for transmission of data to earth. All equipment was functioning normally, the service reported. (Tass Intl. Service in Russian, Feb 7/85)

June 21: The USSR today launched a secret rocket from its Tyuratam launch site that may have been an antisatellite weapons test or the first launch of a new Soviet rocket that used liquid hydrogen fuel instead of kerosene, the *Washington Post* reported. The rocket later broke into three pieces, the largest of which was three feet long, the North American Air Defense Command (NORAD) said. One piece fell out of orbit June 24 and burned up in the atmosphere; the two other pieces came down June 28.

The objects were in an orbit 121 miles high at the lowest point and 215 miles at the highest. The orbit was inclined at 64.4°, a course only slightly off the one the Soviets use to test new rockets and satellites, because it passed directly over a highly instrumented corridor in the Soviet Union.

U.S. intelligence sources said they were baffled by the small size of the three pieces. One source suggested the launch may have been a test of a new Soviet rocket that used liquid hydrogen fuel that resulted in the launch vehicle exploding with most of the debris falling to earth out of radar contact. Another source speculated it might have been a test of a new antisatellite weapon that failed or was deliberately blown up.

The June 21 mystery launch followed that same day the launches of *Cosmos 1663* and *Progress 24*. The Soviets launched *Cosmos 1664* June 26, resuming their numbered Cosmos series after skipping the June 21 "no-name" launch. NORAD gave the secret launch the designation 1985–53–A. (The "53" stood for the 53rd object put into space that year; "A" meant NORAD classified it as a payload, not a rocket launcher. NORAD gave the suffix "B" to launch vehicles that went into orbit.) (*W Post*, July 5/85, A2)

Astronautics and Aeronautics, 1985

Space Program

January 25: FBIS reported that the USSR planned to build before the end of the century an orbiting power plant equal in size to a small town to provide electricity by solar energy. The craft would also carry panels of solar batteries. Soviet cosmonauts had already tested in outer space assembly methods for such a plant. (FBIS, Moscow World Service in English, Jan 25/85)

March 15: Soviet cosmonaut Aleksey Leonov said in an interview with a Tass correspondent that continued advances in space would be inconceivable without extravehicular activities, FBIS Tass in English reported. Leonov enumerated the accomplishments of cosmonauts during extravehicular operations, but pointed out that the benefit of space research was not confined to applied tasks. "Cosmonautics engenders a host of ideas the solution of which holds out a promise of immense benefit to mankind," he said.

He added that he hoped the space age would contribute to the unification of all people and states so they could concentrate on common problems. "It seems to me that it would be reasonable that all countries should pool their efforts and resources, which are unfortunately now used for military purposes, for the lofty cause of peaceful uses of outer space," Leonov concluded. (FBIS, Tass in English, Mar 15/85)

April 1: The Department of Defense's 1985 edition of *Soviet Military Power* pointed out that the Soviets were developing a version of the U.S. Space Shuttle, a space plane, and directed-energy weapons and were engaged in military-related experiments abroad the *Salyut 7* space station, the *Washington Times* reported. The book concluded that the USSR's grand strategy was to attain global supremacy "by means short of war—exploiting the coercive leverage inherent in superior forces, particularly nuclear forces." (*W Times*, Apr 1/85, 1A)

April 9: In remarks at a meeting marking the 25th anniversary of the Gagarin Space Training Centre, Gen. Georgiy Beregovoy, center commander, said that it had become an international space academy where 58 USSR cosmonauts and participants in international space programs from the socialist countries and from France and India trained, FBIS, Tass in English reported.

Beregovoy also said results of space research were used in virtually every sector of the national economy, with more than 800 USSR institutions and organizations using the research results. He said that during missions, cosmonauts were concerned with problems of agriculture; radio and electronics engineering; metallurgy; welding; studies of sea currents, offshore areas, and bottom sedimentation in river estuaries; and compiling maps of shoals and coastal areas.

In other remarks at the meeting, Leonid Kizim, a participant in the longest space mission, which had lasted 237 days, said that during its 25 years the training center had become a major research institution capable of resolving most of the USSR's complex scientific and technical problems. (FBIS, Tass in English, Apr 9/85)

May 27: U.S. Air Force's imaging reconnaissance spacecraft had observed the USSR's 200-foot tall oxygen/hydrogen-powered rocket hardware, to be used in both their space shuttle and new unmanned heavy booster programs, being frequently mounted and removed from its Tyuratam launch pad, suggesting Soviet dissatisfaction with the ground test results and leading U.S. space experts to believe the troubles could greatly slow the Soviet's ability to launch large new space station elements or space weapons such as laser battle stations, *Aviation Week* reported.

The Soviets had not test flown any of the new heavy space shuttle or new unmanned-booster hardware, and serious delays in their flight test schedules would undoubtedly arise from the ground test problems.

However, U.S. reconnaissance had also discovered the Soviet shuttle program recently had added a significant new element—a second large space shuttle orbiter vehicle—observed sitting partially in a hangar at the Soviet Ramenskoye Flight Test Center east of Moscow. The other Soviet heavy orbiter was sitting outside the hangar. The orbiters were almost identical in size and design to the U.S. Space Shuttle orbiter, and the discovery of the second vehicle showed a forward program direction. (*AvWk*, May 27/85, 21)

June 6: The USSR launched at 10:40 a.m. Moscow time today the *Soyuz T–13* carrying cosmonauts Col. Vladimir Dzhanibekov, mission commander and a veteran of four previous space missions, and Viktor Savinykh for a rendezvous with the *Salyut 7* space station in the first Soviet manned launch in nearly a year, the *Washington Times* quoted the official news agency Tass as saying. By late evening, *Soyuz T–13*, fitted with new flight controls, was in an orbit ranging from 203 to 182 miles above earth.

On June 8 the Soyuz spacecraft docked with the space station using the new flight controls and an onboard computer in the Soviet program's first known manual docking. The time between launch from the Baikonur cosmodrome in Kazakhstan and the rendezvous with *Salyut 7* was twice as long as in previous Soviet space missions, leading observers to believe the time was spent testing the new flight controls.

In the past, personnel on the ground and automated onboard systems controlled Soviet spacecraft docking with Salyut stations, while cosmonauts on board simply monitored the operation. But since there had been aborted dockings in recent years, the Soviets might have given more control to Dzhanibekov.

The Salyut space station had been mothballed since the previous October when three Russians returned to earth after 237 days in space, a record. (*W Times*, June 7/85, 7A; *NYT*, June 8/85, 4; *W Post*, June 9/85, A27)

During June: Although the Soviet Union in the past had responded to the introduction of every new U.S. weapons system by installing a corresponding system of its own and many observers assumed the same would be true with the Strategic Defense Initiative (SDI), there were increasing indications that the Soviet leadership might decide to restrict itself to taking countermeasures against the U.S. system and renounce the development of any Soviet counterpart to SDI, the *Bulletin of Atomic Scientists* reported.

In 1983 Henry Trofimenko, head of the Foreign Policy Department of the Institute for the Study of the USA and Canada, said that in giving effective answer to the U.S.'s military program "the USSR is not going to match the U.S. in development of every new system of weapons, nor is it going to imitate it."

More recently, a Soviet analysis by a committee of scientific and strategic experts of the implications of SDI stressed that Soviet countermeasures against a prospective U.S. system could be very effective, could be taken quite easily and quickly, and would cost much less than the system against which they were directed.

Possible countermeasures discussed in the analysis included destruction of the space platforms (done by relatively small missiles, by land-based lasers, by armed satellites functioning as "space mines," or by "clouds" of obstacles set in their path; encasing missiles with material capable of absorbing laser beams or of reflecting them; masking missile launches by means of smokescreens; designing a pattern of missile launches over time that would force the lasers constantly to redirect themselves in haphazard fashion, thereby reducing the effectiveness of the system; or launching dummy missiles in order to use up the destructive power of the system.

The Soviet scientists agreed with western investigators that an estimated total cost of SDI could eventually reach $2 trillion, and even at that expenditure SDI could not guarantee that at least a few missiles in a hypothetical Soviet first strike might not get through to their targets.

The *Bulletin* also pointed out that Soviet scientists had noted that the technologies required for countermeasures were in a much more advanced state than those required for SDI itself; in fact, much of the necessary technology already existed. Well before SDI would be in place, "an effective means of counteraction" could be set up and would probably cost only 1 or 2% as much as the system it was designed to counteract. (*Bulletin of the Atomic Scientists*, June/July 85, 38)

July 5: The first month of the flight of cosmonauts Vladimir Dzhanibekov and Viktor Savinhkh aboard *Salyut 7* was drawing to a close during which time the crew had accomplished all planned activities to bring the Soviet space station into a manned flight mode, checked the condition of onboard systems, and carried out a variety of scientific research projects and experiments, FBIS Moscow TASS in English reported.

The cosmonauts had unloaded the Progress-24 transport spacecraft, positioned the delivered equipment, and replaced three sets of storage batteries as well as individual instrument units. They had used oxygen delivered in containers aboard the transport to pressurize the Salyut's living compartments. The cosmonauts had started preparation of the station's united propulsion unit for refuelling by checking for leaks in the fuelling lines. They would then pump compressed nitrogen out of the fuel tanks.

The schedule for July 5 called for the cosmonauts to perform maintenance on a system for the regeneration of water from atmospheric moisture, take medical checks including measurements of body mass and evaluations of the condition of muscles, and exercise on a bicycle ergometer and running track. The cosmonauts would continue geophysical experiments in a research program on earth's natural resources and environment. They also would make visual observations and take photos of individual areas of the Atlantic Ocean. (FBIS Moscow Tass in English, July 5/85)

July 10: The USSR today launched *Cosmos 1667* intended for the continuation of research on the effects of spaceflight on living organisms, TASS in English reported. During the flight, there would be experiments to study processes of adaptation to weightlessness and to investigate opportunities for radiation shielding during spaceflight.

The satellite carried two monkeys, Verny and Gordy, for studies on vestibular and hemodynamic responses of living organisms to weightlessness at the acute period of adaptation. Experiments carried out at that time would take quantitative measures of the excitability of the vestibular apparatus and to note increases in its reactivity. The experiments would also yield direct data on the outflow and inflow of blood to the head.

Experiments with 10 male rats aboard the satellite were intended to determine the influence on all parts of a living organism of the acute period of adaptation to weightlessness and later readaptation.

Ten tritons (mollusks) carried on the satellite previously had a portion of their front limbs and lenses amputated in order to study possibilities of regeneration and division of cells at zero gravity. And a biocalorimeter would monitor 1,500 Drosophila flies aboard the satellite to determine processes of energy exchanges during the emergence of the flies from nymphs and to study the flies' metabolism. Guppy fish, cornseeds, and crocuses were also experimental subjects.

Scientists from the U.S., Bulgaria, Hungary, the German Democratic Republic, Poland, Romania, Czechoslovakia, and France cooperated with Soviet scientists in developing the experimental equipment for the flight and in pre- and postflight examination of the animals and plants abroad the spacecraft. (FBIS, Tass in English, July 10/85, July 13/85)

July 22: During a "Science and Engineering" radio program, Boris Belitskiy responded to a New Zealander's question about the possibility of there someday being a larger Soviet space station, FBIS Moscow World Service in English reported. "A Soviet orbital station of the next generation will be a larger and more comfortable complex with better facilities for research work, pilot production, cosmonaut exercises, recreation, and everyday life generally," Belitskiy said.

When asked whether the Soviets were developing a reusable spacecraft like the U.S. Space Shuttle, Belitskiy responded, "For the present, freighters of the Progress type meet the needs of resupplying Salyut orbital stations. Nevertheless, Soviet space scientists do see a future for reusable transport systems and have been conducting tests in this area. The technological problems involved have, on the whole, been solved, but this cannot be said of the economic problems . . . This being so, Soviet space scientists are interested in quite a few other approaches to the problem as well."

In response to a question about a manned mission to Mars, Belitskiy said, "Some day, undoubtedly, there will be such a mission but that's still a long way down the road. And a vast project of this kind, bound to be very expensive, would be greatly facilitated by more cooperation between the major space powers." (FBIS, Moscow World Service in English, July 22/85)

July 28: Over 700 enterprises and organizations in the Soviet Union used pictures taken from Soviet orbital Salyut space stations, helping to save the equivalent of $8 million, FBIS Moscow World Service in English reported. For example, the photos assisted in selecting routes of tunnels for the Baykal-Amur railway in eastern USSR. And cosmonauts Vladimir Dzhanibekov and Viktor Savinykh, then aboard *Salyut 7*, were working on a new program of geophysical research. (FBIS Moscow World Service in English, July 28/85)

September 18: The *Soyuz T-14* spaceship, after launch September 17, docked today with the *Salyut-7/Soyuz T-13* orbital complex manned by Vladimir Dzhanibekov and Viktor Savinykh, FBIS TASS in English reported. After checking the integrity of the docking compartment, spaceship commander Lt. Col. Vladimir Vasyutin, flight engineer Georgiy Grechko, and research Lt. Col. Aleksandr Volkov entered the station.

During the planned eight-day joint flight, the cosmonauts would conduct geophysical, astrophysical, and medical research as well as technical and biotechnological experiments. Then Savinykh, Vasyutin, and Volkov would continue work on the station, while Dzhanibekov and Grechko returned to earth in the *Soyuz T–13*. (FBIS, Tass in English, Sept 17/85, Sept 18/85)

September 26: Soviet cosmonauts Vladimir Dzhanibekov and Georgi Grechko in the *Soyuz T–13* spacecraft returned to earth September 26 from the *Salyut 7* orbiting complex, making a soft landing 220 km northeast of Dzhezkazgan, the *Washington Post* reported. Vladimir Vasyutin, Viktor Savinykh, and Aleksandr Volkov remained onboard *Salyut 7*. The partial replacement of the crew, FBIS Moscow Domestic Service in Russian reported, for the first time provided for continuous use over a prolonged period of a manned space complex, thus substantially enhancing its efficiency.

The present Soviet space effort was the fourth long-term expedition aboard the *Salyut 7* station since its 1982 launch. The overlapping of crews meant that the *Salyut 7* would not have to be shut down and restarted. Left empty after the previous mission ended in October 1984, the station suffered an electrical failure that created severe and dangerous problems for Dzhanibekov and Savinykh when they docked June 8 with the station.

The Soviet space program emphasized endurance flights in space with the ultimate goal of establishing large permanently manned orbiting complexes. The previous Soviet space mission set a record for human endurance in space when three crew members lived aboard *Salyut 7* for 237 days. (*W Post*, Sept 27/85, A21; FBIS Moscow Service in Russian, Sept 26/85)

October 14: At a news conference to discuss the recent flights of cosmonauts Vladimir Dzhanibekov and Georgiy Grechko, Soviet space officials said today that they expected to have a permanently manned space station by 1990, but that the *Salyut 7* orbital laboratory would not be the spacecraft that hosted the rotating crews, the *NY Times* reported. Oleg Gazenko, head of the health ministry's Biomedical Problems Institute, which oversaw space medicine, said no unresolved problems remained to block development of the permanent station and that the crew rotation on *Salyut 7* showed that incoming cosmonauts performed their research work better when they were spared the start-up operations needed to reactivate the space laboratory.

At about the same time, Soviet officials told a U.S. Congressional delegation headed by Rep. Bill Nelson (D-Fla.) [see U.S. Space Policy/International, Oct. 10] that it was building a new space station that would be ready for launch in 1986, *Defense Daily* reported. Because the Soviets did not want to have two space stations in orbit at the same time, timing of the launch depended on the condition of the orbiting *Salyut 7* station, which was nearly doubled in size in September by the orbiting and deployment of a large new module. Nelson said the Soviets gave no details of the new station, which, if it is designed to be permanently manned, would beat the planned U.S. space station by seven or eight years. (*NYT*, Oct 15/85, C11; *D/D*, Oct 21/85, 259)

October 31: The Soviet Union, which had kept its space program under tight military control since it launched in 1957 the first satellite, announced creation of a civilian space agency, the *NY Times* reported. The agency would be responsible for the design, construction, and use of spacecraft for scientific research, remote sensing applications such as surveys of resources and crops, and joint space programs with other countries.

In making the announcement, the government newspaper Izvestia identified the new agency as the Main Administration for the Creation and Use of Space Technology for the Economy and for Scientific Research, the *Post* said. It would be known as Glavkosmos, an acronym for the Russian terms "main" and "space."

Izvestia did not specify the position of the agency in the Soviet government's table of organization but did identify the head of the agency as A. I. Dunayev.

The only other publicly disclosed space-related agency in the Soviet Union was the Space Research Institute, founded in 1965 and a unit of the Soviet Academy of Sciences. Western officials generally regarded that institute as performing some civilian space coordinating functions. However, in announcing Glavkosmos, Izvestia said, "It goes without saying that the effective use of space technology is of interest to many government ministries and agencies and to scientific organizations. The space program has reached such scope as to require the creation of a special coordinating agency."

According to the announcement, the new agency would consider proposals for projects, work out long-term plans, and administer programs. It would also be responsible for launching satellites on vehicles expected to be supplied by the military and for collection and dissemination of satellite information.

Izvestia said the agency would administer the Intercosmos program and the search and rescue satellite program, an international effort of the Soviet Union, the U.S., Canada, and France. The announcement did not make clear whether the new agency would administer any manned flights; but the emphasis seemed to be on automated space vehicles concerned with remote sensing, which until then had been part of the Cosmos series, a mixture of nearly 1700 civilian and military satellites launched since 1962. (*NY Times*, Oct 31/85, B13)

November 21: The *Soyuz T-14* spacecraft carrying three Soviet cosmonauts made an unscheduled return to earth today because its commander, Vladimir Vasyutin, was ill and needed hospital treatment, the *Washington Post* quoted the official news agency Tass as saying. The illness forced the crew to leave the *Salyut 7* orbiting laboratory, the first time in either the Soviet or U.S. programs that a spaceflight was curtailed because a crew member became sick in orbit, the *Washington Times* said.

Vasyutin, who was making his first flight, landed after spending 65 days in space. He, Viktor Savinykh, and Alexander Volkov, who were both reported feeling well, had been conducting scientific experiments aboard *Salyut 7*. Tass gave no details of Vasyutin's illness, but western space experts said it was unlikely he was suffering from a simple case of space sickness—the inability to acclimatize properly to conditions in space. The Tass report quoted doctors who performed a preliminary examination on Vasyutin as saying his condition was satisfactory.

The unscheduled return of the cosmonauts left the *Salyut 7*, which had been in orbit since April 1982, unmanned for the first time since June.

Almost a week later, the *Post* reported that officials at Johnson Space Center said that cosmonauts who stayed in touch with U.S. astronauts had said Vasyutin caught a cold that grew worse and spread to his sinus cavity and lungs. The sources said Vasyutin had a fever that refused to break and may have come down with viral pneumonia, which was untreatable with the antibiotics the cosmonauts carried. (*W Post*, Nov 22/85, A26, 27/85, A3; *W Times*, Nov 22/85, 2A)

December 31: During 1985 the Soviet Union launched 96 spacecraft, the U.S. 17, the *Washington Post* reported, reflecting what space specialists said was the fact that most Soviet spy satellites burned out within weeks while U.S. satellites often remained in orbit for years. The Soviets launched 33 photo reconnaissance satellites, amounting to more than one-third of the total Soviet launches from January 1 through December 26.

In 1985 the Soviets continued to stress the ability to locate with satellites U.S. ships at sea, Navy officials commented. One type of Soviet ocean surveillance satellite, which had radar beams that could penetrate clouds, sought aircraft carriers and other big ships. The radar provided the location of ships below the satellite, forcing the Navy to resort to new methods to foil radar detection.

Nicholas Johnson, advisory scientist to Teledyne Brown Engineering Co., kept a widely used log of U.S. and Soviet space launches, and he said that the Soviets launched in 1985 five ocean surveillance satellites, two carrying radar and three equipped with electronic eavesdropping gear. He added that Soviet launches included 19 communication satellites, two of a variety never seen before, and seven satellites designed to warn Moscow of a nuclear attack.

Through espionage the Soviets had learned a great deal about U.S. spy satellite capabilities and had taken steps to mask some of their military activities, intelligence officials noted. Space specialists largely agreed that the U.S. was well ahead of the Soviets in the art of spying from space, although they indicated the Soviet's reconnaissance satellites were steadily improving. (*W Post*, Dec 31/85, A4)

Spacecraft

February 28: The USSR's *Vega 1* and *Vega 2* Venus and Halley's Comet exploratory spacecraft had traveled 18.7 and 17.8 million miles from earth, respectively, and were functioning normally, Tass in English reported. The Soviets had had 67 radio communications with the stations since launch to measure flight-path parameters, monitor onboard systems, and to receive scientific and telemetric data.

Stations located in Evpatoriya and Simeiz in the Crimea, Goldstone (U.S.), Jodrell Bank (U.K.), Canberra (Australia), and Onsala (Sweden) received on January 21 and February 18 signals from the spacecrafts' radio transmitters to prepare for research on the circulation of the Venusian atmosphere using balloon probes. The reception of signals tested correlation of the various ground-measuring systems to ensure pinpointing of balloon probe locations during their forthcoming drift in the Venusian atmosphere. (FBIS, Tass in English, Feb 28/85)

April 26: The USSR today launched the automatic "Prognoz-10-Intercosmos" station to study the structure of interplanetary and near-earth shock waves arising from the interaction of solar wind plasma and the earth's magnetosphere, FBIS, Tass in English reported. The station carried scientific equipment designed by USSR and Czechoslovakian scientists under an "intercosmos" program of international cooperation.

The station was in a high elliptic orbit with an apogee of 200,000 km, perigee of 400 km, period of 96 hours and 25 minutes, and inclination of 65°. The report stated all onboard systems and scientific equipment were functioning normally and that the USSR's coordination-computer center and institutes of the Academy of Sciences were processing incoming data. (FBIS, Tass in English, Apr 26/85)

May 29: Finland and Sweden would together build measuring equipment for two Mars space probes for launch in three years by the USSR, FBIS, Helsinki Domestic Service in Finnish reported. The work was significant, the service quoted Risto Pellinen of the Finnish Meteorological Institute as saying, because it was the first time Finns would build equipment to be launched into space.

Altogether 11 countries and the European Space Agency would participate in the program. The Mars probes would study solar winds, the planet itself, its near surroundings, and its two moons (Phobos and Delmos). Finland with Sweden would build equipment to measure the nature and characteristics of space particles.

Participants and funding sources for the Finnish program were the Center for the Development of Technology, the Academy of Finland, the Ministry of

Trade and Industry, the Scientific-Technical Cooperation Committee of the Foreign Ministry, and the Meteorological Institute. (FBIS, Helsinki Domestic Service in Finnish, May 29/85)

June 12: Geoffrey Perry of the United Kingdom's Kettering Group said the radio aboard the Soviet *Salyut 7* space station came on the air June 12, indicating the two cosmonauts who entered it had been successful in repair efforts, *Aerospace Daily* reported. Perry said he had received no radio signals from the Salyut for some time, supporting the belief of Western observers that the space station experienced some kind of electrical problem since the last cosmonauts had left it in October 1984.

Perry noted the cosmonauts were transmitting on June 12 from 7:09 a.m. to 7:16 a.m. GMT on the Soyuz frequency, but by 8:46 a.m. GMT, the next time they were within his radio range, they had switched to the Salyut frequency. (*A/D,* June 13/85, 2)

August 5: Cosmonauts Vladimir Dzhanibekov and Viktor Savinykh, after completing a docking operation, boarded the *Salyut 7* orbital station to discover cold and darkness, FBIS Moscow World Service in English reported. The electricity supply system had broken down, water was frozen, and a crust of ice covered the instrument panels. It took a week and a half of intensive work by the cosmonauts to eliminate the defects and restore electricity and life support systems. Mission planners sent an automatic cargo ferry that delivered instruments, fuel, water, and food supplies. After restoring the station, the crew continued with planned research.

In a report on a Pravda article on the same topic, the *Washington Post* said the failure of two batteries had paralyzed the *Salyut 7* station after the previous three-man crew mothballed the station in October 1984 following a record 238 days in orbit.

A *NY Times* article noted Pravda mentioned the risky docking took 50 hours, during which the cosmonauts had to rely on visual clues instead of the automatic electronic radar pulses exchanged between the Salyut and their Soyuz T-13 spacecraft.

"They've snatched it back from the brink of death," the *NY Times* quoted James Oberg, an American expert on the Soviet space program, saying. "It's a major coup."

Later the cosmonauts made a five-hour space walk in which they replaced two of the ship's solar panels. The excursion gave them a chance to test new semirigid space suits, which were shown on Soviet TV. (FBIS Moscow World Service in English, Aug 5/85; *W Post,* Aug 6/85, A12; *NYT,* Aug 7/85, D19)

October 17: Soviet cosmonauts Vladimir Vasyutin, Viktor Savinykh, and Aleksandr Volkov continued their work aboard the *Salyut 7/Soyuz T-14* orbital station following the October 2 docking with *Cosmos 1686*, FBIS Moscow in English to North America reported. Soviet space officials designed *Cosmos 1686* to function as a heavy transport vehicle, an inter-orbit tug, or as a specialized module for research or production. During docking, the Soviet's flight control center and crew on the orbiting complex guided search for the satellite, approach, and mooring.

Cosmos 1686 delivered equipment and various other cargo for continued functioning of the complex; the cosmonauts would test satellite equipment and elements of its construction and develop methods of controlling orbital complexes of large dimensions and masses.

Soviet science correspondent Boris Belitskiy said of the *Cosmos 1686* mission: "The experience gained in recent years in operating such modular spacecraft with the *Salyut 6* and *Salyut 7* stations showed that such satellites can considerably extend the active life of future orbital complexes and make their operation more fruitful. Satellites of this new type have an advanced computerized control system, a large supply of propellant, and a power system incorporating solar batteries. The satellite's control system makes possible the automatic search for an orbital station, rendezvous, and docking, all in an unmanned mode. The onboard computer enabled mission control to transmit flight assignments that would over several days, if necessary months, assure the autonomous functioning of the satellite's control system.

"Also, the satellite's control system is highly economical," Belitskiy continued. "The control system itself could select an optimal mode dictated by considerations of saving propellant. A previous heavy satellite of this new type, *Cosmos 1443*, was tested in 1983 as a transport craft of large cargo capacity. It was fitted with a recoverable capsule which returned a 350-kg payload to earth. The present *Cosmos 1686* is a modular version. Such a modular design of future extraterrestial colonies is particularly promising because long-term orbiting stations can most reasonably be assembled directly in orbit from separate modules.

"Can such automatic craft make space research significantly more efficient?" Belitskiy continued. "That is the question the flight of *Cosmos 1686* is expected to answer. The total weight of the craft with its payload exceeds 20 tons, and it's almost as long as the *Salyut 7* station. The diameter of its widest part exceeds four meters and it carries over three tons of propellant and five tons of cargo. Docked with the *Salyut 7* station, *Cosmos 1686* has almost doubled the working space available aboard the orbital station and made conditions for the crew more comfortable. Its flight program provides for extensive research for economic needs." (FBIS Moscow in English to North America, Oct 17/85, Oct 2/85)

December 14: Soviet scientists from the Interkosmos Institute in Moscow together with scientists from the European Space Agency (ESA), the Universities of Utrecht (Netherlands), Birmingham (Great Britain), and Tuebingen and the Max Planck Institute for Extra-terrestrial Physics in Garching, West Germany were planning to launch in spring 1986 an observatory to investigate X-rays in space, the magazine *Geo* reported. The observatory would carry four instruments intended to investigate the X-ray stars of the Milky Way, the remnants of dying stars (supernova), and nuclei of other active galaxies. They called the project "Salyut-Hexe," in which Salyut referred to the Soviet space station and Hexe to the English abbreviation for high-energy X-ray experiment.

The magazine noted that Dr. Claus Reppin of the Max Planck Institute said it was not clear whether the observatory would be coupled to the existing Soviet *Salyut 7* space station or to a new space station, *Salyut 8*. Reppin confirmed that the German scientists would not, for reasons of secrecy, be present at the installation and assembly of the instruments. Soviet scientists had spent two weeks in West Germany to familiarize themselves with the instruments so that they would be able to assemble and install them. Reppin also said the West European scientists did not know whether they could be present at the launch, which would probably be at Baykonur, but he indicated it was probably more important that they be at the ground control center in or near Moscow by a few days after launch, when measuring instruments would begin operation.

The experiment, the magazine added, was part of the Soviet effort to set up a permanently inhabited large space station. At the beginning of 1986, the Soviets planned to launch the new station with several coupling connections for attachment of various modules. In this way, a settlement in space would grow gradually according to a building block principle; up to 30 cosmonauts were in training to serve as crew on the new station. (FBIS, Hamburg DPA in German, Dec 14/85)

UNITED KINGDOM

Satellites

February 13: NASA announced that the United Kingdom's contribution to active magnetospheric particle tracer explorers (AMPTE), the *UK Subsatellite* (*UKS*), did not respond to commands when it had passed over Chilton on January 16 or since. Efforts to contact *UKS* by Chilton and the Deep Space Network (with 10 kw-uplink power) were unsuccessful.

British officials said they would continue periodic efforts to contact the satellite, but were not optimistic. In five months of operation, the *UKS* had supported three chemical releases and had met 70% of the UK-project objectives. (NASA announcement, Feb 13/85)

UNITED STATES

Congress

February 17: Sen. Jake Garn (R-Utah) today joined five astronauts and a French pilot for the final two-hour practice aboard the Space Shuttle Challenger, which they would fly in two weeks, the *Washington Post* reported. They had run through launch procedures and checked all spacecraft systems to avoid any surprises at liftoff, ending with a simulated liftoff at 4 p.m.

Garn, who would fly on Challenger as a congressional observer, was chairman of a subcommittee that oversaw NASA funding. The other non-NASA spaceman was French Air Force Lt. Col. Patrick Baudry, who would serve as a payload specialist. (*W Post*, Feb 18/85, A14)

March 20: In response to numerous comments regarding the appropriateness of Sen. Jake Garn's (R-Utah) flight on the Space Shuttle, Sen. Barry Goldwater (R-Ariz.) wrote in a letter to the *Washington Times* that Garn had as good a background in aviation as any member of Congress and his was a long, deep, and sincere interest. "Of all the members of this body that I know, he is the least motivated by any desire for publicity.

"Why is it wrong for a member of Congress, the body that is responsible for authorizing and funding the whole space program, to engage in flight if for no other reason than to get a better idea of how the whole operation works?"

Sen. Goldwater concluded that he would defend the desirability of Sen. Garn's flight, as he would be able "to give the rest of us in Congress who vote the money a better idea of how the whole operation is going." (*W Times*, Mar 20/85, 7A)

July 20: The White House today issued President Reagan's Proclamation 5358 for Space Exploration Day, 1985. In the proclamation, the President said, "Sixteen years ago, on July 20, 1969, American astronauts sent a message to Earth: 'The Eagle has landed.' In a dramatic and compelling moment in history, the first humans had reached solid ground beyond our own planet . . .

"Space exploration is little more than a quarter century old. In that brief period, more has been learned about the cosmos and our relation to it than in all the preceding centuries combined. The ever-increasing knowledge gained from peaceful space exploration, and the uses to which that knowledge is put, potentially benefit all those on Spaceship Earth . . .

"In recognition of the achievements and promise of our space exploration program, the Congress, by Senate Joint Resolution 154, has designated July 20, 1985, as "Space Exploration Day . . .

"I call on the people of the United States to observe the occasion with appropriate ceremonies and activities." (Admin. of Ronald Reagan, July 22/85, 924)

U.S. Air Force

Aircraft

February 8: The EC–18B advanced-range instrumentation aircraft (ARIA), the first of four former 707 commercial aircraft modified by Aeronautical Systems Division's (ASD) 4950th Test Wing, rolled out on January 4, the Air Force Systems Command's (AFSC) *Newsreview* reported. To augment ARIA-mission capabilities, ASD had purchased the Boeing 707–320 aircraft from American Airlines in 1982 as replacements for four of the seven Boeing EC–135s in the current ARIA fleet.

The aircraft would provide worldwide missile- and space-testing support by serving as airborne tracking stations over land where geographical constraints limited ground-tracking stations and over broad ocean areas where no tracking stations existed. The EC–18B's ARIA fleet improvements included more room for mission equipment, increased fuel capacity, and more fuel-efficient engines.

Modification of the 707s had offered major cost savings over purchase of new aircraft; program costs for the four aircraft were $25 million—$6 million for aircraft purchase and $19 million for ARIA conversion. At the time of purchase, Boeing officials had estimated new aircraft would cost $25 million each, not including ARIA-conversion costs.

The most obvious external modification was the large bulbous nose—a 9 ft. radome housing the world's largest airborne-steerable antenna, a 7 ft. dish for telemetry reception. The nose also housed a smaller weather antenna. Other additions included a smaller radome for communications on the aircraft top and wingtip-probe antennas for high-frequency radio transmission and reception.

ASD also outfitted the aircraft with a navigation station, mission-critical cockpit avionics, a modified electrical system, and an improved environmental-control system. (AFSC *Newsreview*, Feb 8/85, 6)

April 25: Ted Stevens (R-Aka.), chairman of the Senate defense appropriations subcommittee, told Air Force officials today that Congress would provide funds in FY 86 to buy a number of Northrop F–20 Tigershark fighters, possibly for the Guard or Reserve, despite what the Air Force decided about the plane, *Defense Daily* reported. "You're going to get some F–20s if you like it or not," he said, although he added if the Air Force could prove to Congress the F-20 could not do the job, that would be a different matter. He also said that no one in Congress was "shilling" for one company or another, but simply wanted to have competition, which would ensure the best product at the lowest price.

Air Force Assistant Secretary Thomas Cooper said the Air Force was very interested in the F-20, but that the question had not been put to the Defense Resources Board and that the Defense Department had made no final decision to buy the F-20 in FY 87.

When Sen. William Proxmire (D-Wis.) asked about cutting the FY 86 F-16 buy from 180 to 161 aircraft and using the money freed to buy 32 F-20s, Cooper answered that the Air Force would not simply buy 32 F-20s without a follow-on commitment.

Northrop had proposed to provide 396 F-20s to the Air Force over a four-year period at a fixed price of $15 million each. However, Cooper said General Dynamics had told the Air Force that it would provide a fixed-price proposal for the F-16 if the Air Force requested it.

Sen. Warren Rudman (R-N.H.) pointed out the "supplier of the F-16 has not had the best record recently in defense contracting, . . . in fact, is in mild disrepute with some sectors" of the Department of Defense and is "guilty of some misconduct." Therefore, he said, the government should not "continue to reward them" if there was a viable competitor who gave them a better price. (*D/D*, Apr 26/85, 321)

May 14: A Northrop F-20 Tigershark crashed today while practicing its routine for the Paris Air Show during a stopover at Goose Bay, Labrador, *Aerospace Daily* reported. The pilot, whose name was withheld pending notification of next of kin, was fatally injured. A Northrup spokesman said an investigation was underway to determine the cause of the crash and that the company "would expect the cooperation of the Air Force in the investigation since they are the executive agency for the F-20."

The only other F-20 was at Edwards Air Force Base, California, and would remain there to continue the flight demonstration program "because of the high interest in the F-20 program in this country," the spokesman continued.

The first F-20 had crashed the previous October during a demonstration flight at Suwon Air Base near Seoul, South Korea, killing Northrop chief test pilot Darrell Cornell. Northrop reported at the time that the crash was "pilot induced" and the "the aircraft and all its systems functioned properly." The fourth Tigershark was on the assembly line.

An attempt earlier in the month to provide funding for the Northrop F-20 failed, *Defense Daily* reported, when an amendment by Rep. James Courter (R-N.J.) to substitute 30 Northrop F-20s for 24 General Dynamics F-16s in FY 86, brought up in a closed door markup by the House Armed Services procurement and military nuclear systems subcommittee, failed by a note of 8-4. An amendment by the panel's chairman, Rep. Samuel Stratton (D-NY), calling for competition between the F-20 and the F-16 in the FY 87 budget, prevailed.

The Stratton amendment specified that the Secretary of the Air Force, in the development of the tactical aircraft fighter program for the Five-Year Defense Plan beginning in FY 87, "shall establish a competition for procurement of tactical fighter aircraft to meet the requirements of the Air Force above a minimum number of F-16 and F-20 aircraft that the Secretary determines appropriate for meeting the requirements of the active and reserve components. Such competition shall be among all suitable aircraft, including the F-16 and F-20 aircraft."

A report drafted for the subcommittee stated that the F-16 and F-20 "are both roughly comparable aircraft. Both have new and modern radar and avionics systems, both have excellent performance as air-to-air combat fighters, and both have highly accurate air-to-ground bombing systems." The report did note that the F-16 had the advantage in range and payload capabilities, while the F-20 has "a significant advantage" in reliability and maintainability factors that affect the combat readiness and operating costs." (A/D, May 15/85, 1; D/D, May 2/85, 9)

May 17: The U.S. Air Force awarded Martin Marietta Corp. an $87-million contract to begin production of the Low-Altitude Navigation and Targeting Infrared System for Night (LANTRIN), the Air Force Systems Command *Newsreview* reported. The award enabled Martin Marietta to buy specialized factory equipment to produce the first two of 700 navigation pods for LANTRIN, which consisted of navigation and targeting pods mounted on tactical aircraft and a head-up display in the cockpit. The Air Force scheduled production of the targeting pod for spring 1986.

Production of the entire system, which would give tactical pilots a day/night under weather navigation and weapon-delivery capability, would cost $3.16 billion for the 700 systems for use on F-15E, F-16, and A-10 aircraft. The remaining 698 systems would come under fiscal year options on the contract over the next eight years.

The only system of its kind, LANTRIN had a terrain-following radar that scanned the horizon while keeping the pilot at a safe altitude, even in poor visibility. The navigation pod turned night into day by distinguishing the difference in temperature of the terrain below. The result was a daylight scene that appeared on the head-up display mounted in front of the pilot. The combination of infrared and radar allowed the pilot to fly safely at a few hundred feet above ground.

The targeting pod also contained a laser designator for delivery of laser guided weapons and an automatic handoff capability to allow acquisition and delivery of Maverick missiles against tactical targets. (AFSC *Newsreview*, May 17/85, 7)

United States

May 31: The U.S. Air Force's Aeronautical Systems Division's 4950th Test Wing unveiled a generic testbed, the airborne digital avionics test system (ADATS), for flight testing newly developed aircraft avionics systems, the Air Force System Command's *Newsreview* reported. The wing's flight test engineering division developed the test pallet in a continuing effort to enhance its mission capabilities at reduced costs.

Air Force personnel could easily load and unload the test pallet from C-141 and C-130 aircraft and would use it to flight test aircraft components that used a Mil-Std-1553B data bus (a format for transferring digitized information to and from various systems and sensors in an aircraft).

The avionics test system pallet provided or simulated all electronic signals the test item received from the aircraft in which it was designed to fly. To accomplish this the pallet had navigation, air data, and time-measuring systems. A mission computer on the pallet controlled the test item and the ADATS's functions. Flight-test engineers instructed the ADATS through a computer console during each test.

Before ADATS, test wing electronic technicians typically built a dedicated test pallet for each test item. With ADATS, only software needed changing. Future planned upgrades for ADATS included a global positioning system for better navigation accuracy, a radar altimeter, and Doppler velocity sensor. Other planned improvements should permit ADATS to satisfy most digital avionics flight-test requirements through the year 2005. (AFSC *Newsreview*, May 312/85, 4)

July 8: In an internal memorandum Assistant Defense Secretary James Wade Jr. asked the Air Force why F-15 and F-16 fighter jets were getting more—rather than less—expensive to build as time went on, the *Washington Times* reported. The cost of an aircraft normally decreased over the years because the high initial costs of design, engineering, and getting an assembly line into operation were no longer charged.

The memo said the price of the F-15—calculated in 1970 dollars—rose from $5.5 million to $7.4 million a plane in 1985. In current dollars, an F-15 cost 26.3 million. The price of an F-16, in 1975 dollars, rose from $3.9 million in 1980 to $4.8 million in 1984. In current dollars, the plane cost $14 million. McDonnell Douglas Corp. had been making F-15s since 1970; General Dynamics Corp. started building F-16s in 1975.

A Pentagon spokesman said the Air Force was working on a response. However, Pentagon documents provided to Sen. Charles Grassley (R-Iowa) showed construction hours for both aircraft were well above engineering predictions. A single F-15 was supposed to take 22,978 manhours to build; during the first quarter of 1984, McDonnell Douglas needed 37,193 manhours to complete a plane, the documents reported. An F-16 was supposed to take 837,300 manhours to build; in the first quarter of 1984, General Dynamics took 1.4 million manhours to build a plane. (*W Times*, July 8/85, 10C)

August 1: Under-Secretary of the U.S. Air Force E. C. Aldridge, Jr. announced in an internal information memorandum that the policy of the U.S. Air Force was to "ensure that the unique capabilities that can be derived from the presence of military man in space shall be utilized to the extent feasible and practical to enhance existing and future missions in the interest of national security objectives." The memorandum implied that U.S. Air Force would be committed to working with the National Aeronautics and Space Administration in its most recent manned space undertaking to deploy in orbit a manned, permanent orbiting space station in the early 1990s. Memorandum for the Vice Chief of Staff, USAF from Under-Secretary E. C. Aldridge, Jr., August 1, 1985.

August 20: Honeywell's Military Avionics Division would soon begin for the McDonnell Aircraft Co. full-scale development, costing about $500 million, of a tactical electronic warfare intermediate support system (TISS) for the F-15 Eagle aircraft program, *Defense Daily* reported. The award to Honeywell was part of a larger contract the U.S. Air Force recently awarded to McDonnell to upgrade the F-15 tactical warfare system (TEWS).

The Honeywell TISS would test the F-15's electronic warfare system to minimize downtime, reduce maintenance costs, and improve overall missile effectiveness.

Stanley Moeschi, Honeywell vice president, noted that, "This contract demonstrates the maturity of Honeywell's automatic test technology and a substantial advancement in state-of-the-art digital and radio frequency electronic warfare testing." (*D/D*, Aug 20/85, 1)

September 5: As part of its FY 87 defense budget planning, the Department of Defense (DOD) approved a competition between the Northrop F-20 Tigershark and the General Dynamics F-16 Falcon as a fighter interceptor aircraft for the air defense fleet, *Defense Daily* reported. However, DOD had not decided how many prototype aircraft it would purchase for the competition.

Although a FY 86 defense authorization bill specified a competition between the two aircraft, some members of the U.S. House wanted to direct the Air Force to set a minimum number of F-16 and F-20 fighters that it would procure in FY 86 and then institute the competition.

The possibility of a competition had set off a price war between Northrup and General Dynamics, with General Dynamics proposing a specially configured F-16C to compete in cost and effectiveness with the F-20. (*D/D*, Sept 5/86, 12)

September 26: A Federal grand jury in St. Louis subpoenaed 11 McDonnell Douglas employees and certain company records as part of its investigation of alleged overcharges on F-15 contracts with the Air Force, *Defense Daily* reported. The Defense Contract Audit Agency alleged that the company furnished inaccurate cost and pricing data to justify $28 million in probable inflation in the cost of manufacturing major pieces of the F-15's equipment.

McDonnell spokesman Gerald Meyer said the subpoenas apparently related to an investigation of pricing elements in FY 80 and 81 F-15 contracts, although McDonnell did not know the full scope of the investigation.

Meyer said McDonnell had "no reason to believe there was any unlawful conduct" on its part and that it would cooperate with the Justice Department and other government agencies participating in the investigation. He added that McDonnell felt the serving of the subpoenas "was inappropriate and unnecessary because it was not preceded by more ordinary channels of communications. All they had to do was ask." (*D/D*, Sept 26/85, 130)

NASA and U.S. Air Force

February 27: As a result of White House prodding, the Air Force had signed an agreement to use NASA's Space Shuttle for at least eight flights a year for 10 years starting in 1988, and the Air Force would get a discount when NASA worked out a new pricing policy for the start of FY 89, the *Washington Post* reported. The Air Force would pay a fixed fee at the start of each fiscal year, then a per-flight charge less than commercial and other government customers would pay.

Under the agreement, NASA would drop opposition to an Air Force plan to buy 10 single-use unmanned rockets to orbit two satellites a year for five years starting in 1988, and NASA and the Department of Defense would work together to ensure a fully-operational and cost-effective Space Shuttle.

NASA hoped the Space Shuttle would begin to break even in 1987, as the agency anticipated making from that time on 24 Space Shuttle flights yearly for 10 years. The new agreement called for the Air Force to use one third of all Space Shuttle flights for the 10 years starting in 1988.

The agreement also permitted the Air Force to pick the single-use rocket it would use to orbit secret satellites too small for economical use of the Space Shuttle. (*W Post*, Feb 27/85, A22)

April 26: Assured access to space is one of the Defense Department's "bottom lines" in fulfilling its mission, said Col. Victor Whitehead, deputy for the Air Force Space Command's (AFSC) Expendable Launch Vehicles (ELV), "and we're quite proud of our record," AFSC's *Newsreview* reported. "We've gone three years in a row without a launch failure. In that time we have launched 30 totally successful ELVs.

"We've learned over the years what makes a successful launch; we don't go until everything is ready."

The Space Command was examining what might be beyond the Space Shuttle and ELVs for space transportation, considering both manned and unmanned systems as a follow-on. "But I don't think you'll see a future scenario in which we will be dependent on a single system," Whitehead said. "There will be alternative ways of flying payloads. Our job is to determine what they are."

The colonel pointed out that the Space Shuttle was a very good system, but that the Air Force recognized that if it stopped using ELVs and was dependent on the Space Shuttle alone, "our access to space would be limited if the shuttle, for some reason, is not available. We therefore found we needed a complementary ELV."

Whitehead noted the follow-on Titan 34D7 would fulfill this alternative need. "This new booster will give us into the mid-1990s a totally separate way to launch a payload," he said. "We determine it should have the same capabilities as the shuttle in terms of performance, payload, and volume. This way, a payload that could fly on the shuttle can also fly on the Titan 34D7." But it was a true complement to the Space shuttle in that it had "no common modes with shuttle. We have a completely separate set of contractors and separate launch facilities for the shuttle and Titan. If one system is down, the other continues on and flies those payloads," Whitehead concluded. (AFSC *Newsreview*, Apr 26/85, 3)

November 20: NASA announced it had scheduled an Air Force AF–16 satellite for launch December 12 aboard a Scout launch vehicle from Wallops Flight Center.

NASA and the Department of Defense entered into agreements in June 1962 for joint use of the Scout launch vehicle. NASA and the Air Force Systems Command continued the agreement under a memorandum of understanding dated April 19, 1977, and amended May 17, 1983. Under the agreement, NASA maintained the Scout launch vehicle system, and DOD used the system for appropriate missions.

The Air Force had requested NASA to provide Scout launch vehicles for the Instrumented Test Vehicle Program. The Air Force would pay in accordance with the existing interagency agreements the costs associated with this launch of a Scout vehicle. (NASA MOR M–490–605–85–01 [prelaunch], Nov 20/85)

December 12: NASA announced that a Scout vehicle launched the Air Force AF–16 satellite [see U.S. Air Force/NASA and U.S. Air Force, Nov. 20] at 9:35 p.m. EST today from the Wallops Flight Facility.

United States

The Scout launch vehicle, S-207C, performed satisfactorily and placed the spacecraft into an orbit with the following parameters: 418 km, apogee; 170 km, perigee; and 37.05°, inclination.

The U.S. Air Force, in a brief statement released that same day, announced that NASA launched aboard a single booster from Wallops Island two instrument test vehicles, or ITVs, into low orbit, the *Chicago Tribune* reported.

"The satellites will be checked out and maintained in orbit for an unspecified time as part of the U.S. antisatellite program," the Air Force said. An Air Force official commented later that it was "safe to assume there will be another [antisatellite] test during the first quarter of 1986." (NASA MOR M-490-605-85-01 [postlaunch], Jan 10/85; *Chi Trib*, Dec 14/85, A4)

U.S. Science and Technology

January 11: Research at LaRC would improve teleoperation capabilities by gradually increasing the automation level of their teleoperation equipment, the *Langley Researcher* reported, as tests focused on the interface between the human operator and the remotely-controlled equipment using a direct-view station and a station where the operator controlled the action while viewing a TV monitor to assess performance loss as a result of the TV link.

Operators used a Unimation Puma manipulator to pick up a peg, depress switches with it, and insert it into a receptacle. In one test, operators individually moved each joint. In another, operators used resolved-rate control (controlling all joints at once, somewhat like the human arm) to move the end effector in attitude and translation in its own axis system.

Use of the smaller of two pegs and resolved-rate control significantly improved performance; however, there was no significant difference between viewing directly or through closed-circuit TV. Researchers would use these results to assess future incremental improvements to the teleoperator man/machine interface. (LaRC *Researcher*, Jan 11/85, 2)

January 16: Marshall Space Flight Center (MSFC) announced completion of a study done by McDonnell Douglas under a 1983 contract that showed undeniable need for both humans and machines in space. MSFC was continuing to refine and validate the data.

In the human role in space (THURIS) study, engineers identified 37 generic tasks requiring human involvement ranging from removing and replacing protective coverings around spacecraft to surgically acquiring tissue samples. The study then established three criteria for allocation of tasks (performance time, relative cost, and technological risk), which became the basis for an objective method to allocate tasks and determine how much of each of the generic tasks would benefit from human involvement.

For each task, researchers rated man's participation from essential to not significant. For example, in a task entitled "compute data," man's role rated not significant. However, for the task "surgical manipulations", man's participation rated essential.

A second THURIS product was an analytical method to define how man should participate in space systems. The method objectively compared humans, machines, and systems requiring both humans and machines and consisted of a short, logical sequence of questions and a set of highly condensed supporting data covering performance times, relative costs for each man/machine combination, and technology availability data for the 37 generic activities.

"The study provided a set of fundamental data for planners that can be used as a guide for many future activities," MSFC's Stephen Hall said. "To me . . . machines and automation allow man to proceed to greater challenges and achievements. . . The THURIS study found that there are tasks for which man is best, and tasks for which machines are best. For most activities, a mixture of both is the best guarantee of success." (MSFC Release 85–1)

August 2: NASA announced that NASA, the Department of Energy (DOE), and the Department of Defense (DOD) selected the reactor thermoelectric power system concept for further design, development, and ground demonstration testing in Phase II of the SP-100 space reactor power program. The SP-100 program required developing and demonstrating a compact nuclear power system that would provide a safe and highly reliable source of hundreds of kilowatts of electric power for a broad range of civilian and military space applications including the Strategic Defense Initiative (SDI) in the early to mid-1990s and beyond.

In the past, lower power nuclear sources provided the electric power for several NASA and DOD missions including Transit, Pioneer, Apollo, Viking, Voyager, and Lincoln Experimental satellites. The Galileo and Ulysses missions would also use nuclear electric power sources.

The selection of the reactor thermoelectric power system concept followed three years of Phase I data collection and technical investigation. The three agencies considered four reference reactor system concepts employing different power conversion techniques: in-core thermionics and out-of-core thermoelectrics, which were both static energy conversion processes, and Stirling and Brayton cycle engines, which were both dynamic energy conversion machines.

Selection criteria included safety; reliability; the capability of the system to meet mission power requirements; the potential for the technology to cover a range of power requirements up to 1000 kw; cost, schedule, and programmatic risks; survivability; and launch operations. The three agencies determined all of the power system technologies were adequate under those criteria; however, they ranked the reactor thermoelectric power system as having the greatest potential to fulfill reliably the overall requirements for future space power needs with minimum performance and schedule risk.

The next steps in the SP-100 program required DOE to prepare the request for proposals and, pending the availability of funds, to select competitively a contractor to design, develop, fabricate, and test during the period FY 1986–1991 the major systems of the selected concept. DOE would select shortly a site for the reactor system program test. (NASA Release 85-116)

Policy

July 17: A Congressional Office of Technology Assessment report on space cooperation, commissioned by Sens. Spark Matsunaga (D-Hi.) and Claiborne Pell (D-R.I.), said the U.S. and USSR could conduct valuable scientific exchanges, but the U.S. must handle any renewal of exchanges carefully in order to protect national security, the *Washington Times* reported. The report, the release of which coincided with the 10th anniversary of the joint Soyuz-Apollo docking mission in space, said the U.S. space program could particularly benefit from Soviet expertise in the life sciences and planetary sciences and that cooperation could lead to "substantive gains" in some areas of U.S. space research and applications.

In 1982 the U.S. cut off scientific contact between the two countries to protest the Soviet's invasion of Afghanistan, the declaration of martial law in Poland, and the exile of Russian human rights activist Andrei Sakharov. However, Congressional attitudes toward joint U.S.-USSR ventures had changed since then.

The report addressed scientific and practical benefits of cooperation, the potential for transferring sensitive military technology and know-how, the foreign policy aspects of space cooperation such as reducing tensions, and perceptions about Soviet motivations and behavior with respect to overall U.S.-Soviet relations.

Angelo Codevilla, an aide to Sen. Malcolm Wallop (R-Wyo.) and an expert on strategic space programs, said the key to cooperation with the Soviets is "to what extent do we give away the whole store" of the U.S. technology advantage. (*W Times*, July 17/85, 2A)

July 24: Presidential Science Advisor Dr. George Keyworth said to the U.S. House aviation subcommittee today that the U.S. was in the forefront of an unprecedented revolution in aeronautical technology, *Defense Daily* reported, and he called for an immediate start by the country on a streamlined, focused, and coordinated program by both government and industry to develop the necessary technology to field—before the end of the century—hypersonic transatmospheric passenger transport and space launch vehicles/military space planes. The latter could provide a two orders of magnitude reduction in the cost of transporting a pound of payload into orbit. Keyworth said he believed it was possible to develop the technology for the follow-ons to current commercial jetliners and the Space Shuttle under one R&D program. Although he was not certain if one vehicle could fill both roles, he predicted that the vehicles would be "very similar."

Keyworth did not give specifics about funding requirements or a timetable but said that the government should in the course of the coming fall's discussions of the FY 87 budget focus on the opportunities available in the hypersonic area. Alluding to the British view that they could build the HOTOL (Horizontal Take-Off and Landing) transatmospheric vehicle by 1997, Keyworth said it was possible for the U.S. to exceed that goal. He noted that the Reagan Administration over the next few months would work with the aerospace industry to develop a program focusing on well-defined goals and added that he hoped NASA and the Defense Advanced Research Projects Agency would appropriately budget funds to make the program stand out. Keyworth said he was not talking about tomorrow's technology in the hypersonic area but technology that "is here" waiting to be assembled.

At the same hearing, Rep. Tom Lewis (R-Fla.) said the British claimed that the HOTOL would be able to put payloads in orbit at one-half to one-fifth the cost of the Space Shuttle, and that they would capture 75% of the commercial launch market by the year 2000. Deputy Assistant Commerce Secretary Crawford Brubaker added that Britain had closed to the outside world all information about the engines being developed for HOTOL, so "they think they have made a breakthrough" on those engines. It was reported that a combination of Rolls-Royce conceived, airbreathing engines using atmospheric oxygen and liquid oxygen/liquid hydrogen rocket engines would power HOTOL. (*D/D*, July 25/85, 129)

Resources

February 4: The Office of Science and Technology Policy, Executive Office of the President, reported that the President's proposed FY 86 budget requested $60 billion for research and development (R&D), of which $40 billion would support non-DOD projects and nearly $8 billion, basic research.

Although obligations for non-DOD R&D would decrease slightly from 1985 to 1986, reflecting determination to reduce the federal deficit, obligations for basic research in the physical sciences and engineering would increase by 7%, and those for all basic research would increase 1%. Actual outlays during 1986 would grow by 5%, permitting some modest real growth.

In testimony before the U.S. House Committee on Science and Technology, Dr. G.A. Keyworth, Science Advisor to the President, said he strongly supported the necessity of slowing the growth of science and technology funding in the short term, but wouldn't hide his concern over the vitality of U.S. science over the long run. "Our real challenges will come in fiscal years 1987 and 1988, when we simply will have to find ways to ensure our ability to pursue—and pursue vigorously—new avenues of research. . . . We have to be prepared to make hard choices to fund new starts for high-priority research facilities under whatever fiscal scenario we face in coming years. What will be at stake will be the scientific leadership that we can't afford to compromise." (Office of Science and Technology Policy release Feb 4/85; U.S. House Comm. on Science and Technology testimony, Feb 5/85)

During February: In its *Federal Scientific and Technical Workers: Numbers and Characteristics, 1973 and 1983*, the National Science Foundation reported that the federal government was the largest single employer of scientific and technical personnel in the U.S.

Government employment of scientists grew by 20% between 1973 and 1983, from 74,000 to 89,000, although such employment declined during that period at NASA and the Department of Defense.

The variation by government agency in the proportion of scientists, engineers, and computer (SEC) specialists with at least a bachelor's degree was not great, ranging from 95% at the Department of Transportation (DOT) to 99% at NASA. The proportion of NASA SEC personnel with advanced degrees increased from 29% to 36% between 1973 and 83, reflecting the higher separation rates of bachelor's degree holders and those without degrees during the long-term cutbacks in SEC employees at NASA. The number of SEC staff at NASA declined 8% between 1973 and 83.

The decline of 1,800 electronics technicians during the period reflected primarily a fall in total white-collar employment levels of 10% at DOT and NASA; DOT's electronics technicians went from 8,800 to 7,800, NASA's from 900 to 600. (NSF 85–312 [final report], Feb 85)

June 21: The National Science Foundation (NSF) in its *Science Resources Studies Highlights* reported that since 1980, annual increases in federal support of industry-performed research and development, primarily as a result of Department of Defense funding, had outpaced growth in company-financed R&D expenditures.

In 1983 federal funding of industrial R&D performance amounted to $20.4 billion, 11% more than the 1982 level (7% in constant 1972 dollars), whereas industry's own R&D spending increased 8%. Federal and company funds together in 1983 were up 9% to a total of $62.9 billion. An NSF projection for 1984 placed total industrial R&D expenditures at $70.5 billion, 12% over the 1983 level.

Companies in aircraft/missiles and electrical equipment industries received more than three-fourths of all R&D funds provided by the government.

Industrial firms spent $42.5 billion of their own funds on R&D in 1983; between 1975 and 1980 the average annual constant-dollar rate of growth in company R&D financing was 6.6%, which slowed to 4.9% over the following 3-year period.

The number of full-time-equivalent (FTE) R&D scientists and engineers in industry rose 3% during 1983 to 538,000. The electrical equipment industry, which employed over one-fifth of industrial R&D scientists and engineers, showed the highest gain—6%. (*Science Resources Studies Highlights*, June 21/85, 1)

July 5: University of Texas scientists recently held a farewell ceremony at the McDonald Observatory in the mountains of west Texas for the Korad laser that had for the past 16 years measured the distance between the earth and the moon to within four inches, the *Washington Post* reported.

The laser, one of the last experiments still in use from the days of the Apollo moon missions, would have as a replacement the McDonald Laser Ranging System, which would compute the constantly changing distance to the moon to within two inches. However, the two lasers worked the same way, bouncing a beam of laser light off a reflector left on the moon by astronauts Neil Armstrong and Edwin "Buzz" Aldrin and measuring its return to the observatory's 107 in. telescope. (*W Post*, July 5/85, A15)

During October: Despite a recent surge of interest among undergraduate students, many university aerospace engineering departments were coping with faculty shortages, uncertain research support, and inadequate funding to operate and maintain their research facilities, the National Research Council's (NRC) *NewsReport* reported. A study prepared for NASA by an NRC committee chaired by Morris Steinberg, vice president for science at Lockheed Corp., said, "Faculty positions today are not especially attractive to ambitious young aerospace engineers." The number of doctorates awarded annually in the field dropped by half in the decade ending with 1983, indicating students were choosing careers in industry rather than academe. The committee concluded that NASA, which depended on the nation's universities for ideas and expertise, should take several steps to help remedy the problem.

NASA should bolster its support of campus research efforts that addressed "long-term fundamental problems whose solutions are likely to have lasting impact," the committee said. NASA also should institute a system of peer review of research proposals, establish Ph.D. fellowships in aerospace engineering, and coordinate its efforts to support university research and teaching in the field. (NRC *NewsReport*, Oct 85, 17)

U.S. Space Policy

During January: The U.S. government released an unclassified version of the National Space Strategy based on the National Space Policy President Reagan unveiled on July 4, 1982, and on his 1984 State of the Union Address, *Space World* reported. The strategy identified selected high-priority efforts and responsibilities and provided for implementation plans for major space-policy objectives.

The document gave new impetus for future manned military-space operations, underscored Administration support for the space station program and establishment of future civilian-space goals, encouraged space commercialization, ordered a joint NASA/Department of Defense (DOD) study of post-1995 launch vehicles, and called for full Space Shuttle cost recovery by October 1, 1988.

Gilbert Rye, director of space programs for the National Security Council (NSC), said of the strategy: "To our knowledge this is the first document of its kind to lay out in any coherent manner a list of priorities that cover the total U.S. space program. It should be useful for the Congress, the private sector, executive branch agencies, and the American people to fully understand the main thrust of the U.S. space program in the years to come."

The strategy authorized DOD to procure a limited number of expendable launch vehicles to complement the Space Shuttle, but did not specify particular civilian space program goals. A Presidential National Commission on Space during 1985 was to identify goals, opportunities, and policy options for the U.S. civil space sector for the next 20 years. (*Space World,* Jan 85, 8)

April 16: USA Today reported that its debate for that day would explore the pros and cons of U.S. manned space missions. In its editorial, *USA Today* said, "Human brains can adapt to change. Machines cannot. Without men and women in space, there would be no one to even try to fix the Navy's stranded satellite" [Navy communications satellite carried on the Discovery 51–D mission]. ". . . Space is our last frontier. Without humans aboard to ride rockets to the stars and beyond, man will never embrace what he has envisioned," the editorial concluded.

USA Today guest columnist Geoffrey Keller, a professor of astronomy at Ohio State University, wrote that one of the values of the manned space program was that it would help to lengthen the useful life of expensive space telescopes (for example, the Hubble Space Telescope), by permitting astronauts to replace worn-out telescope parts and to make telescopes more powerful as new and more efficient cameras were invented and installed.

In an opposing view, James Van Allen, a professor of physics at the University of Iowa, wrote that nearly all of what he considered the really important and durable products of space technology were "achieved by much-less-expensive unmanned spacecraft, operating automatically and under command control from ground stations, often for many years in earth orbit or in the far reaches of the solar system.

". . . The public acclaim for the Apollo program has left a permanent imprint on NASA and has, in effect, committed it to an overriding emphasis on further development of manned space flight," Van Allen continued. "This emphasis is simultaneously NASA's greatest strength and its greatest weakness. Manned space flight, with all its mythological foundations, has assumed the aura of a religion. The space shuttle and the proposed permanently manned space station are primarily embodiments of this religion," he wrote. But, he added, "the relevant results to date are far too meager to justify such hyperbolic expectations and such enormous expenditures. Meanwhile, the proven applications of space technology are languishing for lack of resources." (*USA Today*, Apr 16/85, 8A)

July 26: A new study issued by the Congressional Office of Technology Assessment (OTA) said competition from other nations and several private companies in launching spacecraft put the U.S. under pressure to protect its economic and technological leadership in space, the *NY Times* reported. This protection could be accomplished by reassessing the Space Shuttle's pricing policy, promoting greater private investment in space-related goods and services, and forging a long-term space policy to assure a competitive edge, report recommended.

The European Space Agency, a consortium of 11 western European governments, broke the U.S. monopoly in launch services for the West with its Ariane rocket program. Arianespace, a corporation owned by the French government and European banks and aerospace companies, was aggressively pursuing customers for Ariane's services and had won several contracts that would otherwise have gone to U.S. conventional rockets or the Space Shuttle. France was planning later in the year to use the Ariane to inaugurate the world's first commercial remote-sensing satellite service, competing with U.S. Landsats that surveyed the world's geologic, water, and agricultural resources.

China announced the previous month a new commercial space program using its own satellites, launching rockets, and ground stations. The Chinese had rockets capable of boosting satellites into the high orbit needed for communications satellites.

Japan, emphasizing the export potential of space technology, was developing its own rocket launching capability and was planning to launch the next year the first of a series of ocean and land remote-sensing satellites.

India also had joined the nations launching satellites, and Brazil was building a rocket base with the intention of becoming the first South American launching power.

Although the USSR was apparently tempted to enter the commercial market, western space experts questioned whether the Russians would ever be a major force in commercial space operations. They said the USSR might be reluctant to allow outside scientists and businessmen access to their facilities and other governments would probably not allow advanced communications satellites to be exported to the Soviet Union.

In its report, the OTA said that other nations developed their own space launching capabilities out of a desire to be technologically independent, to gain any economic benefits that derived from space technology, and to be regarded as "space powers." Consequently, the report concluded that the U.S.'s "competitive strategy based on price or superior technology alone will not prevent foreign entry into the launch service business."

At stake, besides prestige, was a share of what by the end of the century could be a $50 billion annual business, according to estimates by some economists in the aerospace field.

The report recommended that the U.S. government investigate new trade and regulatory policies to reduce the risks and uncertainties that hindered private investment in space technology. The study concluded that NASA by itself was "not well-equipped either to promote or to regulate growth in the commercial exploitation of space." The report suggested the regulation of "space industries" should be integrated with the regulation of their counterparts on earth. (*NYT*, July 26/85, A1)

September 6: NASA and the U.S. Air Force announced award of $5 million 26-month contracts to Boeing Aerospace Co., General Dynamics, Martin Marietta, and Rockwell Internatl. to perform studies on space transportation architecture—the total transportation system of flight elements, ground and orbital support systems, and their operational interactions. NASA's Marshall Space Flight Center (MSFC) awarded and would manage the General Dynamics and Martin Marietta contracts; the Air Force Space Division in Los Angeles, the Boeing and Rockwell contracts. The two agencies and four companies would maintain close coordination throughout the award period.

Broad objectives of the studies were to determine the nation's overall space transportation system architectures, including transportation and support systems needed to simultaneously meet mission and operational requirements, while substantially reducing total life-cycle cost; to identify the technologies required for the architectures; and to refine the resulting transportation and support system concept(s) for the mid-1990s if firm requirements were identified.

After the agencies first analyzed and provided projected mission/payload/operational requirements for the mid-1990s to 2010, the companies would analyze mission requirements, develop and analyze architecture approaches, define future transportation system concepts, and identify technologies applicable to transportation system options.

Transportation system architectures and concepts would include launch and upper-stage flight systems, mission control concepts, ground support systems, logistics support systems, and on-orbit operations for both manned and unmanned systems. The resulting transportation architecture and vehicle concepts should outline and define the most promising concepts for improved cost-effectiveness and mission need accommodation for the specified period. (NASA Release 85-126)

November 12: NASA announced that its Marshall Space Flight Center (MSFC) issued a request for proposals to LTV, Martin Marietta, and TRW to compete for a contract to design, develop, and manufacture an orbital maneuvering vehicle (OMV). The three companies worked previously on OMV definition studies and had until December 20, 1985, to respond. NASA expected to award the contract in June 1986 and planned the first OMV flight for early 1991.

The contract would include provisions for testing and hardware flight testing before the OMV's actual operational missions. The company selected would build one vehicle, with NASA having an option to request construction of a second.

Often called a "space tug," the OMV would transfer satellites and other objects between earth orbits and would extend the reach of the Space Shuttle by about 1000 miles. It would have the ability to retrieve satellites from high orbits, bring them back to the Space Shuttle for maintenance and repair, then return them to their operational orbits. The OMV would also be able to reboost satellites as their orbits gradually decayed.

The OMV would be an unmanned spacecraft, 15 feet in diameter and approximately 4 feet long. Its life would be about 10 years with refurbishment and on-orbit maintenance included in the design. NASA expected initially to deploy the OMV from the Space Shuttle for short duration missions; later the OMV would remain in orbit for extended periods for use in both Space Shuttle-based and space station-based modes of operation. (NASA Release 85-151)

Civilian Programs

May 13: The National Space Institute (NSI) announced establishment of "Space Outreach '85," a program to acquire original ideas from the public as to potential uses of space for social and economic benefit. The program was intended to broaden what appeared to be a too narrow debate on a new and clarified set of long-term civilian space goals.

Rules for the program were that ideas could not exceed 750 words; submissions could not include projects already under review by the federal government; and proposals must be creative, innovative, and feasible.

The NSI would present all ideas received during the program to Congress, NASA, and, in particular, the newly created National Commission on Space [see Mar. 30].

Judges for the program, which had the support of the Sophron Foundation of McLean, Virginia, were Walter Boyne, director of the Smithsonian Institution's National Air and Space Museum; Evert Clark, technology editor, *Business Week*; former astronaut Michael Collins; and Robert Cowen, science writer for the *The Christian Science Monitor*.

The NSI would present awards for the outstanding entries, with the writer of the most innovative suggestion receiving an all-expense-paid trip to see a Space Shuttle launch. (NSI Release, May 13/85)

August 9: World amateur radio operators participated in the shuttle amateur radio experiment (SAREX) on Space Shuttle mission 51–F, the *Lewis News* reported. SAREX allowed people connected with amateur radio to talk with Challenger's crew and watch TV transmissions from the flight deck.

The SAREX test, conducted during off-duty hours in cooperation with on-board mission specialists/ham operators Tony England and John-David Bartoe and mission commander Gordon Fullerton, allowed world amateur radio operators access to the Space Shuttle's voice and video transmissions to earth on amateur 2-m band. A video buffer circuit designed, built, and tested by Lewis Research Center (LeRC) ham operators was key to SAREX's success.

The SAREX system worked in two primary modes, one mode at a time. The first allowed voice communications back and forth between radio operators on the ground and spacecraft crew.

The second primary mode was amateur TV. Using a set-up similar to the ham radio system, amateur TV operators and the Space Shuttle crew sent images to one another—the first such use of two-way video in space. Users transmitted color TV pictures in a slow-scan-mode—one picture every 8 to 36 seconds.

SAREX, closely connected to the Young Astronaut Program, was a joint effort of the American Relay League and NASA. A ham radio experiment with mission specialist Owen Garriot conducted in 1983 on STS–9 was generally credited as the genesis for the current SAREX. (*Lewis News*, Aug 9/85, 1)

October 22: In its recently published "1986 Long-Range Program Plan," NASA outlined plans for an evolutionary, permanently manned space station in low-earth orbit; operation by the year 2000 of man-tended platforms in equatorial, polar, and geosynchronous orbit; and routine manned missions on the moon and later Mars by the early 21st century, *Defense Daily* reported. The annually updated plan, summarizing the status of NASA planning as of the end of February 1985, also detailed some 120 ongoing and planned NASA space projects and missions for approximately the next ten years, with some out to the year 2000.

The plan did not discuss budget figures, but for the near-term assumed budgets with the 1% growth promised annually by President Reagan (which had since disappeared in the deficit battle in Congress).

The report did not project a need for more than four Space Shuttle orbiters but did not rule out such a need, while calling for development by 1990 of an orbital maneuvering vehicle (OMV).

Beyond 1990, objectives of the NASA spaceflight program included developing an orbital transfer vehicle complementary to the Space Shuttle for transportation of payloads, to, between, and beyond earth orbits; defining, designing, and providing a second-generation space transportation system including unmanned cargo vehicles and second-generation orbiters; developing and operating on a routine basis, beginning in the mid-1990s, geosynchronous orbit space platforms that were unmanned, permanent, and multifunctional; developing and putting into routine operation by the year 2000 geosynchronous orbit facilities that were permanent, multifunction, and able to be periodically manned; developing technology and techniques to construct, deploy, or assemble such facilities in space and to test and service them in orbit; and encouraging and supporting NASA and industry development of technology to improve concepts for space boosters that significantly reduced launch costs.

The NASA report noted that "achievements in the early 21st century in science, exploration, earth applications, and commercial uses would depend on two trends: first, the increasing capabilities of space systems with regard to accessibility, payloads, stay times, and variety and sophistication of operations; and second, the increasing capability of instruments with regard to detection, resolution, pointing accuracy, and data collection and management made possible by improvement of their power supplies and cooling mechanisms."

In the 21st century, the NASA report said, "automated or human-tended instruments located on the lunar surface will begin complementary observations" with instruments in low-earth orbit and geosynchronous orbit. For manned transportation from the space station, "a cyrogenic version of the orbital transfer vehicle evolutionary family, currently in early stages of preliminary design, is expected to provide by the year 2000 reusability for manned and sortie flights to at least geosynchronous orbit . . . It also should be able to provide the basis for transportation for longer flights to establish a lunar base and for planetary missions such as a Mars sample return . . . Routine access to the lunar surface will make possible the first intensive, systematic study of another major celestial body," the report noted. "Extensive sample collection and scientific traverses conducted by humans and long-term instrument networks installed and managed by humans will help determine the details of the moon's structure, composition, and history. They also will make accessible the record of solar and cosmic ray particle fluxes preserved in the lunar soil . . . Similar scientific activities can be carried out on Mars, either by large automated spacecraft or by a manned mission . . ." (D/D, Oct 22/85, 268)

United States

November 29: NASA's Jet Propulsion Laboratory (JPL) pioneered the concept of a satellite system for mobile users and, because a U.S. mobile satellite industry was emerging, was turning its attention to developing technologies critical for future systems, the JPL *Universe* reported.

Similar to cellular phones popular in urban areas, the mobile satellite system (MSS) would provide voice and data communications for the entire North American continent, employing one or two satellites in geosynchronous orbit to relay communications instead of depending on ground-based, line-of-sight relay towers that celluar phones used.

There were four categories of MSS users: people who needed uninterrupted communications while they traveled over wide geographical areas, for example interstate truckers; those responding to unpredictable events such as medical emergencies or natural disasters; people working at planned but temporary installations such as oil or gas drilling facilities, mining camps, or archeological excavations; and those in very remote areas.

The new technologies JPL was developing included mechanically steerable low-profile and electronically steerable medium-gain vehicle antennas, digital speech compression, digital modems, and multiple-access techniques for integrated voice and data.

NASA was working with industry and, through a joint endeavor agreement, would exchange a Space Shuttle deployment for 15% of the channel capacity for two years to test the system. "But first," said Dr. Firouz Naderi, JPL's mobile satellite experiment project manager, "we must further develop the technology and try to squeeze as many channels as we can from a very narrow frequency allocation. That is the biggest obstacle we face."

Also, JPL must design sophisticated multi-beam antennas that service "spot areas" and create techniques so frequencies could be reused to provide the most efficient use of the limited spectrum.

Recently JPL sponsored a two-day briefing on the state of the industry. Attending were more than 250 representatives of 120 organizations including the Federal Communications Commission (FCC), which would regulate the MSS; 11 of 12 candidate companies that petitioned the FCC for the right to operate the system; communications equipment manufacturers such as General Electric, Harris, and Motorola; Canadian government and industry representatives; banking officials; and potential users of the system. At the briefing, the FCC said it hoped by the middle of 1986 to assign a frequency to the mobile satellite industry as well as license an operator to be in charge of the system.

NASA scheduled for launch in 1990 the first generation satellite with a 5 to 7 m antenna. During the program's second phase, NASA would launch from the Space Shuttle a 20m (65 ft.) antenna. NASA then planned for the program's final phase to construct on the proposed space station a 50m (180 ft.) antenna for deployment into geosynchronous orbit. (JPL *Universe*, Nov 29/85, 1)

During November: Quoting from the *Congressional Record*, NASA reported remarks of Rep. Don Fuqua, (D-Fla.) chairman of the House Science and Technology Committee, who introduced the Federal Science and Technology Revitalization Act of 1985.

The act ". . . would establish an alternative personnel management system for scientific and technical people in the federal government," Fuqua said. "This is something I have been interested in for a long time, and I am hopeful that the proposal I am introducing will be able to help improve the quality of government-operated federal laboratories by encouraging the recruiting and retention of highly qualified scientific and technical individuals."

Major provisions of the bill would permit agencies to include scientific and technical personnel in the new personnel management systems; simplify job evaluation and remove covered positions from the classification requirements of 5 U.S.C., Chapter 51; provide flexibility to develop a salary structure that ensured a competitive position in the labor market and that reflected the hiring and pay policies needed to attract, retain, and motivate a highly qualified scientific and technical work force; increase base pay on performance, not longevity; allow waiver of the pay cap for up to 5% of specially qualified scientific and technical personnel; provide for performance and special awards and remove the pay cap for lump-sum awards; and create a senior scientific and technical personnel service.

In May 1983 the Federal Laboratory Review Panel of the White House Science Council chaired by David Packard reported that federal laboratories had several serious deficiencies and, consequently, a number of them did not meet the quality and productivity standards that might be expected, Fuqua explained. The panel reported that salaries at federal laboratories were noncompetitive with the private sector at entry and senior levels and that federal laboratories had to deal with a personnel management system that was cumbersome and had little flexibility.

As a result, there existed what the panel referred to as an alarming "inability of many federal laboratories—especially those under civil service constraints—to attract, retain, and motivate qualified scientists and engineers."

The panel concluded that administrative and legislative actions should be initiated to create, at government-operated laboratories, a scientific-technical personnel system that was independent of current civil service personnel systems, Fuqua continued.

This "bill is the legislative attempt to deal with these very real problems," Fuqua said. "My own experience . . . is that the requirements for NASA and Department of Agriculture labs may be very different, but attracting and retaining quality personnel is absolutely essential for the space program and for emerging fields in agriculture." (*NASA Activities*, Nov 85, 9)

Commercialization

January 9: NASA Administrator James Beggs, in a January 9 speech in Washington at the Conference on International Business in Space, predicted that early in the 21st century there would be a permanently manned base on the moon to serve a lunar mining industry and as a way-station to other points in the solar system.

He called space the endless frontier and outlined the unique attributes it offers: first, a vantage point for communications and earth observations and, from telescopes beyond the atmosphere, an improved view of the universe; second, a zero gravity environment, which offers unusual manufacturing opportunities; and third, a near-perfect vacuum, which also offers new options for industrial processes.

He noted that the communications-satellite business was one of the fastest growing industries in the world, with an estimated market of more than $3 billion a year in sales potential through the year 2000, but that the other two unique attributes of space were under-used. To remedy this, he reported three major NASA initiatives: new high-technology ventures, new commercial applications of existing technology, and unsubsidized initiatives that would transfer existing space assets on earth to private hands if they could be operated for profit.

Beggs concluded that expanding commerce in space would be difficult, but not impossible, in order ". . . to transform the promise of space into a brighter tomorrow for all mankind." (NASA Release 85-1)

January 17: John Townsend, corporate vice president and president of Fairchild Space Co., said in a January 16 speech to the American Institute of Aeronautics and Astronautics' (AIAA) National Capital Section that those planning commercial space ventures must monitor the uncertainty of NASA's Space Shuttle manifest and orbiter fleet's long-term availability, *Aerospace Daily* reported. "The manifest is at risk," he said and cited difficulties NASA had in "remanifesting" resulting from the orbiter Challenger's thermal-protection tile-adhesion problems. NASA had cut back its planned 13 missions to 12 in 1985, and Townsend expressed concern over the potential problems of a 4-orbiter fleet that could be exacerbated by an accident.

Townsend also expressed concern over the Space Shuttle pricing policy, noting NASA's recommendation of a 22% price increase in Space Shuttle flights from $71 million (1982 dollars) per flight to $87 million, and concluding that NASA's "expendable launch vehicle people have Ariane to face." (A/D, Jan 17/85, 90)

January 25: The Celestis Group, an organization of Florida undertakers, signed a contract with Space Services, Inc., headed by Donald Slayton, one of the original seven U.S. astronauts and based in Houston, to orbit a payload of cremated human remains at 1,900 miles high, the *NY Times* reported. The mission, scheduled for late 1986 or early 1987, would cost about $15 million. The Department of Transportation (DOT), which must approve the contract, indicated it had no immediate objections.

In 1982 Space Services had become the first private company to launch its own rocket, the 36-foot Conestoga, into suborbital flight. The company's original liquid-fuel rocket had exploded on the launching pad a year before. The space-burial mission would use a second-generation Conestoga capable of putting 1,500 lb. into orbit.

John Cherry, who had formed the Celestis Group, said that a Conestoga nose cone could contain as many as 13,000 capsules, each 3/8 inches by 1/4 inches, holding ashes reduced in volume by a Celestis-developed technique. Burial price would be $3,900 a customer. A reflective material would cover the nose cone in the first launch, enabling viewing of the satellite mausoleum as it passed overhead. Later missions would be deep-space burials, in which the nose cone would eject the capsules for dispersion into the cosmos.

The Commercial Space Launch Act of 1984 empowered the DOT to license all commercial space launchings. DOT examined proposals for their effect on public safety, national security, and international treaty obligations. (*NYT,* Jan 25/85, A13)

February 13: The U.S. Department of Transportation (DOT) approved plans of a Florida-based undertakers and engineers' consortium to orbit in 1986 or early 1987 a mausoleum [see U.S. Space Policy/Commercialization, Jan. 25] with the announcement that the plan represented "a creative response to the president's initiative to encourage the commercial use of space," the *Washington Post* reported. The approval was the first granted by the DOT's new licensing authority for commercial space activities and followed checks with the Department of Defense, the State Department, and NASA. The mausoleum's 1,900 mile-high orbit would place the spacecraft in the Van Allen radiation belts, a region of space rarely used by other spacecraft.

The mausoleum would be in the nose cone of a rocket, Conestoga 2, designed and built by Space Service Inc. (SSI), a Houston firm headed by Donald Slayton, one of the original Project Mercury astronauts. Conestoga 1, a one-stage rocket, had made a successful test flight in 1982. The Celestis Group, Melbourne, Florida, would pay SSI $14 million to put the 300 lb. cargo into orbit and charge $3,900 per cremated body.

A Celestis spokesman said that since announcement of its plans, the group had received hundreds of calls from people wanting to sign up. Slayton said seven or eight other companies had approached him about setting up similar businesses. (*W Post,* Feb 13/85, A2)

April 9: Hans Mark, former NASA deputy administrator, Ames Research Center director, and Secretary of the Air Force, on April 9, during part of an MIT dinner lecture series, disparaged government attempts to commercialize space, the *Space Commerce Bulletin* reported. "Now that I'm out of government, I can say that I found nothing funnier than bureaucrats wringing their hands in town around here about how to commercialize something where most of them, myself included, don't have the slightest idea on how to make investments, how to judge markets, how to do all those things that are important for commercialization," Mark said.

He characterized commercial space efforts as being politically popular but backed with little substance, with much of his criticism directed at private sector remote sensing and expendable launch vehicle operations. He said he didn't believe either technology would be a commercial success in the U.S., and he predicted private Space Shuttle operation would never work.

Mark considered true commercial space activity to be technology transfer programs and government-industry relations as set up in the 1950s to build hardware for government programs. He added, however, that government couldn't take credit for technology transfer as being space commercialization because that was not what "intellectuals think commercialization means and we need to do other things. So we need to hire 50 people in NASA headquarters to do commercialization."

Regarding the impact of the Strategic Defense Initiative (SDI) on commercial space, Mark said, "There's no doubt in my mind that important commercial ventures will follow from SDI simply because of the nature of the contractor systems that we use. We will build high-intensity lasers and the first commercial application of those lasers will surely be in welding or in something like that . . . That's what I like to call the commercialization of space," Mark concluded. (*Space Commerce Bulletin,* Apr 26/85, 1)

June 28: NASA Administrator James Beggs announced that 21 teams submitted proposals to establish centers for the commercial development of space, the objective of which would be to stimulate high-technology research in the microgravity environment of space. NASA expected this research to lead eventually to development of new products that either had commercial potential or would contribute to possible commercial ventures.

The research areas proposed by the teams included semiconductor crystal growth, remote sensing, communications technology, and biotechnology.

A panel of technical, managerial, and financial experts would review over the next 45 to 60 days the proposals to identify winning proposals. NASA would fund, beginning around mid-September 1985, between three and six of the centers for up to $1 million per year each for a period not to exceed five years. (NASA Release 85-98)

July 17: NASA and the National Bureau of Standards (NBS) announced the first sales of a product manufactured in space—tiny polystyrene spheres to serve as NBS standard reference material. The spheres could improve microscopic measurements made throughout the economy in electronics, medicine, and other high-technology areas.

A chemical process developed for NASA by Lehigh University produced billions of the 10 micrometers (1/2500th of an inch) in diameter spheres during several space Shuttle flights. Earthbound processes could not yield sufficiently uniform materials in usable quantities; when produced in a low-gravity environment, the spheres grew uniformly in size and shape.

NBS had packaged the spheres into approximately 600 Standard Reference Material (SRM) units priced at $384 per unit. Each unit was a 5-milliliter vial that contained about 30 million spheres in a 0.4% concentration by weight; the remainder was water. NASA and NBS shared sales proceeds equally. (NASA Release 85-106)

August 20: NASA announced it had signed an agreement with Space Industries, Inc. (SII) for the company to construct and operate an industrial space facility (ISF), the first habitable privately owned commercial platform in space, which NASA would deploy from the Space Shuttle. SII also signed a separate agreement with NASA's space station office that would provide for exchange of information during the definition and preliminary design phase of the space station. The agreement laid the groundwork for subsequent discussions and negotiations during the space station development period to begin in mid-1987. SII was the first private company to sign an agreement with NASA's space station office to share information that could result in a commercial facility capable of compatible operations with the space station.

The modular ISF would initially measure 35 by 14.5 feet. Although not intended to be permanently manned, it would be habitable and provide a shirt sleeve work environment for astronauts when docked with either the Space Shuttle or space station.

NASA Administrator James Beggs said, "We hope the ISF will be the first of many such platforms to be funded and built by private industry that will complement the permanently manned space station and lead eventually to an industrial park in space."

Max Faget, president of SII, said, "The facility is scheduled to be deployed in 1989 and will respond to a variety of private research and manufacturing needs. Industry could take advantage of the unique gravity-free environment of space to conduct experiments that cannot be effectively duplicated here on earth."

SII considered the cost of building the ISF to be proprietary. Under the novel agreement, a provision stipulated that SII would reimburse NASA for all costs incurred by the government associated with the deployment of the ISF when it became operational and was revenue generating. (NASA Release 85-119)

August 22: NASA announced selection of teams to establish Centers for the Commercial Development of Space. The Centers were intended to stimulate high-technology research that took advantage of the characteristics of space to develop new products with commercial potential. The five centers were joint undertakings of government, industry, and academic teams and would work closely with NASA field centers.

NASA initially would fund the centers for a period not to exceed five years, at which time they should be self-sustaining. Funding would range from $750,000 to $1.1 million on a year to year basis.

The centers selected were Batelle Columbus Laboratories, Columbus, Ohio, research area: multiphase materials processing; University of Alabama, Birmingham, Alabama, macromolecular cyrstallography in space; University of Alabama, Huntsville, Alabama, materials processing; Institute for Technology Development, National Space Technology Laboratories, Hancock County, Mississippi, space remote sensing; and Vanderbilt University, Nashville, Tennessee, metallurgical processing in space. (NASA Release 85–120)

September 5: NASA announced that Lewis Research Center (LeRC) opened the Microgravity Materials Science Laboratory (MMSL) to aid scientists on earth in determining what was and was not feasible for science experiments in space [see NASA Installations/Lewis Research Center, Aug. 26].

"The MMLS will permit U.S. government, university, and industry researchers to conduct scientific experiments using equipment that functionally duplicates equipment aboard the Space Shuttle," explained Salvatore Grisaffe, chief of the LeRC materials division. "Access to such a laboratory will give U.S. companies a competitive advantage in developing better materials through microgravity research."

Other microgravity research facilities at LeRC included two drop towers in which experiment packages could free-fall up to 500 feet, achieving a weightless condition for up to five seconds, and the Lewis Lear jet, which could fly parabolic trajectories to achieve a microgravity environment inside the plane for up to 22 seconds.

The MMSL was one part of NASA's microgravity science and applications program, which fostered research in the science and technology of processing materials in low gravity. The aims of the program were to obtain a clearer understanding of the factors controlling earth-based processes to guide their improvement and development of new materials that could not be made on earth and procedures to support long-term space operations.

The Space Shuttle then offered up to seven days for microgravity experimentation; the proposed space station eventually would provide a very long-term microgravity research and materials processing capability. However, the most efficient use of such space resources demanded that experimental procedures be based on a firm scientific understanding with extensive prior ground-based examination. (NASA Release 85–123)

September 30: The U.S. Commerce Department announced it signed a contract with Earth Observation Satellite Co. (Eosat) to pay $250 million over the next five years for Eosat to construct two new Landsat satellites (*Landsat 6* and *7*); provide a ground system to operate and process data from the new spacecraft; and operate, process, and market data from *Landsats 4, 5, 6,* and *7,* the *Washington Post* reported. Eosat, a joint venture of RCA Corp., which would build the new spacecraft, and Hughes Aircraft Co., which would provide satellite instrumentation including the thematic mappers, would operate the orbiting *Landsat 4* and *5* for their anticipated lifetime at a cost to the government similar to the government's projected cost. The Commerce Department said the new satellites, *Landsat 6* scheduled for launch in 1988 and *Landsat 7* scheduled for launch when *Landsat 6* neared the end of its approximately five-year lifetime, should provide sharper images and enhance the value of the system for agricultural uses.

The U.S. government would launch the new satellites from the Space Shuttle at an estimated cost of $44.9 million and also pay for any Space Shuttle system modifications necessary for the launches.

Although the contract called for Eosat to market and operate the four Landsats, it allowed the company to terminate date marketing for *Landsat 4* and *5* if cumulative revenue was below 65% of the mutually agreed upon projected revenues for data sales for the two satellites and to terminate operations and data marketing for *Landsat 6* and *7* if total cumulative revenue fell below 60% of projected revenue after launch of *Landsat 6, Defense Daily* reported.

Landsats used cameras and other scanners to produce pictures in various wavelengths for many uses, including agriculture, mineral explorations, fishing, forestry, snow cover and surface water surveys, and land use and city planning. NASA developed the Landsat system and in 1983 transferred it to the Commerce Department. (*W Post,* Oct 1/85, B3; *D/D,* Oct 1/85, 153)

October 29: Society Expeditions, which specialized in offering exotic vacations, announced at an October 29 press conference that on November 15 it would begin taking reservations for regularly scheduled spaceflights, the *Washington Times* reported.

Society Expeditions scheduled its first flight for October 12, 1992, the 500th anniversary of Christopher Columbus's discovery of the new world. A flight would cost $50,000, and there would be a $2000 charge for a three-day orientation. T.C. Swartz, president of Society Expeditions, said more than 3,500 people had already expressed interest in traveling in space and 350 had signed letters of intent and deposited $5,000 to confirm space on the rocket.

He explained that his company signed a five-year $280 million contract to charter the first two commercial reusable spaceships built by Pacific American Launch Systems. Pacific American's president Gary Hudson said it would cost more than $200 million to develop and build a 57-foot long rocket and that each launch should cost about $1 million.

A flight on Pacific American's PHOENIX E, a rocket powered launch and landing vehicle, would begin somewhere in northern California, circle the earth five to eight times, and return to the launch area.

Before certification for commercial use, PHOENIX E would undergo at least 50 test flights and face vigorous examination by the Department of Transportation, Hudson said. Asked if he would be on one of the test flights, Hudson said, "We think it's the ethical responsibility of the designers to go up before the public. But we won't be on the first flight. That's why there's test pilots." (*W Times,* Oct 30/85, 1B)

November 18: The 3M Corp. and McDonnell Douglas joined as a team to produce and market a proprietary drug that McDonnell Douglas produced on board the Space Shuttle to treat patients who had lost their ability to produce red blood cells, *Aviation Week* reported. The ability of the drug erythropoietin to stimulate the body to produce red blood cells could benefit persons who were anemic or had a variety of other disorders in which red blood cell levels were a factor. And it had the potential in a number of medical situations to reduce the need for blood transfusions that carried the risk of complications. Erythropoietin was not widely used because current production techniques could not filter out by-products harmful to the body.

A batch of the material to be produced in an electrophoresis machine on the next flight of Space Shuttle, mission 61–B, would be given to the first human test patients soon. It was expected the drug could be available for sale in 1988 pending Food and Drug Administration approval.

McDonnell Douglas was completing arrangements to bring 3M's Riker Laboratories Div. into the program as the drug company that would market the product in place of Johnson & Johnson's Ortho Pharmaceuticals Div., which recently dropped out of the program in favor of producing the substance through earth-based bioengineering. Ortho believed it would be able to bring the drug to market a year sooner, while McDonnell Douglas believed space processing would be far more efficient and less costly.

McDonnell Douglas was also completing arrangements with a French drug company, Roussel Pharmaceuticals of Paris, to begin processing starting the next summer of a French drug on the Space Shuttle as a second commercial space product. The French product would be purified in the same mid-deck unit that brought erythropoietin to the animal and human test phase. NASA and McDonnell Douglas were completing arrangements to keep the mid-deck system operational for research and to help generate new products under their joint endeavor agreement. (*AvWk,* Nov 18/85, 16)

December 17: NASA announced it had signed with the International Space Corp. (ISC) an agreement for development of the Normal Freezing Furnace, a high-temperature furnace for producing several types of infrared semiconductor crystals in the microgravity environment of spaceflight. The furnace would produce materials through a directional solidification crystal growth process.

The agreement call for NASA to fly the furnace aboard six to eight Space Shuttle missions to perfect the proposed crystal production process, and ISC would make the experimental equipment available to NASA for the agency's exploratory space processing operations.

NASA anticipated that the experiments would lead to new space manufacturing techniques for producing crystal materials, thus enhancing the U.S. electronic industry's position in the highly competitive worldwide semiconductor market. (NASA Release 85-173)

International

February 8: A dispute over application of U.S. laws to foreign companies apparently killed a joint U.S.-W. German project to commercialize remote sensing-equipment data and could affect negotiations over European participation in the planned U.S. space station, *Science* magazine reported. The joint project, known as SPARX, was to finance regular flights on the Space Shuttle of the modular opto-electronic multispectral scanner, an instrument developed by Messerschmitt-Bolkow-Blohm (MBB) under contract with the German Aerospace Research Establishment. NASA said proposals received from SPARX were unacceptable, because data would be available solely on a proprietary basis to SPARX's commercial customers, conflicting with NASA's "open skies" policy mandating nondiscriminatory access to all data obtained from U.S.-launched civilian missions. NASA and MBB were continuing discussions on the possibility of a separate venture using the German equipment that would respect licensing conditions applied to U.S. companies under the Land Remote Sensing Commercialization Act.

Some members of the European space science community—particularly those with reservations about tying Europe's fortunes too closely to those of NASA's proposed space station—were using the apparent conflict as evidence of their concerns, the *Science* article said. The scientists noted the extent to which the regulations and other provisions contained in the legislation's licensing requirements—such as those requiring licensees to deposit any data obtained in a single central archive—would apply to foreign companies becoming NASA's commercial customers. The dilemma hinged on the conflict between companies based outside the U.S. resenting being subject to legislation over which they had no formal control and Congress and U.S.-based companies that would complain if they saw foreign companies getting an advantage by not having to meet domestic-licensing requirements.

Udo Pollvogt of MBB's Washington office said the issues raised by SPARX "definitely need to be resolved over the next two years" as Europeans considered whether participation in the space station would make them unacceptably liable to U.S.-domestic laws. (*Science*, Feb 8/85, 617)

June 12: NASA Administrator James Beggs stirred up a controversy when he reportedly told a Washington gathering of the American Stock Exchange that the Soviets had "turned us down, flat, with no explanation" on the proposal of a joint space mission, the *Washington Times* reported. Later NASA spokesman Miles Waggoner said the Soviets had not rejected the idea, but a joint manned space mission was not likely to happen within the next few years.

President Reagan had offered the Soviets a chance to participate in a cooperative space venture about a year previously. The Soviets responded that they believed it was not the time to fly that kind of mission. "To us, that means the item is still open," Waggoner said. "This time, though, Mr. Beggs decided to look at it from the other perspective."

Other NASA officials, indicating the agency hadn't received any new word from the Soviets about the proposal, speculated Beggs might have said what he did to provoke a reaction from them.

Originally, NASA planners envisioned a joint practice rescue mission in which a U.S. astronaut would use a jet backpack called a manned maneuvering unit (MMU) to fly from a Space Shuttle to a Soviet Salyut space station. The astronaut might also use the MMU to push a cosmonaut back to the Space Shuttle for a brief time. (*W Times*, June 12/85, 4A)

June 20: The House Science and Technology Committee today completed mark-up of legislation authorizing funding to transfer the land remote-sensing system operations to Eosat, a joint venture of Hughes and RCA, *Aerospace Daily* reported. The legislation, H.R. 2800, was identical to a measure reported out by the Senate Commerce, Science and Transportation Committee the previous week. Both bills amended the Land Remote-Sensing Commercialization Act of 1984 that authorized $75 million in FY 85 for such activities. The House bill authorized $295 million for transfer activities in the FY 85-89 period, of which not more than $125 million would be available for FY 85-86.

The Commerce Department and Eosat, which was selected the previous fall to operate the system, had not yet finalized the contract, but a National Oceanic and Atmospheric Administration (NOAA) official told Congress the contract would be signed within two weeks. Eosat was proposing to provide two satellites and a new ground station for a fixed price of $250 million, satellite hardware continuing present capabilities with improvements, greater data processing capabilities at the new ground station, and a program covering 10 years. Government funding would occur in the first five years. (*A/D*, June 21/85, 1)

July 6: Taylor Wang, the first space traveller of Chinese origin who performed materials science and fluids studies in weightlessness during the April 1985 Challenger Space Shuttle flight 51-B, arrived in Beijing, China, for a two-week visit at the invitation of the Astronautics Industry Ministry. Wang, a scientist at the Jet Propulsion Laboratory, was born in China's Jiangxi Province and had not been in China for over 30 years.

Wang said the purpose of his trip was to exchange views with Chinese scholars and to conduct academic exchanges with Chinese specialists in Beijing, Xian, and Shanghai.

During a welcoming ceremony July 8 at the Ministry, Vice-Minister Bao Keming said that China's open policy in recent years had given great impetus to the development of its space technology. He concluded by saying that Wang's visit would further promote the friendship and cooperation between scientists of the two countries and that Sino-American space science and technology collaboration and exchanges were promising. (FBIS, Beijing XINHUA in English, July 6/85, July 7/85, July 8/85)

July 8: Vice President George Bush, after meeting with French President Francois Mitterrand and other French leaders in Paris, said the U.S. Strategic Defense Initiative (SDI) program and the European French-sponsored Eureka technology program were not incompatible, *Defense Daily* reported. Seeking to calm concerns that the Eureka program was in competition with SDI for scientific and technology talent, Bush said that after his talks with the French that "I more firmly believe . . . there is no incompatibility between Eureka. . .and SDI. . .

"They understand our research program, which is strictly related to strategic defense, and I think we understand much more clearly their concept of collective research on broad technology as far as Europe is concerned," Bush commented.

Technology ministers from 16 European countries had scheduled a meeting for July 17 in Paris to discuss projects and areas of cooperation in the Eureka program. Five areas of high technology were identified for joint programs: information technology, robotics, communications, bio-technology, and new materials. (*D/D*, July 8/85, 1)

July 16: On the 10th anniversary of the Apollo-Soyuz Test Project, former astronauts Thomas Stafford, Donald Slayton, and Vance Brand and cosmonauts Alexei Leonov and Valery Kubasov met on July 16 in Washington D.C. at a celebration sponsored by the American Institute of Aeronautics and Astronautics and the Planetary Society, and called on their countries to undertake a joint manned mission to Mars, the *Washington Post* reported.

"People in both countries are already dealing with the technological questions about how to accomplish such a mission," said Leonov, who commanded the Soyuz spacecraft that docked on July 17, 1975 with an Apollo spacecraft. "I know that all big things start with small steps but we can accomplish big tasks, not only in space but on the ground as well. I know we want to work together," he continued.

The conference heard repeated calls for the U.S. and USSR to begin planning a joint manned mission to Mars. Speakers included Carl Sagan of Cornell University; Bruce Murray, former director of the Jet Propulsion Laboratory; former New Mexico Senator Harrison Schmitt, a former astronaut; and Sen. Spark Matsunaga (D–Hawaii), who has sponsored Senate resolutions promoting more cooperation in space between the two countries.

Other conference participants reminded the audience of the difficulties inherent in sending men to Mars. "Our two uppermost concerns are still a large solar flare and the everyday cosmic radiation the Mars pioneers would receive on their two-year round trip," Dr. John Billingham of Ames Research Center said.

"Massive solar flares represent the worst hazard," he explained. "In 1972 a large flare produced a cloud of radiation equal to a dose of 1500 rads and in 1956 an even bigger flare sent out a dose of 2500 rads. Both would have been lethal to men on a trip to Mars.

"We have to find a way to create a kind of bomb shelter inside a ship bound for Mars and for the crew to have their own solar observatory on board to warn them of things like flares," Billingham said.

During the celebration, the Soviet Union released details of its next unmanned mission to Mars—the 1988 launch of a spacecraft and its 1989 landing on Phobos, the larger of Mars's two moons. (*W Post*, July 17/85, A18)

October 10: Members of the House of Representatives, former astronauts, and assorted other government people would depart on October 10 for a five-day stay in Moscow to interest Soviet officials and scientists in a joint manned mission to Mars, the *NY Times* reported. The Soviets invited the group, and the White House cleared the visit.

Heading the 40-member delegation was Rep. Bill Nelson (D–Fla.), who was chairman of the House space science subcommittee and scheduled to fly aboard the Space Shuttle in December. Former astronauts Thomas Stafford and Donald Slayton, also in the group, would join in a celebration in Moscow of the 10th anniversary of the Apollo-Soyuz space mission.

Nelson noted that the formal space cooperation agreement between the two countries expired in 1982 and said that he hoped the Congressional mission would open the way to a "new spirit of scientific cooperation."

Under the sponsorship of the Soviet Academy of Sciences, the group would visit the space research institute, the cosmonaut training center, and the mission control center in Kaliningrad. A U.S. request to visit the Baikonur Astrodrome, the launching facility in Tyuratam outside the city of Leninsk, was pending. (*NYT*, Oct 9/85, B14)

Military Programs

February 7: In taking actions that would likely aggravate the already eroding launch service's market position of the Space Shuttle, *Aerospace Daily* reported that the Department of Defense (DOD) announced plans to fly at least two complementary expendable-launch vehicles per year beginning in 1998, and the National Oceanic and Atmospheric Administration (NOAA) announced in its FY 86 budget submission that "three of their satellites have been committed to the Titan II." (*A/D*, Feb 7/85, 1)

March 3: In his "National Security Launch Strategy" directive, President Reagan told NASA and the U.S. Air Force to begin a study leading to joint development of a bigger and more powerful space shuttle that could begin flying missions just before the year 2000, the *Washington Post* reported. In a major policy change, the Reagan Administration indicated that the Air Force should share with NASA the cost of designing and acquiring the second-generation space shuttle; NASA had borne the $10 billion cost of developing the current Space Shuttle, although the Air Force used it one third of the time.

The directive also covered an agreement by the Air Force to use NASA's Space Shuttle at least eight times a year for the 10 years after 1988 and approved an Air Force decision to buy an improved version of its Titan rocket to supplement the Space Shuttle for military launches starting in 1988 and ending in 1993 [see Feb. 27].

Defense Secretary Caspar Weinberger said in the directive that the Air Force had decided to buy 10 Titan 34D7 single-use rockets for $2.09 billion between then and September 1993 from Martin Marietta Corp. The directive did not resolve the dispute between NASA and the Air Force over Air Force plans to sell at bargain rates obsolete Titan II intercontinental ballistic missiles to launch three weather satellites for NOAA [see Feb 25].

NASA was disappointed in the Air Force decision to buy the Titans, as it had tried to persuade the Air Force to buy an upgraded version of the solid-fuel rocket booster that helped put the Space Shuttle into orbit. Present rocket power limited Space Shuttle cargo weight, flying height, and its cross-range (limiting landings to Kennedy Space Center, Edwards AFB, and the White Sands Missile Range).

The permanently manned space station planned for 1993 would require loads up to 75,000 lb. carried into polar orbit and at least 100,000-lb. loads into near-equatorial orbit at altitudes of at least 500 miles. (*W Post*, Mar 3/85, A12)

March 25: Brig. Gen. Donald Kutyna, director of Air Force space systems and C^3, told the Senate Armed Services Committee the previous week that two-thirds of the FY 86 space budget was earmarked for the military, *Defense Daily* reported. The Air Force would receive about 48% of the national space

United States

budget, NASA 33%, and other Defense Department activities (Defense Advanced Research Projects Agency [DARPA], Navy, and Army) the remaining 18%. Kutyna added that Air Force space activities represented about three-quarters of the total Defense Department space program, with the Defense agencies, mainly DARPA, coming in second. He noted the Air Force budget for FY 86 was $10.95 billion, with "no change percentage-wise" from FY 85 and "very little change among the elements."

Maj. Gen. Carl Beer, deputy chief of staff for plans for the Air Force Space Command, reported to the same committee approval of the Air Force space plan, which was divided into support and operational missions, including the launch, orbit transfer, and on-orbit control of spacecraft and payloads, and the operations necessary "to secure free passage in space and deny the enemy the use of space" when necessary.

Beer confirmed that the requirement for a space-based radar system was established for the Air Force and Navy and the two would meet in the near future to get an R&D program underway. (*D/D*, Mar 25/85, 129)

April 10: The U.S. Navy implemented a plan announced earlier by Navy Secretary John Lehman to abolish the Naval Material Command and to restructure the Naval Electronics Systems Command to integrate space systems, force ships, and aircraft systems into a Space and Warfare Systems Command, *Defense Daily* reported. The actions were part of a two-year-old plan to decentralize acquisition management in the Navy and streamline the decision making process. *Defense Daily* reported the Navy as saying a major impact of the reorganization of Navy acquisition management would be improved accountability by eliminating a reporting layer.

Adm. James Watkins, chief of Naval Operations, said the space aspect of the reorganization was part of a total package intended to bring the Navy "up to speed" in addressing space needs. "We are trying now to put our act together. We have established the Navy Space Command. . .putting in my office the Space and C^3 Directorate, rather than just C^3," Watkins said. "The Space and Warfare Systems Command will be our technical depository, working with the Naval Research Laboratory (NRL) and other laboratories to work our projects in space, of which there are many. . .This is a nice stepping stone on the route to a unified space command," Watkins added.

He also pointed out that the Navy and the Air Force, and perhaps the Army, would rotate into the unified space command that President Reagan would establish on October 1, 1985. Therefore, Watkins said, he saw the reorganization as important not only within the Navy, but also within the unified command system to achieve a higher level of command decision. (*D/D*, Apr 10/85, 225)

August 27: Defense Secretary Caspar Weinberger today cancelled the Army's new battlefield anti-aircraft gun, called the Division Air Defense (Divad) gun, saying that tests showed it would give soldiers no significant improvement over existing weapons, the *Washington Times* reported. Weinberger told a Pentagon press conference that operational tests held from March through June demonstrated the Divad mobile gun system "does not effectively meet the military threat" or justify adding $3 billion to the $1.8 billion already spent for development and production of the first Divads. The cancellation was one of the largest ever made of a weapon in production.

Weinberger said the gun showed poor flexibility and a lack of range against the threat posed by Soviet missile-firing Hind helicopters. The Army declined comment, other than to say it would follow Weinberger's direction. Sources had said, however that the Army fought hard in meetings over the previous several weeks to keep the Divad in production. The Army had planned to buy 615 Divads for a total cost of $4.5 billion and had accepted delivery of 65 units.

Ford Aerospace and Communications Corp. developed and built the weapon, employing 1900 people in its Divad division at Newport Beach, California. Ford Aerospace President Donald Rassier said the company understood the basis for the defense secretary's decision and that the firm had already begun work on new solutions to the growing Soviet threat.

The Divad was based on the 40-millimeter double-barrel Swedish Bofers gun system, designed to be mounted on old M–48 tank chassis that the Army had in stock. Ford Aerospace equipped it with radar from the F–16 fighter for all-weather and nighttime operation.

The Army considered the weapon essential for the defense of quick-moving armored tank divisions, especially in Europe, against both helicopters firing missiles and fixed-wing aircraft. (*W Times*, Aug 28/85, 1A)

September 23: During a ceremony today at Peterson Air Force Base in Colorado Springs, the Department of Defense (DOD) activated its first unified space command, SPACECOM, to oversee all military programs in space, the *NY Times* reported. SPACECOM would oversee and control all Air Force, Army, Navy, and Marine Corps defensive space activities, which consisted of systems devoted to intelligence-gathering, watching for an attack, and facilitating communications and navigation, and would operate separately from the civilian programs run by NASA. However, it was the potential role that the new command might play as headquarters for any space-based defense against enemy missiles that drew most attention, despite official assertions that any such assignment would lie far in the future.

Although the U.S. was committed to the early stage of research for a missile defense system known as the Strategic Defense Initiative, Gen. John Vessey Jr., chairman of the Joint Chiefs of Staff, said after the ceremony that the command's purpose was an extension of the armed services' current strategy of deterrence, and he sought to allay concern that it would be a first step toward war in space.

"There are several things the command will not become," he said. "It is not a force built to escalate the arms race. It is not a force built to achieve dominance for the United States. The command will make its contribution to that fundamental element of United States strategy, the prevention of war."

Rep. Ken Kramer (R–Colo.), who strongly favored development of a missile defense system, said that the Space Command opened a new period in the lives of nations. Noting that he could "speak a little more freely" than the military officials, Kramer said, "I call it the post-nuclear era, in which the people of this nation and potentially all nations will no longer live in this shadow of potential nuclear destruction and nuclear holocaust." Kramer had fought hard to have the military's space functions united and based in his district and said Colorado had become the "military space capital of the free world."

DOD named Gen. Robert Herres, who headed the North American Aerospace Defense Command and the United States Air Force Space Command, to command the new organization. Herres attempted to dampen speculation about SPACECOM's role in strategic defense development. "There is a lot of uncertainty about how that research is going to turn out. That's why we call it research," he said. "I could speculate about what our role might be, but those decisions will depend on how the research turns out. (*NYT*, Sept 24/85, A28)

November 1: The U.S. Air Force's Electronic Systems Div. awarded Grumman Corp. a $657 million contract to build the Joint Surveillance Target Attack Radar System (JSTARS), an airborne radar system for spotting tanks, personnel carriers, or any other enemy vehicle moving or standing still, the Air Force Systems Command (AFSC) *Newsreview* reported. A C–18, a militarized version of Boeing's 707, would serve as the radar's airborne development platform. The JSTAR would include radar equipment and operations and control displays as well as communications links to ground terminals and weapons systems.

Under the fixed-price incentive contract, a Grumman team including Boeing Military Airplane Co. and United Technologies Corp.'s Norden Systems Div. would design, build, and test two full-scale-development airborne radar systems.

The Air Force and Army would test the JSTARS ability to detect, locate, and track enemy ground vehicles and to help ground commanders plan attacks to destroy them. The Air Force expected field testing to begin in March 1990.

Col. Harry Gillogly, JSTARS program manager, said the airborne systems would provide "an electronic high ground to observe maneuvering enemy forces," telling Air Force and Army commanders what forces to apply where and when in order to inflict the most damage at the least cost.

"Both services will be able to attack key targets," Gillogly said, "the Air Force with direct-attack aircraft and missiles, and the Army with artillery, maneuver forces, and ground-launched missiles." (AFSC *Newsreview*, Nov 1/85, 1)

National Space Commission

March 29: In a speech today before the National Space Club, President Reagan urged a greater U.S. effort in the commercial development of space and announced establishment of a 14-member National Space Commission, which he said would "devise an aggressive space agenda to carry America into the 21st century," the *NY Times* reported. He said Thomas Paine, head of a consulting company on high-technology enterprises and former NASA administrator and president and chief operating office of the Northrup Corp., would lead the commission.

Other commission members were Laurel Wilkening, NASA scientist who would be vice chairman; Jeanne Kirkpatrick, former U.S. ambassador to the U.N.; Brig. Gen. Charles Yeager, retired; Neil Armstrong, former astronaut who headed Computer Technology Aviation; Kathryn Sullivan, first American woman to walk in space; Luis Alvarez, Lawrence Berkeley Laboratory physicist; Paul Coleman, Space Research Association president and professor of geophysics and space at the University of California, Los Angeles; George Field, Smithsonian Astrophysical Observatory senior physicist; Lt. Gen. William Fitch, retired, former Marine Corp. deputy chief of staff for aviation; Charles Herzfeld, vice president and director of research and technology at ITT; J. L. Kerrebrock, head of the department of aeronautics and astronautics at the Massachusetts Institute of Technology; Gerard O'Neill, president of Geostar Corp.; and David Webb, a consultant for space development.

President Reagan said the commission would develop long-term goals for civilian space enterprises, but he did not elaborate on the type of commercial space ventures he had in mind. He said only that, "Before the end of the century, many billions of dollars of commercial activity will be taking place in and because of space" and that the U.S. must use the incentives of individual freedom and the profit motive to encourage these commercial uses. (*NYT*, Mar 30/85, 1A)

June 5: Although a mission to put a human colony on Mars seemed unlikely anytime soon, members of the National Commission on Space [see U.S. Space Policy/National Commission on Space, Mar. 29] said it appeared to be only a matter of time until such an undertaking took place, the *Washington Post* reported. The commission was due in May 1986 to present a report outlining what it thought the U.S. space program should look like over the next 20 years.

The U.S. last visited Mars in 1976, when unmanned *Viking 1* and *2* missions landed and began collecting data. The Mars Observer was scheduled for a 1991 launch to study the planet from orbit. NASA officials said that beyond that there was nothing concrete on the drawing boards, although they were considering sending an unmanned craft to land on Mars, collect samples, and return in a manner similar to early moon explorations. (*W Post,* June 5/85, A5)

December 18: Gen. Lawrence Skantze, commander of the Air Force Systems Command, declared that the U.S. military space program was about to begin a new era in which military space systems competed against terrestrial systems to determine which could better handle military situations, *Defense Daily* reported.

Three developments had created the military space program, reaching what Skantze called a "critical mass." These were the establishments of the Unified Space Command that provided for the first time an advocate for U.S. space activities; the Project Forecast II technology study, which would propose among other things "quantum leaps in space capabilities;" and the growing momentum of the Strategic Defense Initiative (SDI).

"This 'critical mass,' merging with the proven combat enhancement roles of space systems and a tight defense budget, would thrust the military into a new era of force structure decision," Skantze said. "A limited budget and its allocation process will place space systems in direct competition with terrestrial systems for resources and solutions to military problems," he commented.

Space systems would have to demonstrate that the benefits they provided were worth more than the terrestrial force addition that would have to be foregone for the space system, Skantze pointed out. In the past, "space systems have often competed for R&D dollars with terrestrial systems that have traditionally done the same or similar jobs," he explained. "At times we have the luxury of being able to afford both. These times are gone. Given the more sophisticated treat, and the corresponding increased complexity and cost of all aerospace systems, tradeoffs among different mission areas are now essential.

"In an era when fielding a single military satellite can cost up to $.5 billion, we need to carefully consider what we are willing to give up to get a space system," he asserted.

Regarding the outlook for a shift to space systems, Skantze said that "future tradeoffs of space systems will be tough for several reasons. One is that right now, space systems are fewer in number, more costly, and subject to different logistics concepts than terrestrial systems. While multiyear contracting can reduce cost, we haven't been able to take advantage of economic order quantities the way we have for missiles, bombs, or even planes. Development and production costs for spacecraft will continue to exceed the flyaway cost of a B-1B or F-15—even with on-orbit satellite repair on the horizon.

"While the Shuttle offers that possibility," he added, "we still have to build space systems reliable, redundant and survivable. For two to ten years at a time, military spacecraft must survive in-space radiation or combat, regulating their own systems or responding to remote controls. Ultra-high reliability is costly, but we can't yet afford squadron-level maintenance of military satellites on-orbit. For now, we're working on new maintenance concepts and better materials, propellants, and hardening to reduce some of the cost of space systems. However, they will remain expensive military alternatives until we find a new way of doing business."

A second reason making tradeoffs between space and terrestrial systems difficult was that "the payoffs are different, and hard to compare," since space systems were "force multipliers," that is, they made weapon systems more capable, Skantze said.

And a final reason that made tradeoffs difficult, he said, was "that a decision for a space system can be all or nothing," meaning the number of fighter wings could be pared and still provide an effective force to some extent, "while the Unified Space Command cannot cut back on satellite numbers and still operate an effective system. The investment question can boil down to doing it all or just not doing it," he commented. (*D/D*, Dec 18/85, 241)

MISCELLANEOUS

May 9: The Coca-Cola Co. had developed the first container capable of dispensing carbonated beverages in weightlessness, and NASA officials said Coca-Cola could be aboard Space Shuttle flights as early as July, the *Washington Post* reported. Coca-Cola President Donald Keough said that including carbonated soft drinks on a Space Shuttle flight reflected NASA's interest in providing home-like comforts for the astronauts.

Because liquids did not pour in weightlessness, astronauts had to sip drinks from straws inserted into plastic containers that collapsed as they were emptied. However, gas in carbonated drinks expanded in weightlessness and low atmospheric pressure and therefore could escape from plastic containers.

A company spokesman said Coca-Cola had solved the problem with a steel supercan equipped with a drinking spout, a screw to adjust beverage flow, and a safety lock to prevent leaks. (*W Post,* May 9/85, A8)

June 20: NASA announced it would test on Space Shuttle mission 51–F, scheduled for mid-July 1985, a Coca Cola Co. technology, developed at its own initiative and expense, to dispense its carbonated beverage in space [see NASA/Technology Transfer, May 9]. Previously it was impossible for astronauts to consume carbonated soft drinks in microgravity because there was no way to dispense the beverages. The test was part of an agreement between NASA and Coca-Cola under which the company would grant NASA a license to use the company's patented technology, a specially designed can, for unrestricted use in dispensing carbonated beverages in space. NASA would also receive the technical information necessary to fabricate its own cans.

NASA added that other companies in the carbonated beverage industry were welcome to propose different technology for the same purpose. (NASA Release 85–96)

June 25: NASA announced that a Pepsi-Cola dispensing technology would fly aboard Space Shuttle mission 51–F, scheduled for launch on July 12, provided time permitted completion of NASA qualification procedures at the Johnson Space Center. The container, developed by Pepsi-Cola USA and Enviro-Spray Systems (a subsidiary of Grow Group) at their own initiative and expense, was capable of dispensing carbonated beverages, as well as other liquids in space, because pressure within the can was automatically renewed, whenever necessary, to maintain a steady flow of liquid.

Pepsi-Cola had granted NASA a license for unrestricted use of the dispensing technology for testing carbonated beverages in space, and NASA would receive the technical information necessary to fabricate its own cans. (NASA Release 85–97)

July 12: NASA scheduled the Space Shuttle Challenger to take today four high-tech cans each of Pepsi-Cola and Coca-Cola, the world's two largest-selling soft drinks, into orbit for the first time, the *Washington Post* reported. Stocking Challenger with both Coca-Cola and Pepsi-Cola resulted from a compromise reached after weeks of highly sensitive negotiations. NASA had not previously flown carbonated drinks in space because no containers were available to control and dispense carbonated liquids in zero gravity.

It took the companies more than two years to develop and test the cans. Pepsi-Cola said its can cost its supplier, Enviro-Spray, $14 million to develop. Pepsi-Cola filled its can with eight ounces of cola and a plastic pouch that expanded when chemicals were mixed inside it to create carbon dioxide gas, which then inflated the pouch and forced the beverage out. Coca-Cola's can was lined with a laminated plastic bag filled with cola that overlaid a second plastic bag containing carbon dioxide under 50 lb. of pressure. The carbon dioxide forced the cola out of the can. A drawback to both cans was that they must be drunk unchilled.

NASA retained the legal right to use whichever container worked best, the *Washington Times* reported, and to fill it with the drink of its choice. (*W Post*, July 12/85, A3; *W Times*, July 8/85, 3A)

August 6: Sometime in 1986 or shortly thereafter a Conestoga rocket might launch a capsule carrying the cremated remains of 15,000 human beings into an orbit 1,900 miles above the earth, Malcom Brown wrote in the *NY Times*, and a few critics had deplored the idea that the U.S.'s first private venture into space should haul so seemingly useless a payload. However, Browne pointed out, the day might come when the space undertakers would orbit human relics of genuine value to future historians, archeologists, scientists, and doctors.

As the living world became more crowded, he wrote, cremation seemed increasingly attractive as a space-saving alternative. However, if the trend continued, scientists in the distant future might encounter a troublesome shortage of human remains representing today's society. Space burial of at least a few representatives of our society would serve future science admirably, he wrote. Safe from grave robbers, souvenir hunters, land developers, and other terrestrial menaces, an orbiting body would remain in the most pristine frozen storage imaginable.

Short of orbiting whole bodies, he added, even the storing of samples of human tissue in space would give future scientists useful time capsules. A system patented by a U.S. inventor, Philip Backman, reduced a human body to 5% of its natural weight by freezing, pulverizing, and vacuum drying it. "A few grams of the resulting powder, taking up no more space in an orbiting mausoleum than cremated remains, would bequeath to future scientists important clues about the deceased person's identity, genetic makeup, pathology, and even style of life," he wrote.

"Space may be the arena of our future wars. It seems fitting that space should also serve as a graveyard, from which our distant descendants could mine the treasure of knowledge," Browne concluded. (*NYT*, Aug 6/85, C3)

August 14: During a news conference today at Johnson Space Center, Space Shuttle mission 51-F commander Gordon Fullerton said the Challenger's crew members could not endorse either the Coca-Cola or Pepsi-Cola they carried into space for the first time, the *Washington Post* reported. Both drinks were warm, fizzy, and full of froth whenever the crew tried to drink them.

"They both failed miserably, mainly because we had no refrigerator. Warm cola is not on anybody's favorite list of things . . . They just weren't at the right temperature and we had no desire to drain the cans the two drinks came in," Fullerton said.

He added that, although the crew suffered no gastric distress from drinking the two colas, "I just can't extrapolate to any great desire to want them."

When reporters asked the seven crew members if they would forsake the fruit juices they usually drank for the colas if the colas were refrigerated, they all shook their heads. "The drinks we have on board now are quite attractive," said astronaut Anthony England. (*W Times*, Aug 15/85, A8)

December 20: NASA announced that a cubic crystal art work, the first nonscientific payload to fly aboard a Space Shuttle, would go into space in the spring of 1986. "The Boundless Aperture," created by artist Lowry Burgess, was one in a series of seven works in a project entitled "Quiet Axis," intended to express through art the scientific observation of order and harmony in the universe.

The De Cordova Museum, the New Works Program of the Massachusetts Council on the Arts and Humanities, and the Massachusetts Artists Foundation funded and would pay for launch of "The Boundless Aperture," which was a six-lb. five-inch sq. cube of bronze-tinted transparent glass. The cube contained such materials as water from 18 of the world's rivers and minute amounts, or an appropriate substitute, of each of the elements in the periodic table.

After flying in a middeck locker of the Space Shuttle orbiter, the work would be placed inside a petrified sycamore tree obtained from the Grand Canyon; and both would hover in a permanent magnetic field inside a rock formation on the grounds of the De Cordova Museum.

Burgess was a professor at the Massachusetts College of Art in Boston and was director of its graduate program in fine arts and design. He was also a fellow and senior consultant at the Center for Advanced Visual Studies of the Massachusetts Institute of Technology. NASA would determine the final cost of flying the work when it arrived at Kennedy Space Center. (NASA Release 85-175)

December 27: NASA announced that its Kennedy Space Center (KSC) awarded a research grant to Tuskegee University to research possibilities of growing food in outer space. In the project, the researchers would select several sweet potato varieties that looked promising and research nutrient delivery and hydroponic systems that appeared best suited to development of the crop.

The sweet potato project was part of NASA's Controlled Ecological Life Support Systems (CELSS) program to search for methods to supply a continuous food source and to regenerate waste during long-duration spaceflights or for proposed lunar colonies. Dr. William Knott, manager of KSC's Life Science Support Facility and technical monitor for the project, said that, if the sweet potato could deliver enough energy conversion efficiency and productivity, NASA would incorporate it into a nearly full-scale "breadboard" of a working CELSS that it was developing. Other plants previously chosen for the CELSS breadboard project included sugar beets, lettuce, snap beans, wheat, soy beans, and white potatoes.

The sweet potato research would focus on three systems: plant growth, food processing, and waste management. The program would begin by studying the plants in a sealed 24 by 12-foot growth chamber with controlled light, food, water, and temperature. Plants would receive food in the chamber by recycling the atmosphere and water that passed through the nutrient system. NASA expected the chamber, once used to test the Mercury spacecraft for flight, might produce enough food for two to three people. A waste management system would treat leftovers from the recycling process, converting the waste products into a nutrient source for the plants.

Knott said a major challenge of the project was to harvest the highest yields possible while using a minimum of space and water to deliver nutrients to the plants. "We will take the edible seeds and fruit out of the chamber for processing and storage. The leaves, stems, and parts of the plant normally not consumed also would be converted into a food material."

Calling the CELSS program "a beyond the year 2000 endeavor," Knott said NASA would test some of the concepts of the program on the proposed space station "just to see if they work." However, he pointed out, "resupply on a long-duration spaceflight, a lunar base, or on a Mars mission . . . would make the 'space farm' . . . much more attractive." (NASA Release 85-180)

APPENDIX A
SATELLITES, SPACE PROBES, AND
MANNED SPACE FLIGHTS, 1985

There were 121 space launches in 1985, of which the USSR conducted 98 or 81%; the U.S. conducted 17 launches. The European Space Agency (ESA) was responsible for three launches; Japan, two; and the Peoples Republic of China, one.

NASA carried out 14 of the U.S. launches. Nine of these were with the Space Transportation System; three used Atlas Centaur vehicles; and two used Scout vehicles. The U.S. Department of Defense sponsored the remaining three launches—one with a Titan IIIB-Agena D and two using Atlas E launch vehicles; all DOD launches were from the Western Space and Missile Center (Vandenberg AFB). During the year NASA used all four of the continental U.S. launch sites: Kennedy Space Center for all Space Shuttle and Atlas Centaur launches; Vandenberg AFB for one Scout launch; Wallops Flight Facility for a Scout launch; and the White Sands Missile Range (New Mexico) for the final launch of an Aerobee sounding rocket.

Soviet space activities in 1985 indicated their program was on the verge of major advances in exploring and exploiting outer space. The USSR tested three new space systems in 1985, established a new network of Molniya 3 satellites, conducted three flights in the inscrutable Cosmos 1603 series, and reactivated the dormant *Salyut 7* station. Also of interest among USSR space advances were two missions, of the type normally supported by the D-1-e vehicle, that demonstrated completely new launch profiles, suggesting either the first variant in 15 years of the basic Proton booster or an entirely new launch vehicle. And for the first time in 20 years the USSR did not acknowledge a space mission (on June 21 from Tyuratam), leading observers to speculate it used either a new launch vehicle or a new class of satellite. The overall USSR launch rate, equalling their rates for the previous five years, resulted from a surge in October that was three times the launch rate of October 1984. In addition, they conducted three launches within eight hours on June 21 and three in less than 11 hours on August 8. The USSR used the Plesetsk facility for 62 launches, Baikonur Cosmodrome at Tyuratam for 35, and the Kapustin Yar complex for one. USSR launch vehicles deployed 119 satellites, of which 71 remained operational by the end of the year. Of the Soviet's 95 identifiable missions, 64 or two-thirds were dedicated government/military missions. And also in 1985 the USSR created Glavkosmos SSR to coordinate civilian space activities in that country.

Two of ESA's three launches deployed communications satellites—during the first launch one satellite for the Arab Satellite Communications Organization and one for Brazil. The purpose of the second mission was to deploy

GSTAR 1 (U.S.) and *Telecom 1B* (France). Their final mission launched *Giotto* for a Halley's comet flyby. Both Japanese launches were associated with data gathering on Halley's Comet; and the Peoples Republic of China launched *PRC-17.*

Sources for these data include *Spaceflight: Satellite Digest; The Soviet Year in Space, 1985;* press releases of NASA, Department of Defense, National Oceanic and Atmospheric Administration, and other government agencies, as well as the Communications Satellite Corporation. Soviet data also derive from statements in the Soviet press, translations from the Tass news agency, international news service reports, and announcements and briefings by Soviet officials. Data on satellites of other nations also come from announcements of their respective governments or national organizations.

Appendix A

Satellites, Space Probes, and Manned Space Flights, 1985

Launch Date	Spacecraft, Country, Int'l Designation, Vehicle, Launch Site	Payload Data	Apogee (km)	Perigee (km)	Period (min)	Inclination (degree)	Remarks
Jan. 7	Sakigake Japan 1985–1A Mu–3S Tanegashima Space Center	Total weight: 141 kg. Objective: Third launch under International Halley Watch originally intended as engineering test before Planet A launch, would pass 1,000,000 km ahead of Halley's comet through expected magnetospheric bow shock. Description: A cylinder 0.7 in length, 1/4 m in diameter, with solar cells covering curved surface.	Heliocentric orbit				Vega 1 and 2 were previous launches under International Halley Watch.
Jan. 9	Cosmos 1616 USSR 1985–2A A–2 Tyuratam	Total weight: 6000 kg.? Objective: "Continuation of outer space investigations." Description: Unavailable.	368	189	90.08	64.9	Spacecraft configuration possibly similar to Cosmos 1648. Intended for photo-reconnaissance. Decayed or recovered after 54 days.

Launch Date	Spacecraft, Country, Int'l Designation, Vehicle, Launch Site	Payload Data	Apogee (km)	Perigee (km)	Period (min)	Inclination (degree)	Remarks
Jan. 15	Cosmos 1617 USSR 1985-3A F-2 Plesetsk	Total weight: 40 kg? Objective: "Continuation of outer space investigations." Description: Unavailable.	1 414	1 414	114.09	82.6	Possibly similar to small satellites normally launched in groups of 8 by C-1, 1 m long, 0.8 in diameter, spheres.
and							
	Cosmos 1618 USSR 1985-3B	Total weight: Unavailable. Objective: "Continuation of outer space investigations." Description: Unavailable.	1 415	1 487	114.0	82.6	
and							
	Cosmos 1619 USSR 1985-3C	Total weight: Unavailable. Objective: "Continuation of outer space investigations." Description: Unavailable.	1 415	1 384	113.76	82.6	

Appendix A

and	Cosmos 1620 USSR 1985–3D	Total weight: Unavailable. Objective: "Continuation of outer space investigations." Description: Unavailable.	1 414	1 390	113.83	82.6
and	Cosmos 1621 USSR 1985–3E	Total weight: Unavailable. Objective: "Continuation of outer space investigations." Description: Unavailable.	1 414	1 395	113.89	82.61
and	Cosmos 1622 USSR 1985–3F	Total weight: Unavailable. Objective: "Continuation of outer space investigations." Description: Unavailable.	1 414	1 400	113.94	82.6
Jan. 16	Molniya 3–23 USSR 1985–4A A–2–e Plesetsk	Total weight: 2000 kg.? Objective: "Continuation of outer space investigations." Description: Unavailable.	39 727	630	717.81	62.87	A comsat for telephone, telegraph and TV links through the "Orbita" system within the USSR and abroad. Spacecraft had cylindrical body and conical motor section at one end deriving power from a "windmill" of 6 solar panels.

451

Launch Date	Spacecraft, Country, Int'l Designation, Vehicle, Launch Site	Payload Data	Apogee (km)	Perigee (km)	Period (min)	Inclination (degree)	Remarks
Jan. 16	Cosmos 1623 USSR 1985–5A A–2 Tyuratam	Total weight: 600 kg.? Objective: "Continuation of outer space investigations." Description: Unavailable.	415	349	92.19	69.99	Spacecraft configuration possibly similar to Comsat 1648. Intended for military photo-reconnaissance. Recovered after 14 days.
Jan. 17	Cosmos 1624 USSR 1985–6A C–1 Plesetsk	Total weight: 700 kg.? Objective: "Continuation of outer space investigations." Description: Unavailable.	808	785	100.9	74.05	Possibly a cylindrical body with domed ends enclosed in a drum-shaped solar array, about 2 m in length and diameter, for military communications using a store-dump technique.
Jan. 18	Gorizont 11 USSR 1985–7A D–1–e Tyuratam	Total weight: 2000 kg.? Objective: "Continuation of outer space investigations." Description: Unavailable.	35 917	35 751	1 435.96	1.49	A cylinder with pair of solar panels and a dish aerial array at one end, 5 m in length, 2 m in diameter. A comsat for continuous telephone, telegraph, and TV links through the "Orbita" system within the USSR and abroad.

Appendix A

Date	Satellite	Details				Notes
Jan. 23	Cosmos 1625 USSR 1985–8A F–1–m Tyuratam	Total weight: Unavailable. Objective: "Continuation of outer space investigations." Description: Unavailable.	392	120	89.62	64.99	Possibly an electronic reconnaissance satellite to cover oceans. Mission failed due to upper rocket stage restart failure.
Jan. 24	Cosmos 1626 USSR 1985–9A F–2 Plesetsk	Total weight: 2200 kg.? Objective: "Continuation of outer space investigations." Description: Unavailable.	664	630	97.67	82.53	Possibly a cylinder 5 m in length, 1.5 m in diameter, with 2 sun-seeking solar panels for electronic intelligence.
Jan. 24	STS–51C U.S. 1985–10A Discovery KSC	Total weight: 70 tons (excluding payload) Objective: To launch classified military payload. Description: Delta-winged vehicle, 37 m long, 24 m across.	Approx 300 km, circula		90	28.5	Crew: T. Mattingly, L. Shriver, E. Onizuka, J. Buchli, and E. Payton (all military) returned Jan. 27, 1985.
and							
Jan. 24	USA–8 U.S/DOD 1985–10B Discovery STS–51C	Total weight: Unavailable. Objective: Development of spacecraft technology and techniques. Description: Unavailable.	Elements not available				Possibly combined electronic reconnaissance with military early warning.

453

Launch Date	Spacecraft, Country, Int'l Designation, Vehicle, Launch Site	Payload Data	Apogee (km)	Perigee (km)	Period (min)	Inclination (degree)	Remarks
Feb. 1	Cosmos 1627 USSR 1985-11A C-1 Plesetsk	Total weight: Unavailable Objective: "Continuation of outer space investigations." Description: Unavailable.	1 018	958	104.87	82.92	Possibly a cylindrical body, domed ends enclosed in drum-shaped solar arrays, about 2 m in length and diameter, for military communications and navigation using a store-dump technique.
Feb. 6	Cosmos 1628 USSR 1985-12A A-2 Plesetsk	Total weight: 6000 kg.? Objective: "Continuation of outer space exploration." Description: Unavailable.	415	355	92.26	72.85	Spacecraft configuration possibly similar to Cosmos 1648. Intended for military photo-reconnaissance.
Feb. 6	Meteor 2-12 USSR 1985-13A F-2 Plesetsk	Total weight: 2200 kg.? Objective: "Continuation of outer space investigations." Description: Unavailable.	961	939	104.05	82.54	Probably a cylinder 5 m in length, 1.5 m in diameter, with 2 sun-seeking solar panels. A meteorological satellite returning scanning radiometer and other remotely sensed data on the earth's surface and cloud cover.

Appendix A

Date	Satellite	Details	Orbit			Incl.	Remarks
Feb. 8	USA–9 U.S. 1985–14A Titan 3B Agena D Vandenberg AFB	Total weight: Unavailable. Objective: Military reconnaissance. Description: Unavailable.	Near-polar, low, sun-synchronous				Launched by U.S. Department of Defense.
Feb. 8 and	ARABSAT I Arab Satellite Communications Organization 1985–15A Ariane 3 (V12) Kourou, French Guiana	Total weight: 1195 kg. (at launch) Objective: A comsat to cover Arab-speaking countries in North Africa and Middle East. Description: A standard design box-shaped body, 2.26 × 1.49 m, with a 2 panel solar array.	35 849	33 911	1 390.1	0.2	Spacecraft built by Aerospatiale of France. Payload had 25 C-band transponder channels and a TV transponder to provide signals to small community aerials. Placed in geostationary orbit over Zaire.
	BRASILSAT 1 Brazil 1985–15B Ariane 3 (V12) Kourou	Total weight: 1140 kg. (at launch) Objective: A national comsat to provide 24 C-band channels (6 GHz uplink, 4 GHz downlink). Description: A spin-stabilized cylinder with a de-spun aerial array at one end. Length was 2.95 m at launch, 7.04 m in orbit; diameter, 2.16 m.	35 840	35 779	1 437.3	0.1	Placed in geostationary orbit above central Brazil at 88 deg. E longitude.

455

Astronautics and Aeronautics, 1985

Launch Date	Spacecraft, Country, Int'l Designation, Vehicle, Launch Site	Payload Data	Apogee (km)	Perigee (km)	Period (min)	Inclination (degree)	Remarks
Feb. 21	Cosmos 1629 USSR 1985–16A D–1–E Tyuratam	Total weight: Unavailable. Objective: "Continuation of outer space investigations." Description: Unavailable.	35 803	35 775	1 436.23	1.37	Possibly similar to Gorizont-Raduga satellites as a developmental, experimental, or military comsat.
Feb. 27	Cosmos 1630 USSR 1985–17A A–2 Tyuratam	Total weight: 6000 kg. Objective: "Continuation of outer space investigations." Description: Unavailable.	336	175	89.61	64.89	Spacecraft configuration possibly similar to Cosmos 1648. Intended for military photo-reconnaissance.
Feb. 27	Cosmos 1631 USSR 1985–18A C–1 Plesetsk	Total weight: Unavailable. Objective: "Continuation of outer space investigations." Description: Unavailable.	512	472	94.45	65.85	Military spacecraft, possibly for radar calibration or electronic reconnaissance.
Mar. 1	Cosmos 1632 USSR 1985–19A A–2 Plesetsk	Total weight: 6000 kg.? Objective: "Continuation of outer space investigations." Description: Unavailable.	267	209	89.26	72.87	Spacecraft configuration possibly similar to Cosmos 1648. Intended for military photo-reconnaissance.

Appendix A

Date	Satellite				Description
Mar. 5	Cosmos 1633 USSR 1985–20A F-2 Plesetsk	Total weight: 2200 kg.? Objective: "Continuation of outer space investigations." Description: Unavailable.	658	637	97.67	82.54	Probably a cylinder, about 5 m in length, 1.5 m in diameter, with 2 sun-seeking solar panels, intended for military reconnaissance and electronic intelligence.
Mar. 13	GEOSAT U.S. 1985–21A Atlas E Vandenberg AFB	Total weight: 635 kg. (spacecraft) Objective: To provide geodetic data on the southern hemisphere and the north Pacific Ocean to replace data lost by premature shutdown of Seasat and to map the earth's ocean surface. Description: Unavailable.	814	757	100.6	108.1	Operated by the U.S. Navy to measure small variations in ocean surface height.
Mar. 14	Cosmos 1634 USSR 1985–22A C-1 Plesetsk	Total weight: 700 kg. (spacecraft) Objective: "Continuation of outer space investigations." Description: Unavailable.	1 011	960	104.82	82.94	A navsat of cylindrical shape about 2 m in length and diameter with domed ends enclosed in drum-shaped solar arrays.
Mar. 20	Cosmos 1635 USSR 1985–23A C-1 Plesetsk	Total weight: 40 kg. (spacecraft) Objective: "Continuation of outer space investigations." Description: Unavailable.	1 513	1 474	115.85	74.06	Single launch of eight probably spheroidal spacecraft about 1 m long and 0.8 m in diameter to provide tactical communications between troops and units in the field.

Launch Date	Spacecraft, Country, Int'l Designation, Vehicle, Launch Site	Payload Data	Apogee (km)	Perigee (km)	Period (min)	Inclination (degree)	Remarks
and	Cosmos 1636 USSR 1985–23B C–1 Plesetsk	Total weight: 40 kg. (spacecraft) Objective: "Continuation of outer space investigations." Description: Unavailable.	1 495	1 475	115.65	74.06	
and	Cosmos 1637 USSR 1985–23C C–1 Plesetsk	Total weight: 40 kg. (spacecraft) Objective: "Continuation of outer space investigations." Description: Unavailable.	1 489	1 465	115.48	74.06	
and	Comos 1638 USSR 1985–23D C–1 Plesetsk	Total weight: 40 kg. (spacecraft) Objective: "Continuation of outer space investigations." Description: Unavailable.	1 481	1 457	115.29	74.06	

Appendix A

	Cosmos 1639 USSR 1985–23E C-1 Plesetsk	Total weight: 40 kg. (space-craft) Objective: "Continuation of outer space investigations." Description: Unavailable.	1 481	1 442	115.13	74.05
and						
	Cosmos 1640 USSR 1985–23F C-1 Plesetsk	Total weight: 40 kg. (space-craft) Objective: "Continuation of outer space investigations." Description: Unavailable.	1 480	1 428	114.96	94.06
and						
	Cosmos 1641 USSR 1985–23G C-1 Plesetsk	Total weight: 40 kg. (space-craft) Objective: "Continuation of outer space investigations." Description: Unavailable.	1 480	1 413	114.81	74.06
and						
	Cosmos 1642 USSR 1985–23H C-1 Plesetsk	Total weight: 40 kg. (space-craft) Objective: "Continuation of outer space investigations." Description: Unavailable.	1 478 1 706	1 400 1 476	114.64 118.0	74.06 74.1

Launch Date	Spacecraft, Country, Int'l Designation, Vehicle, Launch Site	Payload Data	Apogee (km)	Perigee (km)	Period (min)	Inclination (degree)	Remarks
Mar. 22	Ekran 14 USSR 1985–24A D–1–e Tyuratam	Total weight: 2000 kg.? Objective: "Continuation of outer space investigations." Description: Unavailable.	35 789	35 777	1 435.912	0.41	A cylinder 5 m in length, 2 m in diameter, with a pair of boom-mounted solar panels and a flat aerial array at one end, intended to transmit USSR central TV to collective receiving aerials serving remote communities with the USSR.
Mar. 22	Intelsat VA–F10 U.S. 1985–25A Atlas Centaur KSC	Total weight: 2013 kg. Objective: Comsat to operate at C-band (6 GHz uplink, 11 GHz downlink) and Ku-band (14 GHz uplink, 11 GHz downlink). Description: Box-shaped body, 1.66 x 2.10 x 1.77 m, with attached 4 m aerial mast and 15.9 m span solar array.	35 776	35 787	1 398.1	0.4	Satellite increased traffic carrying capacity by 25% over Intelsat 5 to the equivalent of 15,000 telephone channels. Launched by NASA for INTELSAT. First in a series of 6 improved INTELSAT commercial communications satellite.

Appendix A

Date	Designation	Details				
Mar. 25	Cosmos 1643 USSR 1985–26A A–2 Tyuratam	Total weight: 6000 kg.? Objective: "Continuation of outer space investigations." Description: Unavailable.	294	223	89.68	64.77	Spacecraft configuration possibly similar to Cosmos 1648. Third flight of newest photo-reconnaissance model.
Apr. 3	Cosmos 1644 USSR 1985–27A A–2 Tyuratam	Total weight: 6000 kg.? Objective: "Continuation of outer space investigations." Description: Unavailable.	415	349	92.19	70.35	Spacecraft configuration possibly similar to Cosmos 1648. Intended for military photo-reconnaissance. Recovered after 14 days.
Apr. 12	STS–51D U.S. 1985–28A Discovery KSC	Total weight: 70 tons (excluding payload) Objective: To launch 2 satellites, to re-run McDonnell Douglas continuous electrophoresis drug processing experiment, and to fit makeshift tool to remote manipulator arm to attempt reactivation of Syncom IV–3. Description: Delta-winged vehicle, 37 m long, 24 m across.	464	319	92.4	28.5	Crew: K. Bobko, D. Williams, M. Seddon, S. Griggs, J. Hoffman, C. Walker, and E.J. Garn. Touched down on KSC runway on Apr. 19.

461

Launch Date	Spacecraft, Country, Int'l Designation, Vehicle, Launch Site	Payload Data	Apogee (km)	Perigee (km)	Period (min)	Inclination (degree)	Remarks
and	Telesat-I U.S. 1985-28B Discovery STS-51D	Total weight: 1000 kg. Objective: Communications satellite. Description: Cylinder 3 m long (extending to 7 m on deployment of solar array) and 2 m in diameter.	35 798	35 777	1 436.1	2.3	Launched by NASA for Canada; also called Anik C-1.
Apr. 13	Syncom IV-3 U.S. 1985-28C Discovery STS-51D	Total weight: 7500 kg. (including fully fuelled perigee boost motor still attached). Objective: Communications satellite. Description: Cylinder about 3 m long and 4.2 m in diameter.	35 970	35 594	1 435.9	3.2	Deployed successfully, but booster engine failed to ignite. Rendezvous with Discovery successful, but crew unable to operate sequencer lever. Satellite remained inoperable until restarted by crew of STS-51I on Sept. 1.
Apr. 16	Cosmos 1645 USSR 1985-29A A-2 Plesetsk	Total weight: 5 tons? Objective: "Continuation of outer space investigations." Description: Unavailable.	390	215	90.57	62.83	Intended for "materiology" scientific experiments. Recovered Apr. 29.

Appendix A

Date	Name/ID	Details	Col4	Col5	Col6	Description
Apr. 18	Cosmos 1646 USSR 1985–30A F-1-m Tyuratam	Total weight: Unavailable. Objective: "Continuation of outer space investigations." Description: Unavailable.	444	428	93.3	65.06	Intended for electronic reconnaissance over ocean. Phased with Cosmos 1567 and 1588 and later with Cosmos 1682.
Apr. 19	Cosmos 1647 USSR 1985–31A A–2 Plesetsk	Total weight: 6000 kg.? Objective: "Continuation of outer space investigations." Description: Unavailable.	323	169	89.43	67.14	Spacecraft configuration possibly similar to Cosmos 1643. Intended for long-duration, military photo-reconnaissance; recovered June 11.
Apr. 25	Cosmos 1648 USSR 1985–32A A–2 Plesetsk	Total weight: 6000 kg.? Objective: "Continuation of outer space investigations." Description: Unavailable.	327	229	90.07	82.33	Based possibly on the Vostok manned spacecraft, 6 m long, 2.4 m (max) in diameter, with spherical reentry module, instrument unit, and a supplementary package of instruments at the forward end. Intended for photo-reconnaissance. Recovered May 6.
Apr. 26	Prognoz 10 USSR 1085–33A A–2–e Tyuratam	Total weight: 1000 kg.? Objective: "Continuation of outer space investigations." Description: Unavailable.	200 320	421	5 785.11	64.99	Cylinder, 2 m in diameter, 1.5 m long, with instrument package mounted on one face. Joint USSR/Czechoslovakian mission investigating the region of electromagnetic shock between the magnetosphere and the solar wind. Also carried x-ray photometer.

Launch Date	Spacecraft, Country, Int'l Designation, Vehicle, Launch Site	Payload Data	Apogee (km)	Perigee (km)	Period (min)	Inclination (degree)	Remarks
Apr. 29	STS–51B U.S. 1985–34A Challenger KSC	Total weight: 70 tons (excluding payload) Objective: To launch NUSAT–1 and GLOMR and carry Spacelab 3. Description: Delta–winged vehicle 37 m long and 24 m across.	358	345	91.6	57.0	Crew: R. Overmeyer, F. Gregory, D. Lind, N. Thagard, W. Thornton, L. Vandenberg, and T. Wang landed at Edwards AFB on May 6. Global Low Orbiting Message Relay Satellite for DOD failed to leave cannister and was returned to earth.
and							
	NUSAT–1 U.S. 1985–34B Challenger STS–51B	Total weight: 52 kg. Objective: To calibrate air traffic control radar. Description: A 26-facetted polyhedron, 0.48 m in diameter, with 12 faces covered by solar cells.	339	318	91.1	57.0	Northern Utah Satellite.

Appendix A

Date	Satellite				
May 7	GSTAR 1 GTE Spacenet Corp. (U.S.) 1985–35A Ariane 3 (V–13) Kourou, French Guiana	Total weight: 705 kg. (empty) Objective: U.S. domestic comsat. Description: Box shaped, approximately 3.5 m each side.	36 027	201	635.8	7.0
and						
	Telecom 1B France 1985–35B Ariane 3 (V–13) Kourou, French Guiana	Total weight: 690 kg. (empty) Objective: To provide business communications, overseas telephone and TV links, and military point-to-point communications. Description: Box shaped, 2.0 x 1.4 x 1.4 m, with 13.8 m span solar array.	35 960	201	634.5	6.9
May 15	Cosmos 1649 USSR 1985–36A A–2 Plesetsk	Total weight: 6000 kg.? Objective: "Continuation of outer space investigations." Description: Unavailable.	415	356	92.26	72.8	Spacecraft configuration possibly similar to Cosmos 1648. Intended for military photo-reconnaissance. Recovered after 14 days.
May 17	Cosmos 1650 USSR 1985–37A D–1–e Tyuratam	Total weight: 600 kg.? Objective: "Continuation of outer space investigations." Description: Unavailable.	19 171	19 089	675.73	64.82	Cosmos 1650–1652 possibly cylinder, length and diameter about 2 m, with domed ends enclosed in drum-shaped solar arrays. Intended as navigation satellites; Cosmos 1650 the 6th Glonass launch.

Astronautics and Aeronautics, 1985

Launch Date	Spacecraft, Country, Int'l Designation, Vehicle, Launch Site	Payload Data	Apogee (km)	Perigee (km)	Period (min)	Inclination (degree)	Remarks
and	Cosmos 1651 USSR 1985–37B D–1–e Tyuratam	Total weight: 600 kg.? Objective: "Continuation of outer space investigations." Description: Unavailable.	19 145	19 116	675.75	64.83	
and	Cosmos 1652 USSR 1985–37C D–1–e Tyuratam	Total weight: 600 kg.? Objective: "Continuation of outer space investigations." Description: Unavailable.	19 145	19 120	675.84	64.80	
May 22	Cosmos 1653 USSR 1985–38A A–2 Plesetsk	Total weight: Unavailable. Objective: "Continuation of outer space investigations." Description: Unavailable.	273	259	89.83	82.34	Spacecraft configuration possibly similar to Cosmos 1648. Intended for photo-reconnaissance with all or part of payload an earth resources package. Recovered after 14 days.

Appendix A

Date	Satellite	Details				Notes
May 23	Cosmos 1654 USSR 1985–39A A–2 Tyuratam	Total weight: 6000 kg.? Objective: "Continuation of outer space investigations." Description: Unavailable.	343	172	89.66	64.86	Spacecraft configuration possibly similar to Cosmos 1648. Intended for military photo-reconnaissance. Disintegrated June 21.
May 29	Molniya 3–24 USSR 1985–40A A–2–e Plesetsk	Total weight: 2000 kg.? Objective: "Continuation of outer space investigations." Description: Unavailable.	39 886	470	717.80	62.80	Cylindrical body, 4 m long, 1.6 m in diameter, with conical motor section at one end and deriving power from a windmill of 6 solar panels. Comsat intended to provide telephone, telegraph, and TV links through the Orbita system inside the USSR and abroad. Replaced Molniya 3–18; on station June 9.
May 30	Cosmos 1655 USSR 1985–41A C–1 Plesetsk	Total weight 600 kg.? Objective: "Continuation of outer space investigations." Description: Unavailable.	1 016	979	105.06	82.95	Spacecraft configuration possibly similar to Cosmos 1650; spacecraft replaced Cosmos 1447. Intended to aid navigation.
May 30	Cosmos 1656 USSR 1985–42A D–1–e Tyuratam	Total weight: Unavailable. Objective: "Continuation of outer space investigations." Description: Unavailable.	861	807	101.58	71.11	Orbit plane is 45 deg. away from that of Cosmos 1603, suggesting formation of a new satellite system, possibly intended for electronic intelligence gathering. Second Cosmos 1603 type mission.

Astronautics and Aeronautics, 1985

Launch Date	Spacecraft, Country, Int'l Designation, Vehicle, Launch Site	Payload Data	Apogee (km)	Perigee (km)	Period (min)	Inclination (degree)	Remarks
June 6	Soyuz T-13 USSR 1985-43A A-2 Tyuratam	Total weight: 7000 kg.? Objective: To carry crew to Salyut 7. Description: Unavailable.	359	356	91.69	51.64	Near spherical orbital compartment, conical reentry module, and cylindrical instrument unit with solar panels. Intended to deliver V. Dzanibekhov and V. Savimykh to repair Salyut 7, with docking on June 8 at forward docking hatch; undocked Sept. 25.
June 7	Cosmos 1657 USSR 1985-44A A-2 Plesetsk	Total weight: 6000 kg.? Objective: "Continuation of outer space investigations." Description: Unavailable.	274	257	89.82	82.26	Spacecraft configuration possibly similar to Cosmos 1648. Intended for photo-reconnaissance with all or part of payload an earth resources package. Recovered after 14 days.
June 11	Cosmos 1658 USSR 1985-45A A-2-e Plesetsk	Total weight: Unavailable. Objective: "Continuation of outer space investigations." Description: Unavailable.	39 758	592	717.68	62.86	Possibly similar to Molniya-3 (24) and intended for missile early warning. Replaced Cosmos 1481.

Appendix A

Date	Designation	Details				Notes
June 13	Cosmos 1659 USSR 1985–46A A-2 Plesetsk	Total weight: 6000 kg.? Objective: "Continuation of outer space investigations." Description: Unavailable.	415	357	92.27	72.9	Spacecraft configuration possibly similar to Cosmos 1648. Intended for military photo-reconnaissance and recovered after 14 days.
June 14	Cosmos 1660 USSR 1985–47A C-1 Plesetsk	Total weight: 600 kg.? Objective: "Continuation of outer space investigations." Description: Unavailable.	1 526	1 482	116.07	73.63	Spacecraft configuration possibly similar to Cosmos 1650. Possible geodetic mission.
June 17	STS–51G U.S. 1985–48A Discovery KSC	Total weight: 70 tons (excluding payload) Objective: To launch 3 comsats; test laser reflector as part of the SDI program; deploy Spartan and life science and other experiments. Description: Delta-winged vehicle 37 m long, 24 m across.	391	355"	90.2	28.5	Crew: D. Brandenstein, J. Creighton, J. Fabian, S. Lucid, S. Nagel, P. Baudry (France), and S. Abdelazize Al-Saud (Saudi Arabia). Landed at Edwards AFB on June 24.

469

Astronautics and Aeronautics, 1985

Launch Date	Spacecraft, Country, Int'l Designation, Vehicle, Launch Site	Payload Data	Apogee (km)	Perigee (km)	Period (min)	Inclination (degree)	Remarks
and	Morelos 1 Mexico 1985-48B Discovery STS-51G	Total weight: 512 kg. (excluding fuel) Objective: To provide voice, video, data, and facsimile communications using 18 channels at C-band and 4 at Ku band. Description: Hughes HS-376 type, cylindrical, 6.6 m long with solar array fully deployed, and 2.16 in diameter.	35 798	35 776	1 436.1	0.0	Launched by NASA for Mexico.
June 18	Arabsat 1B U.S. 1985-48C Discovery STS-51G	Total weight: 1257 kg. (at launch) Objective: Comsat to cover Arab-speaking countries of North Africa and Middle East. Description: Box shape, 2.26 x 1.64 x 1.49 m with a 2 panel solar array of width 20.7 m.	35 833	35 737	1 436.1	0.0	Launched for the Arab Satellite Communication Organization (ASCO). Payload consisted of 25 C-band transponders and a single TV transponder providing signals to small community aerials.

Appendix A

Date	Name/Designation	Specifications				Remarks
June 18	Cosmos 1661 USSR 1985-49A A-2-e Plesetsk	Total weight: Unavailable. Objective: "Continuation of outer space investigations." Description: Unavailable.	39 761	856	717.61	62.98	Spacecraft configuration possibly similar to Molniya 3(24). Intended as missile early warning. Coplanar with Cosmos 1604 but 40 deg. to west; on station June 27.
June 19	Telstar 3D U.S. 1985-48D Discovery STS-51G	Total weight: 630 kg. (excluding fuel) Objective: Commercial comsat with 30 C-band transponders. Description: Similar to Morelos, except length was 6.83 m with solar array deployed.	35 791	35 788	1 436.1		Launched for AT&T.
June 19	Cosmos 1662 USSR 1985-50A C-1 Plesetsk	Total weight: Unavailable. Objective: "Continuation of outer space investigations." Description: Unavailable.	513	476	94.5	65.84	Minor military Cosmos.
June 20	Spartan 1 U.S. 1985-48E Discovery STS-51G	Total weight: 1008 kg. Objective: To extend capabilities of sounding rocket-class experiments by making observations of the Perseus Cluster, Galactic Center, and Scorpius X-2. Description: Retangular, 126 by 42 by 48 in.	391	355	91.82	28.47	NASA used the Canadian-built remote manipulator system (RMS) to retrieve the U.S. Naval Research Laboratory's Spartan 1 on June 24, 1985, approximately 45 hours after release.

Launch Date	Spacecraft, Country, Int'l Designation, Vehicle, Launch Site	Payload Data	Apogee (km)	Perigee (km)	Period (min)	Inclination (degree)	Remarks
June 21	Progress 24 USSR 1985-51A A-2 Tyuratam	Total weight: Unavailable. Objective: Unmanned ferry to deliver supplies, replacement parts, maintenance equipment, and fuel to Salyut 7. Description: Unavailable.	359	356	91.69	51.65	Spacecraft similar to Soyuz T-13 except reentry module was replaced by a non-recoverable section containing fuel tanks and there were no solar panels. Docked with rear port of Salyut on June 23, undocked July 15, and reentered earth's atmosphere later that day.
June 21	Cosmos 1663 USSR 1985-52A A-2 Plesetsk	Total weight: 6000 kg.? Objective: "Continuation of outer space investigations." Description: Unavailable.	273	259	89.83	82.33	Spacecraft configuration possibly similar to Cosmos 1648. Intended for photo-reconnaissance with all or part of payload an earth resources package; recovered July 5.
June 21	Unknown USSR 1985-53A F (probably) Tyuratam	Total weight: Unavailable. Objective: Unknown Description: Unavailable.	340	195	89.86	64.41	Possibly nuclear-powered ocean radar satellite. Main payload failed to reach orbit and only 3 fragments detected in orbit. All decayed naturally within 7 days.

Appendix A

Date	Name/ID/Location	Details				Notes
June 26	Cosmos 1664 USSR 1985–54A A–2 Plesetsk	Total weight: 6000 kg.? Objective: "Continuation of outer space investigations." Description: Unavailable.	379	224	90.55	72.84	Spacecraft configuration possibly similar to Cosmos 1648. Intended for military photo-reconnaissance. Recovered after 9 days.
June 30	Intelsat VA–F11 U.S. 1985–55A Atlas Centaur Cape Canaveral AFB	Total weight: 2013 kg., 1098 after fuel depletion Objective: Comsat to operate at C-band and L-band to provide equivalent of 15,000 telephone channels. Description: Box shape, 1.66 x 2.10 x 1.77 m with attached 4 m aerial mast and a 15.9 m span solar array.	35 774	34 401	1 400.6	0.1	Launched by NASA for INTELSAT.
July 2	Giotto ESA 1985–56A Ariane 1 (v14) Kourou, French Guiana	Total weight: 960 kg., 512 kg after fuel depletion Objective: To study Halley's Comet, including photos of nucleus during flyby at less than 1000 km. Description: Spin-stabilized cylinder with de-spun aerial reflector, 1.84 min diameter and 1.6 m long (excluding aerial support).	35 420	206	623.46	7.04	Placed into heliocentric trajectory on July 3.
July 3	Cosmos 1665 USSR 1985–57A A–2 Plesetsk	Total weight: 6000 kg.? Objective: "Continuation of outer space investigations." Description: Unavailable.	292	224	89.67	72.84	Spacecraft configuration possibly similar to Cosmos 1648. Intended for military photo-reconnaissance. Recovered after 14 days.

Launch Date	Spacecraft, Country, Int'l Designation, Vehicle, Launch Site	Payload Data	Apogee (km)	Perigee (km)	Period (min)	Inclination (degree)	Remarks
July 8	Cosmos 1666 USSR 1985–058A F-2 Plesetsk	Total weight: 1500 kg.? Objective: "Continuation of outer space investigations." Description: Unavailable.	666	634	97.72	82.53	Probably a cylinder, about 5 m long, 1.5 m in diameter, with 2 sun-seeking solar panels. For electronic intelligence, it replaced Cosmos 1633.
July 10	Cosmos 1667 USSR 1985–59A A-2 Plesetsk	Total weight: 6000 kg.? Objective: "Continuation of outer space investigations." Description: Unavailable.	270	211	89.30	82.35	Spacecraft configuration possibly similar to Cosmos 1648. Biological satellite with payload of 2 Rhesus macaques, white rats, and salamanders. Recovered after 7 days.
July 15	Cosmos 1668 USSR 1985–60A A-2 Tyuratam	Total weight: 6000 kg.? Objective: "Continuation of outer space investigations." Description: Unavailable.	281	230	89.61	70.38	Spacecraft configuration possibly similar to Cosmos 1648. Intended for military photo-reconnaissance. Recovered after 14 days.
July 17	Molniya 3–25 USSR 1985–61A A-2-e Plesetsk	Total weight: Unavailable. Objective: "Continuation of outer space investigations." Description: Unavailable.	39 904	445	717.66	62.83	Spacecraft similar to Molniya 2 (24) and mission to Molniya 3 (24) for communications; established new plane.

Appendix A

July 19	Cosmos 1669 USSR 1985–62A A-2 Tyuratam	Total weight: Unavailable. Objective: To dock with Salyut 7. Description: Unavailable.	357	354	91.65	51.65	Spacecraft configuration similar to Progress 24 but with solar panels. Apparent test flight of new Soyuz/Progress-type vehicle capable of autonomous flight and joint operation with Salyut. Docked with Salyut 7 on July 21; undocked Aug. 29.
July 29	STS–51F U.S. 1985–63A Challenger KSC	Total weight: 70 tons (excluding payload) Objective: To perform Spacelab 2 experiments in infrared and x-ray astronomy; high-energy, solar, and atmospheric physics; life sciences; and fluid properties. Deploy the Plasma Diagnostics Package. Description: Delta-winged vehicle, 37 m long and 24 m across.	321	312	90.5	49.5	Crew: C. Fullerton, R. Bridges, F. Musgrave, A. England, K. Henize, L. Acton, and J. Bartoe. Landed Aug. 6 at Edwards AFB following 1-day mission extension to increase payload observation time. An abort on the pad delayed launch on July 12, followed by a 97-min delay before actual launch.
Aug. 1	Plasma Diagnostics Package U.S. 1985–63B Challenger STS–51F	Total weight: 160 kg. Objective: To perform plasma physics experiments and measure electron emissions from Spacelab experiment. Description: Cylinder 1 m long, 0.7 m in diameter.	Same as orbiter				Flew alongside Challenger for several hours before retrieval on Aug. 2 for further experiments with it attached to the remote manipulator arm. Retrieved for return to earth.

475

Astronautics and Aeronautics, 1985

Launch Date	Spacecraft, Country, Int'l Designation, Vehicle, Launch Site	Payload Data	Apogee (km)	Perigee (km)	Period (min)	Inclination (degree)	Remarks
Aug. 1	Cosmos 1670 USSR 1985–64A F-1-m Tyuratam	Total weight: Unavailable. Objective: "Continuation of outer space investigations." Description: Unavailable.	264	252	89.68	65.01	Probably nuclear powered, radar carrying satellite for reconnaissance over oceans.
Aug. 2	Cosmos 1671 USSR 1985–65A A-2 Plesetsk	Total weight: 6000 kg.? Objective: "Continuation of outer space investigtions." Description: Unavailable.	258	229	89.36	72.88	Spacecraft configuration possibly similar to Cosmos 1648. Intended for military photo-reconnaissance. Recovered after 14 days.
Aug. 3	Oscar 24 U.S. 1985–066A Scout WSMC	Total weight: 55 kg. Objective: Transit-type navigation satellite placed in orbital storage for Stacked Oscar on Scout (SOOS) program. Description: Octagonal prism, 0.25 m long, 0.46 in diameter, with 4 solar panels and gravity gradient stabilization boom.	1 259	1 001	107.9	89.82	Part of U.S. Navy Transit (Navy Navigation Satellite System).

Appendix A

Date	Satellite	Details	Apogee (km)	Perigee (km)	Period (min)	Inclination (deg)	Remarks
and	Oscar 30 U.S. 1985-66B Scout WMSC	Total weight: 55 kg. Objective: Same as Oscar 24. Description: Same as Oscar 24.	1 260	1 000	107.9	89.82	Part of U.S. Navy Transit (Navy Navigation Satellite System).
Aug. 7	Cosmos 1672 USSR 1985-67A A-2 Plesetsk	Total weight: 6000 kg.? Objective: "Continuation of outer space investigations." Description: Unavailable.	273	258	89.82	82.34	Spacecraft configuration possibly similar to Cosmos 1648. Intended for photo-reconnaissance with all or part of payload; also an earth resources package. Recovered after 14 days.
Aug. 8	Cosmos 1673 USSR 1985-68A A-2 Tyuratam	Total weight: 6000 kg.? Objective: "Continuation of outer space investigations." Description: Unavailable.	273	198	89.20	64.79	Spacecraft configuration possibly similar to Cosmos 1648. Intended for military photo-reconnaissance; recovered Sept. 19.
Aug. 8	Cosmos 1674 USSR 1985-69A F-2 Plesetsk	Total weight: 1500 kg.? Objective: "Continuation of outer space investigations." Description: Unavailable.	664	632	97.69	82.53	Probably a cylinder, 5 m long, 1.5 min diameter, with 2 sun-seeking solar panels. For electronic intelligence; replaced Cosmos 1536.
Aug. 8	Raduga 16 USSR 1985-70A D-1-E Tyuratam	Total weight: 2000 kg.? Objective: "Continuation of outer space investigations." Description: Unavailable.	35 831	35 740	1 436.03	1.29	Possibly a cylinder, 5 m long, 2 m in diameter, with pair of solar panels and multi-dish aerial array at one end. Communications satellite stationed at 45 deg. E.

Launch Date	Spacecraft, Country, Int'l Designation, Vehicle, Launch Site	Payload Data	Apogee (km)	Perigee (km)	Period (min)	Inclination (degree)	Remarks
Aug. 12	Cosmos 1675 USSR 1985-71A A-2-e Plesetsk	Total weight: Unavailable. Objective: "Continuation of outer space investigations." Description: Unavailable.	39 732	602	717.35	62.85	Possibly similar to Molniya 1 (64), intended for missile early warning. Replaced Cosmos 1581; on station Aug. 15.
Aug. 16	Cosmos 1676 USSR 1985-72A A-2 Plesetsk	Total weight: Unavailable. Objective: "Continuation of outer space investigations." Description: Unavailable.	347	166	89.63	67.16	Spacecraft configuration possibly similar to Cosmos 1648. Intended for military photoreconnaissance. Recovered Oct. 14.
Aug. 18	Suisei (Planet A) Japan 1985-73A Mu-3S Kagoshima Space Center	Total weight: 141 kg. Objective: To observe the hydrogen corona around the coma of Halley's Comet and to measure solar wind and cometary charge particles. Description: Cylinder, 0.7 long, 1.4 m in diameter, with curved surface covered by solar cells.	151.467 Heliocentric	100.480	282.2 (days)	0.888	Closest Halley's Comet encounter would be March 8, 1986, at 211,000 km from earth. Spacecraft renamed "Suisei" (Japanese for "comet").

Appendix A

Date	Satellite	Details				Description
Aug. 22	Molniya 1 (64) USSR 1985-74A A-2-e Plesetsk	Total weight: 1800 kg.? Objective: "Continuation of outer space investigations." Description: Unavailable.	39 698	654	717.74	62.86	Cylinder, 3.4 m long, 1.6 m in diameter, with a conical motor section at one end and deriving power from a windmill of 6 solar panels. Replaced Molniya 1-61; on station Aug. 28.
Aug. 23	Cosmos 1677 USSR 1985-75A F-1-m Tyuratam	Total weight: Unavailable. Objective: "Continuation of outer space investigations." Description: Unavailable.	263	251	89.65	65.0	Probably weighed several tons and carried nuclear-powered radar for ocean surveillance.
Aug. 27	STS-511 U.S. 1985-76A Discovery KSC	Total weight: 70 tons (excluding payload) Objective: To launch Aussat 1, ASC 1, and Syncom IV-4 and to repair dormant fuel-laden Syncom IV-3. Description: Delta-winged vehicle, 37 m long, 24 m across.	190	190	90	28.45	Crew: J. Engle, R. Covey, J. van Hoften, W. Fisher, and J. Lounge. Landed Sept. 3 at Edwards AFB. All satellites launched successfully. Syncom IV-3 retrieved and repaired from payload bay and subsequent EVA and released on Sept. 1.

Launch Date	Spacecraft, Country, Int'l Designation, Vehicle, Launch Site	Payload Data	Apogee (km)	Perigee (km)	Period (min)	Inclination (degree)	Remarks
and	Aussat 1 Australia 1985-76B Discovery STS-51I	Total weight: 655 kg. (excluding fuel) Objective: To provide communication links within Australia, particularly TV to remote communities. Description: Hughes HS-376-type spin-stabilized satellite. Cylinder 2.82 m long, 2.16 m in diameter, covered with solar cells.	35 795	35 779	1 436.2	0.0	Launched by NASA for Australia's National Satellite Company.
and	ASC 1 American Satellite Co. 1985-76C Discovery STS-51I	Total weight: 1300 kg. (fully fuelled) Objective: Comsat operating at C-band and Ku-band. Description: Three-axis-stabilized box, 3.2 x 1.6 x 1.6 m, with 14 m span solar array and an aerial array on a face.	35 791	35 782	1 436.1	0.1	Launched by NASA for the American Satellite Company.

Appendix A

Date	Satellite	Description				Incl.	Notes
Aug. 29	Syncom IV-4 U.S. 1985-076D Discovery STS-51I	Total weight: 1400 kg. (excluding fuel) Objective: Comsat. Description: Cylinder, 3 m long, 4.2 m in diameter, with curved surface covered by solar cells.	35 805	35 776	1 410.9	0.2	Launched by NASA for the U.S. Navy as replacement for FleetSatCom spacecraft. Although reaching correct geosynchronous orbit, ceased functioning.
Aug. 29	Cosmos 1678 USSR 1985-77A A-2 Plesetsk	Total weight: 6000 kg.? Objective: "Continuation of outer space investigations." Description: Unavailable.	272	258	89.81	82.33	Spacecraft configuration possibly similar to Cosmos 1648. Intended for photo-reconnaissance with all or part of the payload; also an earth resources package. Recovered after 14 days.
Aug. 29	Cosmos 1679 USSR 1985-78A A-2 Tyuratam	Total weight: 6000 kg.? Objective: "Continuation of outer space investigations." Description: Unavailable.	343	173	89.67	64.86	Space configuration possibly similar to Cosmos 1648. Intended for military photo-reconnaissance. Recovered Oct. 18.
Sept. 4	Cosmos 1680 USSR 1985-79A C-1 Plesetsk	Total weight: Unavailable. Objective: "Continuation of outer space investigations." Description: Unavailable.	807	784	100.78	74.05	Possibly cylinder, 2 m long, 2 m in diameter, covered with solar cells. Intended for military communications using store-dump technique. Replaced Cosmos 1538.

Launch Date	Spacecraft, Country, Int'l Designation, Vehicle, Launch Site	Payload Data	Apogee (km)	Perigee (km)	Period (min)	Inclination (degree)	Remarks
Sept. 6	Cosmos 1681 USSR 1985–80A A–2 Plesetsk	Total weight: 6000 kg.? Objective: "Continuation of outer space investigations." Description: Unavailable.	226	219	89.95	82.33	Spacecraft configuration possibly similar to Cosmos 1648. Intended for photo-reconnaissance with all or part of payload an earth resources package. Recovered Sept. 19.
Sept. 17	Soyuz T–14 USSR 1985–81A A–2 Tyuratam	Total weight: 7000 kg.? Objective: To carry crew to Salyut 7. Description: Unavailable.	353	337	91.44	51.62	Spacecraft 7.5 m long, 2.2 m in diameter, with near-spherical orbital compartment, conical reentry module, and cylindrical instrument unit with solar panels, intended to exchange crew for Salyut 7. Returned Sept. 26.
Sept. 19	Cosmos 1682 USSR 1985–82A F–1–m Tyuratam	Total weight: Unavailable. Objective: "Continuation of outer space investigations." Description: Unavailable.	443	429	93.30	65.02	Intended for electronic reconnaissance over oceans. Phased with Cosmos 1567 and Cosmos 1616.

Appendix A

Date	Satellite	Details				Description
Sept. 19	Cosmos 1683 USSR 1985–83A A–2 Plesetsk	Total weight: 6000 kg.? Objective: "Continuation of outer space investigations." Description: Unavailable.	414	356	92.26	72.86	Spacecraft configuration possibly similar to Cosmos 1648. Intended for military photo-reconnaissance. Recovered after 15 days.
Sept. 24	Cosmos 1684 USSR 1985–84A A–2-e Plesetsk	Total weight: 2000 kg.? Objective: "Continuation of outer space investigations." Description: Unavailable.	39 762	583	717.58	62.9	Possibly based on Molniya satellite with cylindrical body, 4 m long, 1.6 m in diameter, surmounted by conical motor section and powered by a windmill of 6 solar panels. Intended for missile early warning. Replaced Cosmos 1586; on station Sept. 27.
Sept. 26	Cosmos 1685 USSR 1985–85A A–2 Plesetsk	Total weight: 6000 kg.? Objective: "Continuation of outer space investigations." Description: Unavailable.	416	356	92.27	72.86	Spacecraft configuration possibly similar to Cosmos 1648. Intended for military photo-reconnaissance. Recovered after 14 days.
Sept. 27	Cosmos 1686 USSR 1985–86A D–1-h Tyuratam	Total weight: 20 tons? Objective: "Continuation of outer space investigations." Description: Unavailable.	351	335	91.39	51.3	Cylinder 13 m long, 4.15 m (max) in diameter, with 2-panel solar array. Intended to carry supplies to and for enlargement of Salyut 7. Docked with Salyut's forward hatch Oct. 2.

Launch Date	Spacecraft, Country, Int'l Designation, Vehicle, Launch Site	Payload Data	Apogee (km)	Perigee (km)	Period (min)	Inclination (degree)	Remarks
Sept. 29	Intelsat VA–F12 U.S. 1985–87A Atlas-Centaur KSC	Total weight: 2013 kg. (with fuel) Objective: Comsat to provide 13,500 2-way voice circuits and 2 TV channels. Description: Box, 1.66 x 2.10 x 1.77 m, with 4 m aerial mast and 15.9 m span solar array.	35 805	34 776	1 410.9	0.2	Last Intelsat commercial communications satellite launched by NASA for INTELSAT from KSC. Satellite replaced INTELSAT V F-1.
Sept. 30	Cosmos 1687 USSR 1985–88A A–2–e Plesetsk	Total weight: 2000 kg.? Objective: "Continuation of outer space investigations." Description: Unavailable.	39 710	626	717.40	62.97	Spacecraft configuration possibly similar to Molniya with cylindrical body surmounted by conical motor section with windmill of 6 solar panels. Intended for missile early orwarning. Replaced Cosmos 1409;
Oct. 2	Cosmos 1688 USSR 1985–89A C–1 Kapustin Yar	Total weight: Unavailable. Objective: "Continuation of outer space investigations." Description: Unavailable.	548	347	93.53	50.68	Possibly for military radar calibration.

Appendix A

Oct. 3	Cosmos 1689 USSR 1985–90A A–1 Tyuratam	Total weight: 1500 kg.? Objective: "Continuation of outer space investigations." Description: Unavailable.	657	572	96.99	97.97	Cylinder, 5 m long, 2 m in diameter, with 2 sun-seeking solar panels. Intended for earth resources remote sensing.
Oct. 3	Molniya 3–26 USSR 1985–91A A–2–e Plesetsk	Total weight: Unavailable. Objective: "Continuation of outer space investigations." Description: Unavailable.	39 738	609	717.62	62.88	Communications satellite on station Oct 30; established new plane.
Oct. 3	STS–51J U.S. 1985–92A Atlantis KSC	Total weight: 70 tons (excluding payload) Objective: To launch 2 satellites for the Department of Defense. Description: Delta-winged vehicle 37 m long, 24 m across.		Elements not available			Crew on board: K. Bobko, R. Grabe, D. Hilmers, R. Stewart, and W. Pailes. Landed Oct. 7, 1985 at Edwards AFB. A dedicated Department of Defense mission.
and							
	DSCS 11 U.S. 1985–92B Atlantis STS–51J	Total weight: Unavailable. Objective: Messages relay. Description: Unavailable.		Elements not available			For DOD's Defense Satellite Communications System.

Launch Date	Spacecraft, Country, Int'l Designation, Vehicle, Launch Site	Payload Data	Apogee (km)	Perigee (km)	Period (min)	Inclination (degree)	Remarks
and	DSCS 12 U.S. 1985-92C Atlantis STS-51J	Total weight: Unavailable. Objective: Messages relay. Description: Unavailable.	Elements not available				For DOD's Defense Satellite Communications System.
Oct. 9	Navstar 11 U.S./DOD 1985-93A Atlas E	Total weight: 873 kg. Objective: Intended for Global Positioning System. Description: Unavailable.	20 709	540	367.7	63.0	
Oct. 9	Cosmos 1690 USSR 1985-94A F-2 Plesetsk	Total weight: Unavailable. Objective: "Continuation of outer space investigations." Description: Unavailable.	1 417	1 382	113.77	82.61	Similar to Cosmos 1617–1622.
and	Cosmos 1691 USSR 1985-94B F-2 Plesetsk	Total weight: Unavailable. Objective: "Continuation of outer space investigations." Description: Unavailable.	1 417	1 412	114.10	82.62	Fragmented on Nov. 22.

Appendix A

Cosmos 1692 USSR 1985–94C F–2 Plesetsk	Total weight: Unavailable. Objective: "Continuation of outer space investigations." Description: Unavailable.	1 417	1 389	113.85	82.6
and					
Cosmos 1693 USSR 1985–94D F–2 Plesetsk	Total weight: Unavailable. Objective: "Continuation of outer space investigations." Description: Unavailable.	1 417	1 393	113.90	82.61
and					
Cosmos 1694 USSR 1985–94E F–2 Plesetsk	Total weight: Unavailable. Objective: "Continuation of outer space investigations." Description: Unavailable.	1 417	1 398	113.96	82.62
and					
Cosmos 1695 USSR 1985–94F F–2 Plesetsk	Total weight: Unavailable. Objective: "Continuation of outer space investigations." Description: Unavailable.	1 417	1 405	114.03	82.61

Astronautics and Aeronautics, 1985

Launch Date	Spacecraft, Country, Int'l Designation, Vehicle, Launch Site	Payload Data	Apogee (km)	Perigee (km)	Period (min)	Inclination (degree)	Remarks
Oct. 16	Cosmos 1696 USSR 1985–95A A–2 Tyuratam	Total weight: Unavailable. Objective: "Continuation of outer space investigations." Description: Unavailable.	281	230	89.62	70.37	Intended for photo-reconnaissance; recovered Oct. 30.
Oct. 21	PRC–17 PRC 1985–96A Unknown Unknown	Total weight: Unavailable. Objective: Unavailable. Description: Unavailable.	384	170	90.1	63.0	Decayed Nov. 7, 1985.
Oct. 22	Cosmos 1697 USSR 1985–97A D–class? Tyuratam	Total weight: Unavailable. Objective: "Continuation of outer space investigations." Description: Unavailable.	854	850	101.97	70.99	Intended for electronic intelligence. Similar to Cosmos 1603 but different launch profile.
Oct. 22	Cosmos 1698 USSR 1985–098A A–2–e Plesetsk	Total weight: Unavailable. Objective: "Continuation of outer space investigations." Description: Unavailable.	39 726	605	717.31	62.92	An early warning spacecraft on station Oct. 27. Replaced Cosmos 1541.

Appendix A

Date	Satellite	Details				Notes
Oct. 23	Molniya 1-65 USSR 1985-099A A-2-e Tyuratam	Total weight: Unavailable. Objective: "Continuation of outer space investigations." Description: Unavailable.	39 731	630	717.92	62.97	Communications satellite on station Oct. 27. Replaced Molniya 1-58.
Oct. 24	Meteor 3-1 USSR 1985-100A F-2 Plesetsk	Total weight: Unavailable. Objective: "Continuation of outer space investigations." Description: Unavailable.	1 251	1 227	110.27	82.55	First of a new generation of weather satellites.
Oct. 25	Cosmos 1699 USSR 1985-101A A-2 Plesetsk	Total weight: Unavailable. Objective: "Continuation of outer space investigations." Description: Unavailable.	339	167	89.56	67.14	Intended for photo-reconnaissance; recovered Dec. 23.
Oct. 25	Cosmos 1700 USSR 1985-102A D-1-e Tyuratam	Total weight: Unavailable. Objective: "Continuation of outer space investigations." Description: Unavailable.	35 814	35 755	1 425.98	1.43	Communications satellite stationed at 95 deg. E. Possibly first data relay satellite.
Oct. 28	Molniya 1-66 USSR 1985-103A A-2-e Plesetsk	Total weight: Unavailable. Objective: "Continuation of outer space investigations." Description: Unavailable.	39 587	508	717.99	62.81	Communications satellite on station Nov. 21. Replaced Molniya 1-56.

Launch Date	Spacecraft, Country, Int'l Designation, Vehicle, Launch Site	Payload Data	Apogee (km)	Perigee (km)	Period (min)	Inclination (degree)	Remarks
Oct. 30	STS–61A U.S. 1085–104A Challenger KSC	Total weight: 70 tons (excluding payload) Objective: To launch Glomr and carry Spacelab D1. Description: Delta-winged vehicle 37 m long, 24 m across.	333	321	91	57.0	Crew on board: H.W. Hartsfield, S.R. Nagel, B.J. Dunbar, J.F. Buchli, G.S. Bluford, E. Messerschmid (W. Germany), R. Furrer (W. Germany), and W. Ockels (Netherlands). Landed at Edwards AFB on Nov. 6.
and	Glomr U.S. 1985–104B Challenger STS–61A	Total weight: 150 lb. Objective: Messages relay. Description: Sixty-two sided polyhedron.	333	318	91.0	57.0	NASA launched the Global Low Orbiting Message Relay for DOD.
Nov. 9	Cosmos 1701 USSR 1985–105A A-2-e Plesetsk	Total weight: Unavailable. Objective: "Continuation of outer space investigations." Description: Unavailable.	39 719	619	717.45	63.03	Early warning satellite on station Nov. 12. Replaced Cosmos 1675.

Appendix A

Date	Designation	Description				Notes
Nov. 13	Cosmos 1702 USSR 1985–106A A–2 Plesetsk	Total weight: Unavailable. Objective: "Continuation of outer space investigations." Description: Unavailable.	414	356	92.26	72.87	Photo-reconnaissance satellite; recovered Nov. 27.
Nov. 15	Raduga 17 USSR 1985–107A D–1–e Tyuratam	Total weight: Unavailable. Objective: "Continuation of outer space investigations." Description: Unavailable.	35 798	35 791	1 436.52	1.37	Communications satellite stationed at 37 deg. E.
Nov. 22	Cosmos 1703 USSR 1985–108A F–2 Plesetsk	Total weight: Unavailable. Objective: "Continuation of outer space investigations." Description: Unavailable.	666	635	9773	82.51	Electronic intelligence satellite; replaced Cosmos 1674.
Nov. 26	STS–61B U.S. 1985–109A Atlantis KSC	Total weight: 70 tons (excluding payload) Objective: To launch Morelos-B, Aussat-2, Satcom KU–2, and Dex Target and to perform experiments in extravehicular assembly of structures, in diffusive mixing of organic solutions, and in production of primary mirrors and metallic crystals. Description: Delta-winged vehicle 37 m long, 24 m across.	370	350	91.9	28.5	Crew on board: B.H. Shaw, B.D. O'Connor, M.L. Cleave, J.L. Ross, S.C. Spring, R.N. Vela, and C.D. Walker. Landed at Edwards AFB on Dec. 3.

Launch Date	Spacecraft, Country, Int'l Designation, Vehicle, Launch Site	Payload Data	Apogee (km)	Perigee (km)	Period (min)	Inclination (degree)	Remarks
Nov. 27	Morelos-B U.S. 1985-109B Atlantis STS-61B	Total weight 10 008 lb. Objective: To provide telephone, TV, and wire services to Mexico through 22 transponders. Description: Cylindrical, spin-stabilized. With antenna and solar panels deployed 21.6 m high and 7.2 feet in diameter.	35 801	35 773	1 436.1	2.9	Launched for Mexico's Secretariat of Communications and Transportation to serve as on station space for Morelos A in orbit.
and							
	Aussat-2 U.S. 1985-109C Atlantis STS-61B	Total weight: 1 322 lb. Objective: To provide domestic communications with eleven 12-watt transponders and four 30-watt transponders to Australia. Description: Cylindrical, spin-stabilized. With antenna and solar panels deployed 21.6 m high and 7.2 feet in diameter.	35 790	35 786	1 436.2	0.1	Second in series of 3 communications satellites launched for the Australian National Satellite Communications System.

Appendix A

Date	Satellite	Description				Remarks
Nov. 28	Satcom KU-2 U.S. 1085-109D Atlantis STS-61B	Total weight: 15 929 lb. Objective: To provide domestic communications with 16 operational transponders and 6 spares, each transmitting 45 watts of power. Description: Box-shaped 67 by 84 by 60 in main structure. Spin-stabilized, solar panels on two deployable arms.	35 801	35 774	1 436.2	0.1	Second in series of 3 satellites for RCA American Communications, Inc.
Nov. 30	Dex Target U.S. 1985-109E Atlantis STS-61B	Total weight: Unavailable. Objective: Unavailable. Description: Unavailable.	38	372	92.1	28.5	
Nov. 28	Cosmos 1704 USSR 1985-110A C-1 Plesetsk	Total weight: Unavailable. Objective: "Continuation of outer space investigations." Description: Unavailable.	1 110	965	104.84	82.93	Navigation satellite; replaced Cosmos 1598.
Dec. 3	Cosmos 1705 USSR 1985-111A A-2 Plesetsk	Total weight: Unavailable. Objective: "Continuation of outer space investigations." Description: Unavailable.	415	356	92.27	72.87	Photo-reconnaissance satellite; recovered Dec. 17.

Astronautics and Aeronautics, 1985

Launch Date	Spacecraft, Country, Int'l Designation, Vehicle, Launch Site	Payload Data	Apogee (km)	Perigee (km)	Period (min)	Inclination (degree)	Remarks
Dec. 11	Cosmos 1706 USSR 1985–112A A–2 Plesetsk	Total weight: Unavailable. Objective: "Continuation of outer space investigations." Description: Unavailable.	334	167	89.51	67.16	Photo-reconnaissance satellite.
Dec. 12	Cosmos 1707 USSR 1985–113A F–2 Plesetsk	Total weight: Unavailable. Objective: "Continuation of outer space investigations." Description: Unavailable.	665	634	97.72	82.54	Electronic intelligence satellite; replaced Cosmos 1515.
Dec. 12	AF–16A U.S. 1985–114A Scout Wallops Flight Facility	Total weight: Unavailable. Objective: An instrument test vehicle in low orbit for future antisatellite test. Description: Unavailable.	776	314	95.6	37.1	Launched by NASA for the U.S. Air Force. A dual payload.

Appendix A

and	AF-16B U.S. 1985-114B Scout Wallops Flight Facility	Total weight: Unavailable. Objective: An instrument test vehicle in low orbit for future antisatellite test. Description: Unavailable.	772	314	95.5	37.1	Launched by NASA for the U.S. Air Force. A dual payload.
Dec. 13	Cosmos 1708 USSR 1985-115A A-2 Plesetsk	Total weight: Unavailable. Objective: "Continuation of outer space investigations." Description: Unavailable.	273	257	89.80	82.28	Photo-reconnaissance satellite; recovered Dec. 27.
Dec. 19	Cosmos 1709 USSR 1985-116A C-1 Plesetsk	Total weight: Unavailable. Objective: "Continuation of outer space investigations." Description: Unavailable.	1 013	963	104.86	82.95	Navigation satellite; replaced Cosmos 1610.
Dec. 24	Molniya 3-27 USSR 1985-117A A-2-e Plesetsk	Total weight: Unavailable. Objective: "Continuation of outer space investigations." Description: Unavailable.	46 773	482	735.75	62.87	Communications satellite. Established final plane of new constellation.

Launch Date	Spacecraft, Country, Int'l Designation, Vehicle, Launch Site	Payload Data	Apogee (km)	Perigee (km)	Period (min)	Inclination (degree)	Remarks
Dec. 24	Cosmos 1710 USSR 1985–118A D-class? Tyuratam	Total weight: Unavailable. Objective: "Continuation of outer space investigations." Description: Unavailable.	19 146	19 096	675.38	64.8	Navigation satellite; new Glonass launch profile.
and							
	Cosmos 1711 USSR 1985–118B D-class? Tyuratam	Total weight: Unavailable. Objective: "Continuation of outer space investigations." Description: Unavailable.	19 151	19 138	676.32	64.84	Navigation satellite.
and							
	Cosmos 1712 USSR 1985–118C D-class? Tyuratam	Total weight: Unavailable. Objective: "Continuation of outer space investigations." Description: Unavailable.	19 153	19 136	676.30	64.85	Navigation satellite.

Appendix A

Dec. 26	Meteor 2-13 USSR 1985-119A F-2 Plesetsk	Total weight: Unavailable. Objective: "Continuation of outer space investigations." Description: Unavailable.	962	939	104.06	82.54	Meteorological satellite.
Dec. 27	Cosmos 1713 USSR 1985-120A A-2 Plesetsk	Total weight: Unavailable. Objective: "Continuation of outer space investigations." Description: Unavailable.	398	216	90.67	62.82	Photo-reconnaissance satellite.
Dec. 28	Cosmos 1714 USSR 1985-121A D-class? Tyuratam	Total weight: Unavailable. Objective: "Continuation of outer space investigations." Description: Unavailable.	853	163	94.78	71.0	Intended for electronic intelligence; same as Cosmos 1697. Launch failed.

APPENDIX B
NASA LAUNCHES, 1985

The following table of NASA launches in 1985 includes payloads carried by the Space Shuttle and rocket launches larger than sounding rockets.

NASA conducted 14 launches with 24 payloads, all of which were successful or partially successful (the booster on *Syncom 3* failed to ignite and was repaired during STS–51J, *Syncom 4* ceased functioning after deployment, and *Glomr* was deployed during a second mission after it failed to leave the cannister on STS–51B). NASA flew for the Department of Defense two dedicated Space Shuttle missions, during which three satellites were deployed, and conducted two Scout launches, each with two payloads. In addition the Space Shuttle flew the first of what likely would be a series of experiments for the Strategic Defense initiative program. The Space Transportation System's Space Shuttles, along with the Payload Assist Module (PAM) D and D–2 and Hughes's inertial upper stage, deployed 11 communications satellites into synchronous orbit for eight organizations and five countries.

NASA in 1985 introduced Discovery, the last and lightest-weight Space Shuttle orbiter; and McDonnell Douglas inaugurated its more powerful upper stage booster, the PAM D–2.

Space Shuttles also carried 11 Get Away Special (GAS) cannisters in 1985; 15 middeck (secondary) payloads for such organizations as 3M Corp. and the IMAX film company; two Shuttle Student Involvement Project experiments; two deployable GAS can satellites (very small satellites in a GAS can); Spacelab 3 with the first large-scale animal colony; Spacelab 2 with the ESA-developed Instrument Pointing System; and Spacelab D1, a dedicated scientific flight purchased by the West German government.

Crew members on Space Shuttle missions in 1985 totalled 52 (four crew members flew twice). Of this total, 19 were scientists performing observations, experiments, or investigations; 27 were U.S. military officers (16 Air Force, five Navy, four Marine Corps, and two Army); and six were payload specialists from foreign countries. A U.S. Senator, E.J. "Jake" Garn, flew on STS–51D.

The following table categorizes vehicle and payload performance as S for successful, P for partially successful, or U for unsuccessful. These categories, which are unofficial, do not take into account that U missions might produce valuable information or that payloads with a long-life design might fail to meet the design requirements, thus becoming officially unsuccessful. Further information on these launches appears in Appendix A and in the text.

Astronautics and Aeronautics, 1985

NASA Launches, 1985

Date	Name (NASA Code)	General Mission	Launch Vehicle (Site)	Performance Vehicle / Payload	Remarks
Jan. 24	Discovery STS–51C	To launch classified military payload.	Discovery (KSC)	S S	Fifteenth flight of Space Transportation System. Crew were commander Thomas Mattingly; pilot Loren Shriver; mission specialists Ellison Onizuka and James Buchli; and payload specialist Gary Payton. Unannounced payload included Inertial Upper Stage.
Jan. 24	DoD	Classified payload.	Discovery	S S	Possibly an early-warning surveillance satellite.
Mar. 22	Intelsat VA–F10	To place spacecraft in geosynchronous orbit to provide 13,500 2-way voice circuits and 2 TV channels.	Atlas Centaur (KSC)	S S	Launched by NASA for Internatl. Telecommunications Satellite Organization (INTELSAT). First in a series of 6 improved Intelsat commercial communications satellites.
Apr. 12	Discovery STS 51–D	To launch 2 communications satellites and complete assigned experiments and objectives.	Discovery (KSC)	S P	Crew were commander Karol Bobko; pilot Donald Williams; mission specialists M. Rhea Seddon, S. David Griggs, and Jeffrey Hoffman; and payload specialists Charles Walker and Senator E. J. "Jake" Garn (first flight of U.S. Senator). Syncom satellite deployed, but it did not reach orbit.

Appendix B

Apr. 12	Telesat-I (Anik C–1)	To provide domestic communications for Canada.	Discovery	S	S	Successfully deployed from Discovery, and apogee kick motor fired 6:30 pm EST, Apr. 14.
Apr. 13	Syncom IV–3 (Leasat 3)	A communications satellite by Hughes and leased to U.S. Navy.	Discovery	S	U	Mission extended by 2 days for an unsuccessful attempt to repair inoperable Syncom IV–3 with impromptu "flyswatter" to activate sequencer lever.
Apr. 29	Challenger STS–51B	To launch NUSAT–1 and Glomr and to carry Spacelab 3.	Discovery (KSC)	S	P	Crew were commander Robert Overmyer; pilot Frederick Gregory; mission specialists Don Lind, Norman Thagard, and William Thornton; and payload specialists Lodewijk Vandenberg and Taylor Wang. Challenger carried Spacelab 3 for 15 scientific investigations, the most intensive science mission to date. Global Low Orbiting Message Relay Satellite (Glomr) failed to deploy and returned to earth with the orbiter.
Apr. 29	NUSAT–1	To calibrate air traffic control radars on the ground.	Challenger	S	S	Northern Utah Satellite had 6-month design life.
June 17	Discovery STS–51G	To deploy 3 satellites, deploy and retrieve Spartan 1, and carry out four experiments.	Discovery (KSC)	S	S	Crew were commander Daniel Brandenstein; pilot John Creighton; mission specialists Shannon Lucid, John Fabian, and Steven Nagel; and payload specialists Patrick Baudry (France) and Sultan Salman Abdelazize Al-Saud (Arabsat). All satellites were deployed and experiments completed.

Date	Name (NASA Code)	General Mission	Launch Vehicle (Site)	Performance Vehicle Payload		Remarks
June 17	Morelos 1	To provide voice, video, data, and facsimile communications for Mexico.	Discovery	S	S	First of two identical satellites launched for Mexico.
June 18	Arabsat 1	A communications satellite to serve Arab-speaking countries in North Africa and Middle East.	Discovery	S	S	Launched by NASA for the Arab Satellite Communications Organization (ASCO).
June 19	Telstar 3D	To provide U.S. domestic communications.	Discovery	S	S	Launched by NASA for AT&T.
June 20	Spartan 1	To extend capabilities of sounding rocket-class experiments by making observations of the Perseum Cluster, galactic center, and Scorpius X–2.	Discovery	S	S	Retrieved June 24 and returned to earth with Discovery.
June 29	Intelsat V–A (F–11)	A communications satellite operating at C-band and L-band to provide 13,500 2-way voice circuits and 2 TV channels.	Atlas Centaur (KSC)	S	S	Second in a series of improved INTELSAT commercial communications satellites launched by NASA.
July 29	Challenger STS–51F	To carry Spacelab 2 and launch and retrieve the Plasma Diagnostic Package (PDP).	Challenger (KSC)	S	S	Crew were commander Charles Fullerton; pilot Roy Bridges, Jr.; mission specialists Karl Henize, Anthony England, and F. Story Musgrave; and payload specialists Loren Acton and John-David Bartow. Orbiter carried igloo-pallet configuration with Spacelab 2 experiments.

Appendix B

Date	Satellite	Launch Vehicle	Purpose		Remarks
Aug. 3	Oscar 24	Scout (WSMC)	Transit-type navigation satellite placed in orbital storage for stacked oscar on scout (SOOS) program.	S S	Intended for U.S. Navy communications as part of Navy Transit (Navy Navigational Satellite System).
and					
	Oscar 30	Scout (WSMC)	Same as Oscar 24.	S S	One of two Navy communications launched for Navy Transit program.
Aug. 27	Discovery STS-51I	Discovery (KSC)	To launch 3 communications satellites, retrieve and repair Syncom IV-3, and complete assigned experiments.	S S	Crew were commander Joe Engle; pilot Richard Covey; mission specialists James van Hoften, William Fisher, and John Lounge. Fisher and van Hoften retrieved and repaired Syncom and released it Sept. 1.
Aug. 27	Aussat 1	Discovery	To provide communications links within Australia.	S S	Deployed for Australia's National Satellite Company.
Aug. 27	ASC 1	Discovery	To provide domestic communications.	S S	Deployed for the American Satellite Company.
Aug. 29	Syncom IV-4	Discovery	Intended as replacement for FleetSatCom spacecraft for U.S. Navy.	S P	Although placed in correct geosynchronous orbit, Syncom 4 (Leasat 4) ceased functioning.

503

Date	Name (NASA Code)	General Mission	Launch Vehicle (Site)	Performance Vehicle	Performance Payload	Remarks
Sept. 29	Intelsat VA (F12)	Communications satellite to provide 15,000 telephone channels.	Atlas Centaur (KSC)	S	S	Last Intelsat commercial communications satellite launched by NASA from KSC.
Oct. 3	Atlantis STS-51J	A classified military mission to launch 2 communications satellites.	Atlantis (KSC)	S	S	The maiden voyage of Atlantis. Crew were commander Karol Bobko; pilot Ronald Grabe; mission specialists Robert Steward and David Hilmers; and payload specialist William Pailes.
Oct. 3	DCSC 11	To improve military communications.	Atlantis	S	S	
Oct. 3	DCSC 12	To improve military communications.	Atlantis	S	S	
Oct. 30	Challenger STS-61A	To carry out Spacelab D1 experiments and deploy Glomr.	Challenger (KSC)	S	S	Crew, largest number ever flown on orbiter, were commander Harry Hartsfield; pilot Steven Nagel; mission specialists Bonnie Dunbar, James Buchli, Guion Bluford; payload specialists Ernst Messerschmid, Reinhard Furrer and Wubbo Ockels.
Oct. 30	Glomr	A message relay satellite for the Department of Defense.	Challenger	S	S	The Global Orbiting Message Relay satellite.

Appendix B

Date	Name	Purpose	Vehicle		Notes
Nov. 26	Atlantis STS-61B	To launch 3 satellites, perform experiments in extravehicular assembly, and carry out scientific experiments.	Atlantis (KSC)	S S	Crew were commander Brewster Shaw; pilot Bryan O'Connor; mission specialists Mary Cleave, Jerry Ross, and Sherwood Spring; and payload specialists Rudolfo Neri Vela (Mexico) and Charles Walker (McDonnell Douglas). Ross and Spring carried out ACCESS (assembly concept for construction of erectable space structures) and EASE (experimental assembly of structures in extravehicular activity), the first flight demonstration of construction of large space structures.
Nov. 27	Morelos B	A communications satellite to serve as on station space for Morelos A.	Atlantis	S S	Second of 2 satellites launched for Mexico's Secretariat of Communications and Transportation.
Nov. 27	AUSSAT 2	To provide domestic communications for Australia.	Atlantis	S S	Second of 3 communications satellites deployed for the Australian National Satellite Communications System.
Nov. 28	Satcom KU-2	A KU-bank domestic communications satellite to operate in the 12 to 14 gigahertz range.	Atlantis	S S	Second of 2 satellites deployed for RCA American Communications, Inc. (RCA Americom).
Dec. 12	AF-16 A & B	To place 2 instrument test vehicles into low orbit for future U.S. antisatellite tests.	Scout (Wallops Flight Facility)	S S	Launched by NASA for U.S. Air Force.

APPENDIX C
MANNED SPACE FLIGHTS, 1985

There were a total of 11 manned spacecraft flights worldwide in 1985: nine Space Shuttle missions conducted by the U.S. and two Soyuz T missions by the USSR to carry crew to and from the *Salyut 7* station.

The U.S. total of manned spacecraft hours in flight in 1985 was 1395 hours 53 minutes; total cumulative man-hours in space in 1985 were 9280 hours 43 minutes. The 1985 USSR total of manned spacecraft hours was 4249 hours 44 minutes; their total cumulative man-hours in space in 1985 were 10,056 hours.

In total, 52 individuals flew aboard Space Shuttles in 1985. Four individuals—three NASA astronauts and one payload specialist—flew twice. The U.S. Air Force flew payload specialists on each dedicated DOD mission; McDonnell Douglas flew the same payload specialist on two missions; U.S. Senator E.J. "Jake" Garn, one of NASA's subcommittee chairman, flew as an observer; a scientist from EG&G Corp. and another from the Jet Propulsion Laboratory flew as part of the Spacelab 3 mission; scientists from Lockheed Palo Alto Research Laboratory and the Naval Research Laboratory flew as part of the Spacelab 2 mission; two scientists from West Germany and one from the Netherlands flew as part of the Spacelab D-1 mission; payload specialists from Saudi Arabia and France flew on a mission together; and a scientist from Mexico flew to perform experiments.

Crew aboard three Space Shuttle missions went on space walks. On the first, astronauts attached an impromptu "flyswatter" to the robot arm in preparation for an attempt to activate *Syncom IV-3*. Later, in a two-part space walk, two astronauts activated that satellite. Finally, in another two-part exercise, two astronauts carried out large-structures assembly techniques in space.

The year of manned space flights for the USSR was marked by the rescue mission of *Salyut 7*, the first partial space station crew rotation; and the emergency return to earth to get an ailing cosmonaut to the hospital. The USSR in 1985 conducted five missions to the manned *Salyut 7* station: two manned Soyuz T's, one Progress resupply ship, one modified Progress flown under a Cosmos designation, and the fourth in a series of enhanced unmanned resupply and support spacecraft.

The year 1985 was the 20th anniversary of the world's first space walk by Alexei Leonov on board Voskhod 2. Two cosmonauts aboard *Salyut 7* conducted one space walk in 1985 to install a third and final main solar panel attachment on the space station and test an improved extravehicular activity suit.

Manned Space Flights, 1985

Date Launched	Recovered	Designation Crew	Weight (kg)	Duration Revolutions	Remarks
Jan. 24	Jan 27	Discovery STS–51C Thomas Mattingly, CDR Loren Shriver, PLT Ellison Onizuka, MS James Buchli, MS Gary Payton, MS		73hr 33min 49 rev.	Fifteenth Space Transportation System (STS) flight and first dedicated Department of Defense Space Shuttle mission; landed at KSC.
Apr. 12	Apr. 19	Discovery STS–51D Karol Bobko, CDR Donald Williams, PLT M. Rhea Seddon, MS S. David Griggs, MS Jeffrey Hoffman, MS Charles Walker, PS E.J. "Jake" Garn, PS		167hr 55min 109 rev.	Deployed two communications satellites and carried first U.S. Senator in space; mission extended by two days for an unsuccessful attempt to repair inoperable Syncom IV–3. Landed at KSC.
Apr. 29	May 6	Challenger STS–51B Robert Overmyer, CDR Frederick Gregory, PLT Don Lind, MS Norman Thagard, MS William Thornton, MS Lodewijk Vandenberg, PS Taylor Wang, PS		168hr 8 min 110.13 rev	Carried Spacelab 3 in cargo bay; Global Low Orbiting Relay Satellite (Glomr) failed to deploy and returned with orbiter. Landed at Edwards AFB.

Appendix C

June 6	Sept. 26	Soyuz T-13 Vladimir Dzhanibekov Viktor Savinykh	7000 (est)	2691hr 12min 1761 rev.	Repair of Salyut-7 uninhabited since conclusion of previous mission in Oct. 1984; Dzhanibekov returned to earth with Grechko (from Soyuz T-14 mission) on Soyuz T-13.
June 17	June 24	Discovery STS-51G Daniel Brandenstein, CDR John Creighton, PLT Shannon Lucid, MS John Fabian, MS Steven Nagel, MS Patrick Baudry, PS Sultan Salman Abdelazize Al-Saud, PS		169hr 39min 112.8 rev	Deployed three communications satellites; tested laser reflector as part of the Strategic Defense Initiative program; deployed and retrieved reusable Spartan-1; first U.S. flight with French and Arabian crew members. Landed at Edwards AFB.
July 29	Aug. 6	Challenger STS-51F Charles Fullerton, CDR Roy Bridges, PLT Karl Henize, MS Anthony England, MS F. Story Musgrave, MS Loren Acton, PS John-David Bartow, PS		190hr 46min 126.5 rev	Carried in cargo bay Spacelab 2 that had an igloo-pallet configuration for experiments; crew launched and retrieved Plasma Diagnostic Package (PDP); mission extended by one day to increase payload observation time. Landed at Edwards AFB.
Aug. 27	Sept. 3	Discovery STS-51I Joe Engle, CDR Richard Covey, PLT James van Hoften, MS William Fisher, MS John Lounge, MS		170hr. 18min. 113.5 rev.	Deployed three communications satellites; retrieved, repaired, and on Sept. 1 released Syncom IV-3. Landed at Edwards AFB.

509

Astronautics and Aeronautics, 1985

Date Launched	Recovered	Designation Crew	Weight (kg)	Duration Revolutions	Remarks
Sept. 17	Nov. 21	Soyuz T-14 Vladimir Vasyutin Georgiy Grechko Aleksandr Volkov	7000 (est)	1557hr. 52min. 1022.2 rev.	Docked with Salyut 7 station; Savinykh, Volkov, and Vasyutin returned to earth on Soyuz T-14 when Vasyutin became ill.
Oct. 3	Oct. 7	Atlantis STS-51J Karol Bobko, CDR Ronald Grabe, PLT Robert Stewart, MS David Hilmers, MS William Pailes, PS		97hr 45min Unavailable	Classified military mission to deploy two communications satellites; Atlantis also carried a civilian experiment to study exposure of crew and orbiter to cosmic rays. Atlantis, on maiden voyage, landed at Edwards AFB.
Oct. 30	Nov. 6	Challenger STS-61A Henry Hartsfield, CDR Steven Nagel, PLT Bonnie Dunbar, MS James Buchli, MS Guion Bluford, MS Ernst Messerschmid, PS Reinhard Furrer, PS Wubbo Ockels, PS		192hr. 45min 127.1 rev.	Challenger carried largest number of crew members ever along with dedicated German Spacelab D1; Furrer and Messerschmid were from W. Germany, Ockels from the Netherlands. Landed at Edwards AFB.

Appendix C

Nov. 26	Dec. 3	Atlantis STS–61B Brewster Shaw, CDR Bryan O'Connor, PLT Mary Cleave, MS Sherwood Spring, MS Jerry Ross, MS Rudolfo Neri Vela, PS Charles Walker, PS	165hr 4min 107.8 rev.		Crew deployed three communications satellites; mission was first flight demonstration of construction of large space structures. Landed at Edwards AFB.

Appendix D

ABBREVIATIONS OF REFERENCES

Listed here are the abbreviations used for citing sources in the text. Not all the sources are listed, only those that are abbreviated.

AAAS Bull	American Association for the Advancement of Science's *AAAS Bulletin*
A&A	American Institute of Aeronautics and Astronautics' magazine, *Astronautics & Aeronautics*
A&A 1985	NASA's *Astronautics and Aeronautics, 1985: A Chronology* (this publication)
ABC	American Broadcasting Company
AEC Release	Atomic Energy Commission news release
Aero Daily	*Aerospace Daily* newsletter
Aero Med	*Aerospace Medicine* magazine
AF Mag	Air Force Association's *Air Force Magazine*
AFHF Newsletter	Air Force Historical Foundation Newsletter
AFJ	*Armed Forces Journal* magazine
AFSC Newsreview	Air Force Systems Command's *Newsreview*
AFSC Release	Air Force Systems Command news release
AIA Release	Aerospace Industries Association of America news release
AIAA Facts	American Institute of Aeronautics and Astronautics' *Facts*
AIAA Release	American Institute of Aeronautics and Astronautics news release
AIP Newsletter	American Institute of Physics *Newsletter*
AP	Associated Press news service
ARC Astrogram	NASA Ames Research Center's *Astrogram*
Astro Journ	American Astronomical Society's *Astrophysical Journal*
Atlanta JC	*Atlanta Journal Constitution* newspaper
AvWk	*Aviation Week & Space Technology* magazine
B News	*Birmingham News* newspaper
B Sun	Baltimore *Sun* newspaper

513

Astronautics and Aeronautics, 1985

Bull Atom Sci	Education Foundation for Nuclear Science's *Bulletin of the Atomic Scientists*
Bus Wk	*Business Week* magazine
C Daily News	*Chicago Daily News* newspaper
C Trib	*Chicago Tribune* newspaper
Can Press	Canadian Press news service
CBS	Columbia Broadcasting System
C&E News	*Chemical & Engineering News* magazine
Cl PD	*Cleveland Plain Dealer* newspaper
Cl Press	*Cleveland Press* newspaper
Columbia J Rev	*Columbia Journalism Review* magazine
ComSatCorp Release	Communications Satellite Corporation news release
CQ	*Congressional Quarterly*
CR	*Congressional Record*
CSM	*Christian Science Monitor* newspaper
CTNS	Chicago Tribune News Service
D News	*Detroit News* newspaper
D Post	*Denver Post* newspaper
DASA Release	Defense Atomic Support Agency news release
DFRC	See FRC.
DJ	Dow Jones news service
DOC PIO	Department of Commerce Public Information Office
DOD Release	Department of Defense news release
DOT Release	Department of Transportation news release
EOP Release	Executive Office of the President news release
ESA Release	European Space Agency news release, use dated (not numbered)
FAA Release	Federal Aviation Administration news release
FBIS—Sov	Foreign Broadcast Information Service, Soviet number
FonF	*Facts on File*
FRC Release	Flight Research Center news release, after 8 Jan. 1976, became Dryden Flight Research Center (DFRC) news release
FRC *X-Press*	NASA Flight Research Center's *X-Press*
GE Forum	*General Electric Forum* magazine
Goddard News	NASA Goddard Space Flight Center's *Goddard News*
GSFC Release	NASA Goddard Space Flight Center news release
GSFC *SSR*	NASA Goddard Space Flight Center's *Satellite Situation Report*
GT&E Release	General Telephone & Electronics news release
H Chron	*Houston Chronicle* newspaper
H Post	*Houston Post* newspaper
INTELSAT Release	Intl. Telecommunications Satellite Org. news release

Appendix D

JA	*Journal of Aircraft* magazine
JPL *Lab-Oratory*	Jet Propulsion Laboratory's *Lab-Oratory*
JPL Release	Jet Propulsion Laboratory news release
JPRS	Department of Commerce Joint Publications Research Service
JSC Release	NASA Lyndon B. Johnson Space Center (Manned Spacecraft Center until 17 Feb. 1973) news release
JSC *Roundup*	NASA Lyndon B. Johnson Space Center's *Space News Roundup*
JSR	American Institute of Aeronautics and Astronautics' *Journal of Spacecraft and Rockets* magazine
KC Star	*Kansas City Star* newspaper
KC Times	*Kansas City Times* newspaper
KSC Release	NASA John F. Kennedy Space Center news release
LA *Her-Exam*	Los Angeles *Herald-Examiner* newspaper
LA Times	*Los Angeles Times* newspaper
Langley Researcher	NASA Langley Research Center's *Langley Researcher*
LaRC Release	NASA Langley Research Center news release
LATNS	Los Angeles Times News Service
LeRC Release	NASA Lewis Research Center news release
Lewis News	NASA Lewis Research Center's *Lewis News*
M HER	*Miami Herald* newspaper
M News	*Miami News* newspaper
M Trib	*Minneapolis Tribune* newspaper
Marshall Star	NASA George C. Marshall Space Flight Center's *Marshall Star*
MJ	*Milwaukee Journal* newspaper
MSFC Release	NASA George C. Marshall Space Flight Center news release
N Hav Reg	*New Haven Register* newspaper
N News	*Newark News* newspaper
N Va Sun	*Northern Virginia Sun* newspaper
NAA *News*	National Aeronautic Association *News*
NAA Record Book	National Aeronautic Association's *World and U.S.A. National World Aviation—Space Records*
NAC Release	National Aviation Club news release
NAE Release	National Academy of Engineering news release
NANA	North American Newspaper Alliance
NAS Release	National Academy of Sciences news release
NAS—NRC Release	National Academy of Sciences—National Research Council news release
NAS—NRC—NAE *News Rpt*	National Academy of Sciences—National Research Council—National Academy of Engineering *News Report*

NASA Actv	*NASA Activities*
NASA anno	NASA announcement
NASA GMR	NASA Headquarters "General Management Review Report"
NASA HHR—39	NASA Historical Report No. 39
NASA Hist Off	NASA History Office
NASA Hq *WB*	NASA Headquarters *Weekly Bulletin*
NASA Int Aff	NASA Office of International Affairs
NASA *LAR,* XIII/8	NASA *Legislative Activities Report,* Vol. XIII, No. 8
NASA Leg Off	NASA Office of Legislative Affairs
NASA MOR	NASA Headquarters Mission Operations Report, preliminary prelaunch and postlaunch report series (information may be revised and refined before publication)
NASA prog off	NASA program office (for the program reported)
NASA proj off	NASA project office (for the project reported)
NASA Release	NASA Headquarters news release
NASA Rpt SRL	NASA report of sounding rocket launching
NASA SP-4019	NASA Special Publication No. 4019
Natl Obs	*National Observer* magazine
Nature	*Nature Physical Science* magazine
NBC	National Broadcasting Company
NGS Release	National Geographic Society news release
NMI	NASA Management Instruction
NN	NASA Notice
NOAA Release	National Oceanic and Atmospheric Administration news release
NRL Release	Naval Research Laboratory news release
NSC Release	National Space Club news release
NSC *News*	National Space Club *News*
NSC *Letter*	National Space Club *Letter*
NSF *Highlights*	National Science Foundation's *Science Resources Studies Highlights*
NSF Release	National Science Foundation news release
NSTL Release	NASA National Space Technology Laboratories news release
NY News	*New York Daily News* newspaper
NYT	*New York Times* newspaper
NYTNS	New York Times News Service
O Sen Star	*Orlando Sentinel Star* newspaper
Oakland Trib	*Oakland Tribune* newspaper
Omaha W-H	*Omaha World-Herald* newspaper
ONR *Rev*	Navy's Office of Naval Research *Reviews*

Appendix D

P Bull	Philadelphia *Evening* and *Sunday Bulletin* newspaper
P Inq	*Philadelphia Inquirer* newspaper
PAO	Public Affairs Office
PD	National Archives and Records Service's *Weekly Compilation of Presidential Documents*
PIO	Public Information Office
PMR *Missile*	USN Pacific Missile Range's *Missile*
PMR Release	USN Pacific Missile Range news release
Pres Rpt 74	*Aeronautics and Space Report of the President: 1974 Activities*
SAO Release	Smithsonian Astrophysical Observatory news release
SBD	*Defense/Space Business Daily* newspaper
Sci Amer	*Scientific American* magazine
Sci & Govt Rpt	*Science & Government Report*, independent bulletin of science policy
SciServ	Science Service News service
SD	*Space Digest* magazine
SD Union	*San Diego Union* newspaper
SET Manpower Comments	Scientific Manpower Commission's *Scientific, Engineering, Technical Manpower Comments*
SF	British Interplanetary Society's *Spaceflight* magazine
SF Chron	*San Francisco Chronicle* newspaper
SF Exam	*San Francisco Examiner* newspaper
Sov Aero	*Soviet Aerospace* newsletter
Sov Rpt	Center for Foreign Technology's *Soviet Report* (translations)
SP	*Space Propulsion* newsletter
Spaceport News	NASA John F Kennedy Space Center's *Spaceport News*
Spacewarn	IUWDS World Data Center A for Rockets and Satellites' *Spacewarn Bulletin*
SR *list*	NASA compendium of sounding rocket launches
SSN	*Soviet Sciences in the News*, publication of Electro-Optical Systems, Inc.
St Louis G-D	*St. Louis Globe-Democrat* newspaper
St Louis P-D	*St. Louis Post-Dispatch* newspaper
T-Picayune	New Orleans *Times-Picayune* newspaper
Tech Rev	Massachusetts Institute of Technology's *Technology Review*
Today	*Today* newspaper
testimony	Congressional testimony, prepared statement
text	Prepared report or speech text

Astronautics and Aeronautics, 1985

transcript	Official transcript of news conference or congressional hearing
UN Reg	United Nations Public Registry of Space Flight
UPI	United Press International news service
USGS Release	U.S. Geological Survey news release
USPS Release	U.S. Postal Service news release
W Post	*Washington Post* newspaper
WFC Release	NASA Wallops Flight Center news release
WH Release	White House news release
WJT	*World Journal Tribune* newspaper
WSJ	*Wall Street Journal* newspaper

INDEX

A-10 aircraft, 406
Abel, Brig. Gen. Richard, 316
ABM. See Antiballistic Missile.
Abrahamson, Lt. Gen. James, 109, 126–127
Abrams M-1 tanks, 24
Abruzzo, Ben, 16–17
A/C. See Atlas/Centaur.
Academy of Finland, 396
Academy of Sciences, USSR. See Soviet Academy of Sciences.
ACCESS. See assembly concept for construction of erectable space structures.
ACFS. See advanced concept flight simulator.
ACLU. See American Civil Liberties Union.
active magnetospheric particle tracer explorers (AMPTE), 77–79, 401
Acton effect, 47
Acton, Loren, 47, 330
ACTS. See Advanced Communications Technology Satellite.
Acuna, Dr. Mario, 79
Adams, Robert McCormick, 237–239
ADATS. See airborne digital avionics test system.
Adelman, Kenneth, 113
ADSF. See automated directional solidification furnace.
Advanced Communications Technology Satellite (ACTS), 169
advanced concept flight simulator (ACFS), 27
advanced fighter technology integration (AFTI), 11, 13–14
advanced medium range air-to-air missile (AMRAAM), 119
advanced range instrumentation aircraft (ARIA), 340, 404
Advanced Tactical Fighter (ATF), 10–11, 117
Advanced Technology Bombers (ATBs-Stealth), 116
Advanced Turboprop Project (ATP), 14–15

Advanced X-ray Astrophysics Facility (AXAF), 39–40
AEDC. See Arnold Engineering Development Center.
Aerobee, 167
—payload, 244
—sounding rocket, 243–244
aerodynamics, 16, 26, 187
Aerojet Electro Systems Co., 73, 132
Aerojet Engineering Corp., 243
Aerojet Strategic Propulsion Co., 368
Aerojet Technologies, 22
Aeronautical Policy Review Committee, 7
Aeronautical Systems Division. See Air Force, U.S., Aeronautical Systems Division.
Aeropropulsion Systems Test Facility (ASTF), 19
Aerospace Safety Advisory Panel, 223
aerospace sales, 1
Aerospatiale, France, 145–146, 159–160, 330
—Space and Ballistic Systems Division, 219
AFE. See American Flight Echocardiograph.
Afghanistan, 413
AFGL. See Air Force, U.S., Geophysics Laboratory.
AFSC. See Air Force, U.S., Systems Command.
AF-16 satellite, 410
AFTI. See advanced fighter technology integration.
Agency for International Development (AID), U.S., 212, 224
Agriculture, U.S. Department of, 179, 424
aggregation of red blood cells, 348
AI. See artificial intelligence.
AIAA. See American Institute for Aeronautics and Astronautics.
AID. See Agency for International Development.
aileron-rudder interconnect (ARI) program, 10

519

AIM-7 radar missile, 119
Air Force, U.S., 2, 6–7, 11, 16, 20, 22–24, 27–29, 33, 81–82, 94, 111–117, 119–120, 122–123, 129–131, 169, 185, 191, 286–287, 292, 294–296, 305–306, 315–318, 321, 330, 335, 344, 364, 371, 389, 404–411, 419, 426, 436–439
—Academy, 357
—AF-16 satellite, 410
—Aeronautical Systems Division (ASD), 10–12, 25, 81–82, 117, 407
—budget, 116–117, 437
—Defense Meteorological Satellite Program, 217
—Deputy for Airlift and Trainer Systems, 82
—Electronics Systems Division, 439
—Flight Dynamics Laboratory, 14
—4950th Test Wing, 404, 407
—Geophysics Laboratory (AFGL), 27–28, 356–357
—Reserves, 404
—Science Advisory Board Task Force on Manned Strategic System Vulnerability, 179
—shuttle assembly task force, 286
—Space Command, 407–410, 437, 439
—Space Division, Los Angeles, 419
—space systems, 436
—space test program satellite P80-1, 317
—Space Technology Center, 132
—Systems Command (AFSC), 12–13, 19, 24–25, 27, 81, 114, 117, 119–120, 122, 124, 132, 201, 221, 286, 404, 406, 439, 441
—3246 Test Wing, 119
—Weapons Laboratory, 176, 179, 182
Air Force One, 81–82
Air India jetliner, 87
Air Injection as an Alternative to Honeycombing experiment, 358
Air Line Pilots Association, U.S., 91
air traffic communications, 141, 339
air traffic control, 86–87, 141–143, 197
Air Transport Association, 85, 141
airborne digital avionics test system (ADATS), 407
Airbus Industrie, 84, 145

Aircraft Accident Investigation Committee. See Japan.
Air-India, 87, 92–93, 96
airline
—deregulation, 95–96
—insurance premiums, 89–90
—passenger costs, 85
—passenger trips, 85
air/skin friction, 4
AKM. See apogee kick motor.
Alabama, 198, 357–358, 382, 429
Alabama A&M University, Normal, AL, 382
Alabama Space and Rocket Center, 357
Alabama, University of, 358
—Birmingham, 429
—Huntsville, 429
Alaska, 75, 220
Al-Bassam, Abdulmohsen Hamad, 280
Albuquerque, NM, 16, 100
Aldrich, Arnold, 304, 312
Aldridge, Edward, Jr., 120, 305, 322, 364, 408
Aldrin, Edwin, 416
Alexander, Joseph, 179–180
Alexandria, VA, 272
Alfven, Dr. Hannel, 60
All Tech Associates, Inc., 357
Allen, Dr. Joseph, 33–34
Allen, Dr. Lew, 214
Alpha Jet, 145
Al-Saud, Sultan Salman Abdelazize, 280, 328
Al-Saud, Sultan bin Salman bin Abdul Aziz, 282
Alvarez, Luis, 440
American Airlines, 404
American Association for the Advancement of Science, 101, 179
American Astronautical Society, 87, 185, 311
American Civil Liberties Union (ACLU), 320–321
American Flight Echocardiograph (AFE), 350–351
American Geophysical Union, 180, 353
American Institute for Aeronautics and Astronautics (AIAA), 101, 179, 311, 434
—National Capital Section, 425
American Institute of Industrial Engineers, 201

American Newspaper Publishers
  Association, 276
American Red Cross, 232
American Relay League, 421
American Rocket Society (ARS), 101
American Samoa, 220
American Satellite Co. (AmSat), 1–2,
  303, 333–334
American Society of Newspaper Editors,
  276
American Stock Exchange, 433
American Telephone and Telegraph
  (AT&T), 300, 330
American University, Washington, DC,
  124
Ames-DFRC. See Ames-Dryden Flight
  Research Center.
Ames-Dryden Flight Research Center
  (Ames-DFRC), 8, 10–11, 13–14, 16,
  98, 294, 365, 367
Ames Research Center (ARC), 5, 12,
  26–27, 37–38, 42, 61, 66, 82–83,
  86, 151, 167, 175, 186–189, 194,
  210, 235, 259, 268, 327, 354, 367,
  374, 377, 426, 435
  –Extraterrestrial Research Division,
    235
  –Space Human Factors Office,
    267–268
  –Space Sciences Division, 37
  –Water on Mars Workshop, 66
AMPTE. See active magnetospheric
  particle tracer explorers.
AMRAAM. See advanced medium range
  air-to-air missile.
AmSat. See American Satellite Co.
Amsterdam, Holland, 62
Anders, Edward, 147
Anderson, Jack, 31
Anderson, Jim, 209
Anderson, Maxie, 17
Andrews AFB, MD, 81
Anik-C (Telesat-1) communications
  satellite, 296, 298–299, 301, 323
Annapolis, MD, 241
Annunzio, Congressman Frank, (D-IL),
  32
Antartica, 100, 218
antennae, 63, 159, 172–173, 231, 423
Antiballistic Missile (ABM) treaty, 130,
  386
antimatter, 55

antisatellite (ASAT)
  –miniature homing vehicle
    (MHV–ASAT), 24
  –pathfinder sensor, 19
  –system, 111–116
  –weapon, 112–114
Anzic, Godfrey, 103
APL. See Applied Physics Laboratory.
apogee boost motor (ABM), 226
apogee kick motor (AKM), 159, 221
Apollo, 31, 34, 102, 138–139, 185,
  307, 310, 412, 416, 418, 434–435
  –*1*, 34
  –*4*, 220
  –*15*, 33
  –*17*, 338
Apollo Soyuz test project, 172, 290,
  294, 434–435
Applications Technology Satellite (ATS),
  220–221
Applied Physics Laboratory (APL), 243
Apt, Dr. Jerome, 33
Arab League, 219
Arab Satellite Communications
  Organization, 220, 280, 330
Arabsat, 136, 219–220, 280, 330
Arabsat Consortium, 300
ARC. See Ames Research Center.
archaeology, 246, 252
Arecibo Observatory, Puerto Rico, 44,
  54, 236, 379
Argus satellite, 315
ARI. See aileron-rudder interconnect.
ARIA. See advanced range
  instrumentation aircraft.
Ariane rockets, 46, 135–136, 138–140,
  145, 213–214, 219, 255, 343–344,
  346–347, 418
Arianespace, 139, 344–347, 418
Arizona, 37–38, 43, 244, 403
Arizona, University of, Tucson, AZ, 37,
  43, 52, 58
  –Lunar and Planetary Laboratory,
    37–38
  –Planetary Sciences Department, 244
Arizona State University, Tempe, AZ,
  244
Arkansas, 81
Arms Control and Disarmament Agency,
  113
Armstrong Aerospace Medical Research
  Laboratory, 13

Armstrong, Neil, 102, 416, 440
Army, U.S., 11, 132, 180, 437–439
  —Audit Agency, 201
  —Division Air Defense (Divad), 180, 184, 437–438
Arnold Engineering Development Center (AEDC), 18–20, 221
ARS. See American Rocket Society.
artificial intelligence (AI), 186
ASAT. See antisatellite.
ASC–1 Satellite, 333–334.
ASD. See Air Force, U.S., Aeronautical Systems Division.
ASEA Robotics, Inc., 193
Aspin, Rep. Les (D-WI), 115
ASRS. See NASA, Aviation Safety Reporting System.
assembly concept for construction of erectable space structures (ACCESS), 341, 355
Association for Education in Journalism and Mass Communication, 276
Association of Government Accountants, 201
Association of Schools of Journalism and Mass Communication, 276
asteroid, 54–55, 61, 142, 148
  —amphitrite, 62
  —flyby, 50–51, 61–62
  —Freia, 50
  —Hedwig, 50
  —impact, 148
  —theory, 147
ASTF. See Aeropropulsion Systems Test Facility.
ASTRO, 60, 244
Astro–1, 56, 278, 283
astrometry, 43
Astro-Electronics, 281
astronauts, 31–35. (See also firsts.)
  —candidates, 32-34, 163
  —naval, 33
  —pilot candidates, 33
  —training, 34–35
Astronomy Survey Committee. See National Academy of Sciences.
AT&T. See American Telephone and Telegraph.
ATAC. See Congress, U.S., Advanced Technology Advisory Committee.
ATB. See Advanced Technology Bombers.

Atchison, Kenneth, 203
ATF. See Advanced Tactical Fighter.
Atlanta, GA, 142
Atlantic City, NJ, 86, 144
Atlantic Ocean, 16–17, 92, 142, 157, 160, 221, 223, 391
Atlantic Research Corp., 368
Atlantis. See Space Shuttle.
Atlas, 72
Atlas/Centaur (A/C) launch vehicle, 18, 20, 62–64, 157, 160, 167, 204, 345
ATMOS. See Atmospheric Trace Molecules Spectroscopy.
atmospheric research, 71–80, 148–149, 192
Atmospheric Trace Molecules Spectroscopy (ATMOS), 383
Atomic Energy Commission, Division of International Affairs, 177
ATP. See Advanced Turboprop Project.
ATS. See Applications Technology Satellite.
AuCoin, Rep. Les (D-OR), 115
auroras, 328, 354
AUSSAT, 333–334, 339
Austin, David, 201
Austin Co., 310
Australia, 44, 69, 71, 129, 131, 171, 185, 220, 235, 255, 300, 303, 333–334, 339, 349, 379
Australian National Satellite System, 339
Austria, 135, 139, 251
automated directional solidification furnace (ADSF), 328, 355–356
Automation and Robotics Panel, 262
autopilot, 117–119
Avco Systems, 112
avionics, 117
awards, 99–104
  —Allan D. Emil Memorial, 311
  —Arthur S. Fleming, 311
  —Federal Design Achievement, 99
  —Lawrence Sperry, 311
  —Louis W. Hill Space Transportation, 311
  —Meritorious Civilian Service, 100
  —NASA Distinguished Service Medal, 176, 311
  —NASA Exceptional Scientific Achievement Medal, 33
  —NASA Exceptional Service Medal, 33, 173, 176, 185

—NASA Group Achievement, 311
—NASA Outstanding Leadership Medal, 185–186, 311
—National Air and Space Museum Trophy, 102
—National Civil Service League Career Service Medal, 176
—Order of Magellan, 102
—Presidential Design, 99
—Presidential Medal of Freedom, 101
—Presidential Rank of Distinguished Executive, 104, 176
—Public Service Medal, 101
—Research & Development magazine's IR-100, 103
—Senate Productivity, 285
—Soviet Geophysical Committee, 100
—Space Act, 99
—Superior Performance, 33
—United States Senate Productivity, 285
—W. Randolph Lovelace II, 311
AXAF. See Advanced X-ray Astrophysics Facility.

B-1B, 116, 441, 444
B-52, 122–123
Babylonia, 43
Bacillus subtilis, 337
Backman, Philip, 444
Bagian, James, 282
Baggett, Blaine, 151
Baikonur Astrodome, USSR, 435
Bain, J.C.D., 97
Baker, Lt. Comdr. Michael, 33
balloons, 16–17, 63, 65, 71–74, 79, 80, 112, 115, 171, 173, 200, 292
Balloon-borne Laser In-situ Sensor (BLISS), 79
Baltimore, MD, 241, 272
Baltimore/Washington International Airport (BWI), 202
Bame, Dr. Samuel, 59
Bangladesh, 224
Bangor, ME, 16
Bankruptcy Court, U.S., 92
bankruptcy laws, 92
Banks, Peter, 248
Barents Sea, 385
barium clouds, 76, 78
Barnes, Richard, 177

Barrett, James, 324, 347
Barsukov, Dr. V. L., 244
Bartoe, Dr. John-David, 330, 421
Batelle Columbus Laboratories, Columbus, OH, 429
Battista, Anthony, 127
Baucom, Lt. Comdr. Chuck, 10
Baudry, Lt. Col. Patrick, 278–280, 296, 403
Bay of Bengal, 224
Bay St. Louis, MO, 207
Baykal-Amur railway, USSR, 392
Baykonur, USSR, 399
Beach, Tom, 246
Beard, Ron, 233
Beaver Creek, AK, 75
Beazley, Mr., 129
Beer, Maj. Gen. Carl, 437
Beggs, James, 34–35, 47, 52, 61, 155, 168–170, 174, 180–182, 184, 260, 263–266, 272, 297, 299, 317, 343–344, 346, 358–359, 368, 425, 427–428, 433
Beijing, People's Republic of China, 434
Belgium, 135, 285, 336
Belitsky, Boris, 392, 396
Bell, Donald Lyndon, 45
Bell Laboratories, 55
Bellows, Randy, 184
Bement, Laurence, 94–95
Bendix Field Engineering Corp., 190, 314
Beregovoy, Georgiy, 388
Beres, Kathleen, 272
Bergemaschi, Silvio, 248
Bering Sea, 220
Bermuda, 172
Beta Pictoris, 39, 41–42
Bethlehem, PA, 100
Bhatt, N.C., 284
big bang theory, 42
Bignier, Michel, 137, 263
Bigot, Charles, 345
Billingham, Dr. John, 235–236, 435
Binzel, Richard, 64, 67
biomedical imaging, 103
Biomedical Problems Institute, USSR, 393
Biorack, 338, 376
Birmingham, England, 398
Black Brant Rockets, 71, 73
Black, Dr. David, 37, 42, 259

black hole, 44, 46, 55-56
Black, Robert, 82
Blaeu, Willem Janszoon, 238
Blaha, John, 278
BLISS. See Balloon-borne Laser In-situ Sensor.
Block 5D-2 DMSP spacecraft, 217
Bluford, Col. Guion, Jr., 335, 338
Bobko, Col. Karol, 278-279, 296, 318-319, 321-322, 324-325, 335
Boeing
  -707, 404, 439
  -727, 27, 143
  -737, 89, 93, 111, 143
  -747, 81, 87-89, 90-93, 240-241, 363
  -747 Space Shuttle Carrier Aircraft (SCA), 360, 362. (See also Space Shuttle Carrier Aircraft.)
  -757-200, 93
  -767, 142
  -EC-18B, 404
  -JT8D engines, 143
Boeing Aerospace Co., 19, 88, 90, 120-121, 190, 258-260, 270, 419
Boeing Aerospace Operations, 192, 314
Boeing Military Airplane Co., 10, 14-16, 439
Boeing Service International, Inc., 313
Bogota, Colombia, 105
Boise High School, Boise, ID, 272
Boland, Rep. Edward (D-MA), 300
Bolden, Charles, Jr., 282, 356
Bolivia, 91
Bolivian Red Cross, 91
bone stiffness analyzer, 210-211
Bonn, Federal Republic of Germany, 255, 265
Bonnet, Roger, 136
Booker T. Washington Senior High School, Houston, TX, 357
Boston, MA, 16, 445
Boulder, CO, 133
Boundless Aperture, the, 445
Bourland, Charles, 274
Boye, Jack, 228
Boyne, Walter, 421
Bradbury, Ray, 109
Brahe, Tycho, 238
Brand, Vance, 278, 282, 434
Brandenstein, Daniel, 279-281, 361
Brandt, Dr. John, 49

Brasilsat, 136, 138
Braunstein, David, 174, 178
Brazil, 73, 136, 418
Bridges, Col. Roy, Jr., 330
Briggs, Dr. Geoffrey, 244
Brigham City, UT, 368
Brinkman, Dr. Albert, 40,
Briscoe, Lee, 331
British Aerospace Corp., 228
British Aerospace Dynamics Group, Space and Communications Division, 263
British Airtours, 89
British Airways, 86, 96
British Civil Aviation Authority, 143
British Museum, the, 43
British Science Research Council, 45
Broome, Dr. Taft, 18
Brown, Dwayne, 203
Brown, Rep. George, Jr. (D-CA), 24, 114
Brown, June Gibbs, 200
Brown, Malcolm, 444-445
Brubaker, Crawford, 414
Bruhweiler, Dr. F.C., 39, 41-42
Bryne, John, 176
Buchli, Lt. Col. James, 282, 295
Bulgaria, 392
Bunch, Dr. Ted, 188
Burch, Michael, 126, 315-316
Burgess, Lowry, 445
Burns, George, 174
Bush, Vice President George, 155, 272-273, 434
Byron, Rep. Beverly (R-MD), 116

C-5 aircraft, 12, 24-26, 116, 375
C-17 aircraft, 19, 24, 116
C-18 aircraft, 439
C-130 aircraft, 407
C-141 aircraft, 407
C. Itoh and Co. Ltd., 346
Caban, Maj. Robert, 33
California, 17, 24, 44, 47, 63-64, 182, 207, 237, 240, 258, 271, 291, 360, 362-363, 367, 405, 431, 438
California Institute of Technology, 133, 179
California Space Institute, 262
California, University of
  -Berkeley, 42, 44, 55-57, 186
  -Kearney Agricultural Center, 249

Index

—Lawrence Berkeley Laboratory, 248
—Los Angeles, 195, 440
—Medical Center, San Francisco, 279
—Santa Barbara, 67
—Santa Cruz, 42
Calio, Anthony, 176, 344, 346
Callahan, Dr. Paul, 327
Cambridge University, England, 18, 40, 45
Canada, 87, 92, 135, 228, 255, 259, 261, 263, 265, 269, 296–297, 300, 357, 394, 423
Canary, Krista, 246
Canaveral Council of Technical Societies symposium, 109
Canberra, Australia, 44, 69, 396
Canizares, Dr. Claude, 40
Canoga Park, CA, 258
Cape Canaveral AFB, FL, 21, 49, 291, 298, 306, 310, 325
—Air Force Station, 124, 157, 160–161, 367
—Range Control Center, 292
—Forecast Facility (CCFF), 292
Cape Henry, VA, 228
capillary pumped loop (CPL) experiment, 330
Carnegie Institution of Washington, 244
Caroline Islands, 220
Carr, Dr. Michael, 66–67
Carter, President James E., 315
Carter, Manley Sonny, 282
Case Institute of Technology, 31, 195
CAT. See clear-air turbulence.
Catholic University of America, Washington, DC, 39, 41, 178
Caussols, France, 55
CCE. See charge composition explorer.
CCFF. See Cape Canaveral Forecast Facility.
CCRES. See Combined Release and Radiation Effects Satellite.
CCSSO. See Council of Chief State School Officers.
CDP. See Goddard Space Flight Center, Crustal Dynamics Project.
Cedars-Sinai Medical Center, Los Angeles, CA, 206–207
Celestis Group, the, 425–426
Cellarius, Andreas, 238
CELSS. See Controlled Ecological Life Support Systems.

Cenker, Robert, 281, 357
Centaur upper stage rocket. See Space Shuttle Centaur.
Center for Blood Research, 358
Center for the Development of Technology, Finland, 396
Centers for the Commercial Development of Space, 429
Central Intelligence Agency (CIA), 124
Centre National d'Etudes Spatiales (CNES). See French National Center for Space Studies.
CEP. See Council on Economic Priorities.
Cerro Central, Peru, 252
Cessna Aircraft Co., 9
Cessna 421 aircraft, 16
CFES. See continuous flow electrophoresis system.
Chaffee, Martha, 33
Chaffee, Roger B., 33–34
Challenger. See Space Shuttle Challenger.
CHAMP. See Comet Halley Active Monitoring Program.
Chang, Dr. Sherwood, 188
Chang-Diaz, Franklin, 356
Chappell, Dr. Charles, 284–285
charge composition explorer (CCE), 77, 79
Charleston County, SC, 56
Charleston, WV, 16
Charon, 63–64
Chattanooga, TN, 16
Chernenko, Chairman Konstantin, 31
Cherry, John, 426
Chevalier, Dr. Roger, 42
Chicago, IL, 16, 85, 142–143
Chicago, University of, Chicago, IL, 56, 147
Chile, 147, 289–290
China. See People's Republic of China.
China Lake Naval Weapons Center, 367
Chincoteague Naval Air Station, 200
Chinese People's Insurance Co., 215
Chretien, Jean-Loup, 41
CIA. See Central Intelligence Agency.
Circumnavigators Club, New York, NY, 102
CISS. See Space Shuttle Centaur, Centaur integrated support structure.
Civil aviation, 82–96

Clark, Evert, 421
clay origin-of-life theory, 188
clear-air turbulence (CAT), 83
Clearwater, FL, 197
Clearwater, Yvonne, 267–268
Cleave, Dr. Mary, 338, 342
Cleveland, OH, 16
clouds. See meteorology.
CLS. See contingency landing sights.
CNES. See French National Space Agency.
Coats, Michael, 278
Coca-Cola Co., 443–445
Cocoa Beach, FL, 109
Codevilla, Angelo, 413
Coleman, Paul, 440
Colladay, Dr. Raymond, 3–4, 175
College of William and Mary, Williamsburg, VA, 180
Collins, Michael, 421
Colombia, 76, 105
Colorado, 114, 133, 439
Colorado Springs, CO, 438
Colorado State University, Boulder, CO, 279
Colorado, University of, 42, 167, 426
 —Boulder, 245, 252
 —Center for Space Law and Policy, 175–176
 —Department of Astro-Geophysics, 180
 —Laboratory for Atmospheric and Space Physics, 58
Columbia. See Space Shuttle Columbia.
Columbus, Christopher, discovery of America, 255, 430
Columbus, OH, 429
Columbus Space Station (Italo-German project), 135, 137, 236, 376
Combined Release and Radiation Effects Satellite (CCRES), 76
comet
 —artificial, 77–79
 —burned-out, 54
 —dynamics, 43
 —gas cloud ejection, 76
 —intercept, 51, 60
 —nucleus, 59
 —penetrator concept systems, 37
 —plasma tail, 50, 60
 —satellite encounter, 43–44
 —surface, 37

Comet Giacobini-Zinner, 43–45, 48–51, 59, 167
Comet Halley, 37, 41, 43, 46, 48–52, 56–58, 61, 63, 65, 109, 135, 137, 163, 173, 178, 225, 283, 396
Comet Halley Active Monitoring Program (CHAMP), 60, 358
Comet Kopff, 37
comet rendezvous/asteroid flyby mission, 51, 55
Comet Tempel 2, 50, 52
Comet Wild-2, 37, 47, 50, 52
Commerce, U.S. Department of, 222, 430, 433
Commercial Space Launch Act of 1984, 426
Committee on the Present Danger, 179
Committee on Solar and Space Physics, 51
communications satellite systems, 103–104, 296
communications traffic processor, 104
Comprehensive Cancer Center, 358
Compton, Dr. Dale, 186
Computer Sciences Corp., 283
Computer Technology Aviation, 440
Comsat Corp., 107, 141
Concord High School, Concord, NH, 272
Concorde Supersonic Airliner, 97, 145, 237, 239
Conestoga rockets, 426, 444
Conference on International Business in Space, 425
Congress of the International Astronautical Federation, the 36th, 149
Congress, U.S., 1, 37, 111–115, 124–125, 182, 239, 260–262, 269, 272, 285, 405, 413, 424, 433, 435
 —Advanced Technology Advisory Committee (ATAC), 262
 —H.R. 2800, 433
 —H.R. 3403, 239
 —House Appropriations Committee, 117, 347, 368
 —House Appropriations subcommittee, 118
 —House Armed Services Committee, 115–116
 —House Armed Services R&D subcommittee, 127

Index

—House aviation subcommittee, 413
—House foreign affairs subcommittee, 113
—House Housing and Urban Development and independent agencies appropriations subcommittee, 275, 299, 368
—House procurement and military systems subcommittee, 405
—House Public Works and Transportation Committee, 95
—House Public Works and Transportation subcommittee on investigations and oversight, 90
—House of Representatives, 51, 171, 435
—House Republican research report on ASAT, 111
—House Science and Technology Committee, 169–170, 270, 273, 414–415, 424, 433
—House Science and Technology transportation aviation and materials subcommittee, 169
—House space subcommittee, 261, 435
—House space science subcommittee, 435
—House subcommittee on space science and applications, 170, 257, 273, 356, 359, 363
—Library of, 122
—Senate, 112, 171, 176, 181, 285
—Senate Appropriations Committee, 117, 260
—Senate armed service strategic and theater nuclear forces subcommittee, 126
—Senate Armed Services, 116, 126, 436
—Senate Commerce, Science and Transportation Committee, 433
—Senate Committee on Rules and Administration, 238
—Senate Joint Resolution 154, 403
—Senate science, technology, and space subcommittee, 169, 270
Congressional Office of Technology Assessment (OTA), 257–258, 261, 413, 418
—report, 413
Connecticut, 350

Connolly, Denis, 103
Conran, Col. Philip, 18
Continental Airlines, 91–92
Continental Telecom Inc., 1
contingency landing sites (CLS), 289
continuous flow electrophoresis system (CFES), 352
Controlled Ecological Life Support Systems (CELSS), 446
controller, 82, 83, 86
Convair C–580 aircraft, 192
Cook Islands, 220
Cooper, Robert, 28–29, 127, 317
Cooper, Thomas, 23, 405
Cooperative Observations of Polar Electrodynamics (COPE), 71
COPE. See Cooperative Observations of Polar Electrodynamics.
Copenhagen, Denmark, 71
Corfu, Greece, 89
Cornell, Darrell, 405
Cornell University, Ithaca, NY, 67, 320
Coronado Airport, NM, 16
cosmonauts, 31, 388, 391–399
Cosmos Satellites, 394,
—*954*, 386
—*1266*, 387
—*1299*, 387
—*1402*, 386
—*1412*, 387
—*1443*, 398
—*1579*, 387
—*1607*, 386–387
—*1616*, 449
—*1625*, 386
—*1648*, 449
—*1663*, 387
—*1664*, 387
—*1667*, 391
—*1686*, 398
COSPAS/SARSAT program, 228
Costa Rica, 246, 356
Council of Chief State School Officers (CCSSO), 271–272
Council on Economic Priorities (CEP), 125
Council on Foreign Relations, 179
counter-rotating pusher propellers propulsion system, 15
Courter, Rep. James (R-NJ), 405
Covey, Col. Richard, 332
Cowen, Robert, 421

527

Cox, John, 320
Coyne, Dr. Lelia, 187
CPL. See capillary pumped loop.
Crab Nebula, 39
Cray-2 supercomputer, 187-189
Creighton, John, 279-280
Creighton University, Omaha, NE, 370
Cremen, Joseph, 327
Crimea, 396
Crippen, Capt. Robert, 34, 278, 322
Cronkite, Walter, 238
Crossfield, Scott, 151
CRRES. See Combined Release and Radiation Effects Satellite.
Culbertson, Philip, 181, 183, 257, 259, 263
Cumberland Elementary School, West Lafayette, IN, 272
Current, Maxwell, 192
Czechoslovakia, 392, 396

Dallas/Ft. Worth International Airport, TX, 88, 96
d'Allest, Frederic, 136, 146
Danish Meteorological Institute (DMI), 71
DARPA. See Defense Advanced Research Projects Agency.
Darmstadt, Federal Republic of Germany, 46
Dartmouth College, Hanover, NH, 177
Dassault-Breguet Aviation Co., 145-146
David Sarnoff Research Center, 281
Davis, Leon, 194
Daytona Beach, FL, 292
DC-9 aircraft, 143
DC-10 aircraft, 81
DDM. See drop dynamics module.
de Bondone, Giotto, 46
De Cordova Museum, 445
de Pater, Imke, 57
DeCair, Thomas, 183, 203
Deep Space Network (DSN), 63, 168, 173, 401. (See also NASA.)
Defense, U.S. Department of (DOD), 3-4, 7, 12, 15, 22, 29, 83, 111-133, 141, 174, 178, 180, 182, 192, 217, 228, 271, 278-279, 294-295, 300, 302, 311, 315-323, 334, 347-349, 364, 388, 405, 408-410, 412, 414-415, 417, 426, 436-439. (See also Air Force, U.S.; Army, U.S.; Marine Corps, U.S.; Navy, U.S.)
—Affairs Division, 320
—budget, 1, 116-117
—early-warning satellite, 315
—Joint Chiefs of Staff, 438
—National Guard, 404
—Office of the Secretary, 177, 179
—Pentagon, 111-115, 126, 131, 182, 217, 302, 315, 318, 326
Defense Advanced Research Projects Agency (DARPA), 5-6, 28-29, 122, 127, 317, 414, 437
Defense Contract Audit Agency, 409
Defense Intelligence Agency, 126, 320
Defense Meteorological Satellite Program, 27
Defense Nuclear Agency, Scientific Advisory Group on Effects, 177, 179
Defense Preparedness Association, 179
Defense Resources Board, 405
Defense Satellite Communications System (DSCS-III), 319, 334
Defense Science Board System Vulnerability Task Force and Associated Task Forces, 179
Defense Systems, Inc., 219
del Valle, Foreign Minister Jaime, 290-291
Delmos, 396
Delory, Greg, 249
Delta Air Lines, 88, 93, 96, 345
Delta rockets, 226, 300, 345
DeMarque, Dr. Pierre, 42
Denmark, 71, 135, 148, 151
Denson, W. John, 288
Denver, CO, 16, 142, 258
Design Verification Flights (Spacelab), 379
Dessler, Alexander, 100
Detroit, MI, 16
DFRC. See Dryden Flight Research Center.
DFVLR. See West German Aerospace Research Establishment.
Dickenson, David, 288
Dicks, Rep. Norman (D-WA), 115
Dicus, Carroll, Jr., 177
digital image processing, 102-103
Dihrab, Saudi Arabia, 219
Diller, George, 305

Index

Dintenfass, Dr. Leopold, 349
Discovery. See Space Shuttle Discovery.
District of Columbia. See Washington, DC.
Divad. See U.S. Army, Division Air Defense gun.
DMI. See Danish Meteorological Institute.
Dobbins AFB, GA, 12, 25
Dobrowolny, Marino, 248
DOD. See Defense, U.S. Department of.
DOE. See Energy, U.S. Department of.
Doi, Takao, 163
Dole, Secretary of Transportation Elizabeth Hanford, 86
Donahey, Lt. Col. Dayl, 25
Doppler radar system, 88
Doppler velocity sensor, 407
Dorfman, Steven, 230, 234
DOT. See Transportation, U.S. Department of.
Double Eagle II, 17
Douglas Aircraft Co., 4, 48
Downey, CA, 258
drop dynamics module (DDM), 374, 382
Dryden Flight Research Center (DFRC), 6–7, 10, 300–301, 338
DSCS. See Defense Satellite Communications System.
DSN. See National Aeronautics and Space Administration, Deep Space Network.
Du Pont Co., 151
Dubin, Maury, 243
Duffy, Capt. Brian, 33
Duigot, Michel, 219
Dulles International Airport, VA, 237, 239–241
Dunayev, A.I. 394
Dunbar, Dr. Bonnie, 282, 380
Durer, Albrecht, 238
Durrance, Dr. Samuel, 283
Dzhanibekov, Vladimir, 389–393, 397

E-2C Hawkey airborne control aircraft, 111
EAA. See Experimental Aircraft Association.
Eagle, moon landing, 403

earth, 33, 47, 51, 53, 61, 65, 71, 74–75, 100–101, 140, 148–149, 226–227, 231, 296, 317, 322, 334, 350–351
—as center of the universe (Ptolemy), 238
—atmosphere, 57, 75, 148–149, 243–244
—horizon, 61
—magnetism, 275
—magnetoscope, 79
—magnetotail, 78
—metallic core, 245
—observation and microgravity program, 135, 137
—orbit, 49, 54, 125, 146, 152, 236, 252–253
—ozone layer, 75
—primitive, 187
—solar system, 48
—troposphere, 74
—vegetation, 249
Earth Observation Mission (EOM), 284–285
Earth Observation Satellite Co. (Eosat), 430, 433
Earth Resources Laboratory, 245
earth station sites, 160
EASE. See experimental assembly of structures in extravehicular activity.
EASE/ACCESS. See experimental assembly of structures in extravehicular activity/assembly concept for construction of erectable space structures.
East Germany. See German Democratic Republic.
Easter Island. See Isla de Pascua.
Eastern Airlines, 91
Eastern Island University, Indonesia, 211
Eastern Space and Missile Center, 72, 123
EC–18B aircraft, 404
ECC. See electrochemical concentration cell.
echocardiography, 301, 328–329, 351
ECS satellites, 139–140
Edelson, Dr. Burton, 148–149, 283, 332
Edwards AFB, CA, 114, 240, 293, 300, 319, 327, 334, 337, 342, 361–362, 404, 436
—Mojave Desert lakebed runway, 332

Edwards, Rep. Mickey (R-OK), 88
EEC. See European Economic Community.
EG&G Energy Management Corp., 375
EG&G Energy Measurements, Inc., 382
Eglin AFB, FL, 119
Egypt, 38, 111, 250
—Great Pyramid of Cheops, 38
EIDI. See Electro-Impulse Deicing System.
El-Asser, Mohammed, 100
El Cichon (volcano), Mexico, 76
electrochemical concentration cell (ECC), 73
Electro-Impulse Deicing System (EIDI), 9–10
Electromagnetic Levitator (EML), 355
Ellington AFB, TX, 360
Elliot, Steve, 246
ELV. See Expendable Launch Vehicle.
emergency evacuation systems, 90, 94–95
Emerson, Terry, 239
EML. See Electromagnetic Levitator.
Energetic Proton Experiment, 59
Energy, U.S. Dept. of (DOE), 133, 201, 208–209, 212, 412–413
Engen, Donald, 142–143
England, Dr. Anthony, 330, 421, 445
England. See United Kingdom.
Engle, Col. Joe, 240, 332–333
Engstrom, Dr. Fredrik, 137
Enterprise. See Space Shuttle Enterprise.
Enviro-Spray Systems, 443–444
EOM. See Earth Observation Mission.
Eosat. See Earth Observation Satellite Co.
Equatorial Communication Co., 2
Erlangen-Nurnberg, University of, Federal Republic of Germany, 109
ERS–1 programs, 139
ESA. See European Space Agency.
ESOC. See European Space Operations Centre.
Estes, Robert, 248
Ethics and Public Policy Center, the, 179
EURECA. See European Retrievable Carrier.
Eureka technology program, 128, 138, 145, 250–251, 434
European Communications Satellites (ECS), 139–140

European Economic Community (EEC), 251
European Retrievable Carrier (EURECA), 137–138, 376
European Space Agency (ESA), 28, 45–46, 48, 51–52, 62, 99, 129, 135–141, 177, 213, 222–223, 252, 255, 259, 261, 263, 265–266, 269, 282–285, 300, 335–336, 338, 343, 373, 377, 380, 396, 398, 418
European Space Operations Centre (ESOC), 45–46, 52
European Space Research Technology Center, 62
European X-ray Observatory Satellite (EXOSAT), 45, 46
EUTELSAT 1, 138
EUVE. See extreme ultraviolet explorer.
EVA. See extravehicular activity.
Evpatoriya, Crimea, USSR, 396
EXOSAT. See European X-ray Observatory Satellite.
Expendable Launch Vehicle (ELV), 409–410
Experimental Aircraft Association (EAA), 85–86
experimental assembly of structures in extravehicular activity (EASE), 17, 341, 355
—assembly concept for construction of erectable space structures (EASE/ACCESS), 341
Extraterrestrial Research Division. See Ames Research Center.
extraterrestrial signals, 235–236
extravehicular activity (EVA), 17, 333
extreme ultraviolet explorer (EUVE), 189

F–8 fly-by-wire research aircraft, 98
F–14 aircraft, 11, 111, 119
F–15 Eagle aircraft, 11, 23, 111–114, 119, 406–409, 411, 441
F–16 Falcon aircraft, 23, 25, 119, 405–408, 438
F–18 aircraft, 119
F–20 Tigershark aircraft, 23, 404–406, 408
F–100 aircraft, 23, 81
F–110 aircraft, 23
F–111 aircraft, 11, 13–14, 16
FAA. See Federal Aviation Administration.

Index

Fabian, Dr. Andrew, 40
Fabian, John, 34–35, 279–282
Faget, Max, 428
Fahd, King of Saudi Arabia, 280
Fairchild Industries, 1–2
Fairchild Space Co., 424
Fairchild Weston Systems, 291
Farquhar, Dr. Robert, 49
FBIS. See Foreign Broadcast Information Services.
FCC. See Federal Communications Commission.
FDA. See Food and Drug Administration.
FEA. See fluids-experiment apparatus.
Federal Aviation Administration (FAA), 9, 15, 27, 82–90, 141–144, 191, 197, 219, 220, 239
 — Aeronautical Center, 87
 — air route traffic control centers, 86–87
 — NASA runway surface traction program, 196
 — regulations, 81, 142, 144
 — Technical Center, 86–87, 144
Federal Communications Commission (FCC), 141, 158, 423
Federal Investigators, Association of, 20
Federal Republic of Germany, 46, 50, 59, 65, 77–79, 119, 135, 145, 152, 159–160, 167, 192, 214, 250–251, 255–256, 263, 284–285, 300, 335–338, 375–376, 380, 399
 — BMFT ministry of research and technology, 380
Federation of American Scientists, 112, 125
Federation of Japanese Economic Organizations, 266
FEE. See French echocardiograph experiment.
Fees, Martin, 45
FEPC. See Flight Equipment Processing Contract.
Ferguson, Graeme, 238
Fernandez, Judge Ferdinand, 184
FES. See fluid experiment system.
Feuerbacher, Berndt, 380
Fichtl, Dr. George, 327, 377, 382–383
Field, George, 440
Finke, Wolfgang, 380–381
Finland, 71, 251, 396–397

first
 — Arab in space, 280, 282
 — black astronaut to take shuttle's controls, 328
 — manned landing on the moon, 310
 — private citizen passenger, 272–277
 — Space Shuttle. See Space Shuttle.
 — to travel twice the speed of sound, 151
first American
 — in space, 151
 — to orbit the globe, 151
 — to pilot the X–15, 151
 — woman as prime payload specialist, 280
 — woman in space, 102, 151, 281
 — woman to walk in space, 32, 440
Fischell, Robert, 99
Fisher, Anna, 278
Fisher, Dr. William, 227, 230, 281, 332–333
Fitch, Lt. Gen. William, 440
Five-Year Defense Plan, 406
fixed service structures (FSS), 287–288
Flight Equipment Processing Contract (FEPC), 190–191, 314
Florensky Memorial Symposium on Venus, 244
Florida, 21, 197, 207–208, 232, 270, 273, 275, 304, 334, 338, 356, 393, 414, 424–426, 435
fluid experiment system, 243, 374
Fluid Physics Module, 338
fluids-experiment apparatus (FEA), 243, 374
Foerster, Robert, 272
Food and Drug Administration (FDA), 431
Ford Aerospace and Communications Corp., 19, 159–161, 438
 — Western Development Laboratories, 224
Ford, President Gerald R., 203
Foreign Broadcast Information Services (FBIS), 105, 137, 163, 213, 215, 387–388, 391–392, 394–396, 398
Forrester, James, 207
Fort A.P. Hill, VA, 195
Fort Davis, TX, 44
Fort Smith, AR, 81
Fowler, Rep. Wyche, Jr. (D-GA), 51
FPE. See French posture experiment.

531

France, 41, 55, 59, 97, 128, 135–138, 144, 146, 159, 161, 167, 191, 214, 228, 251, 255–256, 285, 327–329, 382, 388, 392, 394, 403, 418
Franco/Soviet space cooperation, 41
Franke, John, Jr., 179
Frankel, Edward, 177
Franklin, I.V., 263–264
Freedom Star, the, 371
French Center for Nuclear Studies, 191
French echocardiograph experiment (FEE), 328–329
French Guiana, 46, 129, 139, 219
French materials science experiment, 374
French National Center for Space Studies, 136, 146
French National Space Agency (CNES), 138, 173, 279
French Polynesia, 139
French posture experiment (FPE), 328–329
Friedman, Dr. Herbert, 39, 41
Friendswood, TX, 272
Frimout, Dr. Dirk, 284–285
Fthenakis, Emanuel, 2
FSS. See fixed service structures.
fuel efficiency, 1, 84
Fullerton, Col. C. Gordon, 240, 330, 421, 445
Funk, Scott, 293
Fuqua, Rep. Don (D-FL), 270, 424
Furrer, Reinhard, 284, 335, 338
Future Air Navigation System Committee, 142

G-Star satellites, 136, 138
Gabon, Republic of, 208–209
 —Ministry of Energy and Hydraulic Resources, 208
Gaffney, Dr. F. Drew, 279
Gagarin Space Training Centre, USSR, 388
Galapagos Islands, Ecuador, 329
galaxies, 57–58. (See also Milky Way.)
 —development, 42
 —infrared, 57–58
 —Seyfert, 57
 —starburst, 57

Galileo mission, 20–21, 28, 61–62, 64–66, 169, 178, 197, 214, 252, 261, 280–281, 290, 300, 412
 —propulsion system, 65
Galung-gung (volcano), Indonesia, 76
Gamma-Ray Observatory, 40, 42
gamma rays, 41, 55
Gander, NF, Canada, 96
Gandhi, Prime Minister Rajiv, 155
GAO. See General Accounting Office.
Garching, Federal Republic of Germany, 59, 399
Garcia, Judith, 272
Gardner, Comdr. Dale, 31–32, 278, 322
Gardner, Lt. Col. Guy, 278, 322
Garmire, Dr. Gordon, 40
Garn, Sen. E. J. (R-UT), 273, 275, 277–279, 296, 322–323, 325, 351, 403
Garriot, Owen, 282, 421
GAS. See Get Away Special.
gas turbine engine (convertible), 5–6
Gazenko, Oleg, 393
GD. See General Dynamics Corp.
GD/C. See General Dynamics Corp., Convair Division.
GE. See General Electric Co.
GEC-Marconi, 159–160
Gemar, Capt. Charles, 33
Gemini, 31, 34, 102, 185, 310
General Accounting Office (GAO), 95–96
General Advisory Committee on Arms Control and Disarmament, 177, 179, 185
General Dynamics Corp. (GD), 12, 120, 170, 180, 182, 184, 197, 204, 345–346, 419
 —Convair Division (GD/C), 18, 20–21
General Electric Co. (GE), 5, 11, 15, 23–24, 103, 118–119, 121, 141, 190, 192–193, 260, 319, 359, 423
 —Penn State University, 357
 —Space Systems Division, 258
 —Valley Forge Space Center, Philadelphia, PA, 77
Geneva, Switzerland, 71–72, 112, 229
Geological Society of America, 100
geophysical fluid flow cell (GFFC), 374, 383
geophysics, 147–149

Index

Georgia Institute of Technology, Atlanta, GA, 167
Geostar Corp., 440
Geostationary Operational Environmental Satellites (GOES), 222–224, 226–227, 229–230, 253
geostationary orbit, 105, 136, 140, 220, 222–224, 226, 229–230
geosynchronous orbit, 18–22, 157, 159–160, 168, 171, 214, 220–221, 227, 231–232, 423
geosynchronous space stations. See space stations, geosynchronous.
Gerard, Patty, 246
German Democratic Republic, 392
German ion-release module (IRM), 77–78
German Space Operations Center, Oberpfaffenhofen, Federal Republic of Germany, 78, 375–376, 380–381
Germany, East. See German Democratic Republic.
Germany, West. See Federal Republic of Germany.
Gernigan, Tamara, 33
Get Away Special (GAS), 218–219
 —bridge, 357
 —canisters, 301, 329–330, 357
 —experiments, 56, 247, 356
 —satellites, 326
GFFC. See geophysical fluid flow cell.
Giacconi, Dr. Riccardo, 40
Giacobini-Zinner. See Comet Giacobini-Zinner.
Gibson, Comdr. Robert, 34, 86
Gilbert, Gene, 252–253
Gilbert Islands, 220
Gillogly, Col. Harry, 439
Gilruth, Robert, 102
Giotto interplanetary space probe, 46, 52, 135, 137
Glenn, Sen. John (D-OH), 151
Glickman, Rep. Dan (D-KS), 170
Global Change, 149
Global Low Orbiting Message Relay Satellite (GLOMR), 204–205, 218–219, 374
Global Positioning System (GPS), 122–124, 141, 233–234. (See also Navstar.)
GLOMR. See Global Low Orbiting Message Relay Satellite.

Goddard, Robert, 152
Goddard Laser Tracking Network, 250
Goddard Space Flight Center (GSFC), 37, 39–41, 43, 45, 49, 58–59, 67, 71, 73–79, 86, 112, 152–153, 171, 177–178, 180, 185–186, 189, 199, 204, 208, 218–220, 225–226, 228, 243–244, 247, 250, 258, 330, 334, 356–357
 —Crustal Dynamics Project (CDP), 250
 —Laboratory for Astronomy and Solar Physics, 49
 —Laboratory for Oceans, 252
 —Microwave Sensors and Data Acquisition Systems Branch, 253
 —Wallops Flight Facility, VA, 71, 73–77, 172, 199–200, 244, 411
 —Wallops Range Control Center, 77
Godwin, Dr. Linda, 33
GOES. See Geostationary Operational Environmental Satellites.
Goetz, Robert, 191
Gohagan, Dr. John, 209–210
Goldenberg, Tsvi, 207–208
Goldschmidt, Rudolf, 9
Goldsmith, Philip, 137
Goldstone, CA, 44
Goldwater, Sen. Barry (R-AZ), 102, 239, 403
Goose Bay, LB, Canada, 405
Gorbachev, General Secretary Mikhail, 386
Gordon Bennett Race, 17
Gorton, Sen. Slade (R-WA), 270
Gossamer Condor, 237
Gould 32-87 computer system, 247
GPS. See global positioning system.
Grabe, Lt. Col. Ronald, 278, 281
Graham, Lt. Gen. Daniel, 126, 320
Graham, William, 176–177, 181–183, 191
Gran Patajen, 246, 252
Grand Canyon, AZ, 66, 445
Grassley, Sen. Charles (R-IA), 407
Great Britain. See United Kingdom.
Great Pyramid of Cheops, Egypt, 38
Grechko, Georgiy, 393
Greece, 89, 250
Greely, Dr. Ronald, 244
Green, Jacklyn, 67
Green, Shirley, 183–184
Green Bank, WV, 44

533

Greenbelt, MD, 218
Greenland, 142
—Sondre Stromfjord facility, 71
Gregory, Col. Frederick, 195, 325, 328, 360–361
Gregory, William, 320
Griffin, Gerald, 191, 361
Griggs, S. David, 278–279, 282, 296, 322, 324
Grindeland, Richard, 354
Grissom, Virgil, 34
Grow Group, 443
Grumman Aerospace Corp., 6, 10–11, 111, 165, 190, 192, 270, 439
—Data System Corp., 198
Grundfest, Dr. Warren, 207
GSFC. See Goddard Space Flight Center.
GTE Spacenet Corp., 136
Guam, 233
Guarachi, Bernardo, 91
Gulf of Mexico, 253, 333
Gulf Stream eddies, 353
Gulfstream II aircraft, 15
Gullahorn, Gordon, 248
Gymnastics Federation, U.S., 246

H-II booster rocket, 163
Haise, Fred, Jr., 240
Hall, Donald, 58
Hall, Stephen, 412
Halley's Comet. See Comet Halley.
Halperin, Morton, 320–321
Halter Marine, 371
Hamilton Standard Management Services, Inc., 190, 314
Hammack, Jerome, 222
Hancock County, MS, 429
Hansen, James, 180
Harford, James, 101
Harris, Hugh, 302
Harris Corp., 172, 270, 423
Hart, Terry, 34
Hartigan, James, 93–94
Hartsfield, Henry, Jr., 293, 335, 338
Harvard University, Cambridge, MA, 350
Hat Creek, CA, 44
Hauck, Frederick, 280
Hawaii, 58, 63–64, 129–132, 221, 318, 330

Hawaii, University of, Honolulu, HI, 58, 63
Hawes, Ralph, 180
Hawley, Steven, 282, 300, 356
Hazen Union School, Hardwick, VT, 272
Heath, Dr. Donald, 75
Heise, Ing., 137
Helin, Eleanor, 55
Helsinki, Finland, 385, 396–397
Henderson, Maj. Gen. Donald, 124
Henize, Dr. Karl, 330–331
Hercules, Inc., 367–368
Hermes, 135, 145–146, 255
Herpes Samirii, 350
Herres, Gen. Robert, 364, 439
Herschel, John, 38
Herschel, William, 68
Herzfeld, Charles, 440
Heseltine, Michael, 132–133
hexanitrostilbene (HNS), 95
Hexe. See high-energy X-ray experiment.
Heyerdahl, Thor, 102
HHG-1. See Hitchhiker payload.
Hi Lat Spacecraft, 71
Hieb, Richard, 33
High Frontier system, 126
high precision tracking experiment, 129–130
high-energy X-ray experiment (Hexe), 399
hijackings, 85, 111
Hilmers, Maj. David, 279, 281, 319, 335
Hind helicopters, 438
Hipparcos, 139–140
Hitchhiker (HHG-1) payload, 356–357
HNS. See hexanitrostilbene.
Hodge, John, 185–186
Hoffman, Jeffrey, 278, 296, 322–324
Hogrefe, Arthur, 99
Hokkaido University, Japan, 163
Holland. See The Netherlands.
Holmes, Todd, 81
Honeywell, Inc., Military Avionics Division, 197, 408
Hong Kong, 146, 215
Hoover, President Herbert, 102
Hopkins Ultraviolet Telescope, 283
Horizontal Take-Off and Landing (HOTOL) transatmospheric vehicle, 414

Index

Hosenball, S. Neil, 175–176
HOTOL. See Horizontal Take-Off and Landing.
Houston, TX, 91, 282, 310, 331, 334–335, 357, 425–426
—Chamber of Commerce, 191
Houston Mission Control, 315, 337
Hovestadt, Dr. Dieter, 59
Howard University School of Engineering, Large Space Structure Institute (LSSI), Washington, DC, 17–18
HS–393 Ku-band satellite, 346
Hubble Space Telescope, 40, 42, 48, 53–54, 58, 178, 199, 252, 282, 300, 417
Hudson, Gary, 430–431
Hughes Aircraft Co., 119, 176, 192, 212, 223, 228, 234, 260, 279, 323–325, 334, 339, 430
Hughes Communications, Inc. 222–226, 230–233, 275, 299–300, 334, 346, 360
Hughes-Fulford, Dr. Millie, 279–280
human performance research, 13
human powered aircraft, 96
Human Role in Space, The (THURIS), 411–412
Hungary, 392
Hunt and Co., 187
Huntsville, AL, 198
Hutchinson, Neil, 257
Hyang, Yi Zluo, 214
hydrogen bomb test detonation, 133
hydrogen-fueled scramjets, 4
Hymer, Wesley, 354–355
Hynds, Dr. Robert, 59
Hyperion, 67
hypersonic speed, 4, 7, 29, 97, 199

IBM. See International Business Machines.
IBSE. See Initial Blood Storage Experiment.
ICAO. See International Civil Aviation Organization.
ICBM. See intercontinental ballistic missile.
ICE. See International Cometary Explorer.

Iceland, 142
IHW. See International Halley Watch.
Ikle, Fred, 128
ILC, 190
Illinois, University of, Urbana, IL, 57, 73–74
image processing laboratory (IPL), 103
impact theory, 148
Imperial College, London, England, 59
Independence, the, 371
India, 71, 87, 155, 238, 284, 373, 388
Indian Department of Space, 284
Indian Ocean satellite, 160–161, 213, 386
Indian Space Research Organization, 284
Indonesia, 76, 211–212
industrial space facility (ISF), 428
inertial navigation system (INS), 122–123
inertial upper stage (IUS), 19, 299, 316
Informatics General Corp., 188
Information Agency, U.S., 222
Infrared Astronomy Satellite (IRAS), 39, 41–42, 53, 57–58, 186
Initial Blood Storage Experiment (IBSE), 358
INMARSAT. See International Maritime Satellite Organization.
Innovative Optics, Inc., 186
Inorganic Coatings, Inc., 208
INS. See inertial navigation system.
Institute of Aeronautical Sciences, 101
Institute for the Study of the USA and Canada, Foreign Policy Department, 390
Institute for Technology Development, National Space Technology Laboratories, Hancock County, MS, 429
Instituto Fisica Spazio, 248
instrument pointing system (IPS), 47, 331
Instrument Technology Associates (ITA), 353
Instrumented Test Vehicle (ITV) program, 410
Integrated Systems Analysts, 190
Intelsat. See International Telecommunications Satellites.
intercontinental ballistic missile (ICBM), 132, 217, 344, 347, 436

535

Astronautics and Aeronautics, 1985

Intercosmos Program, 394, 396
 —Prognoz-10-Intercosmos, 396
Interior, U.S. Department of, 200–201
Interkosmos Institute, Moscow, USSR, 398
International Aeronautical Federation, 31
International Air Transport Association, 85
International Association of Geomagnetism and Aeronomy, 100
International Association of Quality Circles, 174
International Astronomical Union, 180
International Business Machines (IBM), 86–87
International Civil Aviation Organization (ICAO), 141, 219
International Cometary Explorer (ICE), 43–46, 48–51, 59–60, 167. (See also International Sun-Earth Explorer.)
 —magnetometer experiment, 60
International Geophysical Year, 243
International Halley Watch (IHW), 109. (See also Halley's Comet.)
International Maritime Satellite Organization (INMARSAT), 136, 227
International Solar Polar Mission, 62, 300. (See also Ulysses.)
international solar terrestrial physics program, 51–52
International Space Corp. (ISC), 431–432
International Sun-Earth Explorer (ISEE-3), 43, 46, 49. (See also International Cometary Explorer.)
International Technology Underwriters, 324, 347
International Telecommunications Satellites (Intelsat), 134, 158–161, 204, 229
International Ultraviolet Explorer (IUE), 38, 41–42, 45
Interplanetary Monitoring Probes, 180
Investigators Working Group (IWG), 285
Ion Composition Experiment, 59
Iowa, 369
Iowa, University of, 417
IPL. See image processing laboratory.
IPS. See instrument pointing system.
IR–IE. See RCA Infrared-Imaging Equipment.
IRAS. See Infrared Astronomy Satellite.
Ireland, 92, 96, 135

IRM. See German ion-release module.
ISC. See International Space Corp.
ISEE-3. See International Sun-Earth Explorer.
ISF. See industrial space facility.
Ishmael, Stephen, 6
Isla de Pascua, 289, 291–292
Israel, 250
ITA. See Instrument Technology Associates.
Italian National Research Council, 247
 —National Space Plan Office (PSN/CNR), 247–248
Italy, 135, 137, 159–161, 250–251, 255, 336
ITT, 440
ITV. See instrument test vehicle program.
IUE. See International Ultraviolet Explorer.
IUS. See inertial upper stage.
IWG. See Investigators Working Group.
Izvestia, 394

Jaguar jet, 145
JAL. See Japan Air Lines.
Japan, 85, 87–88, 93, 96, 109, 159–160, 163–164, 168, 255, 259, 263–266, 269, 283, 285, 418
 —Aircraft Accident Investigation Committee, 88
 —National Space Development Agency, 163
 —Space Activities Commission, 163
Japan Air Lines (JAL), 88–89, 93, 96
Japan Communications Satellite Co. (JCSat), 346
Jarvis, Gregory, 275, 279
Jastrow, Robert, 125
JCSat. See Japan Communications Satellite Co.
Jet Propulsion Laboratory (JPL), 43, 48, 54, 60, 62–65, 77, 79, 102–103, 109, 173, 206–208, 214, 375, 382–383, 422, 434–435
 —Geobotanical Remote Sensing Group, 249
 —JPL/Cedars/Sinai procedure, 206
 —Managed Three-Axis Acoustic Levitator (3AAL), 355-356
 —wide-field/planetary camera (WFPC), 48

Index

JetStar, 4, 8-9, 16
JFK. See Kennedy International Airport.
Jiangxi Province, People's Republic of China, 434
Jirberg, Russel, 104
Jiuquan launching center, 215
John Cabot, the, 87, 92
Johns Hopkins University
—Applied Physics Laboratory, 78-79, 99, 147
—Department of Physics and Astronomy, 283
Johnson, Frank Jr., 183
Johnson, Kathy, 246
Johnson, Scott, 246
Johnson & Johnson, Ortho Pharmaceuticals Divisions, 431
Johnson Space Center (JSC), 22, 31-32, 35, 86, 135, 163, 167, 186, 190-191, 225, 227, 244, 246, 256-259, 265, 268, 271-274, 276-278, 280, 304, 307, 309-310, 312-314, 317-318, 323, 328, 350-351, 358, 361, 370, 395, 443, 445
Johnson Wax Co., 237
Johnston, J. Bennett (D-LA), 286
Joint Surveillance Target Attack Radar System (JSTARS), 439
JPL. See Jet Propulsion Laboratory.
JSC. See Johnson Space Center.
JSTARS. See Joint Surveillance Target Attack Radar System.
JT8D engines, 143
Jupiter, 21, 53-54, 61-62, 64-65, 68, 178, 245, 252, 280, 300, 377
Justice, U.S. Department of, 181, 203, 409

Kagoshima Space Center, Japan, 48
Kaiser, Dr. Michael, 67-68
Kalamazoo, MI, 16
Kaliningrad, USSR, 435
Kanematsu Institute, Department of Medical Research, Sydney, Australia, 349
Kansas State University, Manhattan, KS, 167
Kascak, Thomas, 103
Kazakhstan, USSR, 389
KC-135 tanker aircraft, 365

Keegan, Sarah, 203
Keio University Medical School, Tokyo, Japan, 163
Keller, Geoffrey, 417
Keller, Samuel, 104, 178
Kelly, Judith, 91
Kelly, William, 91
Keming, Vice-Minister Bao, People's Republic of China, 434
Kennedy (John F.) International Airport (JFK), New York, NY, 85
Kennedy Space Center (KSC), 18, 63, 65-66, 120, 157, 159-160, 171, 176, 191-195, 199, 205, 208-209, 220, 226, 240-241, 269, 273, 278, 281, 286-294, 296-297, 299-302, 304-307, 309-310, 312, 316-317, 322, 324-325, 328, 334-335, 338-339, 358, 360-363, 370-371, 374-376, 379, 436, 445-446
—thermal protection system facility, 171
Kentucky Fried Chicken, 248
Kenwood High School, Baltimore, MD, 272
Kenya, 246
Keough, Donald, 443
Kerrebrock, J.L., 440
Kerry, Sen. John (D-MA), 111
Kettering Group, United Kingdom, 397
Keyhole satellites, 124-125
Keyworth, Dr. George, II, 7, 29, 413-414
kinetic-kill vehicle, 126
Kirkpatrick, Ambassador Jeanne, 440
Kirkpatrick, Commodore John, 179
Kitt Peak, AZ, 78
Kizim, Leonid, 389
Knollenberg probe, 9
Knott, Dr. William, 446
Kohrs, Richard, 312
Kondo, Dr. Yoji, 39, 41-42
Konrad, John, 279
Korad laser, 416
Kornfeld, Dale, 100
Korou Space Center, 46, 138-139
Kraft, George, 244
Kramer, Rep. Ken (R-CO), 439
Kramer, Saunders, 87-88
Krasnoyarsk, USSR, 386
Krueger, Dr. Arlin, 76
KSC. See Kennedy Space Center.

Kubasov, Valery, 434
Kupperman, Dr. Charles, 185
Kutyna, Brig. Gen. Donald, 29, 436–437
Kyushu, Japan, 48

La Paz International Airport, Bolivia, 91
Lageos, 250
Lahav, Dr. Noam, 188
Lahore, India, 238
Lake Maracaibo, Venezuela, 76
Lally, Richard, 85
Lamberth, Horace, 312
laminar-flow control, 4–5, 8–9, 16
Lampton, Dr. Michael, 282, 285
Land Management, Bureau of, 200
Land Remote Sensing Commercialization Act, 432–433
landing gear, 195–196
Landsat earth resources satellite, 246, 249, 418, 430
Langford, John, 96
Langley Memorial Aeronautical Laboratory, 102
Langley Research Center (LaRC), 3, 5, 8, 17, 86, 94, 96, 167–168, 175, 184, 194–195, 365–366, 411
 —Aircraft Landing Dynamics Facility, 195
LANTRIN. See Low-Altitude Navigation and Targeting Infrared System for Night.
LAP. See Large-Scale Advanced Propfan.
LaRC. See Langley Research Center.
Large Space Structure Institute (LSSI), 17–18
Large-Scale Advanced Propfan (LAP), 14–15
Larson, Rear Adm. Charles, 33
laser technology, 206–208
 —fiber-optic imaging system, 206
 —JPL excimer laser, 206
 —medical laser, 206
Lathlaen, Peggy, 272
Laudenslager, Dr. James, 207
launch environment instrumentation system (LEIS), 291
Lawless, Dr. James, 188
Lawrence Berkeley Laboratory, 440
Lawrence Livermore National Laboratory, Department of Energy, 133

LDEF. See Long-Duration Exposure Facility.
Le Fevre, Marius, 137
Le Bourget, France, 260
Leakey, Richard, 246
Leakey Foundation, 246
Leasat, 221–223, 225, 230–232, 301, 303–304, 323, 325, 333–334
Lee, Mark, 281
Lee, Thomas, 100
Lees, Frederick, 177
Leestma, David, 32, 102
Lehigh University, Bethlehem, PA, 100, 428
Lehman, Navy Secretary John, 111, 437
LEIS. See launch environment instrumentation system.
Leninsk, USSR, 435
Lennon, Thomas, 246, 252
Lenoir, William, 34
Leonov, Aleksey, 388, 434
LeRC. See Lewis Research Center.
Leslie, Dr. Fred, 376
Lester, Roy, 374–375, 377
Levy, Dr. Eugene, 37, 244
Lewis, Roy, 147
Lewis, Rep. Tom (R-FL), 414
Lewis Lear jet, 429
Lewis Research Center (LeRC), 5, 9–10, 14–15, 18, 20, 26, 31, 85, 96, 103–104, 157, 160, 175, 194, 196–198, 208–209, 211–212, 235, 256–259, 270, 310, 369–370, 421, 429
Li, Wu Ke, 213
Liberty Star, the, 371
Lichtenberg, Dr. Byron, 282, 285
lightning, 74, 191–192
 —location and protection (LLP), 292–293
Lincoln satellite, 412
Lind, Dr. Don, 325–326, 375
Lindstrom, Robert, 302
Linsky, Dr. Jeffrey, 40
lithium, 78–79
Litvack, Dr. Frank, 207
Lloyds of London, 223
LLP. See lightning location and protection.
LLWAS. See low-level, wind-shear alert system.
Lockheed Corp., 12, 15, 287–288, 298, 310, 416

Index

—Georgia Corp., 4–5, 8, 11–12, 25–27
—L-1011-1, 88–89
—Logistics Directorate, 310
—maintenance problems, 25
—Missiles and Space Co., 48, 258
—Palo Alto Laboratory, 330
—Space Flight Co., 190
—Space Operations Co. (LSOC), 205, 288, 294
Loewenthal, Stuart, 103
Long, Sen. Russell (D-LA), 286
Long-Duration Exposure Facility (LDEF), 297–298, 300–301
Long March rockets, 213–215
Lopez, Don, 237
Los Alamos National Laboratory, 59
Los Angeles, CA, 206–207, 419
Louisiana, 253, 285–286
Lounge, John M., 231, 281, 332–333
Lousma, Jack, 34
Low-Altitude Navigation and Targeting Infrared System for Night (LANTRIN), 406
Lowell High School, San Francisco, CA, 248
Lowenthal, Stuart, 103
low-level, wind-shear alert system (LLWAS), 88
LOX/hydrogen cryogenic and hydrocarbon engines, 22
LSOC. See Lockheed Corp., Space Operations Co.
LSSI. See Howard University School of Engineering, Large Space Structure Institute.
LTV, 420
Lucid, Shannon, 279–282
lunar research, 167. (See also moon.)
Lunar and Planetary Science Conference, 16th, 244
Lunney, Dr. Glynn, 310–312
Lust, Gen. Reimer, 263
Luxembourg, 138

MacArthur, Gen. Douglas, 102
MacCready, Paul, 237
MacQueen, Robert, 114
Madrid, Spain, 44
Magnetometer Experiment, 60
magnetosphere, 64, 77–79

magnetotail probe, 78–79
Malinckrodt Institute of Radiology, St. Louis, MO, 209
Manchester International Airport, Manchester, England, 89
Manned Flight Awareness Panel, the, 174
manned maneuvering unit (MMU), 433
Manned Spacecraft Center, 310
Manno, George, 232
Man-Vehicle Systems Research Facility (MVSRF), 26–27
MARECS, 136
Mark, Hans Michael, 176, 364
Mariana, Franco, 248
Mariana Islands, 220
Marine Corps, U.S., 438, 440
Mariner, 180
Marquart, David, 272
Mars, 51, 54–55, 66–67, 149, 235, 244, 365, 392, 396, 421–422, 434–435, 440, 446
Mars Observer, 440
MARSAR II, 12
Marsden, Megan, 246
Marshall Islands, 220
Marshall Space Flight Center (MSFC), 17, 22, 40, 86, 99–100, 167, 171, 175, 198–199, 248–249, 258–259, 267, 284–285, 307–309, 324, 331, 349–353, 358, 366–368, 374–383, 411, 420
—Materials Science Laboratory, 355
—Productivity Enhancement Center, 367
—Research Animal Holding Facility, 383
—Solar Terrestrial Division, Space Science Laboratory, 285
—Space Camp, U.S., 249
—Systems Dynamic Laboratory, 382
Marsik, Stanley, 308
Martin, Dick, 26
Martin Marietta Corp., 2, 258, 260, 267, 285, 349, 406, 419–420, 436
Maryland, University of, College Park, MD, 40
maser clocks, 233–234
Masar II Computer, 12
Mason, Senator, Australia, 129
Massachusetts, 18, 40, 44, 131, 272, 379, 445

539

Massachusetts Artists Foundation, 445
Massachusetts College of Art, Boston, MA, 445
Massachusetts Institute of Technology (MIT), Cambridge, MA, 18, 40, 96, 167, 186, 426
—Center for Advanced Visual Studies, 445
—MIT Monarch, 96
Mataveri Airfield, Easter Island, 289–291
Materials Science Laboratory-2 (MSL-2), 355–356
Matese, John, 38
Matsunaga, Sen. Spark (D-HI), 413, 435
Mattingly, Capt. Thomas, 34, 295, 349
Maui, HI, 129–132, 318
Mauna Kea Observatory, Hawaii, 64
Maverick missiles, 406
MAW. See mission adaptive wing.
Max Planck Institute for Extraterrestrial Physics, Garching, Federal Republic of Germany, 59, 399
MBB. See Messerschmidt-Bolkow-Blohm.
McAuliffe, Sharon Christa, 272–277
McBride, Jon, 278
McCall, ID, 272–273
McCall-Donnelly Elementary School, McCall, ID, 272–273
McCandless, Bruce, 178, 282, 300
McCarthy, Dr. John, 197
McCray, Dr. Richard, 42
McDermid, Dr. Stuart, 207
McDonald, Rebecca, 207
McDonald Laser Ranging System, 416
McDonald Observatory, TX, 416
McDonnell, Sanford, 120–121
McDonnell Douglas Astronautics Co., 9, 11, 116, 120–121, 143, 167, 226, 258, 264, 279, 299, 322, 339, 342, 345, 407–409, 411–412, 430
—continuous flow electrophoresis system, 299
—investigation of alleged overcharges, 409
McDonnell Douglas Technical Services Co., 193, 379
McDougall, Walter, 151
McMurdo Sound, Antarctica, 218
McNair, Ronald, 274, 278
McPherson, David, 180
McPherson, M. Peter, 224
Meade, Capt. Carl, 33

Medea's gradient furnace, 380
medical advances, 206–211
Medina, Ernesto Rodriguez, 105
Mediterranean Sea eddies, 353
Melanesia, 220
Melbourne, Australia, 129
—Overseas Service, 129
—Regional Airport, 292
Melbourne, FL, 273, 426
Melton, Robert, 258
Merbold, Ulf, 284
Mercury, 244
Mercury Program, 102, 176, 184–185, 200. (See also Project Mercury.)
Merino, Admiral Jose Toribio, 289–290
MERIT (interactive data management and enhancement program), 197
Merritt Island, FL, 172
Messerschmid, Dr. Ernst, 284, 335, 337–338, 380
Messerschmitt-Bolkow-Blohm (MBB), 432
—Erno, Space Systems Group of, 137, 159–160, 192
Metcalf, Michael, 272
Meteor-2, 387
meteorites, 148
Meteorological Institute, Finland, 396–397
meteorological interactive data display system (MIDDS), 292–293
meteorological satellites (Metsats), 344, 346
meteorology, 38–42, 71, 76–78, 133, 218, 221, 223–224, 226–227, 238, 333–334
Meteosat, 140, 222–223
Methia, Richard, 272
Metsats. See meteorological satellites.
Meudon Observatory France, 59
Mexico, 75, 167, 220, 300, 329–330, 333, 339
Mexico City, Mexico, 232
Meyer, Gerald, 409
Meyer, Lt. Col. Thomas, 318
MHV–ASAT. See antisatellite miniature homing vehicle.
Miami, FL, 228
Micale, Fortunato, 100
Michigan, 16, 369
Michigan, University of, Ann Arbor, MI, 167, 186

Index

microburst, 88
micro-earth stations, 2
microgravity, environment, 275
Microgravity Materials Science Laboratory (MMSL), 198, 428
microspheres, 100–101
microwave window frequency range, 235
MIDDS. See meteorological interactive data display system.
mid-infrared chemical laser (MIRACL), 131
Military Airlift Command, 12
Milky Way galaxy, 44–46, 55, 329, 399
Millstone Hill, MA, 379
Mil-Std-1553B data bus, 407
Milwaukee, WI, 201, 357
Mineta, Rep. Norman (D-CA), 239–240
Ministry of Astronautics. See People's Republic of China.
Ministry of Trade and Industry, Finland, 397
Minnesota, 369
Minuteman Mark-12 reentry vehicle, 118
MIRACL. See mid-infrared chemical laser.
Mirage Jet, 145
mishap investigation board, 205
missiles, 117–120. (See also individual names.)
mission adaptive wing (MAW), 13–14, 16
mission peculiar equipment support structure (MPESS), 355, 373–374
mission sequence test, 374
Mission Watch, 277
Missouri, 369
MIT. See Massachusetts Institute of Technology.
Mitsubishi Electric Corp., 159–160
Mitsui and Co. Ltd., 346
Mitterrand, President Francois, 128–129, 139, 145, 251
Mixon, Marvin, 323
MLP. See mobile launch platform.
MMS. See multimission modular spacecraft.
MMSL. See Microgravity Materials Science Laboratory.
MMU. See manned maneuvering unit.
Mobile Laser Ranging System (MOBLAS), 7, 250

mobile launch platform (MLP), 287–288
mobile (multiple) satellite system (MSS), 122, 423
mobile subscriber equipment (MSE), 132
MOBLAS. See Mobile Laser Ranging System.
Mode S transponder, 144
Modular Transportation Laser Ranging System, 250
Moeschi, Stanley, 408
Mojave Desert, CA, 361
monkeys in space, 391
monodisperse latex reactor processor, 100
moon, 49, 244, 396. (See also lunar research.)
—rocks, 245
Moore, Jesse, 174, 312–313, 331–332, 334, 337, 343, 359–360, 363
MOP-1, 140
Morelos satellites, 329–330, 339
Morgan, Barbara, 272–274, 277
Mori, Mamoru, 163
Morison, Samuel, 124–125
Morton Thiokol Inc., 221, 367–369
—Wasatch Division, 20, 367
Moscow, USSR, 39, 115, 385, 389, 395, 399, 435
Mosquito Lagoon, FL, 191–192
Moss Point, MS, 371
motion sickness research, 152, 278, 323
Motorola Corp., 104, 423
Mount Illimani, Bolivia, 91
Mount St. Helens, WA, 76
MPESS. See mission peculiar equipment support structure.
MSE. See mobile subscriber equipment.
MSFC. See Marshall Space Flight Center.
MSL–2. See Materials Science Laboratory-2.
MSS. See mobile satellite system.
Mullane, Lt. Col. Michael, 278, 322
Mulloy, Lawrence, 368
multimission modular spacecraft (MMS), 77
Munich, Federal Republic of Germany, 336–337
Muroc, 101
Murray, Bruce, 435
Murray, Dr. Steven, 40
Musgrave, Dr. F. Story, 330
Mushotzky, Dr. Richard, 40

Mutual Broadcasting System, 217–218
—MultiComm, 217
—Mutual Satellite Services division, 217
Mutual Radio Network, 217–218
MV Kreuztrum, 92
MVSRF. See Man-Vehicle Systems Research Facility.
MX missiles, 118

N-II rocket, 163
Naderi, Dr. Firouz, 423
Nagashima, Japan, 17
Nagel, Lt. Col. Steven, 279–282, 335
Naito, Chiaki, 163
NAS. See National Airspace System.
NASA. See National Aeronautics and Space Administration.
NASCAP. See NASA, Charging Analyzer Program.
NASCOM. See NASA, Communications.
Nashville, TN, 429
Natal, Brazil, 73
Nathan, Dr. Robert, 102–103
National Academy of Engineering, 84
National Academy of Sciences, U.S., 39, 61, 75, 100, 149, 179, 259
—Astronomy Survey Committee, 51
National Advisory Committee for Aeronautics, 31, 310
national aeronautics research and development goals, 3

National Aeronautics and Space Administration (NASA)
—Air Force, 408–411
—Air Force Space Transportation Architecture Studies (STAS), 22
—Authorization Bill, 171
—Aviation Safety Reporting System, 82–83
—Board of Contract Appeals, 177
—budget, 168–171, 299
—Charging Analyzer Program (NASCAP), 28
—Communications (NASCOM), 247
—Deep Space Network (DSN), 63–64, 168, 173, 401
—Flight Review Board, 204
—Hardware Contractors Productivity Conference, 175
—National Space Technologies Laboratory (NSTL), 207, 209, 245–246, 252, 306, 309
—Navy program, 97–98
—1986 Long-Range Program Plan, 421–422
—Office of Aeronautics and Space Technology, 183
—Office of Commercial Programs, 173–175, 183, 202
—Office of Inspector General (OIG), 200–201
—Office of Interim Space Station Program, 185
—Office of Management Operations, 174
—Office of Space Flight, 183
—Office of Space Science and Applications (OSSA), 183, 246, 251–252
—Office of Space Station, 183, 185
—Office of Space Tracking and Data Systems, 183–184
—Seasat program, 217
—Scientific and Technical Information Facility (STIF), 201–202
—Solar System Exploration Division, 244
—Sounding Rocket Division, 243
—Space Flight Participant Committee, 272, 275–276
—Space Station Task Force, 185
—Teacher in Space Project, 271–273
National Agrarian University, Lima, Peru, 246
National Air and Space Museum (Smithsonian), 237–241, 251–252, 421
—Trophy, 102
National Airspace System (NAS), 197
National Association of Broadcasters, 276
National Bureau of Standards (NBS), 40, 427–428
National Cancer Institute, 209
—Breast Imaging Project, 209
National Center for Atmospheric Research, 114, 197
National Committee of the International Union of Radio Science, U.S., 180
National Council for Marine Resources and Engineering Development, 100

Index

National Earthquake Information Center, 133
National Geographic Society, 56
National Institutes of Health (NIH), 350
National Museum of American History, 237
National Newspaper Association, 276
National Oceanic and Atmospheric Administration (NOAA), 72–73, 149, 176, 197, 217, 221–228, 230, 253, 292, 344, 346–347, 433, 436
—satellites, 72–73, 217, 224–225
National Oceanic Satellite System (NOSS), 217
National Research Council (NRC), 84, 416
—Committee on Undersea Warfare, 179
National Research Council of Canada, 357
—space sciences budget, 73
National Science Foundation, 44, 149, 415
National Science Teachers Association, 369–370
National Scout Jamboree, 1985, Fort A.P. Hill, VA, 195
National Security Council (NSC), 151, 417
National Security Industrial Association, 317
National Security Launch Strategy directive, 436
National Society of Professional Engineers, 99
National Space Centre. See French National Center for Space Studies.
National Space Club, 440
National Space Commission, 440
National Space Institute (NSI), 420–421
—Space Outreach '85 program, 420
National Space Transportation System (NSTS), 310–312
National Telecommunications and Information Agency, 142
National Tire Modeling Program, 196
National Transportation Safety Board (NTSB), 83, 91
National Weather Satellite Week, 221
Naval Academy, U.S., 33–34
Naval Aviation Logistics Center, 179
Naval Electronics Systems Command, 437

Naval Intelligence Support Center, 124
Naval Material Command, 437
Naval Observatory, 38
Naval Research Laboratory (NRL), U.S., 114, 191, 233–234, 330, 437
Naval Space Command, 32–34
Naval Underwater Systems Center, 350
Navstar (Rockwell), 122–124, 141, 233–234
—Global Positioning System (GPS), 233
Navy, U.S., 10, 23, 111, 119, 124–125, 131–132, 170, 177–180, 183, 217, 228, 230, 234, 243, 297, 303, 323, 326, 333, 395, 437–438
—Bureau of Ordinance, 177
—espionage trial, 124
—Finance Center, 200
—Fleet Numerical Oceanography Center, 217
—Office of General Counsel, 178
—Office of Naval Research, 100
—Remote Ocean Sensing System (NROSS), 180–181, 217
—Space Program Office, 349
NBS. See National Bureau of Standards.
Nelson, Rep. Bill (D-FL), 273, 275, 356, 358–359, 363, 393, 435
Nelson, George, 230, 356
Neptune, 38, 51, 68
Netherlands, The, 53, 62, 135, 167, 251, 335–336, 398
Nevada, 133
Nevado del Ruiz (volcano), Colombia, 76
New Bedford High School, New Bedford, MA, 272
New Hampshire, 272
New Jersey, 86, 258
New London, CT, 350
New Mexico, 44, 100, 131, 182
New Orleans, LA, 16
New Works Program of the Massachusetts Council on the Arts and Humanities, 445
New York, NY, 16, 85, 93, 102, 223, 228
New Zealand, 148, 220, 392
Newfoundland, Canada, 96
Newman, Larry, 17
Newport Beach, CA, 438
Newton, Isaac, 275, 382

543

Nicogossian, Dr. Arnauld, 350
Nicollier, Claude, 282
NIH. See National Institutes of Health.
Nike-Orion, 73–74
Nike-Tomahawk, 71
Nimbus satellites, 77, 253
Nixon, President Richard M., 203
NOAA. See National Oceanic and Atmospheric Administration.
Nomura, Dr. Tamiya, 163
NORAD. See North American Air Defense Command.
Nordsieck, Dr. Kenneth, 283
Normal Freezing Furnace, 431–432
North, Warren, 31
North American Aerospace Defense Command, 439
North American Air Defense Command (NORAD), 87, 100, 364, 387
North American Aircraft Operations, 97. (See also Rockwell International Corp.)
North Carolina, 77
North Carolina State University, Raleigh, NC, 185
Northern Lights (radio waves), 68
Northern Utah Satellite (NUSAT), 205, 218–219, 374
Northrop Services, Inc., 11, 23, 190, 404–405, 440
Northwest Airlines, 93
Northwestern University, Evanston, IL, 42
Norway, 135, 139, 251
NOSS. See National Oceanic Satellite System.
NOVA, 228
NRC. See National Research Council.
NRL. See Naval Research Laboratory.
NROSS. See Navy Remote Ocean Sensing System.
NROSS satellites, 180–181, 217
NSC. See National Security Council.
NSI. See National Space Institute.
NSTL. See National Space Technologies Laboratory.
NSTS. See National Space Transportation System.
NTSB. See National Transportation Safety Board.
nuclear magnetic resonance (NMR), 209
nucleonics, 163

numerical aerodynamic simulation, 187
Nunamaker, Robert, 194–195
Nusantara, P.T. Elektrindo, 213
NUSAT. See Northern Utah Satellite.

Oberg, James, 397
Oberstar, Rep. James (D–MN), 90
Oberth, Herman, 152–153
O'Brien, John, 176–177
Ockels, Dr. Wubbo, 284, 335, 338
O'Connor, Bryan, 281, 338
Ocampo, Augusto Ramirez, 105
Occupational Safety and Health Administration (OSHA), 206
oceanography, 147. (See also National Oceanic and Atmospheric Administration.)
Office of Management and Budget, White House, 217, 269
Office for Official Publications of the European Communities, 138
Office of Science and Technology Policy, White House, 414–415
Ogilvie, Dr. Keith, 59
O'Gorman, George, 287
O'Hare International Airport, Chicago, IL, 85, 143
Ohio, 15, 369
Ohio State University, Columbus, OH, 417
OIG. See NASA Office of Inspector General.
Oklahoma City, OK, 16
Omniplan Corp., 190, 314
OMS. See orbital maneuvering system.
OMV. See orbital maneuvering vehicle.
one-hundredth American to reach space, 281–282
O'Neill, Gerard, 440
Onizuka, Maj. Ellison, 274, 278, 295
Onsala, Sweden, 396
Opal, Chet, 67
optic electronics, 251
opto-electronic multispectral scanner, 432
orbit
 —geostationary, 105, 136, 140, 220, 222–224
 —geosynchronous, 18–22, 157, 159–160, 168, 171, 214, 220–221, 227, 231–232, 423

Index

—north-south, 295
—sun-synchronous, 72
orbital maneuvering system (OMS), 299
orbital maneuvering vehicle (OMV), 267, 420, 422
Orbiter Processing Facility, 288
orbiting astrometric telescope, 37
origin of life theory, 187–188
Orion rocket, 74–75
Orlando International Airport, Orlando, FL, 293
Osaka, Japan, 96
OSHA. See Occupational Safety and Health Administration.
Oshkosh, WI, 85–86
OSSA. See NASA Office of Space Science and Applications.
Ostriker, Dr. Jeremiah, 42
Ostro, Steve, 54
Oswald, Stephen, 33
OTA. See Congressional Office of Technology Assessment.
Ousley, Gilbert, 78
Overmeyer, Col. Robert, 86, 325–328
Owen, David, 288
Owens Valley, CA, 44
ozone, 73, 75, 76, 149
—depletion of, 75
—layer, 75–76
—stratospheric, 75
Ozonesonde, 73

Pacala, Dr. Thomas, 207
Pacific American Launch Systems, 430–431
—PHOENIX E, 431
Pacific Missile Range Facility, 131-132
Pacific Ocean, 16, 53, 77-78, 114, 129, 136, 161, 220-221, 234, 291
Padova University, Instituto Meccanica, 248
Pailes, Maj. William, 319, 335
Paine, Thomas, 440
Palapa B-2, 19, 33, 223
Palestine, TX, 79
Palestinians, 111
Palmer, Patrick, 57
Palmdale, CA, 240, 362
Palomar Observatory, 64
PAM. See Payload Assist Module.
Pan Am World Services, 292

Pan American World Airways, 87
Pan American World Health Organization, 232
Pan Pacific education and communications experiments by satellite (PEACESAT), 220
Panama, 34–35
Papadopoulos, Konstantinos, 248
Paris, France, 17, 157, 173, 434
Paris Air Show, 31, 145, 213, 260, 266, 405
Parise, Dr. Ronald, 283
Parkersburg, WV, 272
Parliament, British, 133
Parma, Ohio, 369
Pasadena, CA, 133
Patrick AFB, FL, 292
Pattie, Geoffrey, 255-256
Payload Assist Module (PAM), 19–20, 167, 228, 299, 303, 339–340, 356
payload ground operations contract (PGOC), 193
Payload Operations Control Center (POCC), 285, 375, 378
Payton, Maj. Gary, 295
PCG. See protein crystal growth experiment.
Peace Corps, U.S., 91
PEACESAT. See pan Pacific education and communications experiments by satellite.
Peale, Stanton, 67
Pedersen, Kenneth, 177
Peebles, OH, 15
Peenemunde, Federal Republic of Germany, 152
Pei, I.M., 99
Pell, Sen. Claiborne (D-RI), 413
Pellinen, Risto, 396
Penaranda, Frank, 173–174
Pennsylvania State University, State College, PA, 40, 258, 354
Pentagon. See Defense, U.S. Department of.
People's Republic of China, 213–215, 434
—Ministry of Astronautics, 213
—State Bureau of Meteorology, 215
—State Council, 213
Pepsi–Cola USA, 443–445
perihelion, 109
Perini Co., 187

545

Perkin-Elmer Corp., 53–54, 357
Perry, Geoffrey, 397
Perry, Leon, 203
Perseus Cluster, 329
Peru, 77–79, 246, 252
Petersen, Ronald, 288
Peterson, Donald, 34
Peterson AFB, Colorado Springs, CO, 438
PFRR. See Poker Flat Research Range.
PGME. See propylene glycol methyl ether.
PGOC. See payload ground operations contract.
Phase Partitioning Experiment (PPE), 351–352
Philadelphia, PA, 258
Philadelphia University City Science Center, Bioprocessing and Pharmaceutical Center, 353
Philadelphia World Affairs Council, 132
Phillips, Dr. Robert, 279
Phobos, 137, 396, 435
PHOENIX E, 431
Photovoltaic power system, 208–209, 212, 270
physical vapor transport of organic solids (PVTOS), 303
Pierce, Dr. William, 271
Pike, John, 112, 125
PIMS. See programmable implantable medication system.
Pioneer, 61–62, 65, 180, 412
pivoting wing, for supersonic aircraft, 97–98
Planet A, 48
Planet X theory, 38
Planetary
 —detection and study, 38
 —exploration, 52, 61–69, 170
 —flyby spacecraft, 51, 61, 180
 —formation, 37, 39, 41, 43
 —image processing, 102
 —science, 149
 —systems, 39, 197
Planetary Society, the, 434
Planning Research Corp. (PRC), 192
plant experiments, 358, 446
Plasma Diagnostics Package, 28, 354, 381
Plasma Wave Experiment, 59
PLSS. See precision/location strike system.

Pluto, 51, 63–64, 68
POCC. See Payload Operations Control Center.
Point Brown, U.S. Coast Guard Cutter, 74
Poker Flat Research Range (PFRR), AK, 75
Poland, 392, 413
POLAR. See Potential of a Large Object in the Auroral Region.
polar cap studies, 71
Pollack, Dr. James, 66–67
Pollock, Dr. Glenn, 188
Pollvogt, Udo, 432
Portugal, 251
Potential of a Large Object in the Auroral Region (POLAR), 27–28
PPE. See Phase Partitioning Experiment.
Pratt & Whitney, 11, 22, 81, 143, 204
Pravda, 397
PRC. See Planning Research Corp.
precision location/strike system (PLSS), 25
President's Committee on Science and Technology, 183
President's Council on International Economic Policy, 33
President's General Advisory Committee on Arms Control and Disarmament, 177, 179, 185
Presidential National Commission on Space, 417
Presidential Proclamation 5358, 403
Presler, Alden, 196
Princeton, NJ, 258
Princeton University, Princeton, NJ, 42
Probst, Albert, 376
Prognoz-10–Intercosmos, 396
programmable implantable medication system (PIMS), 99
Progress-24, 387, 391–392
Project Mercury, 31. (See also Mercury Program.)
Propeller propulsion systems, 14, 97
propfan, 15, 84
Propylene glycol methyl ether (PGME), 8–9
protein crystal growth experiment (PCG), 352–353, 358
Prouty, Clarke, 219
Proxmire, Sen. William (D-WI), 405
Pshak, Mary, 131

Index

PSN/CNR. See Italian National Research Council, National Space Plan Office.
P.T. Elektrindo Nusantara, Indonesia, 212
Pterosaur, 237
Public Broadcasting System, 151
Public Health Service, 220
Puerto Rico, 54, 236, 271, 330, 379
Purdue University, West Lafayette, IN, 185, 248
PVTOS. See physical vapor transport of organic solids.
Pyle, Thomas, 85

Quality Circles (NASA), 174
quasars, 283

R&D. See Research and Development.
R&D Associates, 185
radar, 438–439
 —beacon system, 144
 —space–based radar system, 437, 439
radar ocean–surveillance satellite (RORSAT), 386–387
Radford, Wade, 99
Radhakrishmam, P., 284
Radio Astronomy Explorer satellite program, 180
radio astronomy technique, 63
Radio Technical Commission for Aeronautics, 141
Radio Television News Directors Association, 276
Radio Wave Experiment, 59
radiosonde instruments, 71
RADL. See Robotic Applications Development Laboratory.
Rahn, Debra, 203
Ramenskoye Flight Test Center, USSR, 389
ramjet, 28, 96
Ramsey, Rear Adm. William, 34
Rand Corp., 176, 179, 181
Rand, Maj. Ronald, 321–322
Ranger spacecraft, 103
Rankine, Brig. Gen. Robert, Jr., 130
Rao, U.R., 155, 284
Rassier, Donald, 438
Ratajczak, Tony, 209

RCA American Communications (RCA Americom), 340
RCA Astro–Electronics, 258, 281
RCA Corp., 225, 430, 433
RCA G-STAR satellite, 221
RCA Infrared–Imaging Equipment (IR-IE), 356–357
RCA SAT-COM satellite, 20, 276
 —C-band, 281, 340
 —F-2R, Transponder 13, 276
 —KU-1, 281, 356
 —KU-2, 278, 339–340, 356
Reagan, President Ronald W., 1, 29, 31, 85, 99, 101, 111–113, 126–129, 168–170, 176, 179, 181, 200, 221, 256, 264, 272, 282, 403, 414, 417, 421, 433, 436–437, 440
 —Administration, 7, 97, 112–113, 118, 121, 126–128, 132, 158, 178, 183, 229, 251, 255, 311, 315, 409, 414, 417
 —National Commission on Space, 52, 417
 —National Security Launch Strategy directive, 436
 —National Space Strategy, 417
 —1980 Presidential Campaign, 185
real-time multiprocessor simulator, 26
recorders, flight data, 83, 87, 91
Red Spot, 377
Redondo Beach, CA, 258
Redmond, Lt. Col. Thomas, 312
Redstone rocket, 152
Reeves, Bill, 333
remote manipulator system (RMS) arm, 257
remote sensing technology, 246
Renaissance, 238
rendezvous and docking, 152
Rensselaer Polytechnic Institute, Troy, NY, 18
Renz, David, 104
Reppin, Dr. Claus, 399
Request for Proposal (RFP), 189, 369
Research Animal Holding Facility, 383
research and development (R&D), 174, 414–416, 437
 —budget, 414, 440
 —facilities, 85
 —space-related computer science, 186–189, 198
 —weapons systems, 23

547

Resnick, Judith, 274, 278
Reynolds, William, 175
RFP. See Request for Proposal.
Rhyolite reconnaissance satellite, 315
Richards, Richard, 278
Richelson, Jeffrey, 124–125
Rickman, Doug, 209
Ride, Sally, 102, 151, 281
Rider, Dr. David, 207
Riesenhuber, Heinz, 255
Rinker, Lewis, 200–201
Rio Abiseo National Park, Peru, 245–246, 252
—Research Project, 246
RMS. See remote manipulator system arm.
RMS Associates, 200, 202
RMS Technologies, Inc., 190, 314
Robotic Applications Development Laboratory (RADL), 193–194
—adaptive control, 194
—artificial intelligence, 194
—closed-loop control commands, 194
—control center for robotics, 193
—real-time tracking system, 194
—servo-controlled robot, 194
—smart systems, 193
—3-D graphics and simulated work cells, 194
—touch and mobility systems, 194
Robotic Control Systems, 194
robotic development, 193
rockets. See names of individual rockets.
ROCket OZonesonde (ROCOZ-A), 73
Rockwell International Corp., 11, 97, 103, 141, 191, 240, 243, 258, 298, 362–364, 419
—North American Aircraft Operations, 97
—Rocketdyne Division, 258, 303, 309
—Space Station Systems Division, 258, 264
—Space Transportation Systems Division, 199
Rockwell Shuttle Operations Co., 190, 313
Rodgers, Dave, 48
Rogers, Tom, 261–262
Roland, Alex, 151
Rolls Royce, 84, 414
Romania, 392
Rome, Italy, 135

Rome, University of, Italy, 248
Rome Air Development Center, 122
RORSAT. See radar ocean-surveillance satellite.
Ross, Maj. Jerry, 278, 322, 338–342, 355
rotary power-transfer device, 104
rotary-wing flight, 5–7
Rothmeier, Steven, 93
rotorcraft, 5–7
Roussel Pharmaceuticals of Paris, 431
Royal Greenwich Observatory, England, 45
Rudman, Sen. Warren (R-NH), 405
Rye, Gilbert, 417

S-207C launch vehicle, 411
Sadilek, Albert, 99
Safer, John, 102
Sagan, Dr. Carl, 39, 320, 435
SAGE II, 73
Sagittarius, constellation, 45
Sahm, Peter, 380
St. Louis, MO, 16, 209, 409
St. Mary's Hospital, Laser Laboratory, Milwaukee, WI, 357
Sakharov, Andrei, 413
Salvator, Giorgio, 137
Salyut mission, 329, 392, 397
Salyut space stations, 388–395, 397–399, 433
Salyut-Hexe project, 399
San Jose State University, San Jose, CA, 187
Sanchini, Dominick, 303
SAREX. See shuttle amateur radio experiment.
Sargent, E. Douglas, 288, 294
SARSAT. See satellite carried search and rescue.
Satcol satellite, 105
satellite, 121–124, 159–161, 217–234, 411–412, 417, 430, 441–442.
(See also individual names; Appendix A.)
—debris, reentry, 87
—early warning, 315
—earth-orbiting, 171
—electronic car map, 123
—failures, 233
—keyhole, 124–125

—meteorological, 155
—Navy, 326
—on-orbit repair, 440
—radar-ocean surveillance, 386–387
—reconnaissance, 123, 314–315
—science, 163
—Soviet, 386–399, 417, 419
—spy, 315–316, 395
—weather, 163, 214, 221, 224, 229–230, 436
satellite carried search and rescue (SARSAT), 225
satellite, communications, 33, 103–104, 155, 211, 232, 419
—based communications network, 2
—commercial communications, 19, 159, 419
—international communications, 160, 213
—Navy, 417
—planetary communications, 197
—telecommunications, 140, 157–161
Satellite Communications for Learning Worldwide (SCOLA), 370
Satellite Television Corp., 107
Satellite Transfer Vehicle, 22
Saturn, 45, 51, 65, 67, 68, 214, 307
Saturn V/Apollo launch, 294
Saudi Arabia, 167, 219, 280, 282–283
Savinykh, Viktor, 389, 391–393, 395, 397
SCA. See Shuttle Carrier Aircraft.
Scandinavia, 71
SCAPE. See self-contained atmospheric protective ensemble.
Scarab 2 underwater search and recovery vehicle, 92
Scarburgh, Col. Edmund, 200
Scarf, Dr. Frederick, 59
SCAS. See stability control augmentation system.
Schmitt, Sen. Harrison (R-NM), 435
Schneider, Ed, 10
Schoenewolf, 1st Lt. John, 123
Scholossberg, Stephen, 178
Science Applications, Inc., 248
Scientific-Technical Cooperation Committee of the Foreign Ministry, Finland, 397
scientists, engineers, and computer (SEC) specialists, 415
SCO–X–2. See Scorpius X–2.

Scobee, Francis, 274, 278
Scorpius X–2 (SCO–X–2), 329
Scott AFB, IL
—Rescue and Command Center, 228
Scout Science and Technology, Inc. (SST), 21
Scout launch vehicle, 167, 410–411
—S-207C, 411
scramjet. See supersonic combustion ramjet.
Scranton, University of, PA, 311
Scully-Power, Dr. Paul, 350, 353
SDC-Burroughs, 144
SDI. See Strategic Defense Initiative.
SEADS. Shuttle Entry Air Data System.
Search for Extraterrestrial Intelligence (SETI), 235–236
Seasat Imaging Radar (SIR-B), 147, 217
Seattle, WA, 90
Seattle-Tacoma Airport, 143
SEC. See scientists, engineers, and computer specialists.
Secret Service, 82
Seddon, Dr. M. Rhea, 278–279, 296, 322–324, 351
Selby, Barbara, 203
Selective Service System, 178
Selenia, Italy, 159–161
self-contained atmospheric protective ensemble (SCAPE), 66
semiconductor crystals, 165
—directional solidification, 165
Senior Executive Service designation, 311
Seoul, South Korea, 93, 405
servomechanism, 103
SETI. See Search for Extraterrestrial Intelligence.
Sever, Tom, 252
SFMD. See storable fluid management demonstration.
Shamsin, V.A., 158
Shanghai, People's Republic of China, 93, 434
Shanghai Institute of Satellite Engineering, 214
Shanghai Institute of Technical Physics, 214
Shaw, Lt. Col. Brewster, Jr., 293–294, 339–341
Shepard, Alan B., Jr., 151
Shergotty, Consortium, 244

Shklovsky, Iosif S., 39
short takeoff and landing (STOL) capabilities, 117
Shrewsberry, Dave, 247
Shriver, Loren, 281, 295
Shroter, A.R., 288
shuttle amateur radio experiment (SAREX), 421
Shuttle Carrier Aircraft (SCA), 360, 362. (See also Boeing 747.)
Shuttle Entry Air Data System (SEADS), 364–366
shuttle infrared leeside temperature sensing (SILTS), 364, 366
Shuttle Inventory Management System (SIMS), 310
Shuttle Operations Strategic Planning Group, 311
Shuttle Processing Contract (SPC), 310
Shuttle Student Involvement Project (SSIP), 358
shuttle upper atmospheric mass spectrometer (SUMS), 365–366
Sicily, Italy 111
Siddon, The Honorable Tom (Canadian Minister of Science and Technology), 261, 265
Sieck, Robert, 306, 312
sigint (signals intelligence), 295
Sigonella Air Base, Sicily, Italy, 111
SILTS. See shuttle infrared leeside temperature sensing.
Silveira, Milton, 210
Simeiz, Crimea, USSR, 396
SIMS. See Shuttle Inventory Management System.
single rotation tractor propulsion technology, 15
SIR-B. See Seasat Imaging Radar.
Skantze, Gen. Lawrence, 441–442
sky marshal force, 85
Skylab, 31, 172, 257, 340, 354
Slayton, Donald, 425–426, 434–435
SLBM. See submarine-launched ballistic missile.
SLS. See Spacelab Life Sciences.
Small Business Innovation
—Development Act of 1982, 202
—high technology, 202
—research program, 202–203
SME. See Solar Mesospheric Explorer Satellite.

Smith, Arthur, 81
Smith, Dick, 294, 376
Smith, Dr. Edward, 60
Smith, Dr. L.G., 74
Smith, Michael, 274, 278, 282
Smithsonian Astrophysical Observatory, 40, 248, 440
Smithsonian Institution, 102, 237–241, 251–252, 421. (See also National Air and Space Museum.)
—Associates lecture, 55
—Board of regents, 239
Snead, Leon, 201
Society Expeditions, 430
Society of Experimental Test Pilots, 10
Society of Professional Journalists/Sigma delta chi, 276
solar
—activity, 114
—chronosphere, 47
—dynamic system, 270
—flares, 435
—optical universal polarimeter, 47
—panels, 397
—power satellite, 18
—system, 68, 140, 149, 235, 245
—system, life beyond, 235
—telescope, 47
—terrestrial relations, 100
—ultraviolet rays, 75
—wind, 59–60
Solar Max, 230, 238
Solar Maximum Mission satellite, 223
Solar Mesospheric Explorer (SME) satellite, 73
Solar Optical Telescope, 42
Solar System Exploration Committee, 61
Solar Ultraviolet Universal Polarimeter (SOUP), 378, 381
solid fuel rocket booster (SRB), 367–369
solid rocket fuel, 323
Solwind satellite, 114
Sondre Stromfjord. See Greenland, Sondre Stromfjord facility.
SOOS–1 spacecraft, 228–229
Sophron Foundation, McLean, VA, 421
sound barrier, 151
sounding rockets, 73–74, 76, 132, 200
SOUP. See Solar Ultraviolet Universal Polarimeter.
South America, 231
South Atlantic, 386

Index

South Bend, IN, 75, 77
South Carolina, 56
South Korea, 405
South Pole, 218
Southern California, University of
 —School of International Relations, 185
Southwestern Louisiana, University of, Lafayette, LA, 38
Southwestern Medical School, Dallas, TX, 279
Soviet
 —Academy of Sciences, 39, 100, 136, 394, 435
 —cosmonauts, 388, 391–399
 —Geophysical Committee, 100
 —laser weapons, 316, 390
 —missile system, 126–127
 —rocket, 385
 —secret launch, 387
 —socialist system, 151
 —space agency, 394
Soviet Military Power, 385
Soviet Union. See Union of Soviet Socialist Republics (USSR).
Soyuz T–6, 41, 434
 —T–13, 389, 397
 —T–14, 394, 397
 —frequency, 397
Soyuz-Apollo mission, 413
space
 —activity, history of, 151–153
 —astronomy, 140
 —camp, 249
 —colonies, 39
 —commercialization, 141, 417, 419
 —probe, 46–47
space, daily life in
 —habitability, 267
 —motion sickness, 152, 323, 395
 —movement, 274
 —personal hygiene, 274
 —preparation of food, 274
 —sleep, 274
 —use of leisure time, 274
Space Activities Commission, 163
Space Command (SPACECOM), 410, 438–439
Space Communications Co. (Spacecom), 1–2, 20
Space Congress, 109
Space Exploration Day, 403

Space Flight Participant Committee. See NASA Space Flight Participant Committee.
space goals, United States, 261
 —long-range civilian space goals, 261
Space Human Factors team, 267
Space Infrared Telescope Facility, 40
Space Industries, Inc., 33, 428
Space Outreach '85, 420
space plans, future, 151–152, 167, 195, 440
space platform, 189
space program, manned, 151, 200
space radio astronomy, 180
space reactor power program, SP-100, 412–413
 —Phase I, 412
 —Phase II, 412
Space-related computer-science technology research, 186
Space Research Association, 440
Space Research Chair, 34
Space Research Institute. See Soviet Academy of Sciences.
space sciences
 —food processing research, 446
 —future directions for research, 259
 —plant growth research, 446
 —waste management research, 446
Space Services, Inc., 425–426
Space Shuttle, 17–18, 20, 31–32, 34, 50, 52, 73, 76, 100, 120, 126, 137, 146, 151–153, 155, 214, 221–222, 227, 234, 241, 248, 257–258, 263, 279, 296–297, 299–301, 303–306, 308, 310, 312, 315–319, 321–326, 330–350, 353–355, 357–365, 368–371, 385, 388, 392, 403, 409–410, 414, 417–418, 421, 425, 428–433, 435, 441, 443, 445. (See also shuttle; Space Shuttle missions.)
 —abort, 302, 307–308
 —based photos, 56
 —communications, 2
 —crew, 48, 223, 274
 —deployed satellites, 22, 24, 27, 28, 29, 33
 —docking, 152
 —emergency landing, 290
 —first private citizen passenger, 272–277
 —flight assignment, 243

551

—flight crew equipment, 190
—foreign competition for launches, 313
—launch, 267, 421
—launch countdown, 312
—maintenance and launch preparation, 312
—manifest, 295, 301, 425
—military missions, 315
—National Selection Committee, 276
—operations, 311, 314, 427
—payload, 302, 310, 312
—robot arm, 59
—Student Involvement Project experiments, 275
—thermal protection tiles, 296, 298, 300, 425
—weather officer, 293
—workload, 13
Space Shuttle Atlantis, 167, 278–279, 318–319, 321. (See also Space Shuttle orbiters.)
Space Shuttle Carrier Aircraft (SCA), 360, 362
Space Shuttle Centaur, 170, 192, 197, 204, 300
—Centaur integrated support structure (CISS), 21
—development costs, 169
—G, 169
—rolling beam, 294
—upper stage rocket, 62, 159, 169–170, 281, 290, 294
Space Shuttle Challenger, 31–32, 47, 60, 131, 147, 152, 218, 252, 276–277, 287, 293–298, 301–303, 321, 326–338, 353, 359–362, 366–367, 377–382, 403, 421, 425, 434, 444–445. (See also Space Shuttle orbiters.)
Space Shuttle Columbia, 56, 151, 163, 168, 238, 278, 281, 302, 305, 356, 361–362, 364–366, 376, 444
—U.S.'s first space shuttle, 238
—maiden flight, 278
Space Shuttle Discovery, 33, 121, 129, 131, 205–206, 225, 227, 230–231, 233, 250, 281, 287, 295, 298, 300, 302–304, 315–317, 321–325, 328–329, 332–334, 339, 359–362, 370–371, 415. (See also Space Shuttle orbiters.)

Space Shuttle Enterprise, 240. (See also Space Shuttle orbiters.)
Space Shuttle main engine (SSME), 22, 301
—1, 330–331
—2, 301
Space Shuttle missions, 17, 32–34, 42, 102, 204, 219, 223–227, 230–231, 247, 250, 275–284, 293–309, 319, 321–325, 330–343, 349, 354–358, 363, 370–371, 376, 378–381, 417, 421, 431, 434, 443
Space Shuttle orbiters, 422, 425
—Atlantis, 167, 278–279, 284, 293–294, 300, 309, 319, 334–335, 338–343, 360
—cabin, 99
—Challenger, 296–297, 299, 425
—Columbia, 278
—Discovery, 33, 298–299, 314, 359
—Enterprise, 240, 286
—Marshall Star, 155
space stations, 1–7, 17–18, 26, 109, 135, 199, 259, 262, 264–266, 432–433. (See also Salyut.)
—automation, 186–187, 262
—center of gravity, 268
—Columbus, 263
—efficiency, 260
—geosynchronous, 167
—manned, 135, 168–169, 255, 264–268, 327, 385, 417–418
—Phase B, 261, 265–266
—program, international, 164
—settlements, 17
—Soviet, 388–394, 433
—SPARX, 432
Space Systems Research Chair, 33, 34
Space Task Group, 310
space technology transfers, 206–212
space telescope data capture facility (ST DCF), 247
Space Telescope Science Institute, 40, 247
Space Transportation Architecture studies, 22
Space Transportation Main Engine (STME) study, 22, 301
Space Transportation System (STS), 183, 190–191, 193, 313–314, 343, 348, 355–356

Index

—missions, 32–34, 60, 218, 243, 279, 297–300, 305–306, 334, 351–352, 362, 373
space walk, 397, 440
Space and Warfare Systems Command, 437
SPACECOM. See space command.
Spacecom. See Space Communications Co.
SPACEFLIGHT (Public Broadcasting System), 151
Spacelab, 99, 135, 138–139, 167, 193, 219, 300, 326–327, 335–337, 350, 361, 373–383
—animal flights, 327
—development, 99
—European design, 99
—module, 285
—pallet, 285
Spacelab Life Sciences (SLS) missions
—1, 279, 285, 362, 373
—2, 280, 330–331, 374, 377–379, 381
—3, 285, 325, 328, 373–374, 378
—4, 278
—12, 328
—D-1, 284, 335–338, 375–376, 380
—D-2, 284
Spacelab Payload Project Office, 377
Spacelab Verification Flight Test (VFT), 379
Spacenet Corp. See GTE.
spacesuits, 167, 190
Spacetrack, 121
Spain, 44, 135, 148, 251, 336
Spartan 1, 247, 328–329
Spartan-Halley, 58, 59, 60
SPARX, 432
SPC. See Shuttle Processing Contract.
Speakes, Larry, 113, 181
Sperry Corp., 86
Sperry Flight Systems, 104
Spitznagel, Dr. Ed, 209
Spring, Lt. Col. Sherwood, 339–342, 355
Springer, Lt. Col. Robert, 278
Sputnik 1, 151
SRB. See solid fuel rocket booster.
SRI International, 186, 260
SRM. See Standard Reference Material.
SSIP. See Shuttle Student Involvement Project.

SSME. See Space Shuttle Main Engine.
SS–N–20 missile, 385
SS–NX–23 missile, 385
SST. See Scott Science and Technology, Inc.
SS–24, SS–25 missiles, 385
ST DCF. See space telescope data capture facility.
stability control augmentation system (SCAS), 11–12
Stafford, Thomas, 434–435
Stallings, William, 247
Standard Reference Material (SRM), 428
Stanford University, Stanford, CA, 177, 179, 186, 210, 248
—Invitational Rugby Tournament, 211
star formation, 42–43
Star 30C, 221
Star Wars. See Strategic Defense Initiative.
STAS. See NASA/Air Force Space Transportation Architecture Studies.
State, U.S. Department of, 426
Stearns Catalytic Corp., 306
Steele, Dr. Charles, 210–211
Steers, Louis, 11
Steimle, Hans-Ulrich, 284
Steinberg, Morris, 416
Steinetz, Bruce, 103
Stevens, Rep. Ted (R-AR), 404
Stewart, Col. Robert, 279, 319, 335
STIF. See NASA Scientific and Technical Information Facility.
Stirling cycle engine, 412
STME. See Space Transportation Main Engine.
Stockholm, Sweden, 149
Stofan, Andrew, 104
STOL. See short takeoff and landing capabilities.
Stone, Nobie, 248
Stone, Randy, 296
storable fluid management demonstration (SFMD), 349
Stover, Col. Thomas, 12
Strategic Defense Initiative (SDI), 28–29, 109, 125–133, 145, 151, 251, 318, 390, 412, 434, 441
—budget, 126–128
—commercialization, 427
—organization, 329
—research, 364, 386, 434, 438
—Space Shuttle activity, 311

Astronautics and Aeronautics, 1985

Strategic Defense Office, 109
Stratton, Rep. Samuel (D-NY), 116, 405
Stratton Amendment, 405–406
STS. See Space Transportation System.
STSOC. See Space Transportation System Operations Contract.
Study of Paper Fiber Formation in Microgravity, 358
submarine-launched ballistic missile (SLBM), 385. (See also missiles.)
 —SS–N–20, 385
 —SS–NX–23, 385
submarines, 385
 —Delta IV class, 385
 —Typhoon class, 385
subsonics, 3, 7
Suitland, MD, 124, 223, 253
Sullivan, Dr. Kathryn, 31–32, 102, 282, 300, 440
Sullivan, Thomas, 184
SUMS. See shuttle upper atmosphere mass spectrometer.
sun, 21, 47–48, 55, 65, 68–69, 72, 75, 109, 244, 252
Sundstrand Corp., 270
Sunnyvale, CA, 258
sun-synchronous orbit, 72
super-loki datasonde dart, 73
supercomputer, 187–189, 256
supernovas, 39, 45–46, 399
supersonic combustion ramjet (scramjet), 4, 28, 96–97
supersonics, 3–4, 28–29, 96–98, 199
Suwon Air Base, Seoul, South Korea, 405
Sverdrup Technology, Inc., 221
Swartz, T.C., 430
Sweden, 74, 135, 251, 396
Swedish Bofers, 438
Switzerland, 63, 72, 135, 251, 336
Sydney, University of, Sydney, Australia, 322
Symbolics Inc., 186
synchrotron radiation, 39
Syncom satellite, 227, 297, 361. (See also satellite.)
 —I, 232
 —II, 232
 —III, 232, 234
 —IV, 231, 233, 299
 —IV-3, 221, 323
 —IV-4, 334
 —IV-5, 275

System Development Corp., 190, 200, 314
System Management American Corp., 190, 314

T–38 aircraft, 34
Taam, Dr. Ronald, 42
tactical electronic warfare intermediate support system (TISS), 408
tactical warfare system (TEWS), 408
Taft, William Howard, IV, 23, 124
Takeuchi, Reiichi, 264, 266
TAL. See trans-Atlantic landing sites.
Tasmania, Australia, 131
Tasmania, University of, Australia, 172
Tass News Agency, 137, 387–388, 391–392, 394–396
Taurus-Nike-Tomahawk rocket, 76
Taurus-Orion rocket, 71, 74
Taurus-Tomahawk rocket, 71
Taylor, Edward, 3
TDRS. See Tracking and Data Relay Satellite.
TDRSS. See Tracking and Data Relay Satellite System.
Teal Ruby experimental space-based aircraft detection system, 129, 317
Technology Development of California, 188
Tedesco, Dr. Edward, 64
Teledyne Brown Engineering Co., 395
Teledyne Industries, Inc., 20
Telesat Canada, 296, 299
telescope, orbiting astrometric, 37
Telestar 3-D, 330
Tempel 2, 50–52
Tennessee, University of, Space Institute, Nashville, TN, 19
Terrier-Malemute rocket, 71, 131–132
tethered satellite system (TSS), 247–248
TEWS. See tactical warfare system.
Texas, 44, 64, 79, 207, 272, 416
Texas, University of, 63, 67, 167, 184, 186, 416
 —Health Science Center, Southwestern Medical School, Dallas, TX, 279
Texas A&M, College Station, TX, 167
TF–34 engine, 5
TF–39 engine, 12, 24
Thagard, Dr. Norman, 278, 282, 375

Index

Thatcher, Prime Minister Margaret, 132–133
thermal dust emission, 58
Thomas Jefferson School for Science and Technology, Alexandria, VA, 272
Thomas, Gene, 338
Thomas, John, 378
Thomas, Peter, 67
Thompson, Dr. Roger, 58
Thomson-CSF, 159, 161
Thorne, Lt. Comdr. Stephen, 33
Thornton, William, 375
3AAL. See Managed Three-Axis Acoustic Levitator.
3M Corp., 303, 431
—Material Sciences Laboratory, 3, 278
—Riker Laboratories Division, 431
Thuot, Lt. Pierre, 33
THURIS. See human role in space study, the.
Tiamet rocket, 199
—fighters, 23, 404–406
Tigershark. See F-20 Tigershark.
tilt-rotor aircraft, 11–12, 15. (See also V-22 and XV-15 aircraft.)
Tiros weather satellites, 214
TISS. See tactical electronic warfare intermediate support system.
Titan, 51, 67, 123, 131, 217, 344, 347, 410, 436
Titan-probe/Saturn orbiter, 51
Titan-II intercontinental ballistic missile (ICBM), 217, 344, 347, 436
Titan-II rocket, 131
Titan-34D7, 123, 410, 436
Tokyo, Japan, 87, 93, 96, 163, 264
TOMS. See Total Ozone Mapping Spectrometer.
Torr, Dr. Marsha, 285
Total Ozone Mapping Spectrometer (TOMS), 76
Toulouse, France, 146
Townes, Charles, 55–56
Townsend, John, 425
TPL. See trans-Pacific landing site.
TR-1 aircraft, 25
tracking network, international, 173
Tracking and Data Relay Satellite (TDRS), 168, 247, 277–278, 296–300
Tracking and Data Relay Satellite System (TDRSS), 171

trans-Atlantic landing (TAL) sites, 289
trans-Pacific landing (TPL) sites, 289, 291
transatmospheric vehicles, 3–4, 7, 28–29
Transit satellite, 228–229, 412
Transpace Carriers, Inc., 345
transport-aircraft flight efficiency, 4
Transportation, U.S. Department of (DOT), 95, 185, 199, 345–346, 415, 426
TRMM. See Tropical Rainfall Measuring Mission.
Trofimenko, Henry, 390
Tropical Rainfall Measuring Mission (TRMM), 252–253
Truly, Commodore Richard, 33–34, 240
TRW, Federal Systems Division, 258, 260, 420
Tsiolkovsky, Konstantin, 152
TSS. See tethered satellite system.
Tucson, AZ, 37, 58
Tuebingen, University of, Federal Republic of Germany, 398
turbofan, 10, 14
turboprop, 10, 84
turbulence, 71, 74, 83
Turkey, 250–251
Tuskegee University, AL, 446
Typhoon missiles, 385
Tyuratam, USSR, 387, 389, 435

UARS. See Upper Atmosphere Research Satellite.
Uchinoura, Japan, 163
UCS. See Union of Concerned Scientists.
UDF. See Unducted Fan demonstrator engine.
UKS. See UK Subsatellite.
UK Subsatellite (UKS), 401
Ultimate Field Trip, the, 274
ultraviolet radiation detection, 48, 75–76, 244
Ulysses mission, 20–21, 62, 169, 197, 252, 280–281, 290, 300, 412
UN. See United Nations.
Unducted Fan (UDF) demonstrator engine, 15
Unesco, 222

Unified Space Command, 440, 442
Unimation Puma manipulator, 411
Union of Concerned Scientists (UCS), 125–126
Union of Soviet Socialist Republics (USSR), 23, 41, 100, 109, 112–115, 120–121, 125, 127, 130, 137, 151, 155, 158, 220, 222, 228, 255, 268, 283, 295, 315–316, 319, 329, 385–399, 413, 419
—Academy of Sciences, 100
—Biomedical Problems Institute, 393
—embassy in Helsinki, Finland, 385
—Glavkosmos, 394
—military power, 385
—space plane, 388
—Space Research Institute, 65
—Venera-Halley space probe, 63
United Aircraft, 81
United Airlines, 93–94
United Kingdom, 48, 53, 59, 72, 75, 78, 84, 89, 97, 119, 135, 137, 143, 159, 214, 251, 255–256, 263–264, 398, 401, 414
—Department of Trade and Industry, 133
—Bureau of Ordinance, 177
—Information Technology, 255
—Kettering Group, 397
—Meteorological Office, 72, 159–160
United Nations (UN), 105, 122, 296, 440
—Legal Subcommittee for Outer Space, 105
United States-United Kingdom Joint Working Group on Atomic Weapons, 179
United Technologies Corp., 81
—Chemical Systems Division, 368
—Hamilton Standard Division, 15
—Norden Systems Division, 439
—United Space Boosters, Inc.'s Booster Production Co., 366
upper air wind information, 197
Upper Atmosphere Research Satellite (UARS), 77–78
Uranus, 38, 51, 65, 67–68, 168, 178, 252
Urban, Dr. Eugene, 332, 381
uresco ultrasonic immersion testing system, 3
Usada, Japan, 44

USS Detroit, 228
USS Saratoga, 111
USSR. See Union of Soviet Socialist Republics.
Utah, 275, 277, 296, 367–368
Utrecht, University of (the Netherlands), 40
Utsman, Thomas, 301, 312

V-22 Osprey tilt-rotor aircraft, 12, 15
Valentine, Ed, 17, 355
Van Allen, Dr. James, 243, 417–418
van Allen Belts, 77
van den Berg, Dr. Lodewijk, 325–326, 375
van Hoften, James, 227, 230–231, 332–333
Vandenberg AFB, CA, 27, 114, 120, 123, 192, 199, 228, 240, 278, 287–289, 291, 294–295, 300, 305, 321, 335, 363, 365, 367, 371
—Space Launch Complex-6, 286
—Western Space and Missile Center, 72, 128
Vandenkerckhove, Jean, 213
Vanderhoff, John, 100
Vannier, Dr. Michael, 209–210
vapor crystal growth system (VCGS), 374
VAS. See VISSR Atmospheric Sounder.
Vasyutin, Lt. Col. Vladimir, 392–395, 397
vay Speybroek, Dr. Leon, 40
VCAP. See Vehicle Charging and Potential Experiment.
VCGS. See vapor crystal growth system.
Vega, 53
Vega spacecraft, 65, 137, 173, 396
Vehicle Charging and Potential Experiment (VCAP), 381
Vela, Rudolfo Neri, 339
Vellinger, John, 249
Vendervender Junior High School, Parkersburg, WV, 272
Venus, 21, 41, 61, 63, 65, 137, 149, 173, 244–245
Vernadsky Institute of Sciences, 244
Very Large Array telescope, 57
very-large scale integration (VLSI), 103
very long baseline (VLB), 51
Very Long Baseline Interferometry (VLBI), 63

Index

Vessey, Gen. John, Jr., 438
Veterans Administration Medical Center, 279
VFT. See Spacelab Verification Flight Test.
Vickers-Armstrong, Ltd. 185
Viking spacecraft, 66, 99, 235, 412, 440
Vikmanis, Maris, 13
Vikram Sarabhai Space Center, Trivandrum, India, 284
Virgin Islands, 271
Virginia Polytechnic Institute, Blacksburg, VA, 167
Virginia, University of, Charlottesville, VA, 42
VISSR Atmospheric Sounder (VAS), 226–227
VLB. See very long baseline.
VLBI. See Very Long Baseline Interferometry.
Vlcek, Brian, 369
VLSI. See very-large scale integration.
volcanic eruptions, 76. (See also individual volcanoes.)
Volkov, Aleksandr, 395, 397
von Braun, Werner, 151
Voyager spacecraft, 62, 65, 168, 178, 180, 252, 412

Wade, James, Jr., 407
Waggoner, Miles, 433
Wales, Robert, 220
Walker, C.B.F., 43
Walker, Charles, 279, 322
Walker, David, 280
Walker, Richard, 38
Wallop, Sen. Malcolm (R-WY), 413
Wallops, John, 200
Wallops Flight Facility. See Goddard Space Flight Center.
Wallops Island, VA, 200, 411
Wallops P-3 search aircraft, 74, 147
Wallops Range Control Center. See Goddard Space Flight Center.
Wallops Skyvan aircraft, 75
Wallops Station, VA, 223
Wang, Dr. Taylor, 214, 327, 375, 434
Warner, Sen. John (R-VA), 111, 386
Washington, DC, 16–17, 29, 39, 41, 102, 113, 125, 191, 233, 240–241, 255, 271–272, 432–433

Washington Space Business Roundtable, 270
Washington, University of, Seattle, WA, 167, 220
Washington University Medical Center, St. Louis, MO, 209
wastewater treatment technology, 207–208
Waterville, ME, 75, 77
Watkins, Adm. James, 349–350, 437
Watterson, Maj. John Brett, 120, 322
Wawatobi, Indonesia, 212
Webb, David, 440
Weber State College, Ogden, UT, 219
Webster, Christopher, 80
weightlessness, 152, 248, 350–352
Weinberger, Secretary of Defense Caspar, 113, 115–116, 120, 126, 132–133, 295, 316, 385, 437–438
Weitz, Paul, 33
Wenger, Niki, 272
West German Aerospace Research Establishment (DFVLR), 335, 375–376, 380, 432
West Germany. See Federal Republic of Germany.
West Lafayette, IN, 272
West Virginia, 16, 44, 272
Westar satellites, 19, 33, 223, 276, 278
Western Samoa, 220
Western Space and Missile Center. See Vandenberg AFB.
Western Union, 278
Westinghouse Electric Corp., 144
Westwood Elementary School, Friendswood, TX, 272
Wetherill, Dr. George, 244–245
WFPC. See wide-field/planetary camera.
Wheelon, Albert, 232–233
White, Dr. David, 188
White, Edward, 34
White, Dr. Simon, 43
White, Terry, 337
White House
 —Advisory Committee on Arms Control and Disarmament, 181
 —Office of Science and Technology Policy, 3, 180
 —Press Office, 203
 —Science Council, Federal Laboratory Review Panel, 424
White Mountains, AK, 75

White Sands Missile Range, NM, 119, 131, 172, 243, 247, 318, 436
Whitehead, Victor, 409–410
Whitmire, Daniel, 38
Wichita State University (WSU), Wichita, KS, 9
Wick, Charles, 222
wide-field/planetary camera (WFPC), 48
Widick, Fritz, 321
Widick, H.W., 294
Wilkening, Dr. Laurel, 52, 440
Williams, Clifton C., 33–34
Williams, Comdr. Donald, 282, 296, 322
Williams, Jane, 33
Wilson, Dr. Andrew, 40
wind shear, 83, 88
Winkler, Martin, 21
Wisconsin, 85, 369
Wisconsin, University of, Madison, WI, 167
—Washburn Observatory, 283
Wisconsin Ultraviolet Photopolarimetry Experiment, 284
Wisdom, Jack, 67
Wisniewski, R., 343
WMO. See World Meteorological Organization.
Wolbach, Wendy, 147
Wolverton, Dr. Billy, 207
woman astronaut, first American
—as prime payload specialist, 280
—in space, 102, 151, 281
—to walk in space, 32, 440
Wood, H. William, 184–185
Woodgate, Dr. Bruce, 58
World Administrative Radio Conference, 141, 229

World Meteorological Organization (WMO), 71–72
World Radio Conference, 141
Wornick Corp., 190, 314
Wright Aeronautical Laboratory, 191
Wright Patterson AFB, OH, 14

X–29 aircraft, 6–8, 11, 168
XBQM–106 unmanned vehicle, 117–118
Xian, People's Republic of China, 434
Xiannian, President Li, People's Republic of China, 214
Xichang launching center, People's Republic of China, 215
X-ray timing explorer (XTE), 189
XTE. See X-ray timing explorer.
Xue, Li, 213, 215
XV–15 tilt-rotor research aircraft, 11–12, 15–16
X-wing rotorcraft, 6

Yager, Col. Walter, 286
Yale University, New Haven, CT, 42
Yeager, Brig. Gen. Charles, 101, 151, 440
Yeager, Glennis, 101
Yeomans, Dr. Don, 43
Young Astronauts Program, 31–32, 421
Young Cosmonauts, 31
Young, Dr. Donald, 210–211
Young, John, 91, 278, 282, 322
Yugoslavia, 158

zero gravity, 100, 189, 243, 256, 425

The NASA History Series

HISTORIES

Anderson, Frank W., Jr., *Orders of Magnitude: A History of NACA and NASA, 1915-1980* (NASA SP-4403, 2d ed., 1981).

Benson, Charles D., and William Barnaby Faherty, *Moonpart: A History of Apollo Launch Facilities and Operations* (NASA SP-4204, 1978).

Bilstein, Roger E., *Stages to Saturn: A Technological History of the Apollo/Saturn Launch Vehicles* (NASA SP-4206, 1980).

Boone, W. Fred, *NASA Office of Defense Affairs: The First Five Years* (NASA HHR-32, 1970, multilith).

Brooks, Courtney G., James M. Grimwood, and Loyd S. Swenson, Jr., *Chariots for Apollo: A History of Manned Lunar Spacecraft* (NASA SP-4205, 1979).

Byers, Bruce K., *Destination Moon: A History of the Lunar Orbiter Program* (NASA TM X-3487, 1977, multilith).

Compton, W. David, and Charles D. Benson, *Living and Working in Space: A History of Skylab* (NASA SP-4208, 1983).

Corliss, William R., *NASA Sounding Rockets, 1958-1968: A Historical Summary* (NASA SP-4401, 1971).

Ezell, Edward Clinton, and Linda Neuman Ezell, *On Mars: Exploration of the Red Planet, 1958-1978* (NASA SP-4212, 1984).

Ezell, Edward Clinton, and Linda Neuman Ezell, *The Partnership: A History of the Apollo-Soyuz Test Project* (NASA SP-4209, 1978).

Green, Constance McL., and Milton Lomask, *Vanguard: A History* (NASA SP-4202, 1970; also Washington: Smithsonian Institution Press, 1971).

Hacker, Barton C., and James W. Grimwood, *On the Shoulders of Titans: A History of Project Gemini* (NASA SP-4203, 1977).

Hall, R. Cargill, *Lunar Impact: A History of Project Ranger* (NASA SP-4210, 1977).

Hallion, Richard P., *On the Frontier: FLight Research at Dryden, 1946-1981* (NASA SP-4303, 1984).

Hansen, James R., *Engineer in Charge: A History of the Langley Aeronautical Laboratory, 1917-1958* (NASA SP-4305).

Hartman, Edwin P., *Adventures in Research: A History of Ames Research Center, 1940-1965* (NASA SP-4302, 1970).

Levine, Arnold, *Managing NASA in the Apollo Era* (NASA SP-4102, 1982).

Muenger, Elizabeth A., *Searching the Horizon: A History of Ames Research Center, 1940-1976* (NASA SP-4304, 1985).

Newell, Homer E., *Beyond the Atmosphere: Early Years of Space Science* (NASA SP-4211, 1980).

Pitt, John A., *The Human Factor: Biomedicine in the Manned Space Program to 1980* (NASA SP-4213, 1985).

Roland, Alex, *Model Research: The National Advisory Committee for Aeronautics, 1915–1958* (NASA SP–4103, 1985).

Rosenthal, Alfred, *Venture into Space: Early Years of Goddard Space Flight Center* (NASA SP–4301, 1968).

Rosholt, Robert L., *An Administrative History of NASA, 1958–1963* (NASA SP–4101, 1966).

Sloop, John L., *Liquid Hydrogen as a Propulsion Fuel, 1945–1959* (NASA SP–4404, 1978).

Swenson, Loyd S., Jr., James M. Grimwood, and Charles C. Alexander, *This New Ocean: A History of Project Mercury* (NASA SP–4201, 1966).

REFERENCE WORKS

Aeronautics and Space Report of the President, annual volumes for 1975–1982.

The Apollo Spacecraft: A Chronology (NASA SP–4009, vol. 1, 1969; vol. 2, 1973; vol. 3, 1976; vol. 4, 1978).

Astronautics and Aeronautics: A Chronology of Science, Technology, and Policy, annual volumes 1961–1978, with an earlier summary volume, *Aeronautics and Astronautics, 1951–1960*.

Dickson, Katherine M., ed., *History of Aeronautics and Astronautics: A Preliminary Bibliography* (NASA HHR–29, 1968, multilith).

Hall, R. Cargill, ed., *Essays on the History of Rocketry and Astronautics: Proceedings of the Third through the Sixth History Symposia of the International Academy of Astronautics* (NASA CP–2014, 2 vols., 1977).

Hall, R. Cargill, *Project Ranger: A Chronology* (JPL/HR–2, 1971, multilith).

Looney, John J., ed., *Bibliography of Space Books and Articles from Non-Aerospace Journals, 1957–1977* (NASA HHR–51, 1979, multilith).

Roland, Alex F., *A Guide to Research in NASA History* (NASA HHR–50, 6th ed., 1982, available from NASA History Office).

Skylab: A Chronology (NASA SP–4011, 1977).

Van Nimmen, Jane, and Leonard C. Bruno, with Robert L. Rosholt, *NASA Historical Data Book 1958–1968*, vol. 1, *NASA Resources* (NASA SP–4012, 1976).

Wells, Helen T., Susan H. Whiteley, and Carie E. Karegeannes, *Origins of NASA Names* (NASA SP–4402, 1976).

Recent volumes are available from Superintendent of Documents, Government Printing Office, Washington, DC 20402; early volumes from National Technical Information Service, Springfield, VA 22161.

www.ingramcontent.com/pod-product-compliance
Lightning Source LLC
Chambersburg PA
CBHW081713170526
45167CB00009B/3568